W9-BWM-867

ENCYCLOPEDIA
OF HUMAN
BIOLOGY
VOLUME 8 Index

ENCYCLOPEDIA OF HUMAN BIOLOGY

VOLUME 8 Index

Editor–in–Chief
Renato Dulbecco
The Salk Institute
La Jolla, California

ACADEMIC PRESS, INC. *Harcourt Brace Jovanovich, Publishers*
San Diego   New York   Boston   London   Sydney   Tokyo   Toronto

Index manuscripts were prepared as follows: Robert Boyd, Volume 1 (in part), Volume 4 (in part), Volume 5 (in part), Volume 6, and Volume 7; Alan Greenberg, Volume 3, Volume 5 (in part); Ed Serdziak, Volume 1 (in part), Volume 4 (in part); Egon Stark, Volume 2. Final compilation, editing, and typesetting was done by Linda Fetters. Nikki Fine prepared the "Index of Related Titles."

This book is printed on acid-free paper. ∞

Copyright © 1991 by ACADEMIC PRESS, INC.

Academic Press, Inc.
San Diego, California 92101

United Kingdom Edition published by
Academic Press Limited
24–28 Oval Road, London NW1 7DX

Library of Congress Cataloging-in-Publication Data

Encyclopedia of human biology / [edited by] Renato Dulbecco.
    p. cm.
    Includes index.
    ISBN 0-12-226751-6 (v. 1). -- ISBN 0-12-226752-4 (v. 2). -- ISBN
0-12-226753-2 (v. 3). -- ISBN 0-12-226754-0 (v. 4). -- ISBN
0-12-226755-9 (v. 5). -- ISBN 0-12-226756-7 (v. 6). -- ISBN
0-12-226757-5 (v. 7). -- ISBN 0-12-226758-3 (v. 8)
    1. Human biology--Encyclopedias. I. Dulbecco, Renato, 1914-
    [DNLM: 1. Biology--encyclopedias. 2. Physiology--Encyclopedias.
QH 302.5 E56]
QP11.E53 1991
612'.003--dc20
DNLM/DLC
for Library of Congress   91-45538
                              CIP

PRINTED IN THE UNITED STATES OF AMERICA
91 92 93 94   9 8 7 6 5 4 3 2 1

CONTENTS OF VOLUME 8

CONTRIBUTORS

Numbers in parentheses indicate the pages on which the authors' contributions begin. Volume number is indicated first followed by a colon and the page number in that volume.

Robert Ader (6:319), School of Medicine, University of Rochester, Rochester, New York 14642

Bernard W. Agranoff (5:341), Neuroscience Lab Building, University of Michigan, Ann Arbor, Michigan 48109

Tim J. Ahern (3:411), Genetics Institute, 87 Cambridge Park Dr., Cambridge, Massachusetts 02140

Shin-Ichi Akiyama (4:107), Department of Cancer Chemotherapy, Institute of Cancer Research, Kagoshima University, 1208-1 Usuki-cho, Kagoshima 890, Japan

R. W. Albers (4:557), Laboratory of Neurochemistry Public Health Service, National Institutes of Health, Bethesda, Maryland 20892

Kari Alitalo (5:545), Cancer Biology Laboratory, Haartmanink, 3, Helsinki 00290, Finland

Daniel L. Alkon (2:305), Section on Neural Systems, Laboratory of Cellular/Molecular Biology, National Institute of Neurologic, and Communication Disorders and Stroke, Bethesda, Maryland 20892

Frank D. Allan (3:609), Blue Run Farm, 116 W. Rt. 1, Stanardsville, VA 22973

Milton Alter (7:299), Neurology Department, Medical College of Pennsylvania, Philadelphia, Pennsylvania 19129

Amnon Altman (1), 0.97, (7:745) Division of Cell Biology, La Jolla Institute for Allergy and Immunology, 11149 N. Torrey Pines Road, La Jolla, California 92037

Burton M. Altura (6:345), State University New York, Health Science Center at Brooklyn, Brooklyn, New York 11203

David G. Amaral (4:227), Salk Institute, Box 85800, San Diego, California 92037

Vincent T. Andriole (6:399), Department of Medicine/1090 Lmp, Yale University, 333 Cedar Street, 201 LCI, New Haven, Connecticut 06510

Friderun Ankel-Simons (2:593), Duke University Primate Center, 2518 Lanier Place, Durham, North Carolina 27705

Otto Appenzeller (4:49), Departments of Neurology, School of Meicine, University of New Mexico, Albuquerque, New Mexico 87131

Jacob A. Arlow (6:305), New York University College of Medicine, 120 E. 36th St., New York, New York 10016

Robert Atkinson (1:63), Department of Human Resource Development, University of Southern Maine, Center for the Study of Lives, Gorham, Maine 04038

Daniel E. Atkinson (2:503), Department of Chemistry, University of California, Los Angeles, Los Angeles, California 90034

Mary Ellen Avery (5:699), Harvard Medical School, Department of Pediatrics, The Children's Hospital, Boston, Massachusetts 02115

Jesus Avila (5:893), Centro de Biologia Molecular, Facultad de Ciencias, Universidad Autonoma de Madrid, E-28049 Madrid, Spain

Nathan Back (4:589), Department of Biochemical Pharmacology, State University of New York, Buffalo, New York 14260

George S. Bailey (3:671), Department of Food Science and Toxicology, Oregon State University, Corvallis, Oregon 97331

James B. Bakalar (5:439), Massachusetts Mental Health Center, Boston, Massachusetts 02215

Charles M. Balch (4:887), Department of General Surgery, The University of Texas, M. D. Anderson Cancer Center, Houston, Texas 77030

Janice I. Baldwin (6:865), Department of Sociology, University of California, Santa Barbara, Santa Barbara, California 93106

John Baldwin (6:865), Department of Sociology, University of California at Santa Barbara, Santa Barbara, CA 93106

Kenneth L. Barker (7:737), Texas Tech University, Health Science Center, Room 2B106, Lubbock, Texas 79430

Erik Barquist (2:31), Department of Surgery and Medicine, Brain Research, UCLA, Los Angeles, California 90024

Jerome Barre (1:707), Departement de Pharmacologie, Faculte de Medecine de Paris XII, F-94010 Creteil, 8 Rue du General Sarrail, France

Renato Baserga (2:253), Fels Research Institute, Temple University, School of Medicine, Philadelphia, Pennsylvania 19140

Anthony S. Bashir (7:153), Division of Communication Disorders, Emerson College, Boston, Massachusetts 02116

John N. Bassili (5:785), Department of Psychology, University of Toronto, Toronto, Ontario, Canada M5S 1A1

C. Daniel Batson (1:197) Department of Psychology, University of Kansas, Lawrence, Kansas 66045

Robert C. Baumiller (2:437), Department of Obstet. and Gynec., Georgetown University, School of Medicine, 3800 Reservoir Road N.W., Washington, D. C. 20007

Gary K. Beauchamp (6:715), Monell Chemical Senses Center, University of Pennsylvania, Philadelphia, Pennsylvania 19104

Aldo Becciolini (6:469), Radiation Biology Laboratory, Universita degli Studi di Firenze, Dipartimento Di Fisiopatologia Clinica, Viale Morgagni, 85, 50134 Firenze, Italy

Sidney L. Beck (7:405), Department of Biological Sciences, DePaul University, Chicago, Illinois 60614-3238

Carl Becker (7:511), Department of Pathology and Laboratory Medicine, Medical College of Wisconsin, Milwaukee, Wisconsin 53226

Gurrinder S. Bedi (4:589) Department of Biochemical Pharmacology, State University of New York, Buffalo, New York 14260

Alvin J. Beltz (2:333) Department of Veterinary Biology, University of Minnesota, St. Paul, Minnesota 55108

Rutger Bennet (1:317), Department of Pediatrics, Korolinska Institutet St. Goran's Children's Hospital, Stockholm, Sweden

Jacques Benveniste (6:43), Director of Research, INSERM U-200, Universite Paris-Sub, 32 rue des Carnets, 92140 Clamart, France

Carolyn D. Berdanier (4:999), Department of Foods and Nutrition, University of Georgia, Athens, Georgia 30602

Otto G. Berg (3:117) Department of Molecular Biology, Uppsala University Biomedical Center, Box 590 S-75124, Uppsala, Sweden

George W. Bernard (2:817), Schools of Medicine and Dentistry, University of California, Los Angeles, Department of Anatomy and Cell Biology, UCLA School of Medicine, Los Angeles, California 90024

Giorgio Bernardi (4:251), Laboratoire De Genetique Moleculaire, Institut Jacques monod, Paris, 75005 France

Irwin S. Bernstein (1:113), Department of Psychology, University of Georgia, Athens, Georgia 30602

Gary G. Berntson (5:799), Department of Psychology, Ohio State University, Columbus, Ohio 43120

Ellen Berscheid (4:535), Department of Psychology, University of Minnesota, Minneapolis, Minnesota 55455

Joseph R. Bertino (3:785), American Society Professor of Pharmacology and Medicine, Memorial Sloan-Kettering Cancer Center, New York, New York 10021

Ernest Beutler (3:777), Research Institute of Scripps Clinic, 10666 N. Tomey Pines Rd., La Jolla, California 92037

A. D. Beynon (2:769), Department of Oral Biology, The Dental School, Framlington Place, Newcastle-upon-Tyne NE2 4BW, United Kingdom

Albert E. Bianco (5:645), Molecular Parasitology & Research Group, Imperial College of Science and Technology, London SW7 2BB, England

Mark H. Bickhard (2:547), Department of Educational Psychology, University of Texas, Austin, Austin, Texas 78712

Howard R. Bierman (4:687, 4:839), Institute for Cancer and Blood Research, Loma Linda School of Medicine, Beverly Hills, California 90211

Bruce Birren (6:379), Division of Biology 14775, California Institute of Technology, Pasadena, CA 91125

Oscar A. Bizzozero (5:271), Biochemistry Department, E. K. Shriver Center, Waltham, Massachusetts 02254

Francis L. Black (4:939), Department of Epidemiology and Public Health, Yale University School of Medicine, New Haven, Connecticut 06510

Lars Bläckberg (3:47), Department of Pediatrics, Univesity of Umea, 901 85 Umea, Sweden

Ronald Blackburn (2:709), Department of Psychology, Ashworth Hospital, Parkbourn Maghull, Liverpool L31 1HW, United Kingdom

Richard E. Blackwell (6:555), Department of Obstetrics and Gynecology, Division of Reproductive Biology, University of Alabama at Birmingham, University Station, Birmingham, Alabama 35294

Henriette Bloch (6:819), EPHE-CNRS, URA 315, 41, Rue Gay-Lunce, 75005 Paris, France

Robert M. Bloochin (6:533), Albert Einstein College of Medicine, Bronx, New York 10461

Joseph P. Blount (2:67), Medical Center, 3001 Green Bay Road, Chicago, Illinois 60064

John E. Blundell (6:723), Biopsychology Group, Psychology Department, University of Leeds, Leeds, LS2 9JT, United Kingdom

Richard P. Bobbin (2:521), Department of Otorhinolaryngology and Biocommunication, Louisiana State University Medical School, Kresge Hearing Research Laboratory of the South, New Orleans, Louisiana 70112

Nicholas Bodor (7:101), Center for Drug Discovery, College of Pharmacy, University of Florida, Gainesville, Florida 32610

Dani P. Bolognesi (6:641, 6:641) LaSalle, St. Ext., P.O. Box 2926, Durham, North Carolina 27710

Robert M. Bookchin (6:533), Department of Dermatology, Rush-Presbyterian-St. Luke's Medical Center, 1653 Cedar St., New Haven, CT 06510-8040

Frank W. Booth (3:505), Department of Physiology and Cell Biology, University of Texas Medical School, Houston, Texas 77030

D. A. Booth (3:637), Food and Nutrition Laboratory, School of Psychology, University of Birmingham, Birmingham B15 2TT, England

Maarten C. Bosland (6:177), Institute of Environmental Medicine, NYU Medical Center, Long Meadow Road, Tuxedo, New York 10987

Graham Boulnois (5:1), Department of Microbiology, University of Leicester, Leicester LE1 9HN, England

H. Bouma (5:737), Instituut vorr Perceptie Cnderzoek, 5600 MB Eindhoven, The Netherlands

D. G. Bouwhuis (5:737), Instituut vorr Perceptie Cnderzoek, 5600 MB Eindhoven, The Netherlands

Anne Bowcock (3:135), The University of Texas Southwestern, Medical Center at Dallas, Dallas, Texas 75235

Charles L. Bowden (1:307; 4:993; 5:91), Department of Psychiatry, The University of Texas Health Science Center, San Antonio, Texas 78229

G. Stephen Bowen (1:151) Center for Prevention Services, Center for Disease Control, 1600 Clifton Road, Atlanta, GA 30333

Steven D. Bradway (6:689), Department of Oral Biology, Dental Research Institute, State University of New York, Buffalo, Buffalo, New York 14214

Joseph G. Brand (7:527), Monell Chemical Senses Center, Univ. of Pennsylvania, Philadelphia, Pennsylvania 19104

Monique C. Braude (3:211; 4:905), 2410 Parkway, Cheverly, Maryland 20785

George A. Bray (5:517), Pennington Biomedical Research Center, Louisiana State University, Baton Rouge, Louisiana 70808

S. Marc Breedlove (4:241), Department of Psychology, University of California, Berkeley, California 94720

Kevin A. Briand (1:479), Department of Psychology, University of Oregon, Eugene, Oregon 97403

R. W. Briehl (7:1), Department of Biochemistry and Physiology, Albert Einstein College of Medicine, 1300 Morris Park Avenue, New York, New York 10461

F. Norman Briggs (5:229), Department of Physiology, Virginia Commonwealth University, Medical College of Virginia, Richmond, Virginia 23298

B. R. Brinkley (5:65), Department of Cell Biology and Anatomy, University of Alabama at Birmingham, UAB Station, Birmingham, Alabama 35294

Lynn A. Bristol (2:751), Laboratory of Pharmacologic and Physiologic Studies, N.I.A.A.A., 12501 Washington Ave., Flow Bldg., Rm. 23, Rockville, Maryland 28502

Nathan Brody (5:769), Department of Psychology, Wesleyan University, Middletown, Connecticut 06457

Sam C. Brooks (2:53), Michigan Cancer Foundation, 110 E. Warren Avenue, Detroit, Michigan 48201

Stephen P. J. Brooks (3:423), Department of Biology, Carleton University, Ottawa, Ontario, Canada K1S 5B6

Barbara A. Brooks (3:553), The University of Texas Health Science Center, at San Antonio, Department of Physiology, 7703 Floyd Curl Dr., San Antonio, Texas 78284

Arnold Brossi (1:177), National Institutes of Health, Bethesda, Maryland 20892

Ian R. Brown (2:431), Division of Life Sciences, University of Toronto, Scarborough Campus, West Hill, Ontario, Canada M1C 1A4

G. V. Brown (4:385), The Walter and Eliza Hall Institute of Medical Research, Royal Melbourne Hospital, Melbourne 3050, Australia

R. C. Brown (7:545), Toxicology Unit, Medical Re-

search Council Laboratories, Carshalton, Surrey SM5 4EF, United Kingdom

J. E. Bruni (4:979), Department of Anatomy/Physical Therapy, University of Manitoba, Winnipeg, MB Canada R3E 0W3

Joseph A. Buckwalter (2:201), Department of Orthopedic, University of Iowa College of Medicine, Iowa City, Iowa 52242

David Burke (3:719, 6:157), Department of Clinical Neurophysiology, Institute of Neurological Sciences, The Prince Henry Hospital, P.O. Box 233, Matraville, New South Wales 2036, Sydney, Australia

Kenneth D. Burman (4:351), Endocrine-Metabolic Service and Kyle Metabolic Unit, Walter Reed Army Medical Center, Washington, D.C. 20307-5001

Karen P. Burton (7:353), Departments of Radiology and Physiology, University of Texas, Southwestern Medical Center, Dallas, Texas 75235

Robert K. Bush (7:337), William Middleton Memorial Veteran's Hospital, Madison, Wisconsin 53705

E. R. Buskirk (1:57), Professor of Applied Physiology, College of Health & Human Development, The Pennsylvania State University, University Park, Pennsylvania 16802

Robert M. Butler (3:891), Department of Geriatrics, Mt. Sinai School of Medicine, 100th St & 5th Avenue, New York, New York 10029

Ralph Buttyan (2:245), Department of Urology, Columbia University, College of Physicians and Surgeons, New York, New York 10032

John T. Cacioppo (5:799), Department of Psychology, Ohio State University, Columbus, Ohio 43120

Edward J. Cafruny (3:93), Graduate School Biomedical Science, University of Medicine and Dentistry of New Jersey, 82 Linden St., Millburn, New Jersey 07041

John E. Calamari (1:43), Psychology Service (116-B), Veterans Administration Medical Center, North Chicago, Illinois 60064

William C. Campbell (2:381), Merck Institute for Therapeutic Research, Drew University, New Jersey 07940

Bruce Campbell (5:835), Servier Research and Development, Ltd., Fulmer, Slough, SL3 6HH, United Kingdom

Roderick A. Capaldi (5:55), Institute of Molecular Biology, University of Oregon, Eugene, Oregon 97403

Richard L. Carter (2:109), Haddow Laboratories, Institute of Cancer Research, and Royal Marsden Hospital, Sutton Surrey SM2 5PT, United Kingdom

Anne Carter (4:443), Laboratory Centre for Disease Control, Tunney's Pasture, Ottawa, Ontario, K1A 0L2, Canada

Willard Cates (6:891), Division of STD, Center for Disease Control, Atlanta, Georgia 30333

Curtis Cetrulo (6:581), Department of Maternal Fetal Medicine, St. Margaret's Hospital for Women, Department of Obstetrics and Gynecology, Tufts University School of Medicine, Boston, MA 02125

Siew Yeen Chai (5:415), Department of Medicine, The University of Melbourne, Heidelberg, Victoria, 3084 Australia

C. E. Challice (2:125), Department of Physics and Astronomy, The University of Calgary, Calgary, AB T2N 1N4, Canada

Britton Chance (3:377), Departments of Biochemistry and Biophysics, Johnson Research Foundation, University of Pennsylvania, School of Medicine, 37th & Hamilton Walk, Philadelphia, Pennsylvania 19104-6089

Kwang-Poo Chang (4:679), Department of Microbiology/Immunology, University Health Sciences/Chicago Medical School, North Chicago, Illinois 60064

Dawn Chescoe (3:271), Microstructural Studies, University of Surrey, United Kingdom

David J. Chivers (3:57), University of Cambridge, Department of Veterinary Anatomy; Tennis Court Road, Cambridge CB2 1QS, England

Joseph Chou (2:371) Laboratory of Biochemical Pharmacology, Memorial Sloan-Kettering Cancer Center and Cornell University Graduate School of Medicine Sciences, New York, NY 10021

Ting-Chao Chou (2:371), Laboratory of Biochemical Pharmacology, Memorial Sloan-Kettering Cancer Center and Cornell University Graduate School of Medicine Sciences, New York, New York 10021

Ching Chung Chou (4:547), Department of Physiology, Michigan State University, East Lansing, MI 48824-1101 IRCM, Clinical Research Institute of Montreal, 110 Pine Ave West, Montreal, Quebec, Canada H2W 1R7

Sandra C. Christiansen (1:183) Division of Allergy and Immunology, Scripps Clinic Medical Group, Inc., La Jolla, California 92037

Fun S. Chu (7:659), Department of Food Microbiology and Toxicology, University of Wisconsin, 1925 Willow Drive, Madison, Wisconsin 53706

Bernard Cinader (6:75), Department of Immunology, Medical Science Building, University of Toronto, Toronto, Ontario M5S 1AB, Canada

Roger A. Clegg (3:581), Hannah Research Institute, AyR KA6 5HL, Scotland

Daniel Cohen (6:79), Centre d'Etude du Polymorphisme Humain, Paris

Erica A. Cohen (2:629), Psychology Service (116B), VA Medical Center, North Chicago, Illinois 60064

Nicholas Cohen (6:319), Department of Microbiology and Immunology, University of Rochester School of Medicine and Dentistry, Rochester, N.Y. 14642

Robert D. Cohen (5:821), The Medical Unit, The London Hospital, University of London, London E1 1BB, England

Zanvil A. Cohn (4:811), Rockfeller University, New York, New York 10021

Andrew J. Coldman (4:921) Division of Medical Oncoclogy, Cancer Control Agency, British Columbia, Vancouver, British Columbia, Canada V52 4E6

Nancy C. Collier (4:99), Washington University School of Medicine, Dept. of Molecular Biology, 8230, St. Louis, MO 63110

Gerald F. Combs (6:789), Division of Nutritional Sciences, Cornell University, Ithaca, New York 14853

P. Michael Conn (2:299), Department of Pharmacology, University of Iowa College of Medicine, Iowa City, Iowa 52242-1109

T. A. Connors (7:545), Toxicology Unit, Medical Research Council Laboratories, Woodmasterne Rd., Carshalton, Surrey SM5 4EF, United Kingdom

Jonathan Cooke (3:281), National Institute for Medical Research, The Ridgeway Mill Hill, London, NW7 1AA, England

Howard Cooke (7:401), MRC Mammalian Genome Unit, King's Buildings, Edinburgh EH4 2XU, United Kingdom

K. E. Cooper (1:751), Department of Medical Physiology, The University of Calgary, Calgary, Alberta, Canada T2N 4N1

Stanley Coren (7:829), Department of Psychology, University of British Columbia, 2136 West Mall, Vancouver, BC, Canada V6T 1Y7

Pelayo Correa (3:765), Department of Pathology, Louisiana State University-Medicine, New Orleans, Louisiana 70112

Virginia L. Corson (3:813), The Johns Hopkins Hospital, Prenatal Diagnostic Center, Baltimore, Maryland 21205

Leda Cosmides (6:493), Department of Psychology, Stanford University, Stanford, California 94305

Antonio Coutinho (4:323), Institut Pasteur, Unite d'Immunobiologist, 25, Rue Du Dr Roux, F-75724 Paris Cedex 15, France

Lex M. Cowsert (5:635), Laboratory of Tumor Virus Biology, National Cancer Institute, Bethesda, Maryland 20892

W. Miles Cox (1:43, 116-B, 1:165, 2:67, 2:629), North Chicago VA Medical Center, The Chicago Medical School, Psychology Service 116B, 3001 Green Bay Road, North Chicago, Illinois 60064

Thomas E. Creighton (6:231), Medical Research Laboratory of Molecular Biology, Cambridge University, Cambridge CB2 2QH, England

Mareo Crescenzi (6:523), Instituto di Immunologia Clinica, Universitá "La Sapienza," Roma, Italy

Carol Crowther (7:459), University of Sydney, Sydney, New South Wales 20006, Australia

Joanne Csete (3:561), Department of Nutritional Sciences, University of Wisconsin-Madison, Madison, Wisconsin 53706

Fernand Daffos (1:729), Institut de Puericulture de Paris, Service de Medicine et de Biologie, Paris, France

Alan R. Dahl (5:285), Inhalation Toxicology Research, Lovelace Biomedical and Environmental Research Institute, Box 5890, Albuquerque, New Mexico 87115

Mary d'Alton (6:581), Department of Maternal Fetal Medicine, St. Margaret's Hospital for Women, Department of Obstetrics and Gynecology, Tufts University School of Medicine, Boston, MA 02125

J. W. L. Davies (7:447), Department of Surgery, Glasgow Royal Infirmary, Glasgow, Scotland, United Kingdom

Louise Davies (4:897), Consultant: Gerontology Nutrition, Hampstead, 85A Redington Rd., London NW3 7RR, England

Karen K. De Valois (7:801), Department of Psychology/Physiological Optics, University of California, at Berkeley, Tolman Hall, Berkeley, California 94720

Michael Dean (3:823–833), Biological Carcinogenesis, Development Program, Program Resource, N.C.I. Frederick Cancer, Research Facility, Frederick, Maryland 21701

Larry L. Deaven (2:455), Center for Human Genome Studies, Los Alamos National Laboratory, Life Sciences Division, LS-4 MS M888, Los Alamos, New Mexico 87545

Samir S. Deeb (2:661), Department of Medicine, Division of Medical Genetics, SK-50, University of Washington, Seattle, Washington 98195

Eric Delson (6:135), American Museum of Natural History, New York, New York 10024

Rik Derynck (7:625, 7:631), Department of Development Biology, Genentech Inc., 460 Point San Bruno Boulevard, South San Francisco, California 94080

David L. DiLalla (3:835), Muenzinger Psychology Building, University of Colorado, Boulder, Colorado 80309

Javier Diaz-Nido (5:893), Centro de Biologia Molecular, Universidad Autónoma de Madrid, Madrid, Spain

Anthony H. Dickenson (5:407), Department of Pharmacology, University College (London), London WC1E 6BT, England

Dirck L. Dillehay (4:819), Emory University School of Medicine, Woodruff Memorial Research Building, Atlanta, Box TT, Georgia 30322

Joakim Dillner (3:463), Department of Virology, Karolinska Institute, SBL, S-10521 Stockholm, Sweden

Walter Doerfler (3:151), Institut fur Genetik, University of Cologne, 121 Weyertal 5000 Koln 41, Federal Republic of Germany

Ralph Dornburg (6:653), McArdle Laboratory, University of Wisconsin, Madison, Wisconsin 53706

Kenneth Dorshkind (4:165), Department of Biomedical Sciences, University of California, Riverside, California 92521

John E. Dowling (6:615), Department of Biology, Harvard University, Cambridge, Massachusetts 02138

Denise M. Driscoll (7:233), Department of Psychology, University of California, Santa Barbara, California 93106

Shyam Dube (3:785), Center for Agricultural Biotechnology, College Park, Maryland 20742

Francis A. Duck (7:723), Wessex Regional Health Authority, Regional Medical Physics Service, Royal United Hospital, Combe Park, Bath BA1 3NG, United Kingdom

Kyle B. Dukelow (3:599), Endocrine Research Center, Michigan State University, East Lansing, Michigan 48824

W. Richard Dukelow (3:599), Endocrine Research Center, Michigan State University, East Lansing, Michigan 48824

Jacques Dunnigan (1:417), Faculty of Science, Universite' de Sherbooke, Sherbrooke P.Q. Quebec, Canada J1K 2R1

Scott K. Durum (2:751), Laboratory of Molecular Immunoregulation, Biological Response Modifies Program, Frederick Cancer Research Facility, National Institutes of Health, NCI, Frederick, Maryland 21701-1013

Sijmen Duursma (3:481), University Hospital, Department of Internal Medicine, Research Group for Bone Metabolism, 3508 GA UTRECHT, The Netherlands

Rosemary Dziak (4:33), Department of Oral Biology, Dental Research Institute, State University of New York, Buffalo, Buffalo, New York 14226

Dwain L. Eckberg (2:147), Professor, Medicine and Physiology, Hunter Holmes McGuire Department of Veterans Affairs Medical Center and Medical College of Virginia, Richmond, Virginia 23249

Duane C. Eichler (6:661), University of South Florida, College of Medicine, Dept. of Biochemistry, MDC 7, 12901 Bruce B. Downs Blvd., Tampa, Florida 33612

E. J. M. Eijkman (5:737) Institute Voor Perceptie Onderzoek, P.O. Box 513, 5600 MB Eindhoven, The Netherlands

Randy C. Eisensmith (5:863), Department of Cell Biology and Institute of Molecular Genetics, Howard Hughes Medical Institute, Baylor College of Medicine, One Baylor Plaza T-721, Houston, Texas 77030

Edward Eisenstein (7:711) Department of Molecular and Cell Biology, University of California at Berkeley, % Stanley Donner ASU, 229 Stanley Hall, Berkeley, CA 94720

Jane Penaz Eisner (4:665), Department of Psychology, School of Arts and Sciences, University of Pennsylvania, Philadelphia, Pennsylvania 19104

J. El-On (6:267), Department of Microbiology and Immunology, Ben Gurion University of The Negev, Beer Sheva 84 105, Israel

James T. Elder (3:445), Department of Dermatology, University of Michigan, Ann Arbor, Michigan 48109

Ervin Elias (7:645), Ben Gurion University of The Negev, Rehavat Tefat 24/1, P. O. Box 653, Beer Sheva 84770, Israel

Everett H. Ellinwood (2:511), Department of Psychiatry, Box 3870 Medical Center, Duke University, Durham, North Carolina 27710

Ari Elson (3:175) Department of Virology. The Weizmann Institute of Science, Rehovot, 76100, Israel

Allen C. Enders (4:423), Department of Anatomy, University of California, Davis, School of Medicine, Davis, California 95616

F. L. Engel (5:737), Institute Voor Perceptie Onderzoek, P. O. Box 513, 5600 MB Eindhoven, The Netherlands

Henry F. Epstein (5:161), Departments of Neurology, Biochemistry, and Cell Biology, Baylor College of Medicine, Houston, Texas 77030

Ulf Eriksson (7:853), Ludwig Institute for Cancer Research, Stockholm Branch, S-10401, Stockholm, Sweden

Mario Escobar (6:605), Department of Pathology, Virginia Commonwealth University, Medical College of Virginia, Richmond, Virginia 23298-0106

Diane C. Farhi (5:681), Case Western Reserve University, Institute of Pathology, Cleveland, Ohio 44106

Elaine B. Feldman (3:649), Medical College of Georgia, Augusta, Georgia 30912-3102

David L. Felten (5:373), Department of Neurobiology and Anatomy, University of Rochester School of Medicine, Rochester, New York 14642

Suzanne Y. Felten (5:373), Department of Neurobiology and Anatomy, University of Rochester, Rochester, New York 14642

Gabriel Fernandez (5:503), Department of Medicine, University of Texas, Health Science Center, San Antonio, Texas 78284

Sydney M. Finegold (1:237), Research Service, Veterans Administration Medical Center, West Los Angeles, California 90073

Leif H. Finkel (5:387), Department of Bioengineering, University of Pennsylvania, Philadelphia, Pennsylvania 19104

Barbara L. Finlay (5:305), Department of Psychology, Cornell University, Ithaca, New York 14853

Susan T. Fiske (1:101), Department of Psychology, University of Massachusetts at Amherst, Amherst, Massachusetts 01003

Becca Fleischer (4:1), Department of Molecular Biology, Vanderbilt University, 1702 Station B, Nashville, Tennessee 37235

Irving H. Fox (6:387), Department of Internal Medicine, Division of Rheumatology, University of Michigan Medical Center, Ann Arbor, Michigan 48109

Allen J. Frances (5:777), Cornell Medical Center, New York, New York 10021

Arthur Frankel (4:415), Department of Medicine, Duke University, North Carolina 27706

Fred H. Frankel (7:615), Department of Psychiatry, Beth Israel Hospital, 330 Brookline Avenue, Boston, Massachusetts 02215

Luigi Frati (6:523), Alfredo Pontecorvi, & Marco Crescenzi, Department of Experimental Medicine, Policlinico Umberto 1, Viale Regina Elena, 324, 00161 Rome, Italy

Morris Freedman (2:763), Rotman Research Institute of Baycres Centre, University of Toronto, Toronto M6A 2E1, Ontario, Canada

Arnold J. Friedhoff (2:217), Millhauser Clinic, NYY School of Medicine, New York, New York 10016

Susan J. Friedman (5:321), Laboratory Biology Chemistry, National Cancer Institute, National Institutes of Health, Bethesda, Maryland 20892

Rose E. Frisch (1:741, 1:741), Department of Population Sciences, Harvard School of Public Health, Cambridge, Massachusetts 02138

R. J. M. Fry (6:427), Biology Division, Martin Marietta Energy Systems Inc., P.O. Box 2009, Oak Ridge, Tennessee 37831

Joseph H. Gainer (4:513, 6:411), Division of Veterinary Medical Research, Food and Drug Administration, Beltsville, Maryland 20705

Karen N. Gale (6:799), Department of Pharmacology, Georgetown University Medical Center, Washington, D.C. 20007

James A. Gallagher (1:811), Human Anatomy and Cell Biology, University of Liverpool, Liverpool L69 3BX, United Kingdom

Brian B. Gallagher (3:455), Department of Neurology, Medical College of Georgia, Augusta, Georgia 30912

Brenda L. Gallie (6:633), Hospital for Sick Children, University of Toronto, 555 University Avenue, Toronto, Ontario M5G 1X8, Canada

S. C. Gandevia (6:157, 6:157), The University of

New South Wales, The Prince Henry Hospital, P.O. Box 233, Matraville, New South Wales, Australia 2036

Paul A. Garber (6:127), Associate Professor, Department of Anthropology, University of Illinois, Urbana-Champaign, Urbana, Illinois 61801

Laurence J. Garey (7:835), Department of Anatomy, Charing Cross & Westminster Medical School, London W6 8RF, England

Stanley M. Garn (4:25), Center for Human Growth and Development, University of Michigan, Ann Arbor, Michigan 48109

Suzanne Gartner (5:811), Henry M. Jackson Foundation Research Laboratory, 1500 E. Gude Drive, Rockville, Maryland 20850

Anthony J. Gaudin (1:371, 2:157, 4:259, 5:131, 5:347, 6:559, 6:587, 7:23, 7:33, 7:85, 7:733, 7:767), Department of Biology, California State University, Northridge, Northridge, California 91330

Jack Gauldie (1:25), McMaster University, Department of Pathology, 2N16, Hamilton, Ontario, Canada, L8N 3Z5

Nori Geary (1:343), Department of Psychology, Columbia University, New York, New York 10027

Russell G. Geen (1:585), Department of Psychology, University of Missouri, Columbia, Missouri 65211

Daniela S. Gerhard (1:93), Departments of Genetics and Psychiatry, Washington University School of Medicine, St. Louis, Missouri 63110

Feroze N. Ghadially (5:561), Cox Science Building, Department of Biology, 249118 Coral Gables, FL 33124

Ian L. Gibbins (1:535), Department of Anatomy and Histology and Centre for Neuroscience, School of Medicine, Finders University, South Australia 5042

Geula Gibori (5:551), Department of Physiology and Biophysics, University of Illinois, Chicago, Illinois 60612

C. C. A. M. Gielen (3:691), University of Nijmegen, Department of Medical Physics and Biophysics, Geert Grootplein Noord 21 NL 6525 EZ Nijmegen, The Netherlands

K. J. Gilhooly (6:145), Department of Psychology, University of Aberdeen, King's College, Old Aberdeen, AB9 2UB Scotland

K. J. Gilhooly (7:467), Department of Psychology, University of Aberdeen, King's College, Old Aberdeen AB9 2UB, Scotland, United Kingdom

J. Christian Gillin (7:71), Department of Psychiatry/V-116A, University of California, San Diego, Box 109, La Jolla, California 92093

George G. Glenner (1:209), University of California, San Diego, School of Medicine, Pathology Department, La Jolla, California 92093

James H. Goldie (4:921), Division of Medical Oncology, Cancer Control Agency, British Columbia, Vancouver, British Columbia, Canada V52 4E6

Bernard David Goldstein (3:383), Department of Environmental and Community Medical UMDNJ, Robert Wood Johnson Medical School, Piscataway, New Jersey 08854

Frank J. Gonzalez (2:737), Laboratory of Molecular Carcinogenesis, National Cancer Institute Building 37 Rm. 3E24, National Institute of Health, Bethesda, Maryland 20892

Peter J. Goodhew (3:271), Materials Science and Engineering, University of Liverpool, Liverpool L69 3BX, England

Steven P. Goodman (2:41), Structural and Cellular Biology, University of Alabama, College of Medicine, Mobile, Alabama 36688

Wm. Jennings Goodwin (1:251), Southwest Foundation for Biomedical Research, San Antonio, Texas 78227

John Gorham (5:675), Agricultural Resources Service, Washington State University, USDA, Pullman, Washington 99164

Paul Graves (3:495), Department of Archaeology, Southampton University, S09 5NH Highfield, 36 New Road, Littlehampton, West Sussex, United Kingdom

Philip Green (3:869), Genetics Department, Washington University, School of Medicine, St. Louis, Missouri 63110

Frank Greenberg (1:697), Institute for Molecular Genetics, Baylor College of Medicine, Houston, Texas 77030

Jane M. Greene (3:649), Medical College of Georgia, Georgia Institute of Human Nutrition, School of Medicine, Department of Medicine, Augusta, Georgia 30912

Robert J. Gregor (5:173), Department of Kinesiology, University of California, Los Angeles, Los Angeles, California 90024

Owen W. Griffith (3:907), Department of Biochemistry, Cornell University Medical College, New York, New York 10021

O. Hayes Griffith (5:903), Institute of Molecular Biology and Department of Chemistry, University of Oregon, Eugene, Oregon 97403-1229

Stern Grillner (4:769), The Nobel Institute for Neurophysiology, Karolinska Institute, Neurofysiologiska Avdelningen, Box 60400, S-10401 Stockholm, Sweden

Lester Grinspoon (5:439), Massachusetts Mental Health Center, 74 Fernwood Rd., Boston, Massachusetts 02215

Yoram Groner (3:175), Department of Virology, The Weizmann Institute of Science, Rehovot 76100, Israel

Lawrence Grossman (6:547), The Johns Hopkins University, Biochemistry Department, Baltimore, Maryland 21205

Sebastian P. Grossman (7:473), Department of Psychology, University of Chicago, Chicago, Illinois 60637

Bennett S. Gurian (1:129), Harvard Medical School, Massachusetts Mental Health Center, Boston, Massachusetts 02115

Paul H. Guth (3:755), Department of Surgery, W112, West Los Angeles VA Medical Center, Los Angeles, California 90073

Bernard Haber (6:829), Department of Neurology, University of Texas Medical School, 200 University #519, Marine Biomedical Institute, Galveston, Texas 77550

David A. Hafler (5:143), Multiple Sclerosis Research, Brigham and Women's Hospital, Harvard Medical School, Boston, Massachusetts 02115

Lori D. Hager (1:825), Department of Anthropology, University of California, Berkeley, Berkeley, California 94720

Randi J. Hagerman (3:709), Child Development Unit B148, The Children's Hospital, Denver, Colorado 80218

Theo Hagg (5:333), Department of Biology, M001, UCSD Medical Center, La Jolla, California 92093

George M. Hahn (4:295), Department of Radiation Oncology, Stanford University Medical Center, Stanford, California 94305-5468

Roger Hainsworth (2:137), University of Leeds, Leeds LS2 9JT, United Kingdom

Brian K. Hall (1:781), Department of Biology, Life Sciences Center, Dalhousie University, Halifax, Nova Scotia, Canada B3H 4J1

Roberta L. Hall (6:845), Department of Anthropology, Oregon State University, Corvallis, Oregon 97331

David L. Hamilton (7:233), Department of Psychology, University of California, Santa Barbara, California 93106

Barbara A. Hamkalo (2:465), Department of Molecular Biology and Biochemistry, University of California, Irvine, Irvine, California 92717

Nabil Hanna (4:469), IDEC Pharmaceuticals Inc., 11099 N. Torrey Pines Rd., La Jolla, California 92037

Yusuf A. Hannun (7:179), Division of Hematology/Oncology, Department of Medicine, Duke University, Durham, North Carolina 27710

John A. Hanover (5:451), Laboratory of Biochemistry, and Metabolism, NIDDK, National Institutes of Health, Bethesda, Maryland 20892

Lawrence V. Harper (1:573), Division of Human Development, Department of Applied Behavioral Sciences, University of California, Davis, Davis, California 95616

John B. Harris (7:587), University School of Neurosciences, Muscular Dystrophy Group Research Lab., Newcastle General Hospital, Newcastle upon Tyne NE4 6BE, England

Richard M. Harrison (7:165), Delta Regional Primate Research Center, Tulane University, Covington, Louisiana 70433

J. T. Hart (5:737), Instituut vorr Perceptie Cnderzoek, 5600 MB Eindhoven, The Netherlands

Ernest Hartmann (3:191), Lemuel Shattuck Hospital, Boston, Massachusetts 02130

John H. Hash (1:287, 1:325), Department of Microbiology, School of Medicine, Vanderbilt University, Nashville, Tennessee 37232

John P. Hatch (1:661), Department of Psychiatry, University of Texas, San Antonio, Health Science Center, San Antonio, Texas 78284

John N. Hathcock (7:559), 603 S. Columbus Street, Alexandria, Virginia 22314

Dieter Haussinger (5:871), Medizinische Universitatsklinik, D-7800 Freiburg, Federal Republic of Germany

A. Wallace Hayes (5:259), Duke University Medical Center, Reynolds Boulevard, Winston-Salem, North Carolina 27102

Barton F. Haynes (7:477), Department of Immunology, Box 3258 Medical Center, Duke University, Durham, North Carolina 27710

Anne M. Haywood (4:453), Departments of Pediatrics and Microbiology, University of Rochester, Box 777, Rochester, New York 14642

Anne M. Heacock (5:341), Neuroscience Lab Building, University of Michigan, Ann Arbor, Michigan 48109

Karen K. Hedberg (5:903), Institute of Molecular Biology and Department of Chemis-

try, University of Oregon, Eugene, Oregon 97403-1229

Sverre Heim (2:445), Department of Clinical Genetics, University Hospital, S-221 85 Lund, Sweden

Renata J. Henneberg (2:805), Department of Anatomy and Cell Biology, University of Cape Town Medical School, Observatory 7925, South Africa, and Department of Anatomy and Human Biology, University of the Witwatersrand Medical School, 7 York Road, Parktown 2193, South Africa

Olle Hernell (3:47), Department of Pediatrics, University of Umea, S-901 85 Umea, Sweden

Evan M. Hersh (1:675), Arizona Cancer Center, The University of Arizona, Tucson, Arizona 85724

Herbert Heuer (5:149), Fachbereich Psychologie, Philipps Universitat Gutenbergser, 18, D-3550 Marburg/Lahn, Federal Republic of Germany

Andrew J. Hill (3:589), Lecturer in Behavioural Sciences, Department of Psychiatry, University of Leeds, 15 Hyde Terrace, Leeds LS2 9LT, United Kingdom

Joseph G. Hirschberg (5:561), Cox Science Building, Department of Biology, P.O. Box 749118 Coral Gables, FL 33124

Mark Hnatowich (1:81), Department of Pharmacology, University of Toronto, Toronto, Ontario M5S 1A8 Canada

Myron A. Hofer (2:869), Department of Neuroscience/Psychiatry, Columbia University, College of Physicians and Surgeons, New York, New York 10032

Celia Holland (4:113), Department of Zoology, Trinity College, University of Dublin, Dublin 2, Ireland

Gert Holstege (4:711, 5:43), Department of Anatomy and Embryology, Rijksuniversiteit Groningen, Oostersingel 69, 9713 EZ Groningen, The Netherlands

William D. Hopkins (2:351), Department of Psychology, Georgia State University, University Plaza, Atlanta, Georgia 30303

Michael Horton (1:761), Imperial Cancer Research Fund, Department of Haematology, St. Bartholomew's Hospital London E4A 7BE U.K.

Roger W. Horton (2:841), Department of Pharmacology and Clinical Pharmacology, St. George's Hospital Medical School, University of London, London SW17 ORE, U.K.

A. J. M. Houtsma (0), 1.11, (5:737), Institute voor

Perceptie, Onderzoek, P.O. Box 513, 5600 MB Eindhoven, The Netherlands

Bruce A. Houtchens (7:127), Department of Surgery, The University of Texas, Health Science Center at Houston, Houston, Texas 77030

Colin M. Howles (3:627), Serono Laboratories (UK) Ltd., 99 Bridge Road East, Welwyn Garden City, Hertshire AL7 1BG, England

Wei-Jen W. Huang (1:165), North Chicago VA Medical Center, The Chicago Medical School, Psychology Service 116B, Veterans Administration Medical Center, North Chicago, Illinois 60064

Ulrich Hübscher (3:439), Department of Pharmacology and Biochemistry, Universitat of Zürich-Irchel, Institut fur Pharmakologie und Biochemie, 8057 Zurich, Switzerland

Patrick L. Huddie (2:305), NIH, NINDS, Laboratory of Molecular and Cellular Neurobiology, Park 5, Rm. 431, Bethesda, MD 20892

James M. Hughes (7:639), Infectious Diseases (C-12), Centers for Disease Control, Atlanta, Georgia 30333

J. C. Hutson (7:419), Department of Cell Biology and Anatomy, Texas Tech. University, Health Science Center, Lubbock, Texas 79430

R. J. Hutz (7:345), Department of Biological Sciences, University of Wisconsin-Milwaukee, Milwaukee, Wisconsin 53201

Louis J. Ignarro (3:371), Department of Pharmacology, University of California, Los Angeles, School of Medicine, Los Angeles, California 90024

L. S. Illis (7:221), Clinical Senior Lecturer in Neurology, University of Southampton Medical School, Southampton, Hampshire 509 4XY, United Kingdom

Tadashi Inagami (1:467), Department of Biochemistry, Vanderbilt University School of Medicine, Nashville, Tennessee 37232

Thomas Innerarity (6:23), Gladstone Foundation, Laborator of Cardiovascular Diseases, University of California, San Francisco, San Francisco, California 94140

Tina R. Ivanov (2:431), Scarborough College, University of Toronto, Scarborough Ontario M1C 1A4, Canada

Susuma Iwasa (5:81), Research and Development Division, Takeda Chemical Industries, Ltd.

Ravi Iyengar (7:15), The Mount Sinai Medical Center, Mount Sinai School of Medicine, New York, New York 10029

David B. Jack (3:199), FIDIA Research Laboratories, 35031 Abano Terme, Italy

Craig M. Jackson (4:177), Blood Services Research Laboratory, American Red Cross, P.O. Box 33351, Detroit, Michigan 48232-5351

G. A. Jamieson (6:53), American Red Cross, Research and Development Administration, 15601 Crabbs Branch Way, Rockville, Maryland 20855

Malcolm Jeeves (4:129), Department of Psychology, University of St. Andrews, St. Andrews, Fife KY16 9JU, Scotland, United Kingdom

Derrick B. Jelliffe (3:17), School of Public Health, University of California, Los Angeles, Los Angeles, California 90024

E. F. Patrice Jelliffe (3:17), School of Public Health, University of California, Los Angeles, Los Angeles, California 90024

Walter Jennings (3:749), University of California, Davis, California 95616

Arthur H. Jeske (2:783), Department of Pharmacology, University of Texas Dental Branch at Houston, 7447 Cambridge, #99 Houston, Texas 77054

Göte Johansson (1:355), Department of Biochemistry, Chemical Center, University of Lund, S-221 00 Lund, Sweden

Patricia V. Johnston (3:567), Department of Food Science and Division of Nutritional Sciences, University of Illinois at Urbana-Champaign, Urbana, Illinois 61801

Johannes Jones (6:581), Department of Maternal Fetal Medicine, St. Margaret's Hospital for Women, Department of Obstetrics and Gynecology, Tufts University School of Medicine, Boston MA 02125

Kenneth C. Jones (6:559, 6:587, 7:23; 7:85), Department of Biology, California State University, Northridge, Northridge, California 91330

L. B. Jorde (4:431; 7:733; 7:767), Department of Human Genetics, University of Utah School of Medicine, Salt Lake City, Utah 84132

Janet E. Joy (2:489), Public Health Service, Alcohol, Drug Abuse, and MHA, National Institute of Mental Health, Saint Elizabeths Hospital, Washington, D.C. 20032

David C. Joy (6:735), Department of Zoology, University of Tennessee, Knoxville, Tennessee 37996

Jon H. Kaas (7:119), Department of Psychology, Vanderbilt University, Nashville, Tennessee 37240

Dagmar K. Kalousek (3:293), Department of Pathology, University of British Columbia, Children's Hospital, Vancouver, B.C. V6H 3V4

Ismet Karacan (7:77), Baylor College of Medicine, Sleep Disorders and Research Center, Houston, Texas 77030

Morley R. Kare (5:751, 7:527), Monell Chemical Senses Center, University of Pennsylvania, Philadelphia, Pennsylvania 19104

Neal F. Kassell (2:357, 2:629), Psychology Service (116B), VA Medical Center, North Chicago, Illinois 60064

Frederick H. Kasten (5:561), Cox Science Building, Dept. of Biology, P.O. Box 249118, Coral Gables, FL 33124

Martin M. Katz (1:607), Division of Psychology, Department of Psychiatry, Albert Einstein College of Medicine, Montefiore Medical Center, Bronx, N.Y. 10467

Kenneth H. Kessler (2:629), North Chicago VA Medical Center, The Chicago Medical School, Psychology Service 116B, 3001 Green Bay Rd., North Chicago, IL 60064

Patrick M. Kelley (1:641), Department of Biology, Wayne State University, Detroit, Michigan 48202

Joseph Kennedy, Jr. (6:581), Department of Pediatrics, St. Margaret's Hospital for Women, Tufts University School of Medicine, Boston, MA 02125

Robert Kennison Department of Obstetrics and Gynecology, Tufts University School of Medicine, Boston, MA 02125

Barbara Kent (3:891), Department of Geriatrics, Mt. Sinai School of Medicine, Annenbert 13-30, 1 Gustave L. Levy Place, New York, New York 10029

M. Gabriel Khan (2:167), University of Ottawa, Ottawa, Canada

Felipe Kierszenbaum (7:681), Department of Microbiology, and Public Health, Michigan State University, Giltner Hall, East Lansing, Michigan 48824-1011

Robert B. Kimsey (7:891), Department of Tropical Public Health, Harvard University, School of Public Health, Boston, Massachusetts 02115

Christopher Kirby (7:143), Department of Biochemistry, University of Arizona, Tucson, Arizona 85724

Jan Klein (4:379), Max-Planck-Institut für Biologie, Abteilung Corrensstrasse 42, Im-

mungenetik, D-7400 Tubingen, Federal Republic of Germany

Renate F. Klein (7:511), Cornell University Medical College, New York, NY 10021

Hynda K. Kleinman (4:623), Laboratory of Developmental Biology, National Institute of Dental Research, National Institutes of Health, Bethesda, Maryland 20892

Arthur Kleinman (7:323), Harvard University, Department of Anthropology, Cambridge, Massachusetts 02138

Norman R. Klinman (1:563), Department of Immunology-IMM-6, Scripps Clinic and Research Foundation, La Jolla, California 92037

Willem J. Kolff (1:385, 1:401), Department of Surgery, University of Utah, Salt Lake City, Utah 84112

George P. Knight (1:595), Department of Psychology, Arizona State University, Tempe, Arizona 85287

Cahide Kohen (5:561), Cox Science Building, Department of Biology, P.O. Box 249118, Coral Gables, FL 33124

Elli Kohen (5:561), Cox Science Building, Department of Biology, P.O. Box 249118, Coral Gables, FL 33124

Luci Ann P. Kohn (2:689), Department of Anatomy and Neurobiology, Washington University School of Medicine, St. Louis, Missouri 63110

Bryan Kolb (2:1), Department of Psychology, University of Lethbridge, Lethbridge, Alberta Canada T1K 3M4

Jerry Kolins (1:737), Medical Director, Community Blood Bank of North County and Associate Director of Laboratories, Palomar Pomerado Health System, Escondido, California 92025

Ludwig Kornel (7:257), Professor of Medicine and Biochemistry, and Senior Attending Physician, Rush Medical College and Rush College of Graduate Studies, Rush Presbyterian-St. Luke's Medical Center, 1653 W. Congress Parkway, Chicago, Illinois 60612

Ertugrul Koroglu (7:77), Baylor College of Medicine, Sleep Disorders and Research Center, Houston, Texas 77030

E. P. Köster (5:737), Instituut voor Perceptie Cnderzoek, 5600 MB Eindhoven, The Netherlands

Joachim Krebs (2:89), Department of Biochemistry, Swiss Federal Institute of Technology, Zurich, Switzerland

K. H. E. Kroemer (3:473), Industrial Ergonomics Lab., Virginia Polytechnic Institute, and State University, Blacksburg, Virginia 24061

J. H. A. Kroeze (5:737), Instituut voor Perceptie Cnderzoek, 5600 MB Eindhoven, The Netherlands

George Krol (4:863), Division of Neuroradiology, Memorial Sloan-Kettering Cancer Center, 1275 York Avenue, New York, New York 10021

Lawrence Kruger (7:793), Department of Anatomy, 73323 Chs, UCLA, Los Angeles, California 90024

Susanne Krüger-Kjaer (3:859), The Danish Cancer Registry, Institute of Cancer Epidemiology, Copenhagen, Denmark

Patrick C. Kung (7:395), T-Cell Sciences, Inc., 38 Sidney Street, Cambridge, Massachusetts 02139-4135

Sheree KwongSee (1:119), Department of Psychiatry, McMaster University, Hamilton, Ontario L8N 375, Canada

Bert N. La Du (5:829), Department of Pharmacology, University of Michigan, Ann Arbor, Michigan 48109

Ole Didrik Laerum (3:621), University of Bergen, Department of Pathology, The Gade Institute, Haukeland Hospital, N-5021 Bergen, Norway

Eric Lai (6:379), University of North Carolina, Chapel Hill, North Carolina 27599

David G. Laing (5:759), CSIRO Food Research Laboratory, Box 52, North Ryde, New South Wales 2113, Australia

Joan Lakoski (6:829), Department of Pharmacology, University of Texas Medical Branch, 10th & Market St., J-31, Galveston, Texas 77550

Michael E. Lamm (5:681), Case Western Reserve University, Institute of Pathology, Cleveland, Ohio 44106

Andre Langaney (6:79), Laboratoire de Genetique et Biometrie, Universite de Geneve, Musse de L'homme, 17 Place du Trocadero, Paris, France 75116

J. Larner (93), Department of Pharmacology, Diabetes Research and Training Center, University of Virginia, Charlottesville, Virginia 22901

Zvi Laron (6:339), Beilinson Medical Center, Petah TIQVA, 49100 Israel

Simon K Lawrence (3:475), Scripps Clinic, La Jolla, CA 92037

Robert Leader (5:675), Michigan State University, Center for Environmental Toxicology, East Lansing, Michigan 48824

Robert L. Leahy (2:853), Center for Cognitive Therapy, 30 E. 60 St. Suite 1007, New York, New York 10022

Robert W. Ledeen (3:737; 3:743), Department of Biochemistry, Albert Einstein College of Medicine, 1300 Morris Park Avenue, Bronx, New York 10461

Susane Lederman (7:51), Department of Psychology, Queens University at Kingston, Kingston, Ontario, Canada K7L 3N6

Tong H. Lee (2:511), Department of Psychiatry, Medical Center, Duke University, Box 3870, Durham, North Carolina 27710

Cheang-Kuan Lee (7:361), Department of Chemistry, National Institute of Singapore, 10 Kent Ridge, Singapore 0511

Marjorie B. Lees (5:271), Biochemistry Department, E. K. Shriver Center, Waltham, Massachusetts 02254, and Department of Neurology, Harvard Medical School, Boston, MA 02115

Robert J. Lefkowitz (1:81), Department of Medicine and Biomedicine, Howard Hughes Medical Institute, Duke University Medical Center, Durham, North Carolina 27710

H. Jürgen Lenz (7:285), School of Medicine, Building for Cellular and Molecular Medicine, University of California at San Diego, La Jolla, CA 92093-0648

Donald A. Leopold (5:689), Department of Ortholaryngology, State University Hospital, 750 East Adams Street, Syracuse, NY 13210

Fernand Leroy (3:305), Human Reproduction Research Unit, Free University of Brussels and, Saint Pierre Hospital, St. Pierre Hospital, 1000 Bruxelles, Belgium

Michael D. Levin (6:99), Department of Anthropology, University of Toronto, Toronto, Ontario, Canada M5S 1A1

Michael J. Levine (6:689), Department of Oral Biology, Dental Research Institute, State University of New York, Buffalo, Buffalo, New York 14214

Max Levitan (3:819), Department of Cell Biology and Anatomy, Mt. Sinai School of Medicine, New York, New York 10029

Alexander Levitzki (2:321), Department of Biological Chemistry, Institute of Life Sciences, The Hebrew University of Jerusalem, Givat Harm, Jerusalem, 91904 Israel

Jay A. Levy (1:11), Cancer Research Institute, University of California, San Francisco, School of Medicine, San Francisco, California 94143

Virgilio L. Lew (6:533), Physiological Laboratory, University of Cambridge, Cambridge, United Kingdom

Richard C. Lewontin (6:99), Museum of Comparative Zoology, Harvard University, Cambridge, Massachusetts 02138

Karen Li (7:459), University of Sydney, Sydney, New South Wales 20006, Australia

Robert P. Liberman (6:755), University of California at Los Angeles, Brentwood VA Medical Center, and Camarillo State Hospital, Camarillo, California 93010

Philip Lieberman (4:641), Department of Cognitive and Linguistic Sciences, Brown University, 141 Elton Street, Providence, Rhode Island 02912

Edward H. Livingston (3:755), Department of Surgery, West Los Angeles VA Medical Center, Los Angeles, California 90073

Margaret Lock (4:697), Department of Humanities and Social Studies in Medicine, McGill University, Montreal, Quebec H3G 1Y6, Canada

I. S. Longmuir (5:617), Department of Biochemistry, North Carolina State University, P.O. Box 7622, Raleigh, North Carolina 27650

Daniel S. Longnecker (7:567), Department of Pathology, Dartmouth-Hitchcock Medical Center, Hanover, New Hampshire 03756

Vanda R. Lops (2:403), UCSD Assistant Clinical Professor, Department of Reproductive Medicine, La Jolla, California 92037

Farid Louis (6:581), Department of Pathology, St. Margaret's Hospital for Women, Department of Obstetrics and Gynecology, Tufts University School of Medicine, Boston, MA 02125

Diane L. Lucas (4:177), Division of Blood Diseases and Resources, National Heart, Lung and Blood Institute, National Institutes of Health, Bethesda, Maryland 20892

Henry C. Lukaski (1:667), United States Department of Agriculture, Agricultural Research Center, Human Nutrition Center, Grand Forks, North Dakota 58202

James W. Maas (6:769), Department of Psychiatry, The University of Texas Health Science, Center at San Antonio Medical School, San Antonio, Texas 78284

Robert Macrae (4:205, 4:215), Department of Chemistry, The University, Hull, HU6 7RX, England

Neil B. Madsen (3:919), Department of Biochem-

istry, University of Alberta, Edmonton Alberta T6G 2E1, Canada

James H. P. Main (5:611), Department of Oral Pathology, University of Toronto, Toronto, Ontario, Canada M5G 1G6

Anita C. Maiyar (7:859), Biomedical School, University of California, Riverside, 900 University Avenue, Riverside, California 92521

Frederick D. Malkinson (6:643), Department of Dermatology, Rush-Presbyterian-St. Luke's Medical Center, 1653 W. Congress Parkway, Chicago, IL 60612-3864

Marston Manthorpe (5:333), Department of Biology, M001, UCSD Medical Center, La Jolla, California 92093

John J. Marchalonis (6:505), Department of Microbiology, College of Medicine, University of Arizona, Health Sciences Center, Tucson, Arizona 85724

Alexander R. Margulis (3:1), Department of Radiology, University of California, San Francisco, San Francisco, California 94143

Renato Mariani Costantini (6:523), Institute of Human Pathology, Univerita "G. D. Annuncio," Chieti, Italy

Deborah B. Marin (5:777), Cornell Medical Center, Payne Whitney Clinic, 525 E. 68th St., New York, New York 10021

Ronald W. Maris (7:327), University of South Carolina, Columbia, South Carolina 29208

Jonathan Marks (3:881), Departments of Anthropology and Biology, Yale University, New Haven, Connecticut 06511

Alice Maroudas (1:365), Technion Israel Institute of Technology, Department of Biomedical Engineering, Technion City, Hafia 32000, Israel

Joseph B. Martin (5:393), School of Medicine, University of California, San Francisco, San Francisco, California 94143

Joe L. Martinez (4:673), Department of Psychology, University of California, 3210 Tolman Hall, Berkeley, California 94720

David Mary (2:137), University of Leeds, Leeds LS2 9JT, United Kingdom

Edward J. Masoro (5:477), Department of Physiology, The University of Texas, Health Science Center at San Antonio, San Antonio, Texas 78284

James W. Mass (6:769), Department of Psychiatry, The University of Texas Health Science Center at San Antonio Medical School, San Antonio, Texas 78284

Rajamma Mathew (6:345), New York Medical College, Valhalla, New York 10595

Christopher K. Mathews (5:461), Department of Biochemistry and Biophysics, Oregon State University, Corvallis, Oregon 97331

Rafael Mattera (4:13) Department of Physiology and Biophysics, School of Medicine, Case Western Reserve University, 2109 Adelbert Road, Cleveland, OH 44106

Richard Mattes (5:751), Monel Chemical Senses Center, University of Pennsylvania, Philadelphia, Pennsylvania 19104

Norbert Maus (3:683), University of Gottingen, Department of Psychiatry, 3400 Gottingen, Federal Republic of Germany

Edward R. B. McCabe (5:433), Baylor College of Medicine, Houston, Texas 77030

Linda McCabe (5:433), Baylor College of Medicine, Houston, Texas 77030

Peter P. McCann (6:67), Merrell Dow Research Institute, Merrell Dow Pharmaceuticals, Inc., 2110 Galbraith Rd, Box 156300, Cincinnati, Ohio 45215

Roger O. McClellan (7:575), Chemical Industry Institute of Toxicology, Box 12137, Research Triangle Park, North Carolina 27709

Donald B. McCormick (2:527), Department of Biochemistry, Emory University, 4001 Rollins Research Center, Atlanta, Georgia 30322

Edith G. McGeer (1:231, 4:269, 5:255, 5:667), Department of Psychiatry, Kinsmen Laboratory of Neurological Research, University of British Columbia, Faculty of Medicine, Vancouver, B.C. V6T 1W5, Canada

Jerry R. McGhee (5:137), Department of Microbiology, University of Alabama, UAB Station; BHS 392, Birmingham, Alabama 35294

Henry M. McHenry (3:487), Department of Anthropology, University of California Davis, Davis, CA 95616

L. Jane McNeilage (1:513), Institutes of Medical Research, P.O. Royal Melbourne Hospital, Victoria 3050, Australia

Jeanne M. Meck (2:437), Department of Obstetrics and Gynecology, Division of Genetics, Georgetown University Medical Center, School of Medicine, Washington, D.C. 20007

Robert J. Meier (3:85), Department of Anthropology, Indiana University, Bloomington, Indiana 47405

Leopoldo de Meis (1:53), Department de Bioquimica, Universidade Federal Do Rio De

Janeiro, Instituto de Ciencias Biomedicas, Cidade Universitaria 21910, Rio De Janeiro, RJ Brasil

Frederick A. O. Mendelsohn, (5:415), Department of Medicine, The University of Melbourne, Heidelberg, Victoria, 3084 Australia

Robert Meny (7:315), Pediatrics and Biological Chemistry, The University of Maryland, School of Medicine, Baltimore, Maryland 21201

Carol N. Meredith (2:101, 6:229), Division of Clinical Nutrition, School of Medicine, University of California, Davis, Davis, California 95616

Carl R. Merril (6:191), 2 Winder Court, Rockville, Maryland 20850

Gary B. Mesibov (1:505), Division TEACCH, Department of Psychiatry, University of North Carolina, Chapel Hill, NC 27599

Sarah A. Meyers (6:801), Department of Psychology, University of Minnesota, Twin Cities, Minneapolis, Minnesota 55455

Richard P. Michael (4:929), Department of Psychiatry, Emory University, School of Medicine, Atlanta, Georgia 30332

Noach Milgram (6:149), Department of Psychology, Tel-Aviv University, Ramat-Aviv University, Israel

J. F. A. P. Miller (4:829), Thymus Biology Unit, The Walter and Eliza Hall Institute of Medical Research, Post Office Royal Melbourne Hospital, Victoria 3050, Australia

Dennis D. Miller (5:483), Department of Food Science, Cornell University, Stocking Hall, Ithaca, NY 14850

Walter L. Miller (7:243), Department of Pediatrics, University of California, San Francisco, San Francisco, California 94143

W. Mills (4:791), Center for High Latitude Health Research Studies, University of Alaska, Anchorage, Alaska 99501

Patricia M. Minnes (7:285), Department of Psychology, Queen's University, Kingston, Ontario Canada K7L 3N6

G. F. Mitchell (4:385), The Walter and Eliza Hall Institute of Medical Research, Royal Melbourne Hospital, Victoria 3050, Australia

Felix Mitelman (2:445), Department of Clinical Genetics, University Hospital, Lund, S-221 85 Lund, Sweden

Peter B. Moens (4:949), Department of Biology, York University, Downsview Ontario, Canada M3J 1P3

Edward P. Monaghan (4:241), Department of

Psychology, University of California, Berkeley, California 94720

Kevin T. Morgan (5:689), Chemical Industry Institute, of Toxicology, Box 12187, Research Triangle Park, North Carolina 27709

F. Moriarty (3:251), Queens Orchard, Little Eversden, Cambridge CB3 7HD, United Kingdom

Robert J. Morin (4:783), Harbor-UCLA Medical Center, Torrance, California 90509

Patrice Morliere (5:561), Cox Science Building, Department of Biology, P.O. Box 249118, Coral Gables, FL 33124

Helen Morris (1:311), xczxz Research Institute, Sandoz Pharmaceuticals Corp., East Hanover, New Jersey 07936

Robin D. Morris (2:351), Department of Psychology, Georgia State University, University Plaza, Atlanta, Georgia 30303

Sherie L. Morrison (3:801), Department of Microbiology, Molecular Biology Institute, UCLA, Los Angeles, California 90024

Randall H. Morse (3:125), Laboratory of Cellular and Developmental Biology, NIDDK, NIH, Bethesda, MD 20892

Jacopo P. Mortola (6:567), Department Physiology, McGill University, Montreal, Quebec, Canada H3G 1Y6

B. M. Mueller (4:957), Scripps Clinic and Research Foundation, 10666 N. Torrey Pines Rd., La Jolla, California 92037

Kim T. Mueser (6:755), Assistant Professor of Psychiatry, Medical College of Pennsylvania, Eastern Pennsylvania Psychiatric Inst., Philadelphia, Pennsylvania 19129

Eugenio E. Muller (2:11), Department of Pharmacology, Chemotherapy and Toxicology, University of Milan, Via Vanvitelli 32, Milan, Italy 20129

Martin Muller (2:721), Laboratory for EMI, Institute of Cell Biology, ETH-Zurich

Gregory R. Mundy (4:275), University of Texas, Health Science Center, San Antonio, Texas 78284-7877

Karl Münger (5:635), National Cancer Institute, National Institutes of Health, Bethesda, Maryland 20892

Bryce L. Munger (7:671), Department of Anatomy, The Milton S. Hershey Medical Center, The Pennsylvania State University Medical School, 500 University Drive, P.O. Box 850, Hershey, Pennsylvania 17033

Paul L. Munson (5:657), Department of Pharmacology, School of Medicine, University of North Carolina, Chapel Hill, North Carolina 27599

Masami Muramatsu (3:107), Department of Biochemistry, The University of Tokyo Faculty of Medicine, Hongo, 7-3-1-, Bunkyo-ku Tokyo 113, Japan

Frederick A. Murphy (7:639), Infectious Diseases (C-12), Centers for Disease Control, Public Health Service, 1600 Clifton Road, NE, Atlanta, Georgia 30333

N. B. Myant (2:411), Medical Research Council, Lipoprotein Team, Hammersmith Hospital, Ducane Road, London W12 OHS U.K.

Francene Myers-Steinberg (1:427, 7:845), College of Agricultural and Environmental Sciences, University of California, Davis, Davis, California 95616

Padmanabhan P. Nair (1:623), U.S.D.A. Human Nutrition Research Center, Beltsville, Maryland, 20705, and The Johns Hopkins University School of Hygiene and Public Health, Baltimore, Maryland 20705

Saran A. Narang (3:169), National Research Council of Canada, Division of Biological Sciences, 100 Sussex Drive, Ottawa K1A 0R6, Ottawa, Ontario, Canada

Mitsuhide Naruse (1:467), Department of Internal Medicine, Tokyo Women's Medical College, Shinjuku-Ku, Tokyo, Japan

D. P. Nayak (4:479), Department of Microbiology and Immunology, Jonsson Comprehensive Cancer Center, UCLA School of Medicine, Los Angeles, California 90024-1747

Winifred G. Nayler (2:81), Department of Medicine, University of Melbourne, Austin Hospital, Heidelberg, Victoria 3084, Australia

David S. Nelson (4:853),[1] Kolling Institute of Medical Research, Royal North Shore Hospital of Sydney, St. Leonards NSW 2065, New South Wales, Australia

W. James Nelson (6:59), Institute for Cancer Research, Fox Chase Cancer Center, Philadelphia, Pennsylvania 19111

Buford L. Nichols, Jr. (4:893), Texas Children's Hospital, Baylor College of Medicine, Houston, Texas 77030

Forrest H. Nielsen (7:603), Department of Agriculture, Agricultural Research Service, Grand Forks Human Nutrition Research Ctr., Grand Forks, North Dakota 58202-7166

Marcel E. Nimni (2:559), Professor of Biochemistry, Medicine and Surgery, University of Southern California School of Medicine, Los Angeles, Los Angeles, California 90033

David Njus (1:641), Department of Biological Sciences, Wayne State University, Detroit, Michigan 48202

Julie A. Nordlee (7:337), Department of Food and Technology, University of Nebraska, Lincoln, NB 68583

Anthony W. Norman (7:859), Biomedical School, University of California, Riverside, 900 University Avenue, Riverside, California 92521

Thomas W. North (2:395), University of Montana, Division of Biological Sciences, Missoula, Montana 59812

Stata Elaine Norton (2:341), Department of Pharmacology and Toxicology, University of Kansas Medical Center, Kansas City, Kansas 66103

Robert Nout (3:661), Department of Food Science, Division of Industrial Microbiology, Wageningen Agricultural University, EV Wageningen, The Netherlands

David P. O'Brien (2:619), Department of Psychology, Baruch College, City University of New York, New York, New York 10010

M. John O'Brien (7:315), Pediatrics and Biological Chemistry, The University of Maryland School of Medicine, Baltimore, Maryland 21201

Brian F. O'Dowd (1:81), Department of Pharmacology, University of Toronto, Toronto, Ontario M5S 1A8 Canada, and The Addiction Research Foundation, 33 Russell Street, Toronto, Ontario, Canada M5S 2S1

Irene M. O'Shaughnessy (7:345), Department of Medicine, Division of Endocrinology and Metabolism, Medical College of Wisconsin, Milwaukee, Wisconsin 53226

John E. Obrzut (2:859), Program in Educational Psychology, College of Education, University of Arizona, Tucson, Arizona 85721

Sasumu Ohno (6:837), Department of Biology, Beckman Research Institute of The City of Hope, 1450 E. Duarte Rd. 10, Duarte, California 91001

Chihiro Ohye (7:437, 7:65), Department of Neurosurgery, Gunma University School of Medicine, 3-39 Slowa machi, Maebashi 371 Gunma, Japan

---

1. Deceased.

David M. Ojcius (4:811), Laboratory of Cellular Physiology and Immunology, Rockfeller University, New York, New York 10021

Kathryn C. Oleson (1:197), Department of Psychology, University of Kansas, Lawrence, Kansas 66045

Rafael Oliva (2:475), Berkeley Human Genome Center, Lawrence, Berkeley, CA 94720

S. Jay Olshansky (5:99), Argonne National Laboratory, 9700 S. Cass Ave., Building 301, Argonne, Illinois 60637

Joost J. Oppenheim (2:751), Laboratory of Molecular Immunoregulation, Biological Response Modifies Program, Frederick Cancer Research Facility, National Institutes of Health, NCI, Frederick, Maryland 21701-1013

Grigori Orlavski (4:769), The Nobel Institute for Neurophysiology, Karolinska Institute, Neurofysiologiska Avdelningen, Box 60400, S-104 01 Stockholm, Sweden

Alicia Osimani (2:763), Department of Neurology, Soroka Medical Center, University of Bluburian, P.O. Box 151, Beer Sheba, Israel

Chung Owyang (5:713) University of Michigan Medical Center, 1500 E. Medical Center Dr., Ann Arbor, MI 48109-0368

Eric D. Ozkan (1:641), Department of Biological Sciences, Wayne State University, Detroit, Michigan 48202

Thomas D. Palella (6:387), Department of Internal Medicine, Division of Rheumatology, University of Michigan Medical Center, Ann Arbor, Michigan 48109

A. Michael Parfitt (1:771), Bone and Mineral Research Laboratory, Detroit, Michigan 48202

Lawrence N. Parker (1:71), Professor of Medicine, University of California at Irvine, Endocrinology Section, Veterans Administration Medical Center, Long Beach, California 90822

Robertson Parkman (1:791), Division of Research Immunology/Bone Marrow Transplantation, Children's Hospital of Los Angeles, Los Angeles, California 90027

Toni G. Parmer (5:551), Department of Physiology and Biophysics, University of Illinois, Chicago, Illinois 60612

Jane R. Parnes (2:225), Division of Immunology, Department of Medicine, Stanford University Medical Center, Stanford, California 94305

John A. Parrish (7:71), Wellman Laboratories of Photomedicine, Department of Dermatology, Massachusetts General Hospital, Harvard Medical School, 435 East 70th Street, Apt. 20F, Boston, Massachusetts 10021

Jitendra Patel (6:201), ICI Pharmaceutical Group, ICI Americas, Inc., Senior Research Pharmacologist, Concord Pike, Wilmington, Delaware 19897

John Paul (7:705), 115 Kelvin Drive, Glasgow G208QL, United Kingdom

Robert Pauley (2:53), Michigan Cancer Foundation, 110 E. Warren Avenue, Detroit, Michigan 48201

Antonio Pecile (1:799), Pharcol, Chemother & Medical Toxicol, Milan University, 32 via Vanvitelli, 20129 Milan, Italy

Peter L. Pedersen Department of Biological Chemistry, Johns Hopkins University School of Medicine, Baltimore, MD 21205

Anthony E. Pegg (6:67), Merrell Dow Research Institute, Merrell Dow Pharmaceuticals, Inc., Cincinnati, Ohio 45215

Jay A. Perman (4:611), Johns Hopkins University, Department of Pediatrics, Brady 303, Baltimore, Maryland 21205

Richard E. Perry (5:799), Ohio State University, Columbus, Ohio, 43120

M. F. Perutz (4:135), MRC Laboratory of Molecular Biology, Cambridge CB2 2QH, England

Wallace Peters (4:873), Pulridge House East, Little Gaddesden, Berkhamsted, Herts HP4 1PN, England

Steven F. Petersen (4:195), Departments of Neurology and Neurological Surgery, Washington University Medical School, St. Louis, Missouri 63110

Ronald Pethig (3:9), Institute of Molecular Biomolecular Elec., University of Wales, Bangor Dean Street, Bangor, Gwynedd, LL57 1UT, United Kingdom

Richard E. Petty (5:799), Department of Psychology, Ohio State University, Columbus, Ohio 43120

Herbert Pfister (3:859), Institut für Klinische und Molekulare Virologie, Friedrich-Alexander Universität Erlangen, Federal Republic of Germany

William C. Phelps (5:635), Division of Virology, Burroughs Wellcome Co., Research Triangle Park, North Carolina 27709

James O. Pickles (3:229), Vision, Touch and Hearing Research Center, Department of Physiology and Pharmacology, The University of Queensland, St. Lucia, Qld. 4072, Australia

Glenn F. Pierce (7:499), Department of Experimental Pathology, Amgen, Inc., Thousand Oaks, California 91320

Quentin J. Pittman (4:303), University of Calgary, Neuroscience Research Group, and Department of Medical Physiology, Faculty of Medicine, Calgary, Alberta, T2N 491, Canada

Alexander Pollatsek (6:481), Department of Psychology, University of Massachusetts, Amherst, Massachusetts 01003

Alfredo Pontecorvi (6:523), Institute of Internal Medicine, Universitá "G. D. Annuncio," Chieti, Italy

Annelise A. Pontius (3:215), Harvard Medical School, Boston, Department of Psychiatry, 37 Cumner Street, Hyannis, Massachusetts 02601

Fran Porter (5:625), Department of Pediatrics, Washington University, School of Medicine, 400 S Kingshighway, P.O. Box 14871, St. Louis, Missouri 63178

James S. Porterfield (7:521), Green Valleys, Goodleigh, Barnstaple, Devon EX32 7NH, England

Michael I. Posner (1:479), Department of Psychology, University of Oregon, Eugene, Oregon 97403

George Poste (4:469), Research and Development, Smith Kline & French Laboratories, King of Prussia, Pennsylvania 19406

R. S. Pozos (4:791), Center for High Latitude Health Research, School of Health Science, University of Alaska, 1544 Hidden Lane, Anchorage, Alaska 99501

Ananda S. Prasad (2:649), Department of Internal Medicine, UHC-5-C, Wayne State University Medical School, Detroit, Michigan 48201

Ross L. Prentice (5:599), Division of Public Health Sciences, Fred Hutchinson Cancer Research Center, Seattle, Washington 98104

Theresa P. Pretlow (5:315), Institute of Pathology, Case Western Reserve University, Cleveland, Ohio 44106

Thomas G. Pretlow (5:315), Institute of Pathology, Case Western Reserve University, Cleveland, Ohio 44106

Morton P. Printz (4:283), Department of Pharmacology, University of California, San Diego, Hypertension Specialized, Center of Research, La Jolla, California 92093

Kenneth P. H. Pritzker (2:639), Pathologist-in-Chief, Mount Sinai Hospital, Toronto, Ontario, Canada M5G 1X5

C. Ladd Prosser (1:37), Department of Physiology and Biophysics, University of Illinois, Urbana, Illinois 61801

Volker Pudel (3:683), University of Gottingen, Department of Psychiatry, 3400 Gottingen, Federal Republic of Germany

Anthony Pugsley (6:245), Institut Pasteur, Unite de Genetique Moleculaire, 75724 Paris Cedex 15, France

Helen Pynor (7:459), University of Sydney, Sydney, New South Wales 20006, Australia

Tom Pyszcynski (1:615), Department of Psychology, University of Colorado, Colorado Springs, Colorado 80933-7150

Richard H. Quarles (5:271), National Institute of Neurological Disorders and Stroke, National Institutes of Health

Jose Quintans (4:313), Department of Pathology, Box 414, University of Chicago, Chicago, Illinois 60637

Efraim Racker (3:931), Department of Biochemistry, Molecular and Cell Biology, Cornell University, Ithaca, New York 14853

Ewa Radwanska (1:1), Infertility/Endocrinology, Christ Hospital, Ambulatory Care Unit, Oak Lawn, Illinois 60453

Sharath C. Raja (3:533), Eye Research Laboratories, University of Chicago, Chicago, IL 60637

Stanley I. Rappoport (1:715), National Institute on Aging, National Institutes of Health, Bethesda, Maryland 20892

Howard Rasmussen (3:317), Yale University School of Medicine, Department of Internal Medicine, 333 Cedar Street, New Haven, Connecticut 06510

Kathleen R. Rasmussen (7:737), Biochemistry Department, Texas Tech University, Health Science Center, Lubbock, Texas 79430

Jonathon I. Ravdin (1:217), Case Western Reserve University, School of Medicine, Cleveland, Ohio 44106

Keith Rayner (6:481), Department of Psychology, University of Massachusetts, Amherst, MA 01003

Charles A. Reasner (4:275), University of Texas, Health Science Center, San Antonio, Texas 78284

Jane Mitchell Rees (3:243), Division of Adolescent Medicine, University of Washington, CDMRC WS-10, Seattle, Washington 98195

Arnold Reif (4:337), Mallory Institute of Pathology, Boston University School of Medicine,

Boston City Hospital, Boston, Massachusetts 02118

R. A. Reisfeld (4:957), Scripps Clinic and Research Foundation, 10666 N. Torrey Pines Road, La Jolla, California 92037

Dale Reisner (6:581), Department of Maternal Fetal Medicine, St. Margaret's Hospital for Women, Department of Obstetrics and Gynecology, Tufts University, Boston, MA 02125

Michael Reiss (4:581), Department of Medicine, Yale University, New Haven, Connecticut 06477

Russel J. Reiter (6:1), Department of Cellular and Structural Biology, The University of Texas, San Antonio, San Antonio, Texas 78284-7762

Oween M. Rennert (7:879), Pediatric/Molecular Kinetrics, Georgetown University School of Medicine, 3800 Reservoir Road, N. W., Washington, D.C. 20007

Arnold D. Richards (6:305), New York University College of Medicine, 120 E. 36th St. New York, NY 10016

Darryl Rideout (2:371), Laboratory of Biochemical Pharmacology, Memorial Sloan-Kettering Cancer Center and Cornell University Graduate School of Medicine Sciences, New York, NY 10021

C. R. Robbins (4:39, 7:41), Colgate Palmolive Company, 909 River Rd., Piscataway, New Jersey 08854

Jacob Robbins (7:483), Department of Health and Human Services, National Institutes of Health, NIADDK, Bethesda, Maryland 20892

Jon Robertus (4:415), Clayton Foundation Biochemical Institute, Department of Chemistry, University of Texas, Austin, Austin, Texas 78712

Peter J. Robinson (1:715), Human and Environmental Safety Division, The Procter & Gamble Company, Miami Valley Laboratories, Cincinnati, Ohio 45239

William S. Robinson (4:755, 6:), Division of Infectious Diseases, Department of Medicine, Stanford University, School of Medicine, Stanford University Medical Center, Stanford, California 94305-5107

Sara Rockwell (6:435), Department of Therapeutic Radiology, Yale University School of Medicine, New Haven, Connecticut 06510-8040

Enrique Rodriguez-Boulan (6:59), Department of Cell Biology and Anatomy, Cornell University Medical College, New York, New York 10021

Peter J. Rogers (6:723), PsychoBiology Group, Institute of Food Research, AFRC Institute, Shinfield RG2 9AT, United Kingdom

Bernard Roizman (4:177), The Majorie B. Kovler Viral Oncology Laboratories, The University of Chicago, Chicago, Illinois 60637

Frank M. Rombouts (3:661), Department of Food Science, Division of Industrial Microbiology, Wageningen Agricultural University, 6700 Wageningen EV, The Netherlands

Mary Ann Romski (4:633), Language Research Center, 3401 Pantherville Rd., Decatur, Georgia 30034

Noel R. Rose (1:519), Professor and Chairman of Immunology and Infectious Diseases, The Johns Hopkins University, School of Hygiene and Public Health, Baltimore, MD 21205

Allan Rosenfield (4:913), Dean, School of Public Health, Columbia University, New York, New York 10032

Gordon D. Ross (2:601), Department Microbiology and Immunology, University of Louisville, School of Medicine, Louisville, Kentucky 40292

Paolo Rossi (6:523), Instituto di Clinica Pediatrica, Universitá Tor Vergata, Rome, Italy

Arthur H. Rossof (4:749), Department of Internal Medicine, Rush Medical College, Section of Medical Oncology, Chicago, Illinois 60612

Carl F. Rothe (7:757), Department of Physiology and Biophysics, Indiana University School of Medicine, 635 Barnhill Drive, Building 374, Indianapolis, Indiana 46223

M. A. Rouf (6:701), Department of Biology/Microbiology, University of Wisconsin-Oshkosh, Oshkosh, Wisconsin 54901

J. A. J. Rouffs (5:737), Institute voor Perceptie Cnderzoek, P.O. Box 513, 5600 MB Eindhoven, The Netherlands

Francois Rouyer (6:283), Unite de Recombinaison et Expression Genetique, Institut Pasteur, 75724 Paris, Cedex 15, France

Robert B. Rucker (1:427, 3:261, 7:845, 7:873), Department of Nutrition #2750, College of Agricultural and Environmental Sciences, University of California, Davis, Davis, California 95616-8669

Duane M. Rumbaugh (2:351, 4:633), Department of Psychology, Georgia State University, University Plaza, Atlanta, Georgia 30303

Janet B. Ruscher (1:101), Department of Psychology, University of Massachusetts at Amherst, Amherst, Massachusetts 01003

Ellen Bouchard Ryan (1:119), Department of Psychiatry, and Office of Gerontological Studies, McMaster University, Hamilton, Ontario L8N 375, Canada

Jose M. Saavedra (4:611, 4:611), Johns Hopkins University School of Medicine, Department of Pediatrics, Baltimore, Maryland 21205

David H. Sachs (4:357), Chief, Immunology Branch, National Cancer Institute, National Institutes of Health, Bethesda, Maryland 20892

Juan C. Saez (2:267), Department of Neuroscience, Albert Einstein College of Medicine, Bronx, New York 10461

Petri Salvén (5:545), Departments of Pathology and Virology, University of Helsinki, 00290 Helsinki 29 Finland and German Cancer Research Center, Im Nevenheimer Feld, Heidelberg, Federal Republic of Germany

Willis K. Samson (6:9), Department of Anatomy, University of Missouri School of Medicine, M304 Health Sciences Center, Columbia, Missouri 65212

Merton Sandler (3:393), Department of Chemical Pathology, University of London, Queen Charlotte's and Chelsea Hospital, W6 OXG United Kingdom

Rene Santus (5:561), Cox Science Building, Department of Biology, P.O. Box 249118, Coral Gables, FL 33124

Vicki R. Sara (4:503), Department of Pathology, Karolinska Institute, Box 60500, 10401 Stockholm, Sweden

Bernard G. Sarnat (2:697), University of California, Los Angeles, School of Dentistry, Center for The Health Sciences, Los Angeles, California 90024

Tomio Sasaki (2:357), Department of Neurosurgery, University of Tokyo Hospital, 7-3-1 Hongo, Binkyo-ku, Tokyo (113), Japan

Tomi K. Sawyer (5:725), The Upjohn Company, 301 Henrietta St., Kalamazoo, Michigan 49001

Anthony J. Sbarra (6:581), St. Margaret's Hospital for Women, Director of Medical Research and Labs., Tufts University School of Medicine, Boston, Massachusetts 02125

Richard Scanlan (3:671), Department of Food Science and Toxicology, Oregon State University, Corvallis, Oregon 97331

H. K. Schachman (7:711), Department of Molecular Biology, University of California, Berkeley, Berkeley, California 94720

Robert S. Schenken (3:361), Department of Obstetrics/Gynecology, The University of Texas, Health Science Center, San Antonio, Texas 78284

Detlev Schild (5:523), Zentrum Physiologie und Pathophysiolgie, der Universitat Gottingen, D-3400 Gottingen, Federal Republic of Germany

Paul Schimmel (5:75), Department of Biology, Massachusetts Institute of Technology, Cambridge, Massachusetts 02139

Milton J. Schlesinger (4:99), Department of Molecular Microbiology, Washington Univ. Medical School, P.O. BOX 8230, St. Louis, Missouri 63110-1093

Richard A. Schmidt (5:121), Department of Psychology, University of California, Los Angeles, Los Angeles, California 90024

Lawrence J. Schneiderman (1:655), Division of Health Care Science, School of Medicine, University of California, San Diego, La Jolla, California 92093

Jean Schoenen (7:191), Department of Neurology and Clinical, Neurophysiology, Institute of Medicine, University of Liege, B 4000 Leige, Belgium

Clifton Schor (1:627), University of California at Berkeley, School of Optometry, Berkeley, California 94720

Larry E. Schrader (1:139), College of Agriculture and Home Economics, Washington State University, Pullman, Washington 99164

Martin C. Schultz (7:153), Department of Communication Disorders and Sciences, Southern Illinois University, Carbondale, Illinois 62901

Manfred Schwab (5:545), Departments of Pathology and Virology, University of Helsinki, 00290 Helsinki 29 Finland and German Cancer Research Center, Im Nevenheimer Feld, Heidelberg, Federal Republic of Germany

Robert K. Scopes (3:403), Centre for Protein and Enzyme Technology, La Trobe University, Bundoora, Victoria 3083, Australia

Clay W. Scott (6:201), ICI Pharmaceutical Group, ICI Americas, Inc., Senior Research Pharmacologist, Concord Pike, Wilmington, Delaware 19897

G. Richard Scott (2:789), Department of Anthropology, University of Alaska Fairbanks, Fairbanks, Alaska 99775

Frederick J. Seil (6:37), Office of Regeneration Research, Veterans Administration Medical Center, Portland, Oregon 97201

Ermias Seleshi (6:769), Department of Psychiatry, The University of Texas Health Science Center at San Antonio Medical School, San Antonio, Texas 78284

Martin E. P. Seligman (4:655), Department of Psychology, School of Arts and Sciences, University of Pennsylvania, Philadelphia, Pennsylvania 19104

Daniel P. Selivonchick (3:671), Department of Food Science and Toxicology, Oregon State University, Corvallis, Oregon 97331

Stewart Sell (4:401), Department of Pathology and Laboratory Medicine, Medical School, University of Texas, Houston, Texas 77030

Robert Rasiah Selvendran (3:35), AFRC Institute of Food Research, Norwich Laboratory, Norwich, NR4 7UA, United Kingdom

Barry J. Sessle (5:361), Department of Physiology, University of Toronto, Faculty of Medicine, Toronto, Ontario M5G 1G6, Canada

Rose A. Sevcik (4:633), Language Research Center, Georgia State University, University Plaza, Atlanta, Georgia 30303

Eldon A. Shaffer (3:71), Head, Division of Gastroenterology, Associate Dean (Clinical Services), University of Calgary, Calgary, Alberta Canada T2N 4N1

Christo Shakr (6:581), St. Margaret's Hospital for Women, Tufts University School of Medicine, Boston, MA 02125

Prem M. Sharma (5:205), Salk Institute of Biological Studies, 10010 N. Torrey Pines Road, La Jolla, California 92186-5800

Christoper R. Shea (7:65), Wellman Laboratories of Photomedicine, Department of Dermatology, Massachusetts General Hospital, Harvard Medical School, 435 East 70th Street, Apt. 20F, Boston, Massachusetts 10021

David Shemin (6:117), 33 Lawrence Farm Road, Woods Hole, Massachusetts 02543

Thomas H. Shepard (7:411), Department of Pediatrics, RD-20, School of Medicine, University of Washington, Seattle, Washington 98195

C. J. R. Sheppard (6:747), School of Physics, The University of Sydney, Sydney, NSW 2006, Australia

Irwin W. Sherman (1:261), Department of Biology, University of California, Riverside, Riverside, California 92521

Priyattam J. Shiromani (7:71), Department of Psychiatry/V-116A, University of California, San Diego, La Jolla, California 92093

Stephen B. Shohet (6:537), MacMillan-Cargill Hematology Research Laboratory, Box 0128, University of California, San Francisco, California 94143

Becky A. Sigmon (6:877), Department of Anthropology, University of Toronto, in Mississauga, Mississauga, Ontario L5L 1C6, Canada

Roy A. Sikstrom (6:87), IRCM, Clinical Research Institute of Montreal, Canada H2W 1R7

Artemis P. Simopoulos (5:533), Director, The Center for Genetics, Nutrition and Health, 2001 S. St. NW, Suite 530, Washington, DC 20009

Nich. R. Stc. Sinclair (1:297), Department of Microbiology and Immunology, University of Western Ontario, London, Ontario, Canada, N6A 5C1

S. Eva Singletary (4:887), Department of General Surgery, The University of Texas, M. D. Anderson Cancer Center, Houston, Texas 77030

Pierre C. Sizonenkó (3:337), Division of Biology of Growth and Reproduction, Department of Pediatrics, University of Geneva Medical School, Geneva, Switzerland, Hopital Cantonal Universitaire, 30, Bld de la Cluse, 1211 Geneva 4, Switzerland

Philip Skehan (5:321), Developmental Therapeutics Program, National Cancer Institute, Bldg. 37-Rm. 5012 NIH, Bethesda, MD 20892

D. N. Skilleter (7:545), Toxicology Unit, Medical Research Council Laboratories, Carshalton, Surrey SM5 4EF, United Kingdom

Donald Small (4:725), Boston University School of Medicine, Boston, Massachusetts 02118

Cassandra L. Smith (2:475), Human Genome Center, Lawrence Berkeley Laboratories, 1 Cyclotron Road, Mailstop 74-3110, Berkeley, California 94720

Clark W. Smith (5:725, 7:879), The Upjohn Company, 7240-209-6, Kalamazoo, Michigan 49001

Mark Snyder (6:801), Department of Psychology, University of Minnesota, Twin Cities, Minneapolis, Minnesota 55455

Joseph Sorge (3:777), Stratagene, 11099 N. Torrey Pines Road, La Jolla, California 92037

Harold C. Sox, Jr. (4:61), Dartmouth-Hitchcock Medical Center, Hanover, New Hampshire 03756

John D. Speth (2:189), University of Michigan, School of American Research, P.O. Box 2188, Santa Fe, NM 87504-2188

Andrew I. Spielman (7:527), Monell Chemical

Senses Center, University of Pennsylvania, Philadelphia, PA 19104

Andrew Spielman (7:891), New York College of Dentistry, Department of Oral Medicine and Pathology, 345 E. 24th Street, N.Y., N.Y. 10010

John Spika (4:443), Laboratory Centre for Disease Control, Tunney's Pasture, Ottawa, Ontario, Canada

Harlan J. Spjut (7:695), Department of Pathology, Baylor College of Medicine, 1200 Moursund Avenue, Houston, Texas 77030

David C. Spray (2:267), Department of Neuroscience, Albert Einstein College of Medicine, Bronx, New York 10461

Larry R. Squire (5:401), Department of Psychiatry, University of California, San Diego, San Diego, California 92161

George R. Stark (3:101), Imperial Cancer Research Fund, Lincoln's Inn Fields, London WC2A 3PX, United Kingdom

Theodore L. Steck (4:969), Department of Biochemistry and Molecular Biology, The University of Chicago, Chicago, Illinois 60637

Bertil Steen (3:899), Department of Geriatric Medicine, Gothenburg University, Sweden, Vasa Hospital, S-411 33 Gothenburg, Sweden

Gary S. Stein (4:235), Department of Cell Biology, University of Massachusetts Medical School, Worcester, Massachusetts 01655

Janet L. Stein (4:235), Department of Cell Biology, University of Massachusetts Medical School, Worcester, Massachusetts 01655

Catia Sternini (7:793), UCLA, Los Angeles, California 90024

G. B. Stokes (5:471), Expert Access Pty Ltd., 61 Cargill St., Victoria Park, Western Australia 6100

George C. Stone (4:69), University of California, San Francisco, Graduate Program in Health Psychology, Center for Social & Behavioral Sciences, 1350 Seventh Avenue, San Francisco, California 94143-0844

Trevor W. Stone (5:5), Department of Pharmacology, University of Glasgow, Glasgow G12 8QQ, Scotland, England

Thomas P. Stossel (5:427), Hematology-Oncology Unit, Massachusetts General Hospital, Department of Medicine, Harvard Medical School, Boston, Massachusetts 02114

J. Wayne Streilein (4:391), Department of Microbiology and Immunology, University of Miami School of Medicine, Miami, Florida 33133

Steven J. Stroessner (7:233), Department of Psychology, University of California, Santa Barbara, California 93106

Osias Stutman (5:291), Immunology Program, Memorial Sloan Kettering Institute, New York, New York 10021

Siva Subramanian (7:879), Pediatric/Molecular Kinetrics, Georgetown University School of Medicine, 3800 Reservoir Road, N. W., Washington, D.C. 20007

Keith E. Suckling (1:447), SmithKline Beecham Pharmaceuticals, The Frythe, Welwyn, Hertfordshire AL6 9AR, England

Yukio Sugino (5:81), Central Research Division, Takeda Chemical Industries Limited, 17-25 Jusholtonmachi Z-chrome Yodogawa-ku, Osaka 532, JAPAN

H. Eldon Sutton (3:791, 3:847, 5:243), Department of Zoology, University of Texas, Austin, Texas 78712

A. Suzuki (3:725), The Tokyo Metropolitan Institute of Medical Science, Honkomagome 3-18-22, Bunkyo-ku Tokyo 113, Japan

Thomas Ming Swi Chang (1:377) Director, Artificial Cells and Organs Research Center, Departments of Physiology, Medicine and Biomedical Engineering, MRC Career Investigator, Faculty of Medicine, McGill University, Montreal, PQ, Canada H3G 1Y6

Stephan Swinnen (5:105), Institut voor Lichamelijke Opleiding, Groep Medische Wetenschappen, Catholic University of Leuven, 3030 herverlee, Belgium

Megan Sykes (4:357), Chief, Immunology Branch, National Cancer Institute, National Institutes of Health, Bethesda, Maryland 20892

Yvette Tache (2:31), Department of Medicine, University of California, Los Angeles, Medical Center and Brain Research Institute, Los Angeles, California 90073

George E. Taffet (3:515), Department of Medicine, Baylor College of Medicine M-320, Houston, Texas 77030

Eng M. Tan (1:331), Scripps Clinic and Research Foundation, La Jolla, California 92037

Tadatsugu Taniguchi (4:527), Institute for Molecular and Cellular Biology, Osaka University, 1-3 Yamada-Oka, Suita, Osaka, 565, Japan

Charlotte A. Tate (3:515), Department of Medicine, Baylor College of Medicine, Houston, Texas 77030

Ian Tattersall (6:135), American Museum of Natural History, New York, New York 10024

Mehdi Tavassoli (4:123), Veterans Administration, University of Mississippi School of Medicine, Jackson, Mississippi 39216

Aubrey E. Taylor (5:31), Department of Physiology, University of South Alabama Medical College, Mobile, Alabama 36688

Charles W. Taylor (1:675), Arizona Cancer Center, University of Arizona, Tucson, Arizona 85724

Steve L. Taylor (7:337), Department of Food and Technology, University of Nebraska, Lincoln, Lincoln, Nebraska 68583

Beverly Taylor Sher (4:1007), 3808C E. Steeplechase Way, Williamsburg, Virginia 23185

Henry Tedeschi (2:235, 2:291), Department of Biological Science, State University of New York, Albany, Albany, New York 12222

Davida Y. Teller (2:575), Department of Psychology, NI-25, University of Washington, Seattle, Washington 98195

Howard M. Temin (6:653), McArdle Laboratory, University of Wisconsin, Madison, Wisconsin 53706

Fred C. Tenover (1:551), Antimicrobics Investigative Branch, Centers for Disease Control, Atlanta, GA 30333

Ana-Zully Teran (1:691), Department of Physiology and Endocrinology, Medical College of Georgia, Osteoporosis Research Institute, Augusta, Georgia 30904

Ruedi F. Thoeni (3:1), Associate Professor of Radiology, Chief, Section Gastrointestinal Imaging, University of California, San Francisco, San Francisco, California 94143

Forrest D. Tierson (3:31, 6:123), Department of Anthropology, University of Colorado, Colorado Springs, Colorado 80933-7150

J. Tyson Tildon (7:315), Pediatrics and Biological Chemistry, The University of Maryland, School of Medicine, Baltimore, Maryland 21201

Jean-Paul Tillement (1:707), Department de Pharmacologie, Faculte de Medecine, F-94010 Creteil, 8 Rue du General Sarrail, France

Donald H. Tinker (3:261), Department of Nutrition #2750, University of California, Davis, California 95616

Donald J. Tipper (1:271), Department of Molecular Genetics/Microbiology, University of Massachusetts Medical School, Worcester, Massachusetts 01655

Marc E. Tischler (7:143), Department of Biochemistry, University of Arizona, Tucson, Arizona 85724

John Tooby (6:493), Department of Psychology, Stanford University, Stanford, California 94305

P. Tothill (7:537), Department of Medical Physics and Medical Engineering, University of Edinburgh, Western General Hospital, 15 Craiglockhart Terrace, Edinburgh, EH1 1A, United Kingdom

Harry C. Triandis (1:485), Psychology/329 Psych., University of Illinois, Champaign, Illinois 61820

Brenda J. Tripathi (3:533), Eye Research Laboratories, University of Chicago, Chicago, Illinois 60637

Ramesh C. Tripathi (3:533), Eye Research Laboratories, University of Chicago, Chicago, Illinois 60637

Lap-Chee Tsui (2:731), The Hospital for Sick Children, University of Toronto, 555 University Ave., Toronto, Ontario M5G 1X8 Canada

Edward G. D. Tuddenham (4:151), Clinical Research Centre, Haemostasis Research Group, In Assoc. with Northwick Park Hospital, Harrow Middlesex, HA1 3UJ, England

Hely Tuorila (1:497), Department of Food Chemistry and Technology, University of Helsinki, Viikki SF-00710 Helsinki, Finland

Anne Uecker (2:859), Department of Psychology, College of Arts and Sciences, University of Arizona, Tucson, Arizona 85721

Jill Urban (1:365), University Laboratory of Physiology, Oxford University, Parks Road, Oxford OX1 3PT England

F. L. Van Nes (5:737), Instituut voor Perceptie Cnderzoek 5600 MB Eindhoven, The Netherlands

Dennis E. Vance (883), Department of Biochemistry, University of Alberta, Edmonton, Alberta T6G, Canada

Steven G. Vandenberg (3:835), Muenzinger Psychology Building, University of Colorado, Boulder, Colorado 80309

Silvio S. Varon (5:333), Department of Biology, School of Medicine, UCSD Medical Center, La Jolla, California 92093

Inder M. Verma (3:701), The Salk Institute, Molecular Biology and Virology Laboratory, 10010 N. Torrey Pines Road, La Jolla, California 92037

Antonia Vernadakis (1:433), Department of Psychiatry, University of Colorado School of Medicine, Denver, Colorado 80262

A. Versprille (6:357), Erasmus Universiteit Rotterdam, 3000 DR Rotterdam, Netherlands

Marc Verstraete (2:179), Center for Thrombosis and Vascular Research, Katholieke Universiteit Leuven, Herestraat 49 B-3000 Leuven, Belgium

Claude A. Villee (6:681), Department of Biological Chemistry, Harvard Medical School, Boston, Massachusetts 02115

Adrian O. Vladutiu (4:573), Departments of Pathology, Microbiology, and Medicine, School of Medicine and Biomedical Sciences, State University of New York, Buffalo, New York 14203

Georgirene D. Vladutiu (4:573), The Buffalo General Hospital, Buffalo, New York 14203

Carl-Wilhelm Vogel (4:367), Department of Biochemistry and Molecular Biology, Department of Medicine, and Vincent T. Lombardi Cancer Center, Georgetown University School of Medicine, Washington, D.C. 20007

Gunnar Von Heijne (3:163; 6:237), Department of Molecular Biology, Karolina Institute Center for Biotechnology 5-141-52, Huddinge, Sweden

James L. Voogt (4:611), Department of Physiology, University of Kansas School of Medicine, Lawrence, Kansas, 66045

Robert S. Wallerstein (6:293), Department of Psychiatry and, Langley Porter Psychiatric Institute, 401 Parnassus Ave., San Francisco, California 94143

Margareta Wallin (5:11), Department of Zoophysiology, Comparative Neuroscience Unit, University of Goteberg, S-400 31, Goteberg, Sweden

Adrian R. Walmsley (2:279), Department of Biochemistry, The Adrian Building, University of Leicester, Leicester LE1 7RH, United Kingdom

David Walsh (7:447), Department of Veterinary Sciences Building, University of Sydney, Sydney, New South Wales 20006, Australia

Harry Walter (1:355), Laboratory of Chemical Biology, Veterans Affairs Medical Center, Long Beach, California 90822

Lawrence Warwick Evans (6:329), Department of Psychology, University of Southampton, Southampton SO9 5NH, United Kingdom

Paul M. Wassarman (6:505), Department of Cell and Developmental Biology, Roche Institute of Molecular Biology, Nutley, Kingsland Street, New Jersey 07110

Meryl E. Wastney (7:873), Pediatric/Molecular Kinetrics, Georgetown University Medical Center, Washington, D.C. 20007

David Weatherall (4:145, Institute of Molecular Medicine, John Radcliffe Hospital, Nuffield Department of Clinical Medicine, University of Oxford, Oxford OX3 9DU, United Kingdom

Benjamin S. Weeks (4:623), Lab. for Developmental Biology and Anomalies, National Institute of Dental Research, National Institutes of Health, Bethesda, Maryland 20892

Howard L. Weiner (5:143), Multiple Sclerosis Research, Brigham and Women's Hospital, Boston, Massachusetts 02115

Claire E. Weinstein (7:309), Department of Educational Psychology, University of Texas, Austin, Austin, Texas 78731

Carol D. Weiss (1:11), Cancer Research Institute, University of California, San Francisco, School of Medicine, San Francisco, California 94143

Bruce Werness (635), Laboratory of Tumor Virus Biology, National Cancer Institute, Bethesda, Maryland 20892

Gary R. West (6:595), Center for Prevention Services, Center for Disease Control, 1600 Clifton Road, Atlanta, GA 30333

John B. West (6:595), Department of Medicine M-023A, University of California, San Diego, School of Medicine, La Jolla, California 92093

Scott Wetzler (1:607), Department of Psychiatry, Bronx Municipal Hospital, Bronx, New York 10461

Sandra M. Wheely (7:737), Biochemistry Department, Texas Tech University, Health Science Center, Lubbock, Texas 79430

Ian Q. Whishaw (2:1), Department of Psychology, University of Lethbridge, Lethbridge, Alberta, Canada T1K 3M4

David O. White (7:771), University of Melbourne, Department of Microbiology, Parkville, Victoria 3052, Australia

Senga Whittingham (1:513), Burnet Clinical Research Unit, Institutes of Medical Research, P.O. Royal Melbourne Hospital, Victoria 3050, Australia

Thomas A. Widiger (5:777), Department of Psychology, University of Kentucky, Lexington, Kentucky 40506

Jonathan Widom (2:419), Departments of Chemistry, Biochemistry, and Biophysics, and Beckman Institute, University of Illinois, 505 South Mathews Avenue, Urbana, Illinois 61801

Thomas N. Wight (6:257), Department of Pathol-

ogy SM-30, School of Medicine, University of Washington, Seattle, WA 98195

Hans Wigzell (4:319), Department of Immunology, Tjänste, Statens Bakteriologiska, Laboratorium, Karolinska Institute, 105-21 Stockholm, Sweden

David A. Wilder (7:89, 7:89), Department of Psychology, Rutgers University, New Brunswick, NJ 08903

Lee Willerman (6:855), Department of Psychology, University of Texas, Austin, Austin, Texas 78712

David E. Williams (3:671), Department of Food Science and Toxicology, Oregon State University, Corvallis, Oregon 97331

R. J. P. Williams (5:47), Department of Inorganic Chemistry, University of Oxford, Oxford OX1 3QR, England

Stephen Williams (7:383), Division of Neuroscience, Baylor College of Medicine, Houston, TX 77030

Paul Willner (2:831), Department of Psychology, City of London Polytechnic, London, E1 7NT, England

Brian C. Wilson (5:587), Hamilton Regional Cancer Center, The Ontario Cancer Treatment and Research Foundation, 711 Concession Street, Hamilton, Ontario, Canada L8V 1C3

Hugh R. Wilson (7:815), Department of Ophthalmology and Visual Science, University of Chicago, Eye Research Laboratories, 939 E 57th Street, Chicago, Illinois 60637

Savio L. C. Woo (5:863), Howard Hughes Medical Institute, Department of Cell Biology and Institute of Molecular Genetics, Baylor College of Medicine, Houston, Texas 77030

Michael Woodruff (7:687), The Bield, Juniper Green, University of Edinburgh, Edinburgh EH14 5DH, Scotland, United Kingdom

Gayle E. Woodson (4:647, 4:653), University of California, San Diego, Veterans Administration Medical Center, San Diego, California 92161

Ira G. Wool (6:671), Department of Biochemistry, University of Chicago, Chicago, Illinois 60637

Ronald G. Worton (5:191), Genetics Department and Research Institute, The Hospital for Sick Children, 555 University Avenue, Toronto, Ontario, Canada M5G 1X8

Yue Wu (2:475), Human Genome Center, Lawrence Berkeley Laboratories, Berkeley, CA 94720

Rosalyn S. Yalow (6:455), Veteran Administration Hospital Medical Center, Bronx, New York 10468

Kenneth M. Yamada (3:523), National Institutes of Health, Building 30, Room 414, NIH, Bethesda, Maryland 20892

Tadataka Yamada (5:713), Gastroenterology Research Laboratory, University of Michigan Medical Center, Ann Arbor, Michigan 48109

T. Yamakawa (3:725), The Tokyo Metropolitan Institute, of Medical Science, 18-22 Honkomagome 3 Chome, Bunkyo-ku Tokyo 113, Japan

Ashley J. P. Yates (4:275), University of Texas, Health Science Center, San Antonio, Texas 78284

John Ding-E Young (4:811), Rockfeller University, New York, New York 10021

Maurizzio Zanetti (7:745), Department of Medicine, University of California Medical Center, San Diego, San Diego, California 92103

Lue Ping Zhao (5:599), Division of Public Health Sciences, Fred Hutchinson Cancer Research Center, 1124 Columbia Street, Seattle, WA 98104

Karl Zilles (2:673), Anatomisches Institut, Universität zu köln, D-5000 Köln 41, West Germany

Warren E. Zimmer (2:41), University of South Alabama, College of Medicine, Structural and Cellular Biology Department, 2042 Medical Science Building, Mobile, Alabama 31688

Ari Zimran (3:777), Shaare Zedek Medical Center, Jerusalem Departement de Pharmacologie, Faculte de Medecine de Paris XII, F-94010 Creteil, 8 Rue du General Sarrail, France

R. A. Zoeller (4:725), Boston University, School of Medicine, Boston, Massachusetts 02118

Linda Zuckerman (4:313), Committee on Immunology, University of Chicago, 5841 S. Maryland, Chicago, IL 60637

Marvin Zuckerman (6:809), Department of Psychology, University of Delaware, Newark, Delaware 19716

D. Zumpe (4:929), Department of Psychiatry, Emory University School of Medicine, Room 504N, Georgia Mental Health Institute, 1256 Briarcliff Road NE, Atlanta, GA 30306

Hans H. J. Zwart (4:89), 33 West Rahn Road, Suite 202, Dayton, OH 45429-2219

SUBJECT INDEX

A

Abbott, Maude, **5**:703
Abdomen
    computed tomography of, **3**:6
    magnetic resonance imaging of, **3**:6
    plain film radiography of, **3**:2–3
Abdominal neoplasms, radiation induction of, **6**:479
Abdominal wall muscles, **4**:263
Abduction, as joint movement, **5**:133
Abnormalities, **1**:697–705, **7**:411–18. *see also* Chromosome abnormalities; Teratogens
    achondroplasia, **1**:699
    animal models, **1**:257
    caffeine and, **2**:76
    causes of, **1**:698–701, **7**:413
    and cell death, **2**:250–51
        in cleft palate, **2**:251
        in pseudohermaphroditism, **2**:251
        in syncactylism, **2**:251
    chromosomal aberrations and, **3**:821
    chromosome abnormalities, **1**:699–700
    common, prevalence of, **1**:702(t)
    congenital malformation, **2**:76
    defined, **7**:406, 412
    deformations, **1**:698
    dermatoglyphics and, **3**:92
    DiGeorge anomaly, **1**:698
    disruptions, **1**:697–98
    dose of teratogen and, **7**:414
    Down's syndrome. *see* Down's syndrome
    dysplasias, **1**:698
    embryonic crown-rump length and, **3**:296
    environmental causes, **1**:700–701
    epidemiology, **1**:701–2, 704
    field defects, **1**:698
    hydrocephalus, X-linked, **1**:699
    from imbalances of naturally occurring peptides, **5**:734
    isotretinoin teratogenicity syndrome, **1**:698
    Meckel-Gruber syndrome, **1**:699
    multiple anomalies, **1**:698
    pathogenesis, **1**:697–98
    polygenic/multifactorial causes, **1**:700
    prenatal diagnosis of, **5**:703
    prevention of, **1**:703–4

Abnormalities *(Continued)*
    rate of, during prenatal life, **7**:412–13
    research on, **1**:704–5
    risk factors, **1**:701
    screening for, **5**:433–38
    sequences, **1**:698
    single-gene defects, **1**:698–99
    sources of information on, **7**:417
    species specificity and, **7**:414
    sporadic, **1**:701
    surgery for, **5**:707
    syndromes, **1**:698
    teratogenic causes. *see* Abnormalities, drug-induced
    as teratogenic defect, **7**:406
    treatment of, **1**:702–3
    trisomy 13, **1**:699–700
    trisomy 18, **1**:699
    trisomy 21. *see* Down's syndrome
    in twins, **3**:313
    of unknown etiology, **1**:701
    VACTERL association, **1**:698
Abnormalities, drug-induced, **7**:405–10
    conditions for teratogenicity, **1**:701
    functional disorders associated with, **7**:408
    mechanisms and pathogenesis of, **7**:418
    mechanisms causing, **7**:409
    phocomelia, **1**:701
    specificity of agents causing, **7**:408
    stage of organ development as factor in, **7**:408–9
    structural abnormalities as, **7**:406
    structural anomalies that represent, **7**:407
    susceptibility to, **7**:409
    teratogenic conditions, **1**:701
    thalidomide-related defects, **1**:701
    thalidomide use and, **7**:405
ABO blood system, **4**:362, 380
Aborigines, Neolithic face pattern and, **3**:226
Abortion
    defined, **3**:293
    genomic mutations and, **5**:247
    maternal mortality and, **4**:916–17, 920
Abortion, spontaneous, **1**:1–10, **2**:441
    causes, diagnosis, and treatment of, **1**:4–10, **3**:294
    chromosomal defects and, **3**:302, 303
    definition of, **1**:1

Abortion, spontaneous (*Continued*)
  endometriosis in, **1**:8
  genetic factors in, **1**:5–7
  immune factors in, **1**:5
  infections in, **1**:8
  luteal deficiency in, **1**:4–5
  physiology of, **1**:1–4
  pregnancies after infertility in, **1**:8–10
  symptoms and types of, **1**:4
  and trisomy-16, **2**:438
  uterine factors
    cervical incompetence, **1**:7
    congenital uterine anomalies, **1**:7–8
    intrauterine adhesions, **1**:8
    uterine fibroids, **1**:8
Absence seizures
  in epilepsy, **3**:456
  generalized tonic-clonic seizures and, **3**:459
Absinthe, **2**:346, 347
Absorption, intestinal. *see* Intestinal absorption
Absorption, spectrophometry, atomic, **2**:649
Absorptive pressure, capillary. *see* Capillary absorptive
    pressure
Abuse of older adults. *see* Elder abuse
ABVD antineoplastics, **4**:849
Acallosals, **4**:133
Acantholysis bullosa. *see* Epidermolysis bullosa
Acardiac fetus, twinning and, **3**:312
Acceptance triangle, in food acceptance measurement,
    3:642
Acceptor control ratio, mitochondrial, **1**:459
Accessory cells, macrophages as, **4**:859
Accident prevention, and longevity, **4**:788–89
Acclimatization
  in adaptational physiology, **1**:37, 39
  in regulation of body temperature, **1**:756
ACE inhibitors, **2**:176
  beta-blockers, **2**:176
  calcium antagonists, **2**:176
  centrally acting drugs, **2**:177
  diuretics, **2**:176
  vasodilators, **2**:177
Acebutolol, **2**:171
Acedapsone, specificity of action, **1**:327
Acesulfam-K, as sweetener, **7**:377–78
Acetaldehyde
  in fifth-stage glycolysis, **3**:938
  formation of, in alcohol intoxication, **1**:171
Acetaminophen
  in dentistry, **2**:786
  hepatotoxicity, **2**:748
  mercapturic acid and, **3**:915

Acetazolamide, counterindications for, **3**:96–97
Acetic acid, **3**:662, 663
Acetoin, **3**:938
Acetone bodies. *see* Ketone bodies
Acetyl coenzyme A, **2**:240, 241, 414, 415, 504–10, 571
  metabolism, **2**:415
  oxidation of, **5**:474
Acetyl glyceryl ether phosphorycholine. *see* Platelet ac-
    tivating factor
Acetyl transferase, **2**:118
Acetylcholine, **2**:5, 13, 14, 18, 23–25, 33–36, 136, 147,
    186, 285, 301, 313, 322, 326, 329, 337, 346, 365,
    368, 493, 686, 831, 834, 850
  in amyotrophic lateral sclerosis, **1**:232
  in autonomic neurons, **1**:540
  cat eye blink response and, **2**:314
  cotransmission with vasoactive intestinal peptide in
    secretory tissue, **1**:546
  deficiencies in Alzheimer's disease, **1**:214–15
  defined, **2**:321
  endothelium-derived relaxing factor and, **3**:371–72
  gastric acid secretion and, **3**:76
  importance of its discovery, **2**:322
  ion channels and, **5**:416
  levels in Huntington's disease, **4**:271
  mood and, **2**:850
  muscle contraction and, **7**:26
  myasthenia gravis and, **2**:301
  as neuroactive substance in retina, **6**:629
  as neurotransmitter, **2**:521–23, **7**:216
  sleep-wake cycle and, **3**:194
  spinal cord function and, **7**:205
  vagus nerve and, **5**:715, 716
Acetylcholine receptors. *see* Receptors, cholinergic
Acetylcholinesterase, **2**:13, 14, 337, 850
  effect of pesticides on, **7**:548
  helminth, cell invasion and, **5**:648
  nerve growth factor and, **5**:337
  neurotransmitter removal and, **7**:386
  organophosphorate insecticides and, **3**:253
Acetylcoenzyme A, **1**:458
Acetyleholinesterase, **1**:180
Acetylglucosamine, **1**:276–77; 279, **2**:561
  lysozyme and, **3**:434
  O-linked
    defined, **5**:451
    in nuclear pores, **5**:452, 454
N-acetylmuramic acid, **1**:276–77, **3**:434, 279
Acetylsalicylic acid, **2**:169, 785
  hypercalcemia of malignancy treatment, **4**:281
  stroke treatment with, **7**:306
Acetyltransferase, **6**:46

Achalasia, esophageal, **2:**87
Achilles tendon, **5:**187–88
Achondroplasia, **1:**698, 699
Achuelian assemblages, **3:**497
Acid-base equilibrium
  disorders of, in children, **5:**708
  kidney and, **5:**828
Acid-base imbalance, pH$_e$ homeostasis and, **5:**877
Acid-base pairs
  conjugate, **5:**871
  weak
    nonionic diffusion of, **5:**873
    in pH assessment, **5:**872
Acid-fast stain. *see* Zichl-Neelsen stain
Acid phosphatase, phenotype in HeLa cells, **4:**108
Acid rain, **1:**144
Acid secretion, **3:**755
Acidic noodle. *see* alpha-helix
Acidic xylans. *see under* Xylans
Acidosis
  acetazolamide and, **3:**97
  ammonium ion excretion and, **5:**828
  course of, **3:**575
  defined, **5:**821
  in glutathione metabolism disorders, **3:**916
  hypothermia, **4:**801
  myocardial contraction and, **5:**824
  phosphofructokinase inhibition and, **5:**825, 874
  renal ammoniagenesis failure and, **5:**879–80
  tubular bicarbonate and, **5:**878
Acidosis, lactic
  diabetic ketoacidosis and, **5:**827
  exercise and, **5:**826, 877
  fructose infusions and, **5:**827
  phenformin and, **5:**827
  in shock, **5:**826
Acidotropic agents, **5:**875
Acids
  defined, **5:**871
  intracellular, cell response to, **5:**873–74
  in membrane transport, **5:**823
  thermal injuries and, **7:**448
Acids, free fatty. *see* Fatty acids, nonesterified
Acidulants, food. *see* Food acidulants
Acinar cells
  anatomy of, **6:**690
  defined, **3:**317
  function of, **7:**567–68
  in hormone synthesis, **3:**327–28
  response to toxic agents, **7:**568–69

Acne
  defined, **5:**710
  treatment, **7:**415
Aconitase, **2:**655
Acoustic nerve
  anatomy, **3:**238
  physiology, **3:**238–39
Acoustic transmissions. *see* Sound
Acoustic trauma. *see* Hearing loss
Acousticolateral hair cells. *see* Hair cells
Acquired immunodeficiency syndrome (AIDS), **4:**450, 837. *see also* HIV infections
  AIDS epidemic (Public health), **1:**151–62
  animals models and pathology of, **5:**676–77
  antibody-antigen complexes in, **1:**304
  antibody testing for, **1:**737–38
  AZT and, **2:**399
  bone marrow transplantation for, **1:**797
  CD4 and, **2:**232
  changes in infected cells, **1:**11–23
  chemotherapy of, **2:**395, 399–401
  cofactors, **1:**22
  costs of treatment, **1:**152
  cytopathological effects of HIV infection, **1:**18–19
  depletion of T cells, **1:**18–20
  and 2',3'-dideoxycytidine, **2:**400
  and 2',3'-dideoxyinosine, **2:**400
  disease models, animal, **1:**256
  drug use and, **1:**152
  health education messages on, **6:**899–900
  HIV-1 and, **2:**399–401
  HIV-1 receptor and, **2:**232–33
  HIV proteins and, **1:**21
  immune destruction of T cells, **1:**19–20
  immunosurveillance, **4:**345–46
  impact of, **1:**151–53
  lithium effects, **4:**752
  newborn screening test for, **5:**438
  and nucleoside analogues, listed, **2:**400
  pathogenesis, **1:**18–22
  prevention of, **1:**153–56
  recombinant subunit vaccines against, **7:**750
  RNase H and, **2:**400
  stem cell arrest in, **1:**20
  tissue damage in, **1:**21–22
Acquired language disorders, **4:**638
Acquisition, in classical conditioning, **2:**631
Acquisition phase
  confounded effects in, **5:**122
  defined, **5:**121
Acral-lentiginous melanoma, **4:**890
Acrodermatitis, **7:**890

Acromegaly, **2:**21, **5:**707, 24, 28, 698, 799
Acrosomal membrane, **3:**603
Acrosome, **3:**599, **7:**166
Acrosome reaction
    defined, **3:**599
    mechanics of, **3:**603–4
    species variation in, **3:**604
Acrosome stabilizing factor, **3:**601
Acrylamide, **2:**346
ACTH. *see* Adrenocorticotropic hormone
ACTH-releasing factor. *see* Corticotropin-releasing hor-
        mone
Actin-binding proteins, **5:**429
Actinin, **2:**44, **5:**169, 270
Actinomyces, morphology, **1:**239
Actinomycosis, oral, **5:**614
Actins, **2:**42, 46–48, 126–28, 132, 172, 181, 182, 241–43,
        271
    alpha-actin
        cardiac, **5:**170
        gene, promoter region of, **5:**166
    beta-actin
        age and, **5:**515
        diet and, **5:**515
    alpha-actinin binding of, **5:**223
    brain actin-binding protein, **2:**42
    classes of, **5:**216–17
    defined, **2:**41, **5:**173
    filament, **2:**271
    fluorescence visualization of, **5:**907
    in fruitfly, **5:**169
    function of, **5:**175
    -motor, **2:**243
    multiple, **5:**165
    in muscle contraction, **5:**209
    in muscle development, **5:**165–66
    in myofibril organization, **5:**168
    in polarized epithelial cells, **6:**62
    polymerization of, **5:**169
    in red cell membrane, **6:**542–43
    in skeletal muscle, **7:**25
    in smooth muscle, **5:**170
    ubiquitousness of, **5:**164
    in walk-along muscle contraction theory, **5:**226
Action potential, **2:**5, 125, 134–36
    defined, **2:**124, **5:**341, 523
    depolarization of membrane, **2:**134–35
    ion flow, **2:**134–35
    myocardial, **2:**134–36
    origin, **2:**134–36
    period, refractory, **2:**134
Action stage, of attitude change, **1:**493

Activated charcoal, in artificial cells, **1:**378
Activated killer cells. *see also* Killer cells
    characteristics of, **5:**302
        defined, **5:**293
Activation reactionsn, immune system, **4:**406–8
Activation system, behavioral, **2:**711
Active site-directed enzyme inhibitors, **3:**398
Active sites, enzyme, **3:**393, 919
Active transport. *see* Biological transport, active
Activin secretion, **7:**432
Actomyocin, **2:**181, 241
Actomyosin, **5:**233
Acupuncture analgesia, **5:**732
Acute amoebic colitis. *see* Colitis
Acute brain syndrome. *see* Toxic psychosis
Acute phase, proteins, **1:**31–33
    physiological role of, **1:**31
Acute phase response, **1:**25–35
    cells involved in, **1:**29–30
    general characteristics of, **1:**25–26
    models for, **1:**26
    molecules involved in, **1:**26–29
    physiological mechanisms in, **1:**30–34
    vasoactive mediators involved in, **1:**27
Acyclovir, **2:**397–98
    for *chronis mononucleosis*, **3:**465
    control of virus infection, **7:**790–91
    2-'deoxycytidine-, **2:**397
    and herpes virus
        DNA polymerase, **2:**397–98
        treatment, **2:**398
    metabolism, **2:**398
    monophosphate, **2:**397–98
    triphosphate, **2:**397–98
Acyl-carnitine, **4:**748
Acyl carrier protein, **2:**571, 573
    phospholipid synthesis and, **5:**883
    reactivity of, **3:**437
    sythetase, **2:**573
Acyl-coenzyme A, **4:**748
Acyl-coenzyme A-cholesterol acyltransferase, **2:**412; 417;
        418, **3:**583
Acyl-D-alanyl-D-alanine, **1:**273
Acyl enzymes
    clavulanic acid and, **3:**400
    elastase and, **3:**397
Acylation, **3:**170
Acylglycerols
    behavior and distribution, **4:**743
    characterization, **4:**741–42
    metabolism, **4:**742–43
Ad12-SV40 hybrid virus, **4:**585

Adaptation
  acclimatory changes and, **1**:39
  to aerobic exercise, **3**:512–13
  capacity, **1**:37, 38
  of cellular signaling, **2**:329
  classes of, **1**:37, 38
  cultural, **1**:340–41
  defined, **3**:505
  emotion-focused and problem-focused, **4**:74
  and evolution, **1**:40
  human, **1**:37, 40–41
  ontogentic, **2**:862, 869, 872
  physiological, **1**:37–41
  resistance, **1**:37–38
  temperature, **1**:38–39
  time course of, **1**:39
Addiction. *see* Substance dependence
Addiction, alcohol. *see* Substance dependence
Addison's disease, **1**:77, **2**:500
  causes of, **3**:334
  and cortisol level, **2**:501
Additives, food. *see* Food additives
Adducin, **2**:46, 48
  calmodulin-binding, **2**:48
  in erythrocyte, **2**:48
Adduction, as joint movement, **5**:133
Adenine, **2**:117
Adenine diphosphate, **5**:472
Adenocarcinoma
  of cervix, **3**:863, **5**:607
  of gastrointestinal epithelium, **3**:769
  of large bowel, **3**:775
  neu oncogene amplification and, **5**:548–49
Adenocorticotropic hormone
  big, **3**:334
  cortisol and, **3**:325
  ectopic, **3**:334
  multiple effects of, **3**:330
Adenohypophysial hormones, **5**:398
Adenohypophysis, **5**:393, 394
Adenoma
  colonic, **3**:770–71
  gastric, familial polyposis coli and, **3**:771
  of gastrointestinal epithelium, **3**:769
  growth hormone secretion and, **2**:21
  hepatic cell, oral contraceptives and, **5**:607
Adenomatous polyps, **3**:770, 771
Adenosine, **2**:366
  functions of, **5**:425
  in gastric blood flow regulation, **3**:760
  oocyte maturation inhibition and, **5**:556–57
  as platelet receptor, **6**:53

Adenosine cyclic monophosphate, **2**:19, 49, 58, 90, 238,
    241, 271, 276, 314, 317, 325, 328, 330, 503, 509,
    532, 844, 848
  acyl-coenzyme A: cholesterol acyl transferase and,
    3:583
  adenylyl cyclase and, **4**:16
  as cellular messenger, **7**:20
  cytidine triphosphate and, **3**:583
  defined, **3**:581
  effect of G protein-linked receptors on, **1**:84
  elastin and, **3**:265
  in fatty acid uptake, **3**:588
  in glycogen cascade, **3**:925
  in oocyte maturation, **5**:555
  oocyte maturation induction and, **5**:557
  pancreatic hormones and, **3**:586
  phosphodiesterase, oocyte maturation inhibition and,
    5:556
  in protein phosphorylation, **6**:202
  and regulation of phosphatidylcholine synthesis, **5**:887
  in response of retina to light, **6**:630
  as second messenger, **4**:975
  sperm capacitation and, **3**:601
  sperm hyperactivation and, **3**:604
Adenosine cyclic monophosphate messenger system,
    3:331, 332, 333
Adenosine deaminase, **2**:397
  and 2'-deoxycoformycin, **2**:397
  and erythro-9-(2-hydroxy-3-nonyl)adenine, **2**:397
  phenotype in HeLa cells, **4**:108
Adenosine diphosphate, **2**:238, 240, 503, 505, 507, 509,
    532
  ADP/ATP transport system, **1**:462
  in exercise, **3**:512
  exotoxins causing ADP-ribosylation, **5**:2–3
  as metabolic controller, **3**:379, 380
  rephosphorylation of, **5**:227
  in respiration control, **3**:377–78
  -ribosyl transferase, **2**:532, 533
Adenosine monophosphate
  in glycogen degradation, **3**:923
  nerve growth factor and, **5**:336
  phosphorylase *a* and, **3**:928
  phosphorylase and, **3**:924–25
  phosphorylase *b* and, **3**:929
  vasoactive intestinal polypeptide and, **5**:722
Adenosine triphosphatase, **2**:42, 96, 366
  in glycolysis, **3**:931
  H/u+/U-translocating
    in adrenal medulla, **1**:646–48
    structure of, **1**:647–48
  kinesin and, **5**:20

Adenosine triphosphatase
    myosin and, **5:**165
    organelle transport and, **5:**900
    RNA transport and, **5:**459
    synthesis, **3:**932
    transport, **4:**974
Adenosine triphosphatase, calcium, **2:**96
Adenosine triphosphatase isoenzymes, **5:**237
Adenosine triphosphate, **1:**53–55, **2:**92, 96, 127, 172, 182,
    240, 503, 505–7, 509, 532, 731
    5-adenosylmethionine decarboxylase, **2:**534
    in autonomic neurons, **1:**540
    defined, **5:**205
    dependent ion pumps, **4:**559–60
    and energy transduction, **6:**225–26
    in exercise, **3:**512
    in fourth-stage glycolysis, **3:**937–38
    free energy from hydrolysis of, **1:**54
    fructose phosphorylation and, **5:**827
    functions of, **3:**326–27
    in glycogen degradation, **3:**921
    in glycolysis, **5:**824
    hydrolysis
        in biospheric energy cycle, **5:**472
        and energy transduction, **1:**53–54
        muscle contraction and, **5:**225, 226–27
        myosin and actomyosin-catalyzed, **5:**232–33
        myosin head and, **5:**211
        translocation of thick and thin filaments and, **5:**233
    as membrane energy source, **4:**973
    in mRNA transport, **5:**459
    muscle contraction and, **3:**379–80, **7:**26–28
    nuclear uptake and, **5:**456–57
    phosphofructokinase and, **5:**821
    phosphorylase *b* and, **3:**929
    in respiration control, **3:**377–78
    sodium transport and, **3:**95
    structure of, **1:**53
    as substrate, **3:**436–37
    synthase complex, mitochondrial, **1:**462–63
    synthesis, **1:**853–76, **4:**558
        in adrenal medulla, **1:**645–46
        electron transfer and, **5:**61–62
        inhibitors of, **1:**464–65
        mechanism, **1:**463–64
        mitochondrial respiratory chain and, **5:**55
        by oxidative phosphorylation, **4:**974
        proton motive force and, **5:**871
        type 1 proton pumps, **4:**563–64
        uncouplers of, **1:**464–65
    synthetase, **2:**238, 239

Adenosine triphosphate
    in third-stage glycolysis, **3:**936–37
    yield calculation of, **5:**474–75
Adenosine triphosphate, calcium, **5:**164
S-Adenosylmethionine
    DNA methyltransferases and, **3:**155
    functions of, **3:**437–38
    metabolism, in adrenal medulla, **1:**646
S-Adenosylmethionine decarboxylase, **6:**69–70
Adenoviral genomes. *see* Genes, adenoviral
Adenoviridae, **4:**520
Adenovirus E1A, **4:**585
Adenovirus promotors, **3:**156–57
Adenyl cyclase, **2:**13, 223, 238, 844, 848
    activation by salmonella, **6:**707
    activity, cell receptors and, **6:**514
    adrenal gland and, **1:**74
    adrenergic receptors and, **1:**81
    adrenocorticotropic hormone and, **5:**729
    cellular signal transduction and, **7:**19
    in cyclic AMP messenger system, **3:**331
    energy charge, **2:**503
    exotoxin activity and, **5:**3
    in *fos* gene induction, **3:**705
    gonadoliberin and, **5:**732
    induction by serotonin, **6:**831
    inhibition of, **5:**732, 733
    in lipid regulation, **1:**60
    melatonin synthesis and, **6:**3
    MSH and, **5:**729
    ovarian steroid biosynthesis and, **7:**263–64
    parathyroid hormone and, **5:**662
    platelet activation and, **6:**53–54
    regulation of bone and calcium metabolism, **1:**803
    in sperm capacitation, **3:**601
    thyroid activity and, **6:**18
    vasoactive intestinal polypeptide and, **5:**722
Adenylate cyclase. *see* Adenyl cyclase
Adenylate kinase, **2:**238, **4:**108
Adenylic acid. *see* Adenosine monophosphate
Adenylyl cyclase, **2:**325–30
    activity of, **4:**16
    brain G protein mRNA and, **4:**22–23
    in cyclic nucleotide regulation, **5:**417
    defined, **5:**415
    guanosine triphosphate and, **4:**14
    pertussis toxin and, **4:**18, 21, 22
Adenylyl 5'imidodiphosphate, **5:**20
Adenylyl transferase, **2:**532
Ader, **5:**374
Adherens junction
    cell-cell, **2:**270

Adherens junction (Continued)
cell-substrate, 2:270
cellular, 2:268
in fibroblast, 2:270
proteins, 2:270
Adhesion, focal, 2:270, 271
and cAMP, 2:271
molecule of neural cell, 2:50
Adhesion, in polymorphonuclear neutrophils, 5:428-29
Adhesion molecules, 4:475-76
Adhesion receptors, 4:474
Adhesions
intrauterine, and spontaneous abortion, 1:8
pelvic, and endometriosis, 3:364
Adinazolam, 1:310
Adipose tissue, 1:57-62/B, 2:639, 645
aging and, 3:904
autonomic innervation of, 1:543
characteristics of, 1:57
components of, 6:437
as determinant of minimal weight for restoration of
menstrual cycles, 1:745-46
development of, 1:58
distribution, 1:61
elemental composition of, 6:437
as energy reserve, 3:576
estrogen production and, 3:482
fat, as determinant of minimal weight for first menstrual
cycle, 1:745-46
fatty acid uptake and, 3:582; 588
features distinguishing brown and white, 1:62
female reproduction and, 1:743
function, in fertile females, 1:742-43
history of, 1:57-58
lipid composition of, 1:60-61
lipolysis in white adipocytes, 3:574-75
mass density of, 6:437
measurement of, 1:58-59; 667-73, 5:517-18
measurements of regional distribution of, 5:518
menarche and fertility, 1:741-50
methods to determine body fat, 1:667-73
nutritional regulation of lipoprotein lipase in, 3:587
storage and release of, 1:60
tracylglycerol in, 3:583
ultrastructure of, 1:59-60
white, characteristics of, 1:57
Adjusted retention time (gas chromatography), 3:750
Adjustment disorders, 4:998
Adjuvants, immunologic, 1:684-86
adjuvant contact immunotherapy, 4:347
adriamycin, 1:686

Adjuvants, immunologic (Continued)
antiradiation effects of, 6:423
cyclophosphamide, 1:685-86
diethyldithiocarbamate, 1:684
levamisole hydrochloride, 1:684
muramyl tripeptide phosphatidylethanolamine, 1:684
nonmicrobial stimulators as, 7:754
and synthetic vaccines, 7:753
synthetic vaccines and, 7:753
vaccines and, 6:643, 7:752-53
Admixture, evolutionary, 3:85
Adnexitis, oral contraceptives and, 5:608
Adolescence, 1:63-69. see also Puberty
biological changes during, 1:65-66
biological development during, 1:65-66
cognitive changes during, 1:66-67
cultural background, 1:63-64
as developmental stage, 1:63-65
disturbance during, 1:64-65
growth during, 4:29-30
historical background, 1:64
psychosocial changes during, 1:67-68
religious/spiritual beliefs during, 1:67-68
self concept during, 1:67-68
sexual behavior in, rules of, 6:884-85
sexual maturation in, 3:356
suicide rate of adolescents, 7:330
turmoil during, 1:64-65
universal aspects of, 1:68-69
well-adjusted, 1:69
Adoptees
adoption studies, and criminality, 2:712
in behavioral genetic research, 3:837, 841-42
in complex inheritance studies, 3:856-57
in criminality studies, 3:844-45
ADP. see Adenosine diphosphate
Adrenal cortex
control of hormone secretion, 1:73-74
development and structure of, 1:71-72
diseases, in children, 3:352-53
disorders of, 1:77-78
fetal
ACTH and, 3:343-44
development of, 3:344-45
hormonal action, 1:72-73
pubescent, 3:355
steroid hormone synthesis in, 7:246-52
Adrenal cortex hormones, 2:23; 247, 834
in acute phase response, 1:33
aging and, 3:896
biosynthetic pathways of, 7:260
control of secretion, 7:258-64

Adrenal cortex hormones (*Continued*)
  lactase activity effects, **4:**617
  maintenance of lactation, **4:**606
  mammogenesis, **4:**604
  metabolic transformation of, **7:**277-78
  secretion rates and plasma concentrations of, **7:**261
  section, adrenocorticotropic hormone and, **7:**259-62
  in treatment of allergy, **1:**195
  in treatment of myasthenia gravis, **5:**258
  urinary excretion products, determination of, **7:**278
Adrenal corticotropin hormone. *see* Adrenocorticotropic
    hormone
Adrenal gland diseases, **1:**77-79
  in children, **3:**352-53
Adrenal glands, **1:**71-79
  adrenocorticotropic hormone and, **5:**729
  ascorbic acid concentrations, **1:**430
  control of hormone secretion, **1:**73-74
  cortisol secretion and, **2:**492
  development and structure of, **1:**71-72
  disorders, **1:**77-79
  fetal, at midgestation, **3:**619
  general characteristics of, **1:**71
  histology, **7:**261
  hormonal action, **1:**72-73
  steroid biosynthesis in, **7:**258-62
Adrenal hyperplasia, congenital, **2:**738, **7:**249, 744
  cause of, **5:**708
  characterization, **4:**244
  as fatal disease, **2:**738
  medical treatment for, **1:**704
  newborn screening test for, **5:**437
  virilizing
    abnormal testosterone secretion and, **3:**342
    bone age in, **3:**351
    cortisol and, **3:**352
    screening for, **3:**349
Adrenal hyposecretion, primary. *see* Addison's disease
Adrenal medulla
  autologous transplantation, in Parkinson's disease,
      5:672
  bioenergetics and, **1:**641-53
  at birth, **3:**350
  chromaffin cells, **1:**538
  development and structure of, **1:**72
  disorders, in children, **3:**353
  disorders of, **1:**78-79
  fetal, development of, **3:**346-47
  functions of, **1:**641
  hormones affecting, **1:**73
  matrix composition of, **1:**642
  pheochromocytoma, **4:**290

Adrenal medulla (*Continued*)
  secretion in, **1:**75, 651-53
  secretory vesicles of
  ascorbate regeneration in, **1:**649-51
  energetics of, **1:**646-47
  H^+-linked transport in, **1:**648, 649
  H^+-translocating ATPase of, **1:**646-47
  mechanism of catecholamine transport in, **1:**648
  pharmacology of monoamine transpoater in, **1:**648-49
  structure of, **1:**642-43
Adrenal reticularis. *see* Zona reticularis
Adrenaline. *see* Epinephrine
Adrenarche
  defined, **3:**350
  premature, **3:**357
Adrenergic agents
  for autoimmune diseases, **5:**383-84
  in vitro studies of, **5:**382
Adrenergic beta receptor blockaders, **2:**141; 171, **5:**383,
      176
  acebutolol, **2:**171
  atenolol, **2:**171
  drug design of, **7:**114
  for hypertension, **4:**294
  metoprolol, **2:**171
  nadolol, **2:**171
  propranolol, **2:**171
  sotalol, **2:**171
  timolol, **2:**171
Adrenergic constrictor system, **3:**519
Adrenergic receptors. *see* Receptors, adrenergic
Adrenocorticotropic hormone, **2:**19, 23, 26-28, 141, 144,
      414, 686, 744, 756, 834, 851
  in adrenal gland development, **1:**72
  circadian rhythms and, **5:**396, 398
  control of secretion of, **1:**73-74
  defined, **5:**393
  fetal, **3:**342-44
  function, **6:**92
  functions of, **5:**727, 729
  immunity and, **5:**375
  as neuroregulator, **2:**22-23
  production and function of, **6:**16-17
  regulation of, **5:**398
  regulation of steroid hormone synthesis, **7:**254
Adrenodoxin, **2:**740, **7:**246-48, 743
Adrenodoxin reductase, **7:**246-48
Adrenogenital syndrome. *see* Adrenal hyperplasia, con-
      genital
Adrenoreductase, **2:**740, 743
Adriamycin. *see* Doxorubicin

Adsorbents, **3:**403

Adsorption chromatography, **4:**207

Adult polycystic kidney disease. *see* Kidney, polycystic

Adult relationship development theories, **4:**543-44

Adult T-cell leukemia. *see* Leukemia, T-cell

Adult T cell leukemia virus type I. *see* HTLV-I

Adults

    language disorders, acquired, **7:**162

    zinc metabolism in, **7:**889

Adverse effects, defined, **3:**93

Aegyptopithecus, in Fayum fossils, **3:**489

Aequorin, as fluorescent probe, **5:**577

Aerobic exercise. *see* Exercise

Aerosols, abuse of, **5:**443-44

Affect, **2:**831, 833. *see also* Emotions

    category view of, **1:**101-11

    cognitive foundations, **1:**105-8

    cognitive structures and, **1:**106-7

    cycling, **2:**833

    in depression, **2:**831

    dimensional approach to, **1:**101-2

    dimensions of, **5:**91-92

    disturbances, and personality disorders, **5:**781

    early theories of, **1:**102-3

    effects on judgment and decision-making, **1:**110

    electric amygdala stimulation and, **2:**833

    in feeding behavior research, **3:**596

    frontal cortex and, **2:**833

    influences on cognition, **1:**108-11

    kindling, defined, **6:**781

    memory and, **1:**109-10

    network model of, **1:**110

    persuasion and, **1:**110-11

    physiological theories of, **1:**103-5

    sad, characteristics of, **5:**92

    separate systems view, **1:**111

    social behavior and, **1:**108-9

    social foundations, **1:**105-8

Affective disorders, **5:**91-97. *see also* Depression

    biological factors in, **5:**95

    cannabis and, **5:**445

    characterization, **4:**997

    depression as, **1:**94

    in the elderly

        endocrinopathy and, **1:**132

        epidemiology, **1:**132

        etiology, **1:**132-33

        genetic determinants, **1:**132

        neurotransmitter abnormalities, **1:**132

    epidemiological factors in, **5:**95

    genetic aspects of, **1:**95-96, **3:**843-44

Affective disorders (*Continued*)

    genetic basis of, **1:**95-99

    genetic markers, **1:**93-100

    linkage analysis in, **1:**96

    mania, **1:**94-95

    mania as, **5:**94-95

    other illnesses, **1:**95

    treatment of, **1:**95, **5:**95-97

Affective psychosis, bipolar. *see* Bipolar disorder

Afferent connections, **2:**333, 335-36

Afferent fiber, **2:**678, 680, 683-85

Afferent motor dysphasia. *see* Aphasia, Broca

Afferent nerves

    from the heart, **2:**138-40

    visceral, **3:**763, **7:**794

Affiliation, interpersonal attraction and, **4:**537-38

Affinity chromatography, **4:**209

Affinity extractions/partitioning, biospecific, in aqueous two-phase systems, **1:**357, 361

Affinity purification, in mouse monoclonal antibody applications, **5:**82-83

Aflatoxicosis

    in domestic animals, **5:**260-61

    signs of, **5:**264-65

Aflatoxin B, **2:**111, 114, 121

Aflatoxin B1, **3:**675, **5:**264

Aflatoxin M1, **5:**265

Aflatoxins, **7:**592-93

    discovery of, **5:**260

    diseases attributable to, **5:**264

    drought and, **3:**666

    in food, **3:**675

    food analysis, **4:**225-26

    structure of, **7:**589

African great apes. *see* Apes

African lymphoma. *see* Burkitt's lymphoma

African sleeping sickness. *see* Trypanosomiasis

African trypanosomes. *see* *Trypanosoma*

Afropithecus, later hominoids and, **3:**489

Afterdrop, hypothermia, **4:**804

Afterhyperpolarization, **2:**311

Afterload, **2:**179

Afterload, ventricular, **3:**518

Agammaglobulinemia, **4:**837

Agar dilution, in antimicrobial susceptibility testing, **1:**559

Age

    defined, **3:**899

    dementia. *see* Dementia, senile

    distribution, and sudden infant death syndrome, **7:**317

    at puberty, **3:**353

    role in dietary protein needs, **6:**233-34

Age, chronological
   defined, **3**:899
   functional age and, **3**:900
Age, developmental, **3**:293
Age, gestational. *see* Gestational age
Age, maternal. *see* Maternal age
Age, menstrual. *see* Gestational age
Age determination by skeleton, **3**:350–51
Age factors
   in barbiturate abuse, **5**:442
   in death rates, **5**:100–101
   in gene expression, **5**:515
   longevity and, **4**:786–88
   in membrane fluidity, **5**:507
   in mutational risk, **5**:254
   in obesity, **3**:684
   in T cell proliferative response, **5**:506–7
Age-I mutant, in *Caenorhabditis elegans*, **3**:895
Age-related diseases, **3**:891, 894
   defined, **3**:891
   hyperadrenocorticism in, **3**:897
Aged
   abuse of. *see* Elder abuse
   changes in drug kinetics in, **5**:853
   communication disorders, **1**:121–24
   demography, **1**:129
   diet for, **3**:659
   epidemiology of illnesses in, **1**:129–30
   expressive language, **1**:120–21
   hypothermia, **4**:802–3
   language differences, **1**:119–21
   language performance variability, **1**:124–27
   malnutrition of, **4**:897–903
   mortality of, **5**:102–3
   nonverbal communication, **1**:121
   nutritional requirements, **4**:899–900
   receptive language, **1**:120
   Recommended Daily Allowances, **4**:899–900
   sexuality, **1**:136
   zinc metabolism in, **7**:889
Agenesis, defined, **7**:406
Agent, toxic. *see* Toxic agents
Agglutinins, wheat germ. *see* Wheat-germ agglutinins
Aggregation, of denatured enzymes, **3**:413, 415
Aggregation, cell. *see* Cell aggregation
Aggressive behavior, **1**:113–18
   aggression as punishment, **1**:116
   agonistic behavior, **1**:114–15
   antisocial behavior, **1**:116
   arousal, media effects, **1**:4, 1058
   causes, **1**:115–16
   competition, **1**:116

Aggressive behavior (*Continued*)
   concept of, **1**:113–15
   defensive behavior, **1**:114–15
   definition, **1**:113
   evolutionary perspectives, **1**:117
   functions, **1**:116–17
   genetic causes, **1**:115
   hormonal effects, **4**:246–47
   hormonal influences, **1**:115
   injury as component of, **1**:114
   intentions, **1**:114
   likelihood of raping variable, **1**:5, 1058
   media priming of, **1**:9, 1058
   mutualism and, **1**:117
   observational learning, **1**:1–2, 1058
   ontogenetic experiences and, **1**:116
   physiological causes, **1**:115
   pornography effects, **1**:4–5, 1058
   predation, **1**:114
   as protection against external threats, **1**:117
   psychological causes, **1**:116
   punishment as, **1**:116
   rape myth, **1**:5, 1058
   sequential inferences, **1**:113–14
   sex differences in response to observed violence, **1**:3–4, 1058
   social comparison processes, **1**:3, 1058
   social uses of, **1**:116
   socializing function of, **1**:116–17
   submissive behavior, **1**:115
   threats, **1**:114
   violent scripts, **1**:8–9, 1058
   xenophobia and, **1**:117
Aging, **4**:697–98
   age-associated disease as criterion for, **5**:478–79
   age-associated physiological changes, **5**:478
   ageism, **1**:136
   animal models, **1**:258
   astrocyte changes with, **1**:445–46
   body tissue composition and, **6**:436
   bone fracture related to, **1**:778–80
   cardiovascular disease and mental aging, **3**:900
   cell death and, **2**:250–51
      in Alzheimer's disease, **2**:251
      in benign prostatic hyperplasia, **2**:251
      in Parkinson's disease, **2**:251
   collagen and, **2**:570
   connective tissue and, **2**:646–47
      accumulation of molecules, **2**:647
      changes, **2**:646–47
      crosslinking, **2**:647
      dehydration, **2**:647

Aging
  connective tissue and *(Continued)*
    loss of functional capacity, **2:**647
    loss of tissue bulk, **2:**647
  crosslinking theory and, **2:**570
  current state of knowledge of, **5:**477-78
  defined, **3:**891, 899
  determinant theories of, **3:**892-93
  effects of, **4:**786-88
  elderly hypothermia, **4:**802-3
  experience of, **4:**698
  extrapyramidal function and, **3:**901
  factors influencing, **5:**477-78
  general theories of, **3:**891-92
  genetic aspects of, **3:**892-93
  glucose intolerance and, **2:**105-6
  historical negativism associated with, **1:**130
  identity and, **4:**705-6
  insulin and, **2:**106
  lactose and, **2:**105-6
  lactose intolerance and, **2:**105-6
  language and, **1:**119-27, **4:**637
  language competence in older adults, **1:**124-27
  longevity as criterion for, **5:**478
  modulation of, criteria indicating, **5:**477-79
  nutrition and, **5:**479
  nutritional modulation of, **5:**477-81
  omega-2 fatty acids in, **5:**538-39
  pigmented neuron loss in, **5:**668
  polymorphism of immune system, **6:**75-78
  prostate cancer and, **6:**183
  psychiatric aspects. *see* Geropsychiatry
  rodent
    food restrictions and, **5:**479-80
    relevance to human situation, **5:**480-81
  serotonin and, **6:**834-35
  skin, **7:**48-49
  slowing of, **5:**479
  social status of women, **4:**708
  stochastic theories of, **3:**893-95
  theoretical causes of, **4:**785-86
  vitamin E and, **7:**877
Agnosia, **2:**7
  defined, **3:**215
  tactile, **2:**683
  unilateral spatial, **3:**223
  visual, **3:**223, 224
Agnosic alexia. *see* Alexia, agnosic
Agonistic behavior. *see* Aggressive behavior
Agonists, **2:**305, **3:**455, **4:**13
Agoraphobia, **2:**857
Agranulocytosis, **1:**794, **5:**259

Agraphia, **2:**9, 684
Agricultural chemistry. *see* Chemistry, agricultural
Agricultural products
  aflatoxins in, **5:**264
  mycotoxins in, **5:**261
Agriculturalists, customary diets of, **3:**20-21
Agriculture
  antibiotics for, **1:**216-17
  in developing countries, **3:**26
  in Ethiopia, **3:**562
  food composition tables, U.S., **3:**656
  food technology and, **3:**21-22
  Western diets and chemistry in, **3:**26
Agriculture, U.S. Department of
  on contaminated grain, **5:**263
  Continuing Survey of Food Intakes by Individuals, **3:**650
  Dietary Guidelines, **3:**651
Agrins, in neuromuscular junction, **5:**171
Agronomy, **1:**139-49
  crop sciences and, **1:**141-45
  definition of, **1:**139, 140
  history of, **1:**140-41
  soil sciences and, **1:**145-48
  training and job opportunities in, **1:**148-49
AH antigen, **4:**963
AIDS. *see* Acquired immunodeficiency syndrome (AIDS)
Ainu origin, based on teeth, **2:**798
Air
  abdominal, in plain film radiography, **3:**3
  airborne allergens, **1:**191-93
  airborne toxicants, **7:**577
  alveolar, oxygen transport and, **5:**619
  dead space, calculation of, **5:**619
Air hypothermia, **4:**802
Air pollutants. *see also* Pollutants
  bronchoconstriction and, **3:**387
  concern about, **7:**575-77
  inhalation-associated headaches, **4:**57
  lung disease and, **3:**388
  in upper respiratory tract disease, **5:**694-95
Air quality standards, **7:**585
Airway
  autonomic innervation of, **1:**542
  infection, in children, **5:**706
  smooth muscle, autonomic innervation, **1:**542, 546-47
Akakabi-byo. *see* Red-mold disease, cereal
Akinesia, **2:**682
Akithsia, and schizophrenia, **6:**776
Alanine
  in elastin, **3:**261
  fetal glucagon and, **3:**346

Alanine aminotransferase, **1:**739

Alar, toxicity, **7:**565

Alarmones, defined, **5:**470

Alaska, dermatoglyphics of, **3:**91

Albendazole, **2:**387, 390
  ascariasis treatment with, **1:**270
  trichuriasis treatment with, **1:**270

Albright, **3:**481

Albumin, as drug binder in blood, **1:**708

Albumins
  avian, antimicrobial proteins in, **3:**663
  in immunoassays, **3:**207
  nonesterified fatty acids and, **3:**585
  primate, **3:**488
  thermal injury and, **7:**455

Alcohol
  beneficial effects of, **1:**166
  blackout, defined, **1:**165
  blood levels of, **1:**165
  chronic toxic effects of, **1:**171–76
    cardiovascular system, **1:**174–75
    central nervous system, **1:**171–73
    endocrine system, **1:**174
    fetus, **1:**175–76
    gastrointestinal tract, **1:**175
    liver, **1:**173–74
    pancreas, **7:**570
  consumption, in United States, **1:**166
  distillation, **1:**165
  effects on anxiety disorders, **1:**135
  hypothermia and, **4:**803
  nutrient composition of alcoholic beverages, **5:**497
  toxicology, **1:**165–76

Alcohol, ethyl, **2:**346, 348
  drugs of abuse and, **3:**211–14
  in fifth-stage glycolysis, **3:**938
  in food spoilage, **3:**663
  and pathology due to, **2:**348
  solvent abuse and, **5:**444

Alcohol, methyl, **2:**346

Alcohol dehydrogenase, **7:**885

Alcohol withdrawal delirium, **5:**443
  definition of, **1:**165
  symptoms of, **1:**171

Alcoholic intoxication
  acute effects of
    blackouts, **1:**171
    central nervous system, **1:**170
    hangovers, **1:**170–71
    physical dependence and withdrawal, **1:**171
  barbiturates and, **5:**442

Alcoholic intoxication (*Continued*)
  factors affecting
    characteristics of beverages and how they are consumed, **1:**168
    characteristics of drinker, **1:**168–69
    degree of tolerance, **1:**169
    ethnicity, **1:**169
  hangover and, **1:**170–71
  Naltrexone for, **3:**212

Alcoholics Anonymous
  definition of, **1:**43
  in treatment programs, **1:**49–50

Alcoholism, **1:**43–51, **2:**764
  animal models, **1:**258
  biological determinants of, **1:**47
  congenital defects and, **7:**416–17
  effect on blood interferon levels, **4:**525
  insomnia and, **7:**78–79
  memory and, **2:**74
  personality disorders associated with, **5:**781
  physiological determinants of, **1:**47
  as predictor of suicide, **7:**331–32
  resolution of addiction without treatment, **4:**86
  societal aspects, **4:**84
  sociocultural determinants of, **1:**48
  treatment of, **1:**49–50, **3:**213
  vitamin A activity and, **7:**851

Aldehyde dehydrogenase, **5:**286

Aldehydes
  as enzyme inhibitors, **3:**397–98
  transfer, **2:**528

Aldimine, **2:**534

Aldol condensation product, **3:**266

Aldolase, **4:**576, 578

Aldosterone, **2:**176, 744
  atrial natriuretic factor and secretion, **1:**474
  deficiency of, **3:**353
  effect on ingestive behavior, **4:**248
  functions of, **1:**73, **7:**270–72
  hypertension and, **4:**291
  metabolic transformation of, **7:**278–79
  perinatal, **3:**349
  in potassium-sparing diuretics, **3:**99
  synthesis of, **7:**245

Alert states. *see also* Attention
  maintenance of, **1:**483–84

Aleukia, alimentary toxic. *see* Alimentary toxic aleukia

Aleurone layer, dietary fiber from, **3:**40

Alexia. *see* Dyslexia, acquired

Algae
  omega-3 fatty acids of, **5:**534
  toxins produced by, **7:**590–91

Algae, blue-green, ciguatera and, 3:673
Algae, green, 3:160
Algorithms, and binocular correspondence, 1:637–38
Alicarb, toxicity of, 7:565
Alimentary system. *see* Digestive system
Alimentary toxic aleukia, 5:260, 263, 7:665–66
Alimentation, intravenous. *see* Infusions, intravenous
Alizarin red S, 2:699
Alkaline phosphatase, 2:467, 4:576
 in bone mineralization, 3:481
 function of, 7:885
Alkalines, intracellular, 5:874
Alkaloids
 bisindole, 1:181
 cinchona, 1:181
 ergot, 1:180–81, 7:592
 indole, 1:180–81
 mammalian, 1:179–80
 in medicine, 1:177–82
 morphine, 1:179
 opium, 1:179
 as plant toxins, 7:595
 pyridine, 1:179
 pyrollizidine, 7:549
 tropane, 1:178–79
Alkalosis, respiratory, 5:825
Alkanes
 aliphatic, 7:548–49
 metabolism, 4:736
Alkylating agents, 2:110, 117
 bis-(chloromethyl)ether, 2:110
 chlorambucil, 2:110
 chlornaphazin, 2:110
 ethylene oxide, 2:110
 mustard gas, 2:110
 vinyl chloride, 2:110
Alkylation
 as cause of DNA damage, 6:548–49
 in drug design, 7:105
All-or-none effect, 7:26
Alleles
 allele-specific oligonucleotide, 5:867, 869
 allele-specific oligonucleotide probes, 3:139–40
 allelic exclusion, 3:802
 allelozymes, 4:574–75
 defined, 2:475, 3:135, 5:243, 737, 791, 823, 847
 diversity of, 3:854
 DNA markers and, 3:138
 loss, 2:53
 nonallelism and, 3:792–93
 probes of, 3:182

Allelomorphs. *see* Alleles
Allergens, 1:191–93; 299, 4:410
Allergic encephalomyelitis. *see* Encephalomyelitis, allergic
Allergic response
 classification of, 1:184
 eosinophils in termination of, 5:812
 pathophysiology of, 1:183–90
  amplification of allergic response in, 1:189–90
  cellular effectors in, 1:189
  immediate and late phase responses in, 1:186–88
  immunoglobulin E (IgE) in, 1:183–85
  mast cells and basophils in, 1:185–86
  regulation of immune response in, 1:185
 Type IV, 5:818
Allergic rhinitis. *see* Rhinitis, allergic
Allergies, 1:183–95
 clinical, 1:190
  physical findings in, 1:192
  symptoms in, 1:191
 defined, 4:410
 delayed, 4:411–13
  effect of radiation on, 6:423
  macrophages and, 4:859
 diseases related to aeroallergens in, 1:191–92
 diseases related to ingested or injected allergens in, 1:193
 drug therapy, 1:194–95
 evaluation of, 1:190
 flow cytometry and, 3:624
 laboratory evaluation of
  ancillary investigations in, 1:194
  cytological evaluation in, 1:193–94
  testing for specific IgE in, 1:193
 macrophage activation in, 5:818
 reactions, types of, 4:462
 sulfite, mechanism of, 7:339
 symptoms of, 1:191
 treatment for
  allergen avoidance in, 1:194
  antihistamines in, 1:194–95
  corticosteroids in, 1:195
  crommolyn in, 1:195
  immunotherapy in, 1:195
  sympathomimetics and bronchodilators in, 1:195
 type I, 5:812
Allergies, food. *see* Food hypersensitivity
Alliesthia, in control of meal size and satiety, 1:351
Alloantigen, on antigen-presenting cells, 4:361
Allocation of resources. *see* Resource allocations
Allocortex, 2:675

Alloenzymes, **7**:689
Allograft
    rejection, **4**:362
    sensitization, **4**:360
Allopurinol, **2**:392
Allosteric proteins
    hemoglobin, **4**:138–39
    interaction of binding sites in, **6**:224
Allosteric regulation, in oxidative phosphorylation, **3**:940
Allostery
    defined, **3**:919; 931
    first demonstrated, **3**:923
Allotypes, **4**:314–15, 381–82
Allozygosity, **4**:433
Alopecia, **4**:41–42
Alopecia, androgenetic, **4**:42
Alopecia areata, **4**:42
Alopecia totalis, **4**:42
Alopecia universalis, **4**:42
Alpha 1-acid glycoprotein
    in acute phase response, **1**:32
    as drug binder in blood, **1**:708–9
    in measurements of drug transport, **1**:711–12
Alpha 1-antichymotrypsin, **1**:31
Alpha 1-antiproteinase, elastolysis and, **3**:268
Alpha-1-Antitrypsin, deficiency of, **3**:146
Alpha-1 crystallins, **3**:183
Alpha 1-globin gene, in primate evolution, **3**:882
Alpha-actinin, **5**:221–23
Alpha-amylase, **3**:415
Alpha blockers, **4**:294
Alpha chain disease. *see* Immunoproliferative small intestinal disease
Alpha errors, in randomized trials, **4**:62
Alpha fetoprotein, **1**:704, **4**:341
Alpha-helix, of polypeptides, **6**:220
Alpha-hypophamine. *see* Oxytocin
alpha-Ketoacid
    decarboxylation, **2**:542
    dehydrogenase, **2**:540
alpha-Ketoglutarate, **2**:504–9
alpha-Ketol, **2**:527
alpha-Microglobulin, **2**:118
Alpha-MSH
    in fever, **1**:759–60
    function of, **6**:88
5-Alpha-reductase deficiency, **4**:243–44
Alpha rhythm, **3**:215
Alpha toxin. *see* Phospholipase C
Alphaherpesvirinae, **4**:187
Alphaviruses, **7**:523–24
Alprazolam, **1**:310

Alternatives to animal testing. *see* Animal testing alternatives
Altruism, **1**:197–207
    altruistic personalities, **1**:205
    deductive arguments for, **1**:203–4
    definitions, **1**:198–99
    definitions, implications of, **1**:199–200
    distinguished from egoism, **1**:198–99
    empathy-altruism hypothesis, **1**:205–6
    field observation evidence for, **1**:204–6
    field observation methods, **1**:203–4
    helping, **1**:200
    helping, motives for, **1**:202–3
    implications of, **1**:206–7
    laboratory experiment evidence for, **1**:205–6
    laboratory experiment methods, **1**:204–6
    question of, **1**:197–98
    question of, methods for addressing, **1**:203–4
    reciprocal, logic of, **1**:204
    research questions on, **1**:206
    as resource distribution preference in behavior, **1**:574
Alu repetitive sequences. *see* Repetitive sequences, nucleic acid
Aluminum
    neurological disorders and, **5**:901
    toxicity
        in Alzheimer's disease, **1**:214
        artificial cells in treatment of, **1**:378
Aluminum salts, as vehicles for synthetic vaccines, **7**:754
Alveolar buds, **4**:604
Alveolar elastin
    development of, **3**:267
    in pulmonary emphysema, **3**:268
Alveolar process, **2**:697, 779
Alveolar ventilation
    aging and, **3**:902
    carbon dioxide excretion and, **5**:877
    defined, **6**:596
Alveoli
    abnormal changes in volume, **6**:364–70
    interstitium, structure and function of, **6**:367
    as part of pulmonary tree, **6**:592–93
    primary lobule of, **6**:364
    stability and atelectasis, **6**:367–68
    structure of, **6**:364–66
    surfactant system of, **6**:601
    volume of, and pulmonary pathology, **6**:368
    walls
        diffusion of oxygen across, **5**:619–20
        structure of, **6**:593
Alveus, **4**:228

Alzheimer's disease, **1:**209-16, **2:**686, 764
    abnormal proteins and protein products in, **1:**215-16
    acetylcholine deficiencies in, **1:**214-15
    aluminum toxicity as causative factor, **1:**214
    amyloid fibers, **1:**209
    amyloid precursor protein and, **3:**186
    beta protein and, **1:**215-16
    beta protein precursor and, **1:**215-16
    blood-brain barrier changes and, **1:**214
    cell death and, **2:**251
    central nervous system transmitters in, **5:**412
    cerebrotoxic agents, in pathogenesis, **1:**214
    cerebrovascular amyloidosis in, **1:**209, 211-13
    characterization, **4:**997
    choline acetyltransferase depletion in, **1:**214-15
    cholinergic receptors and, **5:**421
    chromosome 21 and, **3:**180
    clinical diagnosis, **1:**213
    communication effects, **1:**123-24
    computerized tomography with X-ray, **1:**131
    description, **1:**209
    diagnosis, **1:**131-32
    Down's syndrome and, **3:**175
    effects of, **1:**209
    epidemiology, **1:**213
    etiology, **1:**131
    familial, **2:**764
        DNA marker diagnosis of, **3:**148
        genetic locus of, **3:**182
    genetics of, **1:**213
    hippocampal formation and, **4:**232-33
    hyperphosphorylated tau and, **5:**901
    infectious agent theory of, **1:**214
    language disorders associated with, **4:**638
    lesions of, **1:**210-13
    magnetic resonance imaging, **1:**131
    memory pathology, **4:**677-78
    nerve growth factor and, **5:**340
    neurofibrillary tangles, **1:**210-13; 211
    neuroreceptor deficiencies, **1:**214-15
    neurotransmitter deficiencies, **1:**214-15
    pathogenesis of, **1:**215-16
    phosphorylated tau and, **5:**898
    positron emission tomography, **1:**131
    primary dementias, **1:**210
    pseudodementia, **1:**210
    reduction of gray matter in, **1:**213
    secondary dementias, **1:**209-10
    senile plaques, **1:**211-13
    senility and, **1:**209
    single photon emission computerized tomography, **1:**131

Alzheimer's disease *(Continued)*
    theories of, **1:**214-15
    therapeutics, **1:**132
    treatment and management, **1:**213
    vascular changes and, **1:**214
Amacrine cells, **6:**626
alpha-Amanitin, chemoimmunoconjugates with, **4:**370
Amantadine, **2:**399
Amantia species, mushroom poisoning and, **7:**591-92
Amastigote, **7:**681
Amber suppression, and amino acid specificity, **5:**77-78
Amebiasis, **1:**217-30, **2:**382, **6:**268-69, 383
    acute amoebic colitis, **1:**222
    amebic colitis, **1:**222, 229
    amoeboma, **1:**222
    asymptomatic intestinal infection, **1:**222
    brain abscesses associated with, **1:**223
    cell-mediated immune responses to, **1:**228
    complement-mediated resistance to, **1:**227-28
    diagnosis, **1:**229
    fulminant colitis, **1:**222
    genitourinary diseases associated with, **1:**223
    high-risk groups, **1:**219
    invasive
        acquired resistance to, **1:**226-28
        antibody response, **1:**227
    invasive amoebic colitis, **1:**222
    iodoquinol and, **2:**382
    liver abscesses associated with, **1:**222-23
    lung abscesses associated with, **1:**223
    metronidazole and, **2:**382
    nonimmune defense mechanisms, **1:**226
    pathogenesis, **1:**223-26
    polymorphonuclear leukocyte response to, **1:**228
    prevention of infection, **1:**229-30
    symptomatic noninvasive infection, **1:**222
    therapy, **1:**229
    transmission of amoebic infection, **1:**218-19
Amelin, **2:**46, 47
    defined, **2:**41
    localization, **2:**47
Ameloblast, **2:**774, 819-23
Amelogenesis, **2:**821-22
Amelogenesis imperfecta, **5:**611-12
Amelogenin, **2:**822
Amenorrhea, **2:**29, 498
    of athletes and underweight girls, **1:**742-43
    breast feeding and, **4:**608-9
    defined, **5:**393
    delayed menarche and physical exercise, **1:**746-47
    hypothalamic, **2:**28

Amenorrhea *(Continued)*
  hypothalamic/pituitary pathology and, **5:**398–99
  and naloxone, **2:**28
  psychic causes of, **5:**396
  secondary, **2:**28
American Association for Accreditation of Laboratory Animal Care, **1:**255
American Fertility Society, **3:**366
American Joint Committee on Cancer staging system, **4:**890–91
American Pediatric Society, **5:**700
American Psychiatric Association, **5:**445
Ames assay
  advantages of, **5:**251
  defined, **3:**671
  of heterocyclic aromatic amines, **3:**677
Amethopterin. *see* Methotrexate
Amiloride, in potassium-sparing diuretics, **3:**99
Amine hormones
  action mechanisms of, **3:**331–33
  listed, **3:**320
  synthesis, **3:**327–28
Amine oxidase
  extracellular, **2:**564
  inhibitor, beta-aminoproprionitrile, **2:**564
Amines
  aromatic, **2:**110
    4-aminobiphenyl, **2:**110
    benzidine, **2:**110
    2-naphthylamine, **2:**110
  biogenic, pulmonary endothelial cells and, **6:**350
  as neuroactive substances in retina, **6:**629
  neurotransmitters and, **1:**308–9, **3:**322
2-Amino-4-phosphonobutyrate, **2:**523
2-Amino-5-phosphonovalerate, **2:**524
Amino acid sequences. *see also* Leader sequences
  of actin, **5:**169
  alternative RNA splicing and, **5:**166
  of bile salt-stimulated lipase, **3:**50
  calcium pump proteins, **4:**561
  of cytochrome *c* oxidase isoforms, **5:**61
  of dystrophin, **5:**196
  gene specification of, **3:**792
  of human *fos* gene, **3:**703
  hydrophobic, of elastin, **3:**261
  interleukin-2, **4:**529
  in matrix proteins, **3:**528
  of myosin, **5:**169
  of myosin alkali light chains, **5:**216
  of nerve growth factor, **5:**335
  of peptide hormones, **5:**714
  in ribosomal proteins, **6:**672

Amino acid sequences *(Continued)*
  T cell receptor gene sequence, **4:**833–34
  of tropomyosin genes, **5:**218, 219
Amino acids
  abbreviations of, **3:**523
  activation, and protein synthesis, **6:**673
  analogues, pancreatic toxicity and, **7:**569
  availability, in protein, **6:**232
  codons and, **5:**76
  decarboxylase, **2:**534
  deficiency, immunity and, **5:**505–6
  defined, **3:**317
  destruction of, **3:**416, 418
  dietary, commercial supplements of, **6:**232
  digestibility of, **6:**231–32
  enzyme thermostability and, **3:**412
  essential, **6:**230, 628–29
  excitatory, **2:**523
  food analysis, **4:**221–22
  genetic code, **6:**217
  in glutathione degradation, **3:**912
  in hemoglobin, **4:**135
  Ig sequence, **4:**833–34
  integral membrane protein sequence, **4:**972
  as neuroactive substances in retina, **6:**230
  nonessential, **6:**230
  of peptides, **5:**725
  as plant toxins, **7:**594
  and primary structure of proteins, **6:**214–16
  in protein digestion, **3:**80
  proteinase, **4:**183
  pyrolysis products
    defined, **3:**671
    in food, **3:**676–78
  receptors, **2:**339
  recognition of transfer RNA, **5:**75–80
  residues
    in enzyme denaturation, **3:**413
    in histones, **3:**127
  sulfated tyrosine residue 7, **5:**718
  of tachykinins, **5:**720
  transfer, in protein synthesis, **6:**673–74
  transmitter, **2:**339
  transport, across blood-brain barrier, **1:**725
  types of, **3:**431
Amino-hydroxypropylidene bisphosphonate, **4:**280–81
Amino terminal
  defined, **3:**919
Amino-terminal regions, **4:**184
Aminoacyl tRNA synthetase, **5:**76, 77–78
D-alpha-Aminoadipate, **2:**524
p-Aminoazobenzene, **2:**110

4-Aminobenzoic acid, **1:**273

4-Aminobiphenyl, **2:**112

L-alpha-Aminobutyrate, in glutathione synthesis, **3:**910

gamma-Aminobutyric acid. *see* GABA

Aminolevulinate, **2:**534

Aminolevulinic acid, **6:**117–18

Aminonitrile, **2:**571

Aminopeptidase N, **1:**769–70

Aminophosphovalerate, **2:**314

beta-Aminopropionitrile, **2:**564, 569, 571

Aminopropyltransferase, **6:**70

Aminopterin, **5:**704

Aminopyrine, **3:**759

Aminoquinolines, **4:**884

p-Aminosalicylic acid, **1:**327

Aminotransferase, **2:**534

Amiodarone, **2:**173

Amitriptyline, **1:**613–14, **2:**747

Ammonia

from deamidation, **3:**416

excretion of, by kidneys, **5:**879

homeostasis of

bicarbonate and, **5:**877

in liver and kidneys, **5:**879–80

NH4+, elimination of, **5:**879

spirits of (smelling salt), **2:**787

from thermoinactivation of hen egg white lysozyme, **3:**417

Ammoniagenesis, acid-base homeostasis and, **5:**828

Ammoniagenesis, renal. *see under* Ammonium

Ammonium chloride, **5:**827

Ammonium ions, in acidosis, **5:**828

Ammon's horn, **2:**680

Amnesia, **2:**9

anterograde, **2:**9, **5:**402

characterization, **4:**998

defined, **2:**1

epileptic seizures and, **3:**523

functional system damaged during, **5:**404–5

medial temporal, **4:**677

and memory, **2:**765

preserved learning during, **5:**402–3

relationship to autism, **1:**509

retrograde, **2:**9, **5:**403–4

Amnestic dysphasia. *see* Dysphasia, amnestic

Amniocentesis, **1:**703–4

Amnion

defined, **3:**293

functions of, **3:**612

Amnioscopy, **3:**304

Amniotic cavity, formation, **4:**428, 929

Amniotic fluid

defined, **3:**609

fetal swallowing of, **3:**616, 618

fetal thyroid hormones in, **3:**339

Amniotic sac, examination of, **3:**295

Amoebiasis. *see* Amebiasis

Amoebic colitis. *see under* Amebiasis

Amoeboma, **1:**222, **6:**268, 229

Amphetamine, **2:**149, 220, 511, 834, 843

and appetite reduction, **2:**220

dopamine and, **5:**671

epidemics

in Japan (1950-1956), **2:**513

in Sweden (1964-1968), **2:**513

in U.S.A. (1965-1969), **2:**513

studied for first time, **2:**512–13

synthesized in 1887, **2:**512

use in therapy, **2:**513

Amphetamines

abuse of, **5:**440

addiction and, **1:**43–51

alcohol and, **3:**213

barbiturates and, **5:**442

cocaine and, **5:**441

3,4-methylenedioxyamphetamine and, **5:**449

in treatment of autism, **1:**509–10

Amphiarthroses, **1:**371

Amphibians

gas convection in, mechanisms of, **6:**570

olfaction of, **5:**528

poisonous, **7:**598–99

Amphipaths, **4:**970

Amphiphilic substances

defined, **3:**47, 581

fatty acids as, **3:**582

Amphotericin B, **1:**290(t), 292

Ampicillin, **2:**785

Amplicons, **3:**785

Amplification

in cellular signaling, **2:**328

in signal transduction, **4:**15

Amplification, gene. *see* Gene amplification

Amplification control elements, **3:**785

Amplification enhancer regions, **3:**785

Amygdala, **2:**5, 7, 9, 32, 335, 680, **4:**721–23

Amylase, **2:**104, 781, **6:**694

Amyloid beta protein, **3:**182

Amyloid deposits, and kidney dialysis, **1:**409

Amyloid F protein, **3:**262

Amyloid fibers, **1:**209, 211, 215

Amyloid precursor protein, **3:**182, 186, 188

Amyloidosis, cerebral, **1:**210, 215
Amylopectin, **2:**103
Amylose, **2:**103
Amyotrophic lateral sclerosis, **1:**231–35
    animal models for, **1:**234
    defined, **5:**43
    general description of, **1:**232
    as motor neuron disease, **1:**231–35
    nucleus of Onuf motoneurons in, **5:**44
    pathology of
        chemical, **1:**232
        histological, **1:**232
    theories as to the cause of, **1:**233–34
    treatment of, **1:**235
    treatment with interferon, **4:**523
Anabolic steroids
    athletes and, **7:**273–74
    defined, **5:**725
    health and, **7:**273–74
Anabolism, **2:**503
Anaerobes, in skin and wound lesions, **1:**557
Anaerobic exercise. *see* Exercise, anaerobic
Anaerobic infections
    bowel bacterial overgrowth syndrome, **1:**244, 246
    clinical clues to, **1:**244, 248
    common, table of, **1:**244–45
    diagnosis, **1:**246–47
    host defense mechanisms, **1:**242–43
    incidence of, **1:**244
    nosocomial infections, **1:**243
    predisposing conditions, **1:**243
    prognosis for, **1:**249–50
    prophylaxis, **1:**249
    therapy, **1:**247–49
    types of, **1:**244–45
    unique features of, **1:**244, 248
    virulence factors in anaerobes, **1:**241–42
Anaerobiosis, **1:**237–38
Anagen, **4:**41
Anal atresia, **1:**702(t)
Analeptics, **3:**716
Analgesia, **2:**338
    of central gray area by electrostimulation, **2:**338
    descending pain system and, **5:**627
Analgesia, dental, **2:**786
    codeine, **2:**786
    dihydromorphone, **2:**786
    hydrocodone, **2:**786
    meperidine, **2:**786
    methodone, **2:**786
    morphine, **2:**786
    narcotics, **2:**786

Analgesia, dental *(Continued)*
    oxycodone, **2:**786
    pentazocine, **2:**786
    side effects, **2:**786
Analgesics, opioid peptides as, **5:**732
Analytic cytology. *see* Cytology, analytical
Analytical microscopy, **3:**274
Anaphase, **2:**243, 133
    events of, **5:**71
    mitotic spindle phosphorylation and, **5:**902
Anaphora, **6:**490
Anaphylatoxins, **2:**612–14
    assay by radioimmunoassay (RIA), **2:**612
    C 5a, **2:**612, 613
    carboxypeptidase *N* and, **2:**613
    control of, **2:**613
    generation of, **2:**613
    inactivation, enzymatic, **2:**613
    spasmogenicity and, **2:**613
Anaphylaxis, **1:**193, **4:**410–11
    defined, **3:**579
    food hypersensitivity and, **3:**674
Anastomoses, vascular. *see* Vascular anastomoses
Anastomosis
    defined, **2:**357
    leptomeningeal, **2:**362–63
Anatomic pathology
    defined, **5:**681
    in diagnosis, **5:**681–83
Anatomical evolution. *see* Evolution, anatomical
Anatomy
    cardiovascular system, **2:**157–65
    heart, **2:**157–65
    larynx, **4:**653–58
    magnetic resonance images, **4:**869–70
    muscular, **4:**259–67
    vocal tract, evolution, **4:**635–36
Anatomy, comparative, **2:**593–600
    arms, **2:**597–98
    brain, **2:**599
    definition, **2:**593
    face, **2:**599
    feet, **2:**594
    hair, **2:**599
    hands, **2:**598
    head, **2:**599
    inner organs, **2:**597
    legs, **2:**594–95
    neck, **2:**598
    pelvis, **2:**595–96
        as birth canal, **2:**594
    posture, **2:**593–94

Anatomy, comparative *(Continued)*
    reproduction and, **2:**599
    shoulder, **2:**597
    trunk, **2:**596–97
    what makes us human?, **2:**1, 8, 593, 600
Anchor fiber, **2:**132
Anchored PCR. *see* Polymerase chain reaction
Anchorin CII, **2:**207, 211
Ancylostoma, **2:**387
Androgen-binding proteins, **7:**423
Androgen insensitivity syndrome, **4:**243
Androgenic pathway, **7:**260
Androgens
    androstenedione, **2:**744
    biosynthesis of, **7:**263–64
    breast cancer and, **2:**56
    criminality and, **2:**716
    in female genital organogenesis, **3:**342
    female reproduction and, **1:**743
    function of, **7:**272–74
    hair loss and, **4:**42
    masculinization of females and, **2:**744
    metabolic transformation of, **7:**279
    in oocyte maturation, **5:**556
    in pregnancy, biosynthesis of, **7:**275
    in puberty, **3:**355, **7:**273
Anemia
    in cancer, **2:**654
    in chronic disorders, **2:**654–55
    in chronic infections, **2:**654
    as complication from kidney dialysis, **1:**409
    deficiency symptoms and, **2:**650, 654
    hookworm-induced, **4:**116–17
    renal failure and, **2:**654
    of spaceflight, **4:**123–24
    trichuriasis and, **4:**120
    in tuberculosis, **2:**654
Anemia, aplastic, **1:**794
Anemia, hemolytic, **1:**530, **2:**616, **4:**143, **6:**398; 545
Anemia, hypochromic, **2:**653; 654
    in developing countries, **3:**26
    in Uganda, **3:**21
Anemia, pernicious
    and autoimmunity, **1:**528
    role of antibodies in, **4:**408
Anemia, sickle cell, **4:**143, 149
    bone marrow transplantation for, **1:**796
    defined, **5:**259
    diagnosis of
        with polymerase chain reaction, **3:**140
        with restriction fragment-length polymorphisms,
           3:136

Anemia, sickle cell *(Continued)*
    DNA marker diagnosis of, **3:**144–46
    genetic aspects of, cystic fibrosis and, **3:**832
    genetic variation and, **6:**111
    hemoglobin characteristics in, **7:**2
    increased red cell calcium in, **6:**535–36
    newborn screening for, **5:**436
Anemia, sideroblastic, **2:**654–55
    congenital, **2:**654
    dyserythropoietic, congenital, **2:**655
    in females, **2:**654
    hypochromic, microcytic, **2:**650
    in males, **2:**654
    primary, acquired, **2:**654
    secondary, **2:**654
Anemones, toxins produced by, **7:**595–96
Anencephaly
    adrenal cortex growth and, **3:**343, 344
    gonadotropins and, **3:**348
    male, **3:**340
    polygenic/multifactorial causes, **1:**700
    prevalence of, **1:**702(t)
    prolactin and, **3:**347
    prolonged pregnancy and, **3:**338, 344
    in 16-wk fetus, **3:**303
Anesthesia
    from solvents, **3:**386
    space station problems with, **7:**136–37
Anesthesia, dental, **2:**783–84
    amide class, **2:**783–84
    bupivacaine, **2:**784
    ester class, **2:**783
    etidocaine, **2:**784
    lidocaine, **2:**784
    mepivacaine, **2:**784
    prilocaine, **2:**784
    procaine, **2:**784
    and sodium conductance, **2:**783–84
    and toxicity, **2:**784
    and vasoconstriction, **2:**784
    and vasoconstrictor
        epinephrine, **2:**784
        levonordefrin, **2:**784
Aneuploidy, **2:**437
    autosomal, **2:**441
    chemical producers of, **5:**72
    defined, **3:**621, 835
    mental retardation and, **3:**838
Aneurysm, **2:**652
    cerebral, **7:**302
    and copper deficiency, **2:**652
    and stroke, **7:**303

"Angel dust." *see* Phencyclidine

Angina pectoris, **2:**87, 160
    calcium antagonist for, **2:**87
    coronary, **2:**171
    defined, **2:**167
    pathology, **2:**87

Angioedema, hereditary, **2:**617
    and androgen therapy, **2:**617
    kinins and, **4:**599
    and swollen epiglottis, **2:**617

Angiogenesis, **4:**1010–11

Angiogenin, and ribonuclease, **6:**663

Angiography, **1:**449, **3:**1, 2, 3–4

Angioplasty, **1:**454

Angioplasty, coronary, **2:**170

Angiorrhexis, **2:**652

Angiosarcomas, **7:**702

Angiostrongylus costaricensis, **2:**389

Angiotensin-converting enzyme, **3:**396

Angiotensin-converting enzyme inhibitor, **2:**174, 176, **4:**291, 294

Angiotensin II, **2:**20, 27, 38, 144, 174
    as activator of thirst, **7:**475
    functions of, **5:**730, **6:**91
    and hypertension, **4:**291

Angiotensin III, **2:**38

Angtiotensin-forming enzyme. *see* Renin

Angiotensin receptors. *see* Receptors, angiotensin

Angiotensinogenase. *see* Renin

Angiotensins, **1:**473–75
    functions of, **1:**74, **5:**729–30
    in pulmonary blood flow, **6:**598

Angular gyrus. *see* Gyrus, angular

Anhedonia, **2:**511

Animal diseases, infectious, **1:**294–95

Animal feed
    aflatoxins in, **5:**265
    amino acid pyrolysis products in, **3:**677
    antibiotics in, **1:**294–95
    colorectal cancer and, **3:**767
    deoxynevalenol in, **5:**264
    mycotoxins in, **5:**261

Animal intelligence. *see* Thorndike, E. L.

Animal milk. *see* Milk

Animal models. *see* Models, animal

Animal models of disease. *see* Disease models, animal

Animal parasites. *see* Parasites, animal

Animal parasites, diseases caused by. *see under* Protozoal infections, human

Animal testing alternatives, **1:**253. *see also* Animal welfare; Animals, laboratory

Animal welfare, **1:**254. *see also* Animal testing alternatives; Animals, laboratory; Disease models, animal
    Animal Welfare Act (PL89-544), **1:**253
    Health Research Extension Act of 1985 (PL99-158), **1:**253
    Improved Standards for Laboratory Animals Act, **1:**253
    major concerns, **1:**254

Animals
    aquatic, pollutants and, **3:**256
    comparative pathology and induced diseases, **5:**676
    contributions to human behavioral research, **5:**678
    dreaming by, **3:**195
    elastin in, **3:**264
    genetically transmitted metabolic defects of, **5:**678
    major contributions to human conditions, **5:**678–79
    in medical research, **5:**676
    pollutant routes through, **3:**256
    sex differences in, biocultural, **6:**846
    terrestrial, pollutants and, **3:**256
    toxins, **7:**595–98
    transgenic, molecular pathology and, **5:**679

Animals, domestic
    dendrodochiotoxicosis in, **5:**263
    deoxynevalenol and, **5:**264
    ergotism in, **5:**263
    mycotoxicoses in, **5:**260–61
    ochratoxins and, **5:**265
    trichothecenes and, **5:**264

Animals, laboratory. *see also* Animal testing alternatives; Animal welfare; Disease models, animal
    American Association for Accreditation of Laboratory Animal Care, **1:**255
    care of, **1:**254–55
    cocaine and, **5:**441
    Guide for the Care and Use of Laboratory Animals, **1:**255
    medical achievements by use of, **1:**252(t)

Animals, transgenic
    as molecular factories, **5:**680
    molecular pathology and, **5:**679

Anion gap. *see* Acid-base equilibrium

Anion metabolism, **4:**568–69

Anion transporter. *see* Ion transport

Anions
    choloride, body function of, **5:**49–50
    nonmetal, body function of, **5:**49–50

Aniridia-Wilms tumor complex, **2:**439, 440

Anisakiasis, **2:**387

Anisometropia, space perception and, **1:**639

Anisotropy, properties of, **5:**567

Ankles, in bicycling, **5:**187–89

Ankylosing spondylitis, **6:**416–17

Ankyrin, **2:**43, 44, 46, 47, 50
    in brain, **2:**47
    in erythrocyte, **2:**47
    myofibril organization and, **5:**169
Annealing, simulated. *see* Simulated annealing
Annular granules, **5:**452
Annulate lamellae, of oocyte, **5:**553
Anomia. *see* Aphasia, amnesic
Anopheles mosquito, **4:**873, **6:**275-77, 883
Anorexia
    association with hookworm infections, **4:**116
    hypothalamus role, **4:**309
Anorexia nervosa, **2:**498
    behavioral aspects of, **3:**592
    and caffeine, **2:**79
    characterization, **4:**997
    course of, **3:**245-46
    defined, **3:**17, 243
    as eating disorder, **1:**352
Anosmia
    appetite and, **1:**345
    defined, **5:**529
    from head traumata, **5:**524
Anoxia
    ability to withstand and temperature, **4:**792
    acid-base disturbance from, **5:**877
    effect on blood flow, **5:**621
    headaches associated with, **4:**58
    metabolic control and, **3:**380
ANS. *see* Autonomic nervous system
Antagonism, chemotherapeutic, **2:**371-79
Antagonism, drug, **3:**211
Antagonist, **2:**305, **3:**455, **4:**13, 321
Anteater, **2:**707
Anteride, **2:**184
Anterior cerebral artery, **7:**301
Anterior hypothalamus-preoptic area, **4:**719
Anterior intercavernous sinuses, **4:**984
Anterograde axonal transport
    defined, **3:**743
    kinesin in, **5:**19-20
    velocity of, **3:**744
Anthelmintic drugs, **1:**270
Anthropoidea, **6:**139
Anthropology
    changing terminology in, **6:**104-5
Anthropology, dental, **2:**789-804
    dentition, human, **2:**789-91
    environmental stress and, **2:**802-8
    genetics, dental, **2:**793-94
    history inferred from teeth, **2:**791, 792
    phenetics, **2:**791-98

Anthropology, dental *(Continued)*
    phylogeny, **2:**791-98
    population history and tooth size, **2:**794-96
    teeth, **2:**790-91
        observability, **2:**790
        preservability, **2:**790
        variability, **2:**790
    teeth, as indicators
        of age, **2:**791
        of sex, **2:**791
    teeth and behavior, **2:**798-802
Anthropometry
    defined, **3:**17; 649, 473
    of the elderly, **4:**899
    in ergonomic database, **3:**478
Anthroponoses, defined, **7:**891
Anti-GM1 antibodies, **3:**741
Anti-idiotypes, as regulators of immunity, **1:**522
Anti-inflammatory agents
    eicosanoids and, **3:**579
    endogenous antipyresis, **1:**759-60
    nonsteroidal, **2:**785, **4:**477
    steroid action, **1:**311-15
Anti-Müllerian hormones
    Sertoli cell secretion of, **7:**425
    testosterone implants and, **3:**341
Anti-parasitic agents, **2:**390-93
    mode of action, **2:**391-93
    resistance to, **2:**384
Anti-viral agents, **2:**396, 398-99
Antianginal drugs, **2:**171-72
    beta-blocker, **2:**171
    calcium antagonist, **2:**172
    nitrate, **2:**172
Antiarrhythmic drugs, **2:**172-74
    amiodarone, **2:**173, 174
    disopyramide, **2:**173
    and electrocardiogram (ECG), **2:**173
    encainide, **2:**173
    flecanide, **2:**173
    lidocaine, **2:**173, 174
    procainamide, **2:**173
    quinidine, **2:**173
    sotalol, **2:**174
Antiautoantibody-antigen complexes, **1:**301
Antibiosis, **1:**287
Antibiotics, **1:**287-95, 784-85
    adverse reactions to, **1:**292-93
    agricultural uses, **1:**294-95
    ampicillin, **2:**785
    antifungal, **1:**291, 292
    antitumor agents, **1:**288

Antibiotics (*Continued*)
bacterial cell wall inhibitors, **1:**291
bacterial protein synthesis inhibitors, **1:**291
biosynthesis, **1:**289-90
broad-spectrum, **2:**783
cephalexin, **2:**785
for childhood diseases, **5:**703
classification, **1:**288
defined, **5:**699
in dentistry, **2:**784-85
erythromycin, **2:**785
families of, **1:**289(t)
as feed additives, **1:**294-95
fungal, **7:**593
herd medication, **1:**294
history and development, **1:**287-88
inhibitors of bacterial cell wall synthesis, **1:** 271-85
beta-lactam, **1:**288; 290-91
mechanism of action, **1:**290(t); 290-91
medical uses, **1:**292-93
against mycobacteria, **1:**293
mycotoxins as, **5:**260
narrow-spectrum, **2:**783
nucleic acid synthesis inhibitors, **1:**291
resistance to. *see* Drug resistance, microbial
secondary metabolites, **1:**289
selective toxicity, **1:**288
semisynthetic, **1:**288
sensitivity testing of pathogens. *see* Microbial sensitivity tests
tetracycline, **2:**785
Antibiotics, lactam, **1:**288, 290-91
Antibodies, **2:**248-49. *see also* Immunogenetics; Immunoglobulins
antigenic markers, **4:**315
binding
to graft endothelial cells, **4:**362
as technique for protein detection, **6:**196-97
beta-cell repertoire of specificity, **1:**563-65
cellular basis of expression, **1:**564-65
in detection of bacterial pathogens, **1:**553
enzyme-linked, application of, **5:**89-90
fusion proteins, preparation and application of, **5:**89-90
general characteristics of, **1:**563
immunocytochemical analysis of, **3:**277
molecular basis of diversity, **1:**564, 566
polyclonal, preventing graft rejection, **4:**363
response
effect of conditioning on, **6:**324
effect of radiation on, **6:**420-21
to invasive amoebiasis, **1:**227
role in beta cell triggering, **1:**567

Antibodies (*Continued*)
staining of. *see* Immunostaining
tests for infectious diseases, **1:**738
Antibodies, anti-HIV, **1:**20
Antibodies, anti-idiotypic, **4:**314-15, 317, 348-49
Antibodies, antidystrophin, **5:**196, 198
Antibodies, antinuclear
and autoimmune disease, **1:**515-16
in diagnosis of autoimmune disease, **1:**338-39
Antibodies, chimeric
antibody effector functions and, **3:**808-9
antitumor, **3:**810-11
characteristics of, **3:**803-4
mouse/human, **5:**88
antitumor, **3:**810-11
characteristics of, **3:**804
first, **3:**808
potential applications of, **3:**811
production of, **3:**807-8
Antibodies, genetically engineered, **3:801-12**
Antibodies, monoclonal
B cell subsets, **4:**835
biological response modification with, **1:**687-88
bispecific, preparation and application of, **5:**86-87
cancer diagnosis using specific, **5:**83
cancer immunodiagnosis, **4:**346
cancer immunology, **4:**338
chemical modification of, **5:**84
chimeric, preparation and application of, **5:**87-88
cytotoxicity, **4:**348
defined, **3:**671; 801, **5:**81
diagnosis of parasite infections by, **5:**654
in diagnostic pathology, **5:**686
in flow cytometry, **3:**623
and fusion proteins, **5:**89-90
ganglioside structure and, **3:**731
gangliosides and cancer treatment, **3:**733
human, **5:**84-86
application of, **5:**85-86
Epstein-Barr virus transformation to produce, **5:**85
immunotherapy using, **5:**86
immunoconjugates, **1:**688-89
immunotherapy with, **4:**348
isoenzyme analysis, **4:**577
melanoma-associated. *see* Melanoma antigens and antibodies
mouse, **5:**82-84; 335
application of, **5:**82-84
immunodiagnosis using, **5:**83-84
immunotherapy using, **5:**84
preparation of, **5:**82

Antibodies, monoclonal *(Continued)*
    parasite detection, **4:**389
    preventing graft rejection, **4:**363, 365
    protocol for derivation of, **5:**82
    second generation, **5:**86–90
    single-chain, preparation and application of, **5:**88–89
    technology, **5:**81–90
Antibodies, S-IgA, **5:**141
Antibody conjugates
    with cytotoxic drugs, **4:**370
    with protein toxins, **4:**367–69
Antibody-dependent cell cytotoxicity, **4:**340
    carbohydrates and, **3:**805
    defined, **3:**801
Antibody formation
    and aging immune system, **6:**77
    effect of trichothecenes on, **7:**664
    to malaria, **4:**878–79
    retinoids and, **4:**826
    in upper respiratory tract, **5:**692
Antibody variable regions, **1:**564
    defined, **3:**801
    expression of, **3:**806–7
    genetic engineering of, **3:**809
Anticancer agents. *see* Antineoplastic agents
Anticarcinogens, dietary, **3:**674–81
Anticholinesterase agents. *see* Cholinesterase inhibitors
Anticoagulants, **2:**170–71
    and clot formation, **2:**170
    coumarin-type, drug resistance to, **5:**833
    heparin, **2:**171
    in open heart surgery, **1:**395
    warfarin, **2:**171
Anticodon, alanine tRNA, **5:**78–79
Antidepressive agents, **1:**307–10, **2:**220; 833, 837, 842. *see also* specific antidepressive agents
    affecting amine transport, **1:**308–9
    indications for use of, **1:**308
    monoamine oxidase inhibitors, **1:**309–10
    response to, **1:**307–8
    serotonergic receptors and, **5:**422
    standard, **1:**308–9
    table of drugs used as, **1:**309(t)
    in treatment of mood disorders, **5:**95–97
Antidepressive agents, tricyclic, **1:**308–9, 613–14, **2:**833; 837–38, **5:**440, 842
Antidiuretic hormone. *see* Vasopressin
Antiemetics, **4:**909–10
Antiestrogen, **2:**60
    clomiphene, **2:**60
    tamoxifen, **2:**60
    triphenylethylene structure best, **2:**60

Antifungal agents, **1:**327, 328
Antigen-antibody complexes, **1:**297–305, **4:**416
    activation of inflammatory reactions, **1:**298–99
    in AIDS, **1:**304
    allergens, **1:**299
    anti-self responses and, **1:**302
    antiautoantibody-antigen complexes, **1:**301
    antigen-presenting cells, **1:**303–4
    autoimmune disease, **1:**300, 302–3
    B cells
        autoregulatory, **1:**303
        Fc-dependent inactivation, **1:**301
    in chronic infections, **1:**300–301
    complement activation, **1:**298
    eosinophils, **1:**299
    Fc receptors, **1:**298–99
    Fc signaling mechanisms, **1:**301
    immune clearance, **1:**299
    immunization strategies, **1:**299
    induction of immune responses by, **1:**301–2
    K cells, **1:**299
    mechanisms and outcomes, **1:**305(t)
    nature of, **1:**297–98
    phagocytosis and, **1:**299
    reticuloendothelial system, **1:**299
    rheumatoid factor, **1:**302
    role in immunity to infections, **1:**299
    role in regulation of immune response, **1:**301–3
    separation of, by radioimmunoassay, **6:**458
    T cells
        Fc-dependent inactivation, **1:**301
        prevention of Fc signaling, **1:**303
Antigen-presenting cells, **1:**677, **2:**752
    depletion and transplant survival, **4:**361
    Fc signaling and, **1:**312–13
    macrophages as, **4:**859, **5:**818
    major histocompatibility complex, **4:**834–35
    in psoriasis, **3:**450–51
    resident, **4:**396
    T cell sensitization, **4:**361
Antigen-specific receptors, **4:**314
Antigen vaccines, **4:**446
Antigenic determinants. *see* Epitopes
Antigenic drift, viral epidemics and, **7:**782–84
Antigenic variation
    African trypanosomes and, **5:**649, 652–53
    as factor in transmission of infectious disease, **7:**643
    trypanosomiasis and, **7:**684
Antigens, **2:**751
    altered, in autoimmunity, **1:**523–25
    B-cell activation, **2:**754–55

Antigens *(Continued)*
  binding of, to lymphocyte receptors, **6**:520-22
  blood group, **4**:379-81
  carcinoembryonic, **5**:319
  chemically induced, **4**:343-44
  cross reactivity to, **1**:565
  defined, **2**:225, **4**:402
  dermatomyositis and, **1**:336
  detection of bacterial pathogens, **1**:553
  differentiation, **4**:3, 8
  exogenous, detection by immune system, **4**:393-94
  ganglioside, **4**:962-65
  genes and molecules governing recognition, **4**:832-34
  hemopoietic, expressed by osteoclasts, **1**:769
  histocompatibility, in bone marrow transplantation, **1**:792
  immune response. *see* Histocompatibility antigens class II
  immunoparasitology, **4**:386
  labeled, in radioimmunoassay, **6**:457-58
  "late migrating nuclear", in *Xenopus*, **5**:456
  lupus and, **1**:334
  lymphocyte activation, **2**:752
  melanoma-associated. *see* Melanoma antigens and antibodies
  novel, **4**:390
  oncofetal, **3**:733, **4**:341
  oncogene product, **4**:341
  orally administered, *S*-IgA and, **5**:141
  polymyositis and, **1**:336
  presentation
      parasitic diseases, **4**:386-87
      regional immunology, **4**:399
      T cell-beta cell collaboration, **1**:568
  processing, macrophage, **5**:818
  receptors, in beta cell triggering, **1**:567
  regionally located, **4**:394
  release from parasites, **4**:387
  scleroderma and, **1**:337
  self, and autoimmune disease. *see* Autoimmune disease
  self-antigen, **2**:225
  stage-specific embryonic, **3**:730
  surface, T cells and, **5**:506
  T cell receptor, **7**:478-79
  T cell activation, **2**:752-54
  targeting moiety of immunoconjugates, **4**:374-75
  variation by parasites, **4**:387-88
  viral capsid, nasopharyngeal carcinoma and, **3**:465
  virally induced, **4**:341-43
Antigens, CD4, **1**:15-16, 20, **4**:374-75
Antigens, EBV-determined, in Burkitt's lymphoma, **3**:464
Antigens, polyomavirus transforming, **5**:455, 456

Antigens, virally induced, **4**:341-43
Antihemophilic factor. *see* Factor VIII
Antihistamines, ascorbic acid, **1**:432
Antihypertensive agents, **2**:176-77, **4**:293-94
Antimalarial agents, **1**:267, **4**:883-84
Antimalarial vaccines, **4**:884-85
Antimere, **2**:789
Antimicrobial agents, **1**:325-30
  agricultural uses, **1**:329-30
  anti-*Bacteriodes*, **1**:248
  anti-*Fusobacterium*, **1**:248
  antifungals, **1**:327
  antimycohbacterials, **1**:327
  antivirals, **1**:327
  chromosomally mediated resistance to, **1**:328
  classification, **1**:326
  drug resistance. *see* Drug resistance, microbial
  in foodstuffs, **3**:663
  history of, **1**:325-26
  for infants, **1**:317-23
      activity of commonly used, **1**:320
      pharmacology of, **1**:320
      spectrum of, **1**:320
  mechanism of action, **1**:326-27
  medical uses of, **1**:328-29
  pharmacology, **1**:247-48
  plasmid-mediated resistance, **1**:328
  resistance mechanisms in anaerobes, **1**:249
  resistance of specific anaerobes to, **1**:246
  resistance to, **1**:248-49; 328
  side effects, **1**:329
  susceptibility patterns, **1**:247
  synthetic, families of, **1**:326
Antimony-containing drugs, **1**:263
Antimycobacterial agents
  for leprosy, **1**:326, 329
  mechanism of action, **1**:327
  resistance to, **1**:328
  for tuberculosis, **1**:326, 329
Antineoplastic agents, **4**:370
  antibody conjugates with, **4**:370
  immunotoxins, **4**:369
  for lymphoma, **4**:849-50
Antinociception, behavioral, **2**:338
Antinuclear antibodies
  and autoimmune disease, **1**:516
  and autoimmunity, **1**:331-41
  identification of, **1**:332
Antioncogenes. *see* Genes, suppressor, tumor
Antioxidants
  food analysis, **4**:223
  vitamin E as, **7**:875-77

Antiparasitic agents, **4:**121

Antiphosphorylcholine antibodies, **4:**315

Antiplatelet agents, **2:**169–70

    aspirin, **2:**169

    dipyridamole, **2:**169–70

    omega-3-fatty acids, **2:**170

Antiporters

    Cl⁻/HCO3, **4:**565–66

    Na^+/Ca^{2+}, **4:**564

    Na^+/H^+, **4:**564–65

    proton-dependent, **4:**565

Antipsychotic agents. *see* Tranquilizing agents, major

Antirepresentationalism, **2:**553

Antisera, preparation of, **6:**457

Antisocial behavior, **1:**116

Antithrombin III, **2:**171, **4:**182

Antitubercular drugs, **1:**329

Antiviral agents

    mechanism of action, **1:**327

    resistance to, **1:**328

Antral follicles. *see* Follicles, antral

Antral glands, physiology of, **3:**768

Antrum, pyloric. *see* Pyloric antrum

Anus, **3:**82

Anxiety, **2:**340, 854, 857

    behavioral inhibition system and, **5:**774–75

    brain monoamine neurotransmitter systems and, **1:**612–13

    caffeine and, **2:**78–79

    cannabis and, **5:**445

    central nervous system transmitters in, **5:**410–11

    conscience and, **2:**711

    control, pharmalogic, **2:**785–87

    effects of tricyclic drugs, **1:**613–14

    Gray's biological theory of, **5:**774–75

    LSD use and, **5:**448

    in phencyclidine abuse, **5:**446

    role of serotonin in, **6:**832

Anxiety disorders

    alcohol effects, **1:**135

    characterization, **4:**997

    in older adults, **1:**134–35

Aorta, **2:**160, 180, 182,

    effect of radiation on, **6:**414

    surgery, **4:**96

Aorta, thoracic, **3:**263

Apatites, **2:**640–42, 645

Apes

    bipedal, early hominids as, **3:**496–97

    characteristics of, **6:**141

    genetic relations among, **3:**881

    in hominid evolution, **3:**495

Apes *(Continued)*

    Hominidae and, **3:**488

    mating, **4:**934

    Sivapithecus and, **3:**490

Aphasia, **2:**9, 87, 876

    conduction, fiber tract lesions and, **3:**223

    congenital, **7:**160–62

    defined, **2:**763

    language disorders associated with, **4:**638

    motorical, **2:**682

    in older adults, **1:**122

Aphasia, amnesic

    parieto-occipital system lesions and, **3:**219

    right parieto-occipital system lesions and, **3:**223

Aphasia, Broca, **3:**218

    characterization, **4:**638

    in older adults, **1:**122

Aphasia, sensory. *see* Aphasia, Wernicke

Aphasia, Wernicke, **1:**122, **2:**684

    cortical lesions and, **3:**219

    lesioned left angular gyrus and, **3:**220

    parietal lobe impairment and, **3:**219

Apheresis, LDL, **1:**454

Aplastic anemia

    bone marrow transplantation for, **1:**794

    and radiation, **6:**416

Aplysia, **2:**310

    and calcium, **2:**316

    gill-siphon withdrawal reflex, **2:**314–15

    habituation, **2:**316

    learning model, **2:**310

    neuron, sensory, **2:**316, 317

    organs, **2:**310

    reflex habituation, **2:**310

    sensitization, **2:**310

Apnea, and laryngeal reflexes, **4:**649

ApoB. *see* Apolipoproteins, B

Apocrine sweat glands, **7:**44–45

ApoE. *see* Apolipoproteins, E

Apoenzymes, **2:**527

    apo B, **2:**411, 415, 666–71

    apo CII, **2:**667

    apo E, **2:**666

    apo(a), **2:**667–68

    Apolipoprotein, **2:**661, 662

    defined, **2:**527

    genes, **2:**664–65

    hyperlipidemia, familial, **2:**666

    and hyperlipoproteinemia type III, **2:**666

    listed, **2:**664

    metabolism, **2:**664

    receptor ligand, **2:**666

Apolipoproteins, A, **1:**453
Apolipoproteins, B, **1:**452
Apolipoproteins, C, **1:**453
Apolipoproteins, E, **1:**452
   apoE2 form, **1:**453
   genes for, **1:**453
Apomorphine and stereotypy, **2:**223
Apoplexy. *see* Cerebrovascular disorders
Apoproteins, **6:**26, 27
Apoptosis, **2:**752
   cellular, **2:**249–50
   defined, **2:**245
Appearance, personal, and criminality, **2:**713–14
Appendix
   Peyer's patches and, **5:**139
   rabbit, noradrenergic innervation of, **5:**377
Appetite, **1:**343–53
   aerobic exercise and, **3:**508
   control. *see under* Satiation
   and eating disorders, **1:**351–52. *see also* Eating disorders
   food acceptance and, **3:**639
   food preferences and, **1:**345–47
   ingestive behavior and, **1:**344
   regulation, obesity and, **3:**684–85, 686
   and regulation of body weight, **1:**349
   salt, **1:**344, 347
   salt, clinical reports of, **6:**717–18
   specific, **1:**347
   varieties and measures of, **1:**344–45
Appetite triangle. *see* Acceptance triangle
Apple, **2:**106
Apple juice, **2:**106
Apple sauce, **2:**106
Appraisal theories, of emotion, **1:**107–8
Apraxia, **2:**7
   constructional, **3:**223
   defined, **2:**763
   ideational, **2:**766
   ideomotor, **2:**766
   lesioned frontal operculum and, **3:**218
   in older adults, **1:**122–23
Aptotic body, **2:**245
APUD tumors, defined, **3:**768
Aquatic animals. *see under* Animals
Aqueduct of Sylvius, mesencephalic, **2:**333, 334
Aqueductal stenosis, inherited, **1:**698
Aqueous biological materials, **3:**9
Aqueous two-phase systems, **1:**355–64
   advantages of, **1:**356
   affinity partitioning, **1:**357, 361, 364
   apparatuses for, **1:**359–60

Aqueous two-phase systems *(Continued)*
   for bioconversions, **1:**364
   biological particulates, **1:**359
   biotechnological applications, **1:**363–64
   cell population separations, **1:**362
   cells, physical properties, **1:**362–63
   for chiral separations, **1:**364
   for concentration of biomaterials, **1:**364
   countercurrent distribution, **1:**355; 359–60
   dextran-polyethylene glycol systems, **1:**363–64
   hydrophobicity, **1:**356
   membranes, physical properties, **1:**362
   molecular partitioning, **1:**356–57
   multiple extraction procedures, **1:**355, 359–60
   nucleic acids
      physical properties, **1:**362
      separation, **1:**361
   particulate partitioning, **1:**357
   partition coefficient, **1:**355
   phase properties, **1:**357–58
   phase properties, manipulation of, **1:**358
   phase separation, **1:**356
   polymer-salt systems, **1:**356, 359
   preparation of, **1:**358
   procedures for, **1:**358–60
   for protein extractions, **1:**363–64
   protein interaction and binding, **1:**362
   protein separations, **1:**361
   salt partitioning, **1:**356
   separations and fractionations, **1:**361–62
   single-tube partitioning, **1:**358–59
   two-polymer systems, **1:**356, 359
   use of, **1:**356
Aquocobalamine, **2:**537
Ara-A. *see* 9-beta-D-Arabinofuranosyladenine
Arabinans
   complex pectic polysaccharides and, **3:**38
   in dicotyledonous seeds, **3:**39
9-beta-D-Arabinofuranosyladenine, **2:**396–97
9-beta-D-Arabinofuranosylhypoxanthine, **2:**397
Arabinoxylans
   in aleurone layer, **3:**40
   in endospermous tissues, **3:**40
Arachidonic acid, **2:**169, 170, 745
   in acute phase response, **1:**26–27
   arterial metabolism, **2:**367
   defined, **3:**317
   epoxygenase pathway, **2:**745
   G proteins and, **4:**17
   inhibition by steroids, **3:**579
   metabolism, **2:**745, **4:**472–73
   metabolites, in acute phase response, **1:**27

Arachidonic acid *(Continued)*
  mobilization, **4:**472
  nonclassic hormones and, **3:**323
  omega-3 fatty acids and, **5:**509, 533
  phospholipid hydrolysis and, **3:**584
  prostaglandins and, **3:**483
Arachnodactyly. *see* Marfan syndrome
Arachnoid mater
  cranial, **4:**984–86
  spinal, **4:**988–89
  spinal cord, **7:**195
Arachnoid membrane, **1:**716–17
Arachnoid space, **5:**351
Arachnoid villi, **4:**985–86
Arboviruses. *see* Flaviviruses; Togaviruses
Archaebacteria, **1:**274, **2:**543
Archaic reflexes, defined, **6:**821
Arches, dermatoglyphic, defined, **3:**88
Archicortex, **2:**673, 680–81
  of Ammon's horn, **2:**680
  of gyrus,dentate, **2:**680
  and hippocampus, **2:**680
  network modeling of, **5:**388
  parts of, **2:**680
  of subiculum, **2:**680
  volume, **2:**679
Archimedean screw pump, **4:**90
Arcuate fasciculus. *see* Fasciculus, arcuate
Arcuate popliteal ligament, **1:**376
Ardrey, **3:**500
Area tempestas, and convulsive seizures, **6:**781
Arecoline, **1:**178
Argentaffin system. *see* Chromaffin system
Argentation chromatography, **4:**222
L-arginine, nitric oxide synthesis and, **3:**375
Arginine vasopressin, **2:**300
  in fever, **1:**759–60
  hypothalamus control, **4:**311
Argument processing
  cognitive evaluation of, **5:**803
  persuasion and, **5:**800, 805
  vs. peripheral clues, **5:**804
Arm, **2:**597–98
  of human, **2:**597
  of macaque, **2:**597
  radius, **2:**598
  rotation, **2:**598
  ulna, **2:**598
Armadillos, **3:**312
Armstrong, George, **5:**700
Aromagram, of mixture of odors, **5:**761
Aromatase, **2:**57

Aromatic amines, heterocyclic
  chemistry of, **3:**676–77
  mutagen formation and, **3:**678
Aromatic amines, N-hydroxy heterocyclic, **3:**677
Aromatization
  antral follicle growth and, **3:**631
  defined, **3:**627
  of dominant follicle, **3:**632
Arousal, **2:**714
  and criminality, **2:**714
  persuasion and, **5:**807
Arrhythmia, **2:**172–74
  and calcium antagonist, **2:**87
  ectopics, ventricular, **2:**172
  and infarction, myocardial, **2:**172
  supraventricular, **2:**172
  ventricular, **2:**172–74
Arsenic, **2:**111, 112, 114
  biological function of, **7:**606
  deficiency, **7:**605–6
  oxidation states, **7:**552
  toxicity of, **3:**386, **7:**605
Arsine, toxicity of, **3:**386
Arsphenamine, synthesis and use, **1:**326
Art, Upper Paleolithic, **3:**502–3
Artemisia absinthium (wormwood), **2:**347
  and madness, **2:**347
  oil, **2:**347
Artemisinin, **2:**384, 392, **4:**884
Arterial occlusive disease, **5:**604
Arterial pressure
  characterization, **4:**285–87
  mean, determination, **4:**285–87
  normative, **4:**287–88
Arterial venous anastomoses, **4:**796
Arteries, **2:**183–84, 640
  cardiac, **2:**183–84
  coronary, **2:**137
  defined, **3:**755
  elastic, **2:**184
  elastin in, **3:**267
  endothelial, **2:**184
  pulmonary, **2:**183–84
  of spinal cord, **7:**193–95
  structure of, **2:**183–84
  wall-blood barrier, **2:**367, 369
Arteries, gastric. *see* Gastric arteries
Arteriography, coronary, **2:**146
Arterioles
  in aerobic exercise, **3:**510
  in gastric blood flow regulation, **3:**759–60
Arterioles, submucosal. *see* Submucosal arterioles

Arteriosclerosis
    cigarette smoking and, **7**:517
    defined, **7**:511
Arteriovenous malformation, and stroke, **7**:303
Artery, middle cerebral, **7**:301
Artery, splenic. *see* Splenic artery
Arthritis
    role of infectious agents in, **4**:451
    substance P and, **5**:384
Arthritis, rheumatoid, **2**:616, 655
    nutrition and, **5**:509-10
    oral contraceptives and, **5**:608
Arthrodial joints, **1**:374
Arthropods, **2**:390
Articular cartilage and intervertebral disc, **1**:365-70. *see also* Cartilage; Cartilaginous tissues
    matrix
        movement of solutes through, **1**:367
        structure and physicochemical properties of , **1**:365-67
    mechanical deformation of, and fluid flow, **1**:368-69
    osmotic pressure of proteoglycans in, **1**:368
    solute exchange through, routes of, **1**:367-68
Articulation, speech, **4**:662
Articulation disorders, **7**:158
Articulations, **2**:596
    amphiarthroses, **1**:371-76
    bursae, **1**:373
    cartilaginous joints, **1**:371
    diarthroses, **1**:371
    fibrous joints, **1**:371-72
    functional classification, **1**:371; 375(t)
    gomphoses, **1**:372
    of human, **2**:596
    joint cavities, **1**:371
    knee joint, **1**:376
    of macaque, **2**:596
    periodontal ligaments, **1**:372
    structural classification, **1**:371-73, 375(t)
    survey of, **1**:375(t)
    sutures, **1**:371; 372
    symphysis joints, **1**:373
    synarthroses, **1**:371, 372
    synchondroses, **1**:372
    syndesmoses, **1**:372
    synostosis, **1**:372
    synovial cavities, **1**:372
    synovial joints. *see* Synovial joints
Artificial cells, **1**:377-83
    containing adsorbents and immunoadsorbents, **1**:378-80
    containing charcoal, **1**:378

Artificial cells *(Continued)*
    containing enzymes, **1**:382
    containing living biological cells, **1**:382
    as drug carriers, **1**:380, 311
    lipid-membrane, **1**:381
    medical and biotechnological applications of, **1**:378, 383-83
    preparation of, **1**:377-78
        from biodegradable polymers, **1**:381-82
    for treatment of
        acute poisoning, **1**:378
        aluminum or iron overload, **1**:378-79
        kidney failure, **1**:379
        liver failure, **1**:379
Artificial feedback. *see* Feedback, extrinsic
Artificial heart, **1**:385-400. *see also* Heart, mechanical
    blood pump inside chest for, **1**:389-90
    as bridge to transplantation, **1**:398
    control of, **1**:397-98
    cost to society, **1**:398-99
    Dr. Barney Clark as first recipient of, **1**:399
    drive systems for, **1**:389
    durability of, **1**:398
    infection of, **1**:396-97
    intima of
        biolyzed surfaces as, **1**:394
        rough surfaces as, **1**:394-95
    ex vivo and *in vivo* testing, **1**:395-96
    materials for, **1**:398
    need for, **1**:398
    percutaneous (wires) or transcutaneous supply of energy to drive, **1**:397
    power sources for, inside the body, **1**:397
    pulmonary artery-pulmonary artery shunt for testing, **1**:395-96
    smooth surfaces as, **1**:394
    surgical implantation of, **1**:390-91
    testing of, **1**:91-393
    thrombosis affecting, **1**:393
    total, **1**:389
    total orlvad, **1**:387-88
Artificial insemination. *see* Insemination, artificial
Artificial intelligence, **5**:391
Artificial kidney, **1**:401-15. *see also* Dialysis; Hemodialysis; Peritoneal dialysis
    arteriovenous shunt and, **1**:402, 404
    complications from, **1**:408-10
        amyloid, **1**:409
        anemia, **1**:409
        calcium problems, **1**:409
        copper intoxication, **1**:409
        diet restrictions, **1**:409

Artificial kidney
  complications from *(Continued)*
    hemochromatosis, **1:**410
    hypertension, **1:**409
    lack of taste and smell, **1:**410
  convention versus dialysis in, **1:**404–5
  disposable, **1:**406–8
  economic impact from use of, **1:**408
  heart-lung machine and, **4:**93
  historical background of, **1:**402
  home dialysis and, **1:**408
  number of patients using, **1:**408
  poison removal by, **1:**411
  rotating drum, **1:**403
Aryepiglottic muscle, **4:**657
Arylsulfonyl tetrazolides, **3:**170
Arytenoids, laryngeal, **4:**655–56
Asbestos, **1:**417–25, **2:**111, 112, 116
  amosite, **1:**418
  amphiboles, **1:**417
  animal experimentation, **1:**421
  biological potency, **1:**423–25
  chemical composition, **1:**418
  chrysotile, **1:**418
  Convention Concerning Safety in the Use of Asbestos,
    1:425
  crocidolite, **1:**418
  definition, **1:**417
  disease risk and, **1:**422–23
  diseases related to, **1:**420–21
  fiber persistence in tissues, **1:**423–25
  fiber size effects on potency, **1:**423
  fiber types, classification, **1:**418
  ingestion, **1:**422
  inhalation, **1:**422–23
  lung cancer and, **1:**420–21
  lung disease and, **3:**387–88
  mesothelioma, **1:**421
  serpentine, **1:**417
  sources, artificial, **1:**420
  sources, natural, **1:**419–20
  substitutes for, **1:**425
  toxicity, **7:**555–56
  uses of, **1:**419
  workplace exposure, **1:**422
  world production of, **1:**419
Asbestos fibers
  ingestion of, **1:**422
  inhalation, **1:**422
Asbestosis, **1:**420
Ascariasis, **1:**270, **4:**117–18, 121, 618
Ascaris, **2:**386, **5:**474

Ascaris lumbricoides, transmission, **4:**114
Ascending colon, **3:**82
Ascorbic acid, **1:**427–32, **2:**571; 656
  absorption and tissue distribution, **1:**430
  biosynthesis, **1:**429–30
  cellular regulation, **1:**430–31
  characterization, **1:**427–28
  chemical functions, **1:**431
  chemical properties, **1:**428–29
  coenzymatic functions, **1:**431
  deficiency, **2:**571. *see* Scurvy
  as dietary supplement, **1:**432
  food analysis, **4:**220
  functions, **1:**427–28
  gastric carcinoma and, **3:**772
  metabolism, in adrenal medulla, **1:**646
  in nitrosamine inhibition, **3:**679
  occurrence in foods and tissues, **1:**429
  pharmacological uses, **1:**432
  physical properties, **1:**428–29
  properties of, **1:**650
  regeneration, in adrenal medulla, **1:**649–51
  regulatory functions, **1:**431
  supplements, **1:**432
Ascorbic acid reductases, **1:**430
Aseptic meningitis, **4:**461
Aserinksy, Eugene, **3:**193
Asialogangliosides. *see* Glycosphingolipids
Asian great apes. *see* Apes
Asian/Pacific Americans, diet of, **3:**651
Asparagine, **2:**507, **3:**419
Asparaginyl residues, **3:**416, 419
Aspartame, **7:**378–80
Aspartate
  astrocyte uptake of, **1:**440
  in enzyme catalysis, **3:**434
Aspartic acid, **2:**5, 14, 324, 337, 339, 507, 524, 525
  aging and, **3:**419
  in enzyme thermoinactivation, **3:**416, 417
  as signal molecule, **2:**324
Aspergillus, food toxicology and, **3:**674–75
Aspergillus flavus aflatoxin production, **5:**264, **7:**592–93
Aspergillus fumigatus, **5:**262
Aspergillus nidulans, **5:**902
Aspergillus ochraceus, **5:**265
Aspergillus parasiticus, **5:**264
Aspirin. *see* Acetylsalicylic acid
Assembly, protein
  of contractile proteins, **5:**169–70
  defined, **5:**161
  in myofibril organization, **5:**168
  of myosin, **5:**169

Assertion, conditional, **2:**623-24
Assimilation, defined, **3:**71
Association cortex, **2:**7
Associationism, **2:**630
 classical, **2:**547
 and contiguity principle, **2:**630
 and similarity principle, **2:**630
Asthma, **4:**410-11
 in children, **5:**706
 cold-induced, **4:**797
 hypnosis to alleviate, **7:**622-23
 occupational, **3:**387
 as psychophysiological disorder, **6:**337
 role of antibodies in, **4:**408
 sulfite-induced, **7:**338, 339
Astrocytes, **1:**433-46, **2:**341, 343
 biochemical characteristics, **1:**438-41
 functional roles, **1:**441-46
 laminin and, **5:**345
 morphology, **1:**435-38
 perinodal, **1:**437
 pharmacological characteristics, **1:**438-41
 physiological characteristics, **1:**438-41
 protoplasmic characteristics, **1:**435
 Trichinella infection and, **5:**648
Astroglia, **1:**437-38
Astroglial cells, **3:**741
Asymmetry, cerebral, **2:**351, 871, 873-75, 877
 anatomical, **2:**352
 functional, **2:**351-53
ATA box, **4:**147
Ataxia, **2:**649-50
 and copper deficiency, **2:**650-52
 defined, **5:**259
Atebrine, **4:**884
Atelectasis, **6:**367
 from cardiopulmonary bypass, **4:**95
 in infants, **5:**705
Atenolol, **2:**171
ATG, in glucocerebrosidase cDNA, **3:**778
Atherogenesis, **2:**88
Atheroma. *see* Atherosclerosis
Atherosclerosis, **1:**447-55, **2:**171, 661, 663, 665
 abnormal cell proliferation and, **3:**326
 animal models, **1:**450-51
 and artery, schema, **2:**169
 B-mode ultrasound, **1:**449
 characterization, **1:**447
 cigarette smoking and, **7:**517
 consequences of, **1:**447-48
 contributions of animal models to human, **5:**678
 coronary angiography, **1:**449

Atherosclerosis *(Continued)*
 defined, **2:**167, **7:**511
 disease of, **1:**448-49
 efficacy of treatments for, **1:**454-55
 fibrous caps plaque, **1:**449
 genetic factors, **1:**453
 lipoproteins and, **3:**574
 magnetic resonance imaging, **1:**449
 myoblast profileration and, **5:**163
 plaque, **2:**169
 and plasma, lipoproteins, abnormal, **2:**665
 proteoglycans and, **6:**265
 risk factors, **1:**451-52
 scintigraphy, **1:**449
 therapy for, **1:**454-55
 tissue factor and, **7:**498
 vascular muscle cell proliferation and, **5:**170
Atherosclerotic plaques
 development, **1:**449-51
 fibrous caps, **1:**447, 449
 foam cells, **1:**449
 response-to-injury hypothesis, **1:**450
Athletes, **2:**107
 carbohydrate diet, **2:**107
 diet for, **3:**659
 muscle fibers in, **5:**179
 precompetition, **2:**107
Atomic absorption, in drug analysis, **3:**202
Atomic bombs, and incidence of leukemia, **6:**417
Atomic nuclei, **4:**864-65
Atonic seizures, in epilepsy, **3:**456
Atopic reactions, **4:**410-11
ATP. *see* Adenosine triphosphate
ATPase. *see* Adenosine triphosphatase
Atractyloside, **4:**12
Atresia, **2:**245, 247
 defined, **3:**627
 over life span, **3:**628
Atrial fibrillation, **2:**175
Atrial natriuretic factor, **1:**467-78
 central nervous system and, **1:**475
 in disease, **1:**476-77
 endothelin action and, **1:**475
 function of, **6:**91
 gene for, **1:**469-70
 mechanism of action of, **1:**475
 messenger RNA of, **1:**469
 peripheral nerve endings and, **1:**474
 pharmacologic and physiologic actions of, **1:**477-78
 physiological actions of, **1:**473-79
 processing of, **1:**471-72
 receptors, **1:**475-76

Atrial natriuretic factor *(Continued)*
  secretion of, **1:**472–73
  structure of, **1:**468–69
  synthesis of, **1:**470–71
    by atrium, **1:**470
    by extracardiac tissue, **1:**471
    by ventricles of heart, **1:**470–71
Atrioventricular node, **2:**183
Atrium, cardiac. *see* Heart atrium
Atrocities. *see* Violence
Atrophy, **2:**254
  defined, **5:**259
  of muscular disuse, **2:**573
Atropine, **2:**24, 37, 346, 348, 522
  and acetylcholine, **2:**348
  effects of, **1:**178–79
  and hallucination, **2:**348
  pancreatic polypeptide and, **5:**719
Attachment, **2:**857
  as innate behavior, **2:**857
  interpersonal attraction, **4:**542
  and self-development, **2:**857
Attachments, gut
  anatomy, **3:**66, 68
  defined, **3:**57
Attack, vasovagal, **2:**146
Attention, **1:**479–84
  alerting, **1:**483–84
  alertness, **1:**479–80
  approaches to, **1:**479–80
  brain injury effects, **1:**481
  cognitive control of, **1:**482–83
  cognitive neuroscience of, **1:**480
  computerized tomography, **1:**480
  covert shifts of, **1:**481
  crosstalk and, **5:**156
  disorders involving, **1:**484
  executive systems of control, **1:**482–83
  hypotheses of, **1:**480
  to local vs. global objects, **1:**482
  magnetic resonance imaging, **1:**480
  methodological developments, **1:**480
  microelectrode use, **1:**480
  as network of anatomical areas, **1:**480, 484
  norepinephrine effects, **1:**483–84
  orienting, **1:**480–82
  parallel processing, **1:**482
  pattern recognition processes and, **1:**481–82
  positron emission tomographic studies of, **3:**225
  selectivity, **1:**479–80
  sensory interference, **1:**482–83
  unified vs. independent systems of, **1:**483

Attention deficit disorder with hyperactivity, **2:**718, **4:**996
  amphetamines for, **5:**440
  dyslexia and, **3:**216
  as minimal brain dysfunction in child, **2:**342
  symptoms of, **5:**709
Attention stage, of attitude change, **1:**492–93
Attitude
  antecedents to, **1:**488
  attraction as, **4:**536–37
  attributions and, **1:**487
  behavior and, **1:**488–89
  behavior-attitude models, **1:**490–92
  cognitive consistency and, **1:**487
  consequences of, **1:**488–89
  defined, **1:**485–86, **5:**799
  as determinants of food consumption, **1:**497–504
  to food, measurement of, **1:**497–99
  functions of, **1:**486–87
  information and, **1:**487
  life-related, depressive disorder and, **1:**613
  measurement of, **1:**489–90
  models, single vs. multicomponent, **1:**486
  objects of, **1:**485–86
  observational rating scales, **1:**610–11
  persuasion and, **5:**799–800
  self-report tests, **1:**610
  underlying processes of, **1:**487
Attitude change, **1:**485–96
  action stage, **1:**493
  attention stage, **1:**492–93
  audiences, **1:**494–95
  central vs. peripheral processssing, **1:**495
  channels, **1:**494
  compliance, **1:**493
  comprehension stage, **1:**493
  discomfort in value structures and, **1:**495–96
  factors in, **1:**493–95
  identification, **1:**493
  incompatible actions and, **1:**487
  internalization, **1:**493
  messages, **1:**494
  persuasion and, **5:**801–2
  retention stage, **1:**493
  sources, **1:**493–94
  stages, **1:**492–93
  strategies for, **1:**495
  yielding stage, **1:**493
Attraction. *see* Interpersonal attraction
Attractors, network, **5:**391
Attribution theory, **1:**106, **2:**856, **5:**789
Attributional Style Questionnaire, **1:**610
Attributions, attitudes and, **1:**487

Atypical facial pain, **5**:368

Auditory attention deficits, **7**:156

Auditory central nervous system, **3**:239-40

Auditory cortex, **3**:240, 241

Auditory discrimination deficits, **7**:156

Auditory nerve. *see* Acoustic nerve

Auditory perception
    biological basis of, **3**:240-41
    positron emission tomographic studies of, **3**:225

Auditory perceptual disorders, **3**:219

Auditory processing, disorders of, **7**:156

Auditory threshold, **3**:240-41

Auerback, C., **5**:251

Augmented feedback. *see* Feedback, extrinsic

Austin, C. R., **3**:601

Australopithecus, **2**:189
    aethiopicus, **3**:491
    afarensis, **2**:189, **3**:490, **6**:848
    africanus, **2**:189, **3**:491
    anatomy, **3**:496
    as bipedal apes, **3**:497
    boisei, **2**:189
    homo sapiens and, **3**:491
    robustus, **2**:189, **3**:491
    tools and, **3**:498

Authoritarian personality theory, **7**:235

Autism, **1**:505-12
    abnormalities in
        cognitive, **1**:506
            drinking, eating and sleeping, **1**:506
            posture and motor behavior, **1**:506
            self-injurious behaviors as, **1**:506
    characteristics of, **1**:505-8, **4**:997
    defined, **3**:709
    demographics of, **1**:507
    diagnosis of, **1**:506-7
    epidemiology of, **1**:507
    etiology of, **1**:507-8
    fragile X syndrome and, **3**:714
    infantile, symptoms of, **5**:709-10
    and language and hearing impairments, **1**:507
    and mental retardation, **1**:506-7
    neurological factors associated with, **1**:508
    and schizophrenia, **1**:507
    theories of, **1**:508-9
    treatment for, **1**:509-12

Autoantibodies, **1**:513-17
    autoimmune diseases and, **1**:514-17
    dermatomyositis and, **1**:336
    diseases associated with inactivation effects, **4**:408
    to erythrocytes, **1**:514-15
    lupus and, **1**:331-41, 334

Autoantibodies *(Continued)*
    polymyositis and, **1**:336
    to receptors on cell membranes, **1**:515
    scleroderma and, **1**:337
    self-tolerance, **4**:837

Autoantigens, **1**:339, 340

Autocrine, **2**:321

Autocrine effectors
    defined, **3**:318, **5**:714, 445
    epidermal homeostasis and, **3**:449

Autocrine mobility factors, **4**:1013

Autograft. *see* Transplantation, autologous

Autoimmune diseases, **1**:519-33, **2**:299; 616, **4**:324
    characteristics of, **1**:302
    dietary lipids and, **5**:509-10
    discrimination between self and nonself as factor in, **1**:520-26
    DNA marker diagnosis of, **3**:148
    effector mechanisms of, **1**:530-31
    experimental, **1**:526-27
    hemolytic anemia, **2**:616
    human, examples of, **1**:528
    idiotypes, **4**:317
    interleukin-2 and, **4**:532-33
    macrophages role, **4**:862
    measures of immune function in, **5**:144-45
    mechanism of, **1**:300-301
    multiple sclerosis as, **5**:143-44
    myasthenia gravis as, **5**:257-58
    noradrenergic nerves and, **5**:383-84
    organic specific, **1**:527-29
    origin of, in New Zealand mice, **1**:526
    protooncogene expression in, **5**:515
    receptor specific, **1**:530
    regulatory defects in, **1**:302-3
    rheumatoid arthritis, **2**:616
    role of genetics in, **1**:531-32
    serology of, **1**:514-17
    spectrum of, **1**:527-30
    systemic
        characteristics of, **1**:529-32
        nature of, **1**:331-32
    systemic lupus erythematosus, **2**:616
    treatment for
        nonspecific, **1**:532
        replacement, **1**:532
        specific immunosuppression, **1**:532-33
    types of antigens associated with, **1**:528

Autoimmune response, **1**:332-38

Autoimmunity. *see also* Autoantibodies; Autoimmune diseases
    in amyotrophic lateral sclerosis, **1**:234

Autoimmunity *(Continued)*
   antinuclear antibodies and, **1**:331–41
   characterization, **4**:406
   induction, **1**:522–26
      altered antigen and, **1**:523–25
      altered immune response and, **1**:525–26
      methods for, **1**:525
   mechanisms of, **4**:407
Autoimmunity avoidance, **4**:394
Autologous marrow transplantation for leukemia,
   4:695–96
Autologous mixed lymphocyte response, **5**:144–45
Autolysins, bacterial
   and cell wall synthesis, **1**:277–78
   and lethality, **1**:283
Automatism, **2**:718
   in barbiturate abuse, **5**:442
   and criminality, **2**:718
Autonomic dysreflexia, **7**:232
Autonomic motoneurons, **5**:43
Autonomic nervous system, **1**:535–50, **2**:32; 147
   adrenal medullary chromaffin cells, **1**:538
   anatomy of, **5**:355–60
   control of effector tissues, **1**:541–44
   control of orofacial region by, **5**:372
   convergence of synaptic input, **1**:537
   cranial parasympathetic pathways, **1**:538–39
   defined, **2**:147, **3**:317
   divergence of preganglionic neurons, **1**:537
   enteric division, **1**:539–40
   extra-adrenal medullary chromaffin cells, **1**:538
   ganglion cells, **1**:536–37
   of gastrointestinal tract, **3**:73
   hypothalamus role, **4**:307–8, 310–11
   nerve communication in, **5**:409
   neuropharmacology of, **5**:409
   neurotransmitters, **1**:540–41
   organization of, **1**:535–40
   parasympathetic division, **1**:538–39
   parasympathetic system as part of, **5**:359–60
   paravertebral ganglia, **1**:537–38
   persuasion and, **5**:806–7
   postganglionic neurons, **1**:537
   preganglionic and postglandionic fibers of, **5**:358
   preganglionic neurons, **1**:537
   sacral parasympathetic pathways, **1**:539
   splanchnic nerves, **1**:538
   sympathetic chain, **1**:538
   sympathetic division, **1**:537–38
   sympathetic system as part of, **5**:356–59
   synapses, **1**:536
   vagus nerve, **1**:539

Autonomic nervous system *(Continued)*
   varicosities, **1**:536
   voluntary control of, **1**:662
Autonomic nuclei, of gray matter, **7**:204
Autonomous replication. *see* DNA replication
Autooxidation, longevity and, **3**:894
Autophosphorylation, **2**:92
Autopsies, **5**:681, 682
Autoradiography
   to detect radioactive proteins, **6**:197
   in interphase study, **5**:66
   photoelectron microscopy and, **5**:908
   in vitro , **5**:415, 421
Autoreceptor, **2**:323
Autoregulation, **2**:365
   breakthrough of, **2**:365
   of cerebral blood flow, **2**:365
   defined, **2**:357, 365
   diagram of, **2**:365
Autoregulatory escape
   defined, **3**:762
   intestinal blood flow, regulation, **4**:551
Autosomal chromosomes. *see* Chromosomes, autosomal
Autosomal dominant diseases, **3**:140–43
Autosomal dominant inheritance
   characteristics of, **3**:850–51
   disease and, **3**:820
Autosomal dominant traits, mutation rates of, **5**:245
Autosomal recessive diseases
   characteristics of, **3**:820
   DNA marker diagnosis of, **3**:143–44
Autosomal recessive inheritance
   characteristics of, **3**:851
   sib correction method and, **3**:853–54
Autosomes, **2**:437–42
   chromosome number changes, **2**:437–39
   defined, **3**:819, 847
   numbered, **2**:437
   structure changes, **2**:439–41
Autosuggestion, in food acceptance, **3**:646
Autotransplant. *see* Transplantation, autologous
Autozygosity, **4**:433
Availability equivalency. *see* Biological availability
Aversions, dietary. *see* Dietary aversions
Avian erythorblastosis retrovirus. *see* Erythroblastosis
   virus, avian
Avidin, **2**:463, 464
Aviemore model, reciprocal recombination, **4**:952–53
Avogadro's number, defined, **3**:423
Avoidance learning, passive, **2**:711
   and electric shock, as punishment, **2**:715
   in fearful birds, **2**:873

Avoidance learning, passive *(Continued)*
and psychopaths, **2:**715
Awakening. *see* Sleep-wake cycle
Axonal collateral sprouting. *see* Collateral sprouting,
axonal
Axonal flow, in nerve regeneration, **5:**344
Axonal regeneration
nerve growth factor and, **5:**338–39
in peripheral nerve experiments, **5:**344
Axonal transport
in central nervous system, **3:**743–45
defined, **3:**743
ganglioside uptake and, **3:**744
in nerve regeneration, **5:**344
in peripheral nervous system, **3:**745–47
Axonal transport, retrograde, **5:**333
defined, **3:**743
ganglioside turnover and, **3:**747
Axonemes, **5:**18
Axons, 2, **2:**1, 235, 341–44, 677
afferent, of muscle spindles, **3:**721
collateral sprouting, **5:**345, **6:**37–38
defined, **5:**341
efferent. *see also* Beta efferent system; Gamma efferent
system
of muscle spindles, **3:**721
of tibialis anterior muscle, **3:**722
microtubules in, **5:**12, 18
from neocortex, **5:**308–9
skeletomotor, **3:**720
of squid, **3:**10
transport in, **5:**20
Axoplasm, **2:**13
Axotomy, trophic support and, **5:**334
Ayerza's syndrome. *see* Hypertension, pulmonary
Azacytidine
DNA demethylation and, **3:**156
muscle differentiation and, **5:**168
X-localized genes and, **3:**160
Azathioprine, **1:**685, **2:**114
Azidothymidine. *see* Zidovudine
Azo-compound, aromatic, **2:**110
p-aminoazobenzene, **2:**110
p-dimethylaminoazobenzene, **2:**110
AZT. *see* Zidovudine
Azurophilic granules, neutrophil, **5:**812–13

B

B-cell growth factor. *see* Interleukin 4
B-cell lymphoma. *see* Lymphoma, B-cell
B-cell stimulatory factor-1. *see* Interleukin 4

B-cells. *see* B-lymphocytes
B12 coenzymes
chemistry, **2:**536–37
enzyme reactions, **2:**538
functions, **2:**537
interconversions, **2:**538
metabolism, **2:**537
structure, **2:**537
B-lymphocytes, **2:**235, 754–55, 831–32
activation by antigen, **2:**754–55
antigen recognition, **4:**834
autoregulatory, **1:**303
in bone marrow, **2:**754–55
characterization, **4:**165–66, 811, 822–23
chemical sympathectomy and, **5:**383
and cytokines, **2:**754–55
development, **2:**754
difference from T lymphocytes, **4:**335
effect of radiation on, **6:**421–22
Epstein-Barr virus and, **3:**463, 464
Epstein-Barr virus-transformed, **2:**230.457, 658
expression of cell surface proteins, **1:**677
Fc-dependent inactivation, **1:**301
functions, **4:**165–66, 402–3
idiotypes, **4:**315
immunoglobulins, **1:**677
immunoparasitology, **4:**386
migration patterns, **4:**396–97
peripheral, **2:**755
pre-B cell, **2:**754
primary, **1:**566
pro-B cell, **2:**754
production, **4:**331
progenitors, **4:**169–71
receptors, antigen binding to, **6:**520–22
replication, **2:**754
stimulatory factor. *see* Interleukin-4
subsets and cellular interactions, **4:**835–36
T cell action on, **4:**335
targets for immunoconjugates, **4:**373–74
tumor rejection and, **4:**340
Babesiosis, **1:**740, **6:**277
Babinski sign, **7:**219
Baboons, sex differences in, **6:**847–48
Bacillus Calmette-Guérin vaccine. *see* BCG vaccine
Bacillus cereus, **3:**666
Bacitracin, mode of action, **1:**290(t)
Bacitracin A
effect on peptidoglycan biosyntheses, **1:**280
structure of, **1:**273
Back-propagation, **5:**391, 392
Backcross mapping, **3:**827

Baclofen, for Huntington's disease, **4**:274
Bacteria
   antibiotic inhibitors of cell wall, **1**:271-85
   antibody production in, **3**:804-5
   colonic, **3**:43
   detection of, **1**:552-59
   genetic mapping of, **3**:824
   glycolysis and, **3**:933
   intestinal, **3**:78
   oral, **2**:811
   pathogens, versus normal flora, **1**:551-52
   photosynthetic, **2**:238
   polyamine metabolism in, **6**:72
   taxonomy, murein and, **1**:274
   tolerance, antibiotics and, **1**:5283
   toxins produced by, **7**:588-90
   in vaginal canal, **4**:457-58
Bacteria, anaerobic, **1**:237-50
   anaerobiosis, **1**:237-38
   bowel-associated, **1**:241
   characteristics, **1**:238-39
   classification, **1**:238
   colonization resistance by, **1**:319
   common pathogens, **1**:239
   differentiation of genera, **1**:240
   Gram stain, **1**:238
   habitats, **1**:239-41
   infections in humans. *see* Infections, anaerobic
   mechanisms of resistance in, **1**:249
   nose-associated, **1**:239-40
   oral cavity-associated, **1**:240-41
   skin-associated, **1**:239
   stomach-associated, **1**:241
   urethra-associated, **1**:241
Bacterial conjugation. *see* Conjugation, genetic
Bacterial flora, **1**:551
Bacterial infections
   detection of, **1**:551-61
   diagnosis of, **1**:552
   flea-borne, **7**:896
   hosts with strongyloidiasis, **4**:119
   persistence, **1**:283
   as scalar-borne zoonosis, **7**:897
   tick-borne, **7**:894-95
Bacterial toxins
   antibody conjugates with, **4**:368-69
   immunotoxins, **4**:418
Bacteriodes
   activity of various drugs and, **1**:248
   in breast milk feces, **1**:319
   habitats, **1**:240-41
   infections of, **1**:246

Bacteriodes (Continued)
   morphology, **1**:239
   virulence, **1**:241-42
Bacteriology, **1**:554. *see also* Microbiology
Bacteriophage lamdba, **2**:461
   Charon 21A, **2**:461-62
Bacteriophages, in DNA replication study, **3**:441
Bacteriorhodopsin, topographical model of, **1**:86, 89
Baculoviridae, in *Spodoptera frugiperda*, **3**:157
Balance. *see* Equilibrium
Balantidiasis, **6**:275
Balantidium coli, **6**:275
Baldness. *see* Alopecia
Balkan nephropathy, **5**:265-66
Ball-and-socket joints, **1**:374
Balloon pumps, and failing circulation, **1**:386
Bananas, starch digestion and, **3**:43
Bancroft's filariasis, **4**:389
Band, epithelial, **2**:817-18
Barbiturates
   abuse of, **5**:442-43
   in dentistry, **2**:786
   GABA$_A$ receptor and, **5**:423
Barium, in radiology, **3**:4, 5
Barley
   in "basic four" food groups plan, **3**:656
   dietary fiber from, **3**:40
Baroflex, **2**:147
Baroreceptor, **2**:145, 179
Barrington, **5**:44
Barrington, F. J. F., **5**:43
Basal cells
   defined, **3**:445
   of olfactory mucosa, **5**:523
Basal ganglia, **5**:348
   in convulsive seizure propagation and control, **6**:782-83
   Parkinson's disease and, **5**:667
   spelling errors and, **3**:224
Basal layers, of hair, **4**:41
Basal metabolism measurement, **1**:756-57
Basal nucleus, olivary. *see* Olivary nucleus
Base composition
   consensus sequences and, **3**:122
   in oligonucleotide synthesis, **3**:170
Base sequence
   determination of, **3**:877
   of eukaryotic genes, **3**:794
   evolution of, **3**:154
   of human *fos* gene, **3**:703
   methylation and, **3**:153
   mutation and, **5**:244

Basement membrane, **2:**639
   biological activity, **4:**624-26
   cartilage matrix and, **3:**525
   and collagen IV, **2:**566
   components of, **4:**624
   defined, **3:**523
   functions of, **3:**524
   laminin in, **3:**528
   proteoglycans in, **6:**261-62
   structure, **4:**624
   tissue localization, **4:**623-24
Bases (chemistry), **5:**871
Basic fibroblast growth factor. *see* Fibroblast growth factor, basic
"Basic four" food groups plan, **3:**652-63
Basilar membrane
   distortion components of, **3:**237
   tuning curves for, **3:**236
Basilar sinus, **4:**984
Basophils
   in allergic response, **1:**185-86
   characterization, **4:**166-67
   effect of platelet-activating factor on, **6:**47-48
   functions, **4:**166-67
   role in inflammation, **1:**313
Basophils, tissue. *see* Mast cells
Batrachotoxin, **7:**589
Bats, rabies transmission and, **7:**897
Bayliss, **5:**713
BCG vaccine, **1:**676, **2:**759, **4:**447
   immunotherapy with, **4:**347
   and tumor necrosis factor, **2:**759
Beaded hair, **4:**48
Beck Depression Inventory, **1:**610, **2:**842
Becker's muscular dystrophy. *see* Muscular dystrophy
Beck's cognitive model of depression, **2:**855-56
   assumptions, maladaptive, **2:**856
   automatic thoughts, **2:**856
   "if then" statement, **2:**856
   maladaptive assumptions, **2:**856
   schematic processing, **2:**856
   self-schema, **2:**855-56
   thought, automatic, **2:**856
Bed nucleus of the stria terminalis, **4:**721-23
Bedbug, **2:**390
Bedwetting, **7:**82
Beef tapeworm, **2:**389
Beer
   N-nitrosodiethylamine in, **3:**679
   production of, **1:**166-67
Bees, toxins produced by, **7:**597
Beeswing wheat bran, **3:**40

Behavior
   accidental, **5:**790
   affective, **2:**637
   associative network representing information, **5:**795
   attitude-behavior models, **1:**490-92
   attitudes and, **1:**488-89
   attributions of, **5:**792-93
   in biological research, **1:**608-9
   brain relationship. *see* Neuropsychology
   changes in, **2:**629-30
      dementia, **2:**766
      nervous system, **6:**40
   cognitive development and, **1:**579-80
   conditioning, **2:**629-36
   consensus information to explain causes of, **5:**789-90
   consistency information to explain causes of, **5:**789-90
   construct accessibility and, **5:**792
   cooperative, competitive, and indivualistic, **1:**573-83
   criminal. *see* Criminality
   defined, **3:**589
   desirable, **2:**630
   developmental perspectives in, **1:**578
   disorders
      in children, **2:**718, **5:**709-10
      in fragile X syndrome, **3:**713-14
   distinctiveness information to explain causes of, **5:**789-90
   drug effects on, **1:**613-14
   environmental influences and, **5:**773
   expectation fulfillment and memory for, **5:**796
   experience and, **2:**873
   fear, **2:**340
   future directions of research, **1:**582-83
   genetics and, **5:**773
   higher-level neural organization control of, **5:**809
   hippocampal functions, **4:**231-32
   hormonal influences, **2:**870, **4:**241-49
      in courtship, **2:**870
      in mating, **2:**870
      in nursing, **2:**870
   hypothalamal outputs, **4:**307
   immunologic effects on, **6:**325-26
   implications of social categorization for, **7:**93-96
   indirect self-destructive, suicide as, **7:**328
   influence of socialization, **1:**580-81
   integration of cognitive development and socialization in, **1:**581-82
   interpersonal, theory of, **1:**491
   kin selection as sociobiological mechanism in, **1:**575
   learning, **2:**629-36
   marijuana and cannabinoid effects, **4:**907-8
   measurement of, **1:**579, 607-14

Behavior *(Continued)*
    memory and, **2:**340
    motor activity and, **2:**873
    nature of, **3:**590
    nervous system and, **2:**634
    neurotransmitter systems and, **1:**612-13
    objective performance instruments, **1:**611-12
    observation rating scales, **1:**610-11
    olfaction and, **5:**528-29
    pesticides in foods and, **7:**562
    planned, theory of, **1:**491
    preferred interaction styles in, **1:**573
    prenatal, **2:**875
    primate
        food distribution and availability as factor in,
            6:129-30
        infant care as, **6:**130-31
        intrasexual aggression in, **6:**131-32
        kinship and, **6:**130
        mating competition and, **6:**131-32
        predation as factor in, **6:**128-29
    problem-solving, suicide as, **7:**328
    psychobiology of developmental behavior, **2:**869-79
    rage and, **2:**340
    reasoned action, theory of, **1:**490-92
    recipient influences in, **1:**577-78
    relation to health, **4:**72-73
    resource distribution preferences in, **1:**573-74
    risk-taking, **4:**73
    self-report tests, **1:**609-10
    sex and, **2:**340
    situational influences in, **1:**578
    social psychological perspectives in, **1:**575-76
    social value differences in, **1:**577
    sociobiological perspectives, **1:**574-75
    styles of. *see* Self-monitoring
    techniques, in treatment of autism, **1:**510-11
    trait theory and consistencies, **5:**770-71
    understanding causes of, **5:**789-93
    undesirable, **2:**630
Behavior, abnormal, **3:**837-38
    objective performance instruments, **1:**611-12
    observation rating scales, **1:**610-11
    self-report tests, **1:**610
Behavior therapy, **4:**79-80
    defined, **3:**683
    for eating habits, **3:**685
    personality disorders and, **5:**782
Behavioral contracting, **4:**80
Behavioral genetics. *see* Genetics, behavioral
Behavioral inhibition system, **5:**775
Behavioral medicine, **4:**70

Behavioral structure, **3:**590
Behaviorism, **2:**870
Bell, J., **3:**829
Belladona, toxicity, **1:**178
Bendrofluazide, **2:**176
Benign senescent forgetfulness of Kral, **1:**210
Benzene, **2:**111, 112, 340
    biological markers of, **3:**390
    exposure routes of, **3:**385
    gamma-Benzene hydrochloride (Lindane), **2:**390
    inhaled, disposition and effect of, **7:**579
    as neurotoxins, **7:**547
Benzidine, **2:**112
Benzimidazole, **2:**390-92
Benzo(a)pyrene, **2:**110
    assay of, **3:**677
    maximum allowed level in food, **4:**224
Benzodiazepines, **2:**338, 493, 783, 786, 850-51
    abuse of, **5:**443
    as anxiolytic for older adults, **1:**135
    barbiturates and, **5:**442
    ethanol and, **3:**213
    list of, **2:**786
Benzoic acid, food analysis, **4:**224
Benzol. *see* Benzene
Benzopyrene, in marijuana, **4:**906
Benzothiadiazine diuretics. *see* Diuretics, thiazide
Benzothiazepines, **2:**84-85, 96
Benzothiodiazines, induction of glucose intolerance and,
    7:572
Bereavement, **5:**685
Bergmann's fibers, **1:**436
Bergson, **3:**497
Beriberi
    cardiac, **5:**266
    in developing countries, **3:**26
Bernard, Claude, **3:**377, 920, 921
Berson, **5:**713
Beryllium, **2:**111
Besedovsky, **5:**385
beta-(1 → 4)-linked D-mannosyl residues, **3:**39
beta-(1 → 4)-linked D-xylopyranosyl residues, **3:**39
Beta 1-adrenergic receptors. *see* Receptors, adrenergic,
    beta
Beta (beta)-sheets of proteins, **6:**221
Beta-adrenoceptor antagonists. *see* adrenergic beta recep-
    tor blockaders
Beta-amyloid protein. *see* Amyloid beta protein
Beta-blockers. *see* Adrenergic beta receptor blockaders
Beta-carboxyl polypeptide linkages, **3:**419
Beta-carotene deficiency, **3:**772
Beta efferent system, **3:**722

Beta-elimination, in irreversible enzyme thermoinactivation, **3:**417, 418
Beta errors, in randomized trials, **4:**62
Beta-glucosidases, **3:**782
11-Beta-hydroxylase deficiency, **4:**244
Beta-hypophamine. *see* Vasopressins
Beta-lactamases
    effect on peptidoglycan transpeptidase, **1:**282–83
    inactivation in bacteria, **1:**283–85
    inhibition of, **3:**400
Beta-lactams
    acquired resistance to, **1:**283–85
    definition of, **1:**271
    development of, **1:**272
Beta-MSH, function of, **6:**88
beta-N-methylamino-levo-alanine (BMAA), **1:**233
beta-N-oxalylamino-levo-alanine (BOAA), **1:**233
Beta receptors. *see* Adrenergic beta receptor blockaders
Betaherpesvirinae, **4:**187
Betel quid, **2:**114
Betz cells, **2:**682, **5:**309
Beverages
    alcoholic, **1:**166–67
    production, crop sciences and, **1:**142–43
Bezafibrate, **2:**169
Bhang. *see* Cannabis
Bhopal incident, **7:**549
Bias
    in argument processing, **5:**804–5
    in hypothesis testing, **1:**615–22. *see also* Hypothesis
        testing
Bias, protopathic, **2:**67
Biaxial joints, **1:**374, 376
Bicarbonate buffers. *see* Buffers, bicarbonate
Bicarbonates, **2:**34, 37, 39, 282, 539, 540
    gastric, **3:**76
    in gastric blood supply, **3:**763
    hepatic pHdiD and, **5:**827
    homeostasis of
        ammonium and, **5:**877
        in liver and kidneys, **5:**879–80
    lactate and, **5:**825
    mucus-rich, **2:**37
    pHe stabilization and blood, **5:**878
    phosphorylation, **2:**539, 540
    renal function and, **5:**878
    secretin and, **5:**718
    secretion, duodenal, **2:**37
    stress and duodenal secretion, **7:**288
Biceps, **3:**691, 693–94
Biceps brachii muscle, **5:**184
Biceps femoris muscle, **5:**183

Bicuculline, **2:**13, 316
Bicycling, **5:**187–89
Bifidobacterium, in breast milk feces, **1:**319
Biglycan
    defined, **6:**261
    structure and function, **1:**764
Bilateral symmetry. *see* Symmetry
Bilayer membranes, **4:**970
Bile, **2:**414
    from cholesterol, **2:**414
    circulation, enterohepatic, **2:**414
    in digestive process, **3:**78
    in gallbladder, **2:**414
    kinetics of drug excretion in, **5:**846–47
Bile acids and salts, **1:**623–26, **2:**168
    anti-hypercholesterolemic sequestrants, **1:**454
    biological activity of secondary, **1:**625
    biotransformation of, by gut microflora, **1:**625
    circulation, **2:**414
    colipase-dependent lipase and, **3:**51
    defined, **3:**47
    dietary fiber and, **3:**42
    digestive functions of, **1:**625
    in digestive process, **3:**79
    formation of, **2:**414, **4:**747
        abnormalities in, **1:**626
        primary, **1:**624
        secondary, **1:**625
    historical background of, **1:**623
    from LDL-cholestrol, **2:**168
    in lipolysis, **3:**54–55
    metabolism, abnormalities in, **1:**626
    micellar transport and, **3:**54
    microbial role in formation of, **1:**625
    resins binding-
        cholestyramine, **2:**168
        colestripol, **2:**168
        divistyramine, **2:**168
    sequestrants, **2:**168
    solubilized proteins, **4:**971–73
    structural characteristics of, **1:**623–24
    triacylglycerol absorption and, **3:**54
Bile pigments
    metabolism, **6:**117–21
    structure of, **6:**120
Bile salt-stimulated lipase
    lipolysis and, **3:**54
    properties of, **3:**49–50
    sn-2-monoacylglycerol hydrolysis and, **3:**53
    triacylglycerol digestion and, **3:**51
Bile salts. *see* Bile acids and salts
Biliary calculi. *see* Cholelithiasis

Biliary tract diseases, **5:**723
Bilirubin
    disorders of metabolism, **6:**121
    structure of, **6:**120
Biliverdin, **6:**120
Billard, **5:**700
Bilophila wadsworthia, **1:**246
Binary fission and asexual reproduction, **4:**929
Binaural interaction, **3:**240
Binding site, **2:**299-300
Binet, **3:**838
Binge, **2:**511, **3:**589
Binge/gorge, **3:**243
Binocular correspondence, **1:**635-36
Binocular vision. *see also* Space perception
    benefits of, **1:**634-35
    binocular correspondence in, **1:**635-38
    binocular disparity in, **1:**635
    eye movements in, **3:**555
    interocular suppression and binocular rivalry in, **1:**639
    neurophysiology of, **1:**639
    postnatal development of, **1:**640
    role of neuron V1 in, **7:**811
    sensory fusion, stimulus for, **1:**636-37
    space perception and, **1:**627-40
    stereo-depth scaling and, **1:**638-39
    stereopsis and, **1:**638
Bioaccumulation
    defined, **3:**251
    OECD tests for, **3:**257
Bioavailability, **3:**17
Biochemical markers. *see* Biological markers
Biochemistry, basic equations in, **2:**375
    Henderson-Hasselbalch, **2:**375
    Hill, **2:**375
    Michaelis-Menten, **2:**375
    Scatchard, **2:**375
Bioconcentration
    defined, **3:**251
    in fathead minnows, **3:**257
Bioconversions, **1:**364
Biocytinase, **2:**539
Biodegradable microcapsules, **7:**755
Bioelectrical impedance. *see* Tetrapolar bioelectrical impedance analysis
Bioenergetics. *see* Energy metabolism
Bioethics, **1:**655-59
    beginning of life, **1:**655-56
    biological criteria for life's beginning, **1:**656
    Declaration of Helsinki, **1:**658
    distributive justice, **1:**659
    end of life, **1:**656-57

Bioethics *(Continued)*
    euthanasia, **1:**657
    genetic engineering, **1:**658-59
    Hippocratic Oath, **1:**657
    human engineering, **1:**658-59
    human experimentation, **1:**657-58
    Nuremburg Code, **1:**657-58
    persistent vegetative states, **1:**657
    psychological criteria for life's beginning, **1:**656
    rationing and, **1:**659
    religious criteria for life's beginning, **1:**656
    resource allocation and, **1:**659
    social criteria for life's beginning, **1:**656
    Uniform Determination of Death, **1:**656-57
Biofeedback, **1:**661-66, **4:**79
    analog, **1:**664
    binary, **1:**664
    cognitive mediation, **1:**663-64
    experiments in animals, **1:**662
    experiments in humans, **1:**662
    experiments on curarized animals, **1:**662-63
    ideomotor model, **1:**663-64
    immediacy of feedback signal, **1:**664
    knowledge of results and, **1:**664
    learning theory, **1:**661-62
    parameters affecting learning, **1:**664
    somatic mediation, **1:**662-64
    specificity of effects, **1:**663
    training, space travel and, **7:**152
    voluntary control of ANS-mediated responses, **1:**662
Biogenic amines. *see* Monoamines
Biological availability, **3:**17
Biological clocks, **2:**484, 489, 491-95, **4:**785
Biological criteria for beginning of human life, **1:**656
Biological markers
    defined, **3:**891
    in septo-hippocampal test model, **5:**337
    in toxicology, **3:**390
    in transfection, **3:**806
    vital, **2:**697
Biological phenomena, chemical measurement of, **3:**257
Biological pump. *see* Biological transport, active
Biological response modifiers, **1:**675-89
    adherent lymphokine-actived killer cells, **1:**680
    adriamycin, **1:**686
    antibody conjugates with, **4:**370-71
    azathioprine, **1:**685
    cellular, **1:**679-80
    colony-stimulating factors, **1:**686-87
    cyclophosphamide, **1:**685-86
    cyclosporine A, **1:**685
    cytokines, **1:**680-84

Biological response modifiers (Continued)
  diethyldithiocarbamate, **1:**684
  erythropoietin, **1:**687
  growth factors, **1:**686–87
  immunoconjugates, **1:**688–89
  immunomodulatory agents, **1:**684–86
  immunopotentiating agents, **1:**684–86
  immunorestorative agents, **1:**684–86
  immunosuppressive agents, **1:**684–85
  interferon-alpha, **1:**680–81
  interferon-beta, **1:**680–81
  interferon-gamma, **1:**680–81
  interleukin-1, **1:**682–83
  interleukin-2, **1:**683
  interleukin-3, **1:**683
  interleukin-4, **1:**683
  interleukin-5, **1:**683
  interleukin-6, **1:**684
  levamisole hydrochloride, **1:**684
  lymphokine-activated killer cells, **1:**679–80
  monoclonal antibodies, **1:**687–88
  muramyl tripeptide phosphatidylethanolamine, **1:**684
  serological, **1:**687–89
  steroids, **1:**684–85
  thymic hormones, **1:**684
  transforming growth factor-beta, **1:**687
  tumor-infiltrating lymphocytes, **1:**680
  tumor necrosis factor, **1:**681–82
Biological rhythms. see Circadian rhythm; Periodicity
Biological transport. see also Countertransport
  of carbohydrates, **3:**79–80
  in colon and rectum, **3:**82
  defined, **3:**71
  diffusion through membrane bilayers, **4:**971, 973
  of gangliosides, **3:743–48**
  of glutathione, **3:**908, 911
  of hormones, **3:**328
  membrane transport, **4:**972–73
  nonionic
      membrane transport by, **5:**823
      in proton translocation, **5:**873
  of organelles, along microtubules, **5:**18–20
  passive, **2:**279, 280–83
  of pH, **5:**875
  of plasma lipoproteins, **3:**573–74
  of polymeric IgA, **5:**138–39
  reciprocating transporters, **4:**973–74
  redox-driven transport, **4:**974
  in renal function, **3:**95
  in small intestine, **3:**77
  of water, in digestive system, **3:**81

Biological transport, active, **2:**89, 236–37, 279–89, **4:**557–71
  and alpha-helix, **2:**280
  anion metabolism, **4:**568–69
  ATP-dependent, **4:**559–64
  axonal, **2:**49
  in blood-brain barrier, **1:**724
  of calcium, **4:**568–69
  cell pH regulation, **4:**569
  cellular information processing, **4:**558
  cellular volume regulation, **4:**569–70
  chloride for bicarbonate, **2:**282
  Cl^-/HCO_3 antiporter, **4:**565–66
  classes of transporters, **2:**279
  compared to channels, **4:**559
  complex transporters, **4:**567
  cotransporters, **2:**279, 286–87
  epithelial proton transport, **4:**569
  in erythrocyte, **2:**282
  function of transporters, **2:**280–83
  gating, **2:**280
  as glycoprotein, **2:**282
  integration into functional requirements of cells and tissues, **4:**571
  ion binding to transport protein, **4:**560–61
  ion channel, **2:**279, 287, 289
  ion pump, **2:**236–37, 279, 283–86
  ion-transport protein interaction, **4:**560–61
  K^+ metabolism, **4:**568
  in kidneys, **5:**878
  kinins and, **4:**594
  matching to work loads, **4:**559–60
  maximum rate, **4:**559
  membrane-spinning domain, **2:**280
  membranes, **4:**973
  model, **2:**237
  phosphorylation and dephosphorylation reactions, **4:**561–62
  ping-pong mechanism, **2:**283
  primary active transport systems, **4:**562–64
  principles, **4:**558–62
  proton-dependent antiporters, **4:**565
  proton metabolism, **4:**569, **5:**873
  role in physiology, **4:**558
  secondary active transport systems, **4:**564–67
  sodium transport system, **4:**564–68
  stages, **4:**559
  structure of transporters, **2:**280–83
  symporters, **4:**566–67
  topography, diagram of, **2:**283
  total body fluid regulation, **4:**570–71
  transport ATPase, **4:**974

Biological transport, active (*Continued*)
  transporters, **2:**279-83
  uphill, **2:**237
Bioluminescence, measurement of, **5:**580
Biomass, **5:**471
Biomaterials, aqueous two-phase systems for, **1:**364
Biomechanics
  defined, **3:**473
  in ergonomic database, **3:**478
Biomedical research. *see* Research
Biomodulators. *see* Biological response modifiers
Biopsy
  definition of, **1:**57
  in diagnosis of bone tumors, **7:**696-98
  in surgical pathology, **5:**683
Biopsy, chorionic villi. *see* Chorionic villi sampling
Biopsychosocial model, **4:**71
Biosphere
  defined, **5:**471
  energy flow through, **5:**472, 473
Biotin, **2:**467, 539
  amidohydrolase, **2:**539
  and avidin, **2:**467
  as carboxyl group carrier, **3:**437
  carboxylase, **2:**539
  deficiency, immunity and, **5:**513
Biotin-dependent enzymes, **2:**539
Biotinidase, **2:**53
Biotinidase deficiency, **5:**437
Biotinyl-L-lysyl coenzyme, **2:**539
Biotnyl
  chemistry, **2:**537
  functions, **2:**539
  metabolism, **2:**537-39
Biovars, **6:**703
Bipedalism, **2:**593
  of *Australopithecus afarensis*, **3:**490
  digestive system orientation and, **3:**59-60, 68, 69
  of earliest hominids, **3:**497
  evolutionary advantages of, **3:**496
  as factor, in human evolutionary studies of pelvis, **1:**825-26
  in human evolution, **3:**488
Biphenium salts, for hookworm anemia, **1:**264
Biphenyl compounds, polychlorinated, **3:**255
Bipolar cells, **6:**626
Bipolar disorder, **2:**832-34, **5:**92-93, 842
  antidepressant drugs for, **1:**307-8
  endocrine, **2:**834
  genetic, **2:**833-34
  genetic aspects of, **3:**843
  seasonal, **2:**833

Birds
  gas convection in, mechanisms of, **6:**570
  neurological aspects of singing, **5:**307, 313-14
Birefringence, defined, **5:**567
Birnbaumer, **4:**14
Birth. *see* Labor
Birth control. *see* Contraception
Birth control pill. *see* Contraceptive agents
Birth defects. *see* Abnormalities
Biscarboxyethylfluorescein, **5:**872-73
Bisexuality, **4:**930
Bishydroxycoumarin, **5:**266-67
Bismuth-212 isotope, **4:**372
Bisphonates, **4:**280-81
Biston betularia. *see* Peppered moth
Bithionl, **2:**389
Bitter taste
  cell membrane permeability and, **7:**532
  compounds eliciting, **7:**530
  signal transduction for, mechanisms of, **7:**534-35
Black spot disease, cereal, **5:**263
Blacks
  death rate of, **5:**100-101
  diet of, **3:**651
  women
    endometriosis and, **3:**362
    pregnancy pica of, **3:**32
Bladder
  function of, **7:**7634-35
  innervation of, **5:**43-44
  snail fever, **1:**263
  spinal cord dysfunction and, **7:**231-32
Bladder neoplasms, **2:**116
  BCG immunotherapy, **4:**347
  in dogs, **2:**118
Blades, in Upper Paleolithic tool-making, **3:**498
Blalock, Alfred, **5:**703
Blast, **2:**769
Blastocyst
  bovine, **3:**309
  chimeric, **3:**309
  defined, **3:**305
  endometrium preparation for implantation, **4:**429
  transition from free to attached, **4:**424
Blastocytes. *see* Blastomeres
Blastoderm
  in body pattern formation, **3:**286
  diversity in, **3:**288-90
Blastodisc. *see* Blastoderm
Blastomeres
  defined, **3:**305
  in dichorial MZ twinning, **3:**308-9

Blastula. *see* Blastocyst
Blinding, in clinical trials, **4**:62
Blindness, **2**:346
    methyl alcohol induced, **2**:346
    visual substitution aids for, **7**:62
Blix, G., **3**:725
Blocked practice. *see* Practice (Psychology), blocked vs. random
Blood, **2**:639
    alcohol level, **1**:170
    arterioles, **2**:160
    artery, **2**:160–61
    artery wall barrier, **2**:367
    capillaries, **2**:160
    cholesterol and coronary heart disease, **2**:662
    detection of bacterial pathogens in, **1**:554
    dielectric properties of, **3**:13, 15
    drug binding, **1**:709–13
    drug-protein complexes in, **1**:707–8
    drug-protein interaction, **1**:710
    fetal. *see* Fetal blood
    gas transport, **6**:599–600
    hormone transport in, **3**:328
    implications of plasma-bound drugs in, **1**:707
    kidney function and, **3**:94
    lipoproteins, **2**:661–64, **6**:23–25
        as drug binders, **1**:709
    optical properties of, **5**:595–96
    oxygen dissociation curve for, **6**:599
    parathyroid hormone regulation of calcium, **5**:660–61
    purification
        continuous ambulatory peritoneal dialysis as method of, **1**:412–13
        hemoperfusion as method of, **1**:411
        hemosorption as method of, **1**:411
        recirculating peritoneal dialysis as method of, **1**:411–12
        regular peritoneal dialysis as method of, **1**:411
    regulation, **2**:365–66
        chemical constituents during heart surgery, **4**:92
        heart-lung machine complications, **4**:96
        inputs to hypothalamus, **4**:306–7
    regulation, intestinal. *see under* Intestines
    targeting of drugs in, **1**:713
    tests for infectious diseases, **1**:737–40
    trauma caused by artificial heart, **1**:396
    vasopressin secretion and osmolality, **5**:397
    zinc metabolism in, **7**:881–82
Blood, occult. *see* Occult blood
Blood banks, **5**:684
Blood-brain barrier, **1**:715–27, **2**:366–67
    astrocytes and, **1**:445

Blood-brain barrier *(Continued)*
    carrier-mediated transport, **1**:724
    changes in Alzheimer's disease, **1**:214
    cytokines and, **5**:385
    defined, **5**:667
    factors causing disruption, **1**:726–27
    ganglioside penetration of, **3**:739
    general characteristics of, **1**:715–16
    magnetic resonance imaging, **4**:870
    modification of, **1**:726–27
    osmotic opening of, **1**:726–27
    osmotic pressure effects, **1**:726–27
    permeability of
        experimental methods for determining, **1**:721–22
        movement of drugs across, **1**:723
    structure and function of, **1**:716–19
    transport at, **1**:732–26
    water transfer in, **1**:722–23
Blood cells
    changes, space travel and, **7**:148
    classification, **4**:165
    fragility, **7**:876–77
    functions of, **4**:165–67
    microtubules in, **5**:18
    mutation rates in, **5**:248–49
    production regulation, **4**:172–75
    types of, **4**:165–67
    zinc metabolism in, **7**:882
Blood-cerebrospinal fluid barrier, **1**:717–18
Blood circulation. *see also* Fetal circulation; Microcirculation
    in aerobic exercise, **3**:509–10
    cerebral, **2**:365–66, **3**:224–25
        autoregulation of, **2**:365
        chemical control, **2**:365
        metabolic control, **2**:365
        neurogenic, **2**:365
        regulation, **2**:185–86
    circulatory pathway, **2**:161–65
    in dynamic exercise, **3**:519–20
    hepatic, portal, **2**:164–65
    and lupus erythematosus, **2**:601
    of pituitary gland, **5**:395
    pulmonary, **2**:158, 161–62
    regulation by kinins, **4**:594
    in stomach, **3**:755–63
    systemic, **2**:162–63, 165
    tumor cells and, **4**:1011–13
    uteroplacental, **3**:314
    in the venous sinuses, **4**:984
Blood clotting factors. *see* Factor IX through XIII

Blood coagulation, **4:**152-53
    clots, as targets for immunoconjugates, **4:**373-74
    effect of birth control hormones on, **1:**693-94
    extrinsic pathway for, **7:**493-94
    factor VIII deficiency, **4:**153-54
    factor IX deficiency, **4:**154
    factor XI deficiency, **4:**155
    triggering of, **7:**493-95
    von Willebrand's disease, **4:**155
Blood coagulation disorders, **4:**153-55
Blood diseases. *see* Hematologic diseases
Blood flow velocity
    carbonic anhydrase and, **5:**876
    in dynamic exercise, **3:**519
    fatty acid uptake and, **3:**586
    pH and, **5:**824, 876
    space travel influences on, **7:**148
    in stomach, **3:**757-59
Blood glucose
    in exercise, **3:**512
    hormonal regulation of, **3:**323-24
    in hunger and meal initiation, **1:**348-49
    insulin and, **5:**732
    insulin effects, **4:**248
    liver maintenance of, **3:**921
    perinatal, **3:**350
Blood groups
    alleles, differences and similarities in, **6:**109
    antigens, **4:**379-81
    as genetic markers, **3:**828
    genetic polymorphism of, **6:**80
    somatic mutation and, **5:**249
Blood plasma. *see* Plasma
Blood platelets. *see* Platelets
Blood pressure, **2:**144, **4:**284-85. *see also* Hypertension
    in aerobic exercise, **3:**510
    aging and, **3:**901
    angiotensinogen-derived peptides and, **5:**729
    antihypertensive drugs, **4:**293-94
    arterial, **2:**185-86, **4:**285-88
    atrial natriuretic factor and, **1:**473-74
    brain regulation of, **5:**422
    and caffeine, **2:**70
    in children, **5:**708
    control, PAG projections and, **4:**718
    defined, **4:**284
    diastolic, **4:**285
    in dynamic exercise, **3:**518
    during heart surgery, **4:**93-94
    increase, **2:**145
    larynx stimulation effects, **4:**650
    neurotensin and, **5:**721

Blood pressure *(Continued)*
    oral contraceptives and, **5:**602
    regulation, kinins and, **4:**595
    relationship to blood volume, **7:**757-65. *see also* Vascular capacitance
    in static exercise, **3:**520
    systolic, **4:**285
    in vascular compartments, **4:**284-85
    vasopressin secretion and, **5:**397
    vessel response, **2:**141-43
    in weightlessness, **2:**150-51
Blood proteins, **1:**708-9
    as drug binders, **1:**708-9
    in drug distribution and elimination, **1:**711
    in drug transport, **1:**711-12
Blood sugar. *see* Blood glucose
Blood transfusion
    autoimmune disease and, **1:**520
    as immunosuppression, **4:**363-64
    screening tests for, **1:**737-40
    veto phenomenon, **4:**363
Blood vessels
    abnormalities and stroke, **7:**303
    abnormalities in Alzheimer's disease, **1:**211-13, 214
    autonomic innervation, **1:**543, 545
    capacitance. *see* Vascular capacitance
    of eye, **3:**548-49
    of foregut, **3:**64
    of hindgut, **3:**66
    maternal, during implantation, **4:**429
    of midgut, **3:**65-66
    postradiation effects on, **6:**465
    spinal cord, **7:**193
Blood volume
    during cardiopulmonary bypass, **4:**94
    redistribution of
        by active processes, **7:**761-63
        effect of disease on, **7:**763
        effect of stress on, **7:**763
        by passive processes, **7:**760-61
    reduction in astronauts, **2:**151-52
        arterial wall barrier, **2:**367-69
        cerebral vesselS, **2:**357
        sugar, **2:**101, 105
        vein, **2:**160
        venules, **2:**160-61
        vessel, **2:**160-61
    relationship of blood pressure to, **7:**757-65. *see also* Vascular capacitance
    stroke volume and, **3:**518
    techniques for measuring, **7:**763-65
Bloodletting, in pediatrics, **5:**700

Bloom syndrome, and sister-chromatid exchange, 5:251
Blotting, southern
    for muscular dystrophy, 3:146
    for muscular dystrophy detection, 5:195, 201–2
    in papillomavirus infection diagnosis, 3:862
    polymerase chain reaction technology and, 3:139
    for restriction fragment-length polymorphisms, 3:136
    for retinoblastoma gene, 3:143
Blue-green algae. *see* Algae, blue-green
Body composition
    aging and, 3:904
    assessment, 1:669–72
        of fat-free mass, 1:669–70
        in obesity, 1:671
        regional impedence measurements in, 1:672
        of weight loss, 1:671–72
    methods to determine, 1:667
    in puberty onset, 3:355
Body fat. *see* Adipose tissue
Body fluids, mutagen detection in, 5:251
Body hair. *see* Hair
Body mass index
    defined, 3:683
    illustrated, 3:688
Body stalk, defined, 3:293
Body surface area, 5:312–13
Body temperature
    during cardiopulmonary bypass, 4:94
    circadian rhythm of, 1:752
    core, 4:793–95
    and its regulation, 1:751–60
    muscle contraction and, 3:511–12
Body temperature regulation, 2:339. *see also* Thermo-
    tolerance
    aging and, 3:903
    behavior as mechanism for, 1:754
    body temperature set points in, 1:758
    cold, physiological basis, 4:793
    disturbances in, 1:759–60
    evaporative heat loss, 1:755–56
    heat production as mechanism for, 1:756–57
    hominid evolution and, 3:497
    hormonal control, 4:805
    hypothalamus role, 4:310
    in infants, incubators and, 5:705
    nonshivering thermo-genesis and, 1:757
    physical activity and, 1:756–57
    physiological mechanisms used in, 1:754–59
    piloerection as mechanism for, 1:756
    radiation
        as means of body heat exchange, 1:753
        and thermoregulation, 4:794

Body temperature regulation (*Continued*)
    REM sleep and, 3:197
    shivering and, 1:757
    skin circulation as mechanism of, 1:754–55
    temperature controllers in, 1:759
    temperature receptors in, 1:757–58
Body types, described by Hippocrates, 3:474
Body vs. mind, 4:704–5
Body water
    aging and, 3:904
    distribution of, 3:94
    renal reabsorption of, 3:95–96
Body weight. *see also* Adipose tissue; Obesity
    dietary restraint and, 3:689
    educational level and, 3:684
    first menstrual cycle and, 1:744–45
    longevity and, 4:784
    mammalian brain size and, 5:306
    for men, 5:518
    in puberty onset, 3:355
    relative, defined, 3:683
    setpoint, hypothalamus role, 4:309
    for women, 5:518
Bohr effect, 4:138
Boltzmann distribution, 3:235
Boltzmann machine, 5:391
Bombesin, 2:27, 6:89, 35–38
Bonded phase partition columns, 4:217–18
Bonded phases, HPLC, 4:207–8
Bone age measurement. *see* Age determination by skeleton
Bone and bones, 2:639, 7:34, 645
    accumulation, with age, 1:775–76
    alveolar, 2:777, 779, 825
    as an organ system, 1:783–84
    ancient, 1:784
    appositional/resorptive, 2:697–98
    atrophy, 2:697
    of body, 7:34
    bundle-bone, 2:777, 825
    cancellous
        characteristics of, 1:772
        relative permittivity of, 3:12
    cartilage, 2:704
    cartilaginous, 2:697–98
    cell biology of, 1:762
    cell differentiation
        growth factors as signals for, 1:787–88
        hormones as signals for, 1:788–89
        interactions with other tissues as signals for, 1:788
        mechanical factors as signals for, 1:788–89
        signals for, 1:787–89
    cell genes, 1:761–70

Bone and bones *(Continued)*
　cellular heterogeneity in, **1:**817–18
　characteristics of, **1:**782–84
　and collagen, **2:**574
　compact, **2:**640
　　remodeling in, **1:**815
　and computerized tomography, **2:**702
　　of human skull, **2:**702
　　implants as markers of, **2:**699, 701
　　method for, listed, **2:**700
　cortical
　　characteristics of, **1:**772
　　components of, **6:**437
　　elemental composition of, **6:**437
　　mass density of, **6:**437
　coxal, **7:**38
　craniofacial, **2:**697–708
　defined, **2:**817
　deformity, **2:**698
　density
　　and bone strength, **1:**774–75
　　fragility and age-related changes and, **1:**771–80
　endochondral, **2:**705
　estrogen and, **3:**481–83, 485
　estrogen receptors in, **3:**484
　extracellular matrix, **1:**783
　foot, **7:**39
　fossilized, **1:**784
　fractures. *see* Fractures
　fragility, **1:**779–80
　functions of, **1:**811
　general characteristics of, **1:**779–80
　hip, **7:**38
　homeostasis, **1:**774
　hyoid, **7:**34
　hypertrophy, **2:**697
　lamellar, **1:**772–73
　lamina dura, **2:**825
　leg, **7:**39
　long, **2:**215
　　fetal, **2:**202
　　formation. *see* cartilage
　　growth of, **2:**215
　and magnetic resonance imaging, **2:**702
　measurement with dye, **2:**699
　metabolism
　　hormonal regulatory factors in, **1:**800–805
　　regulatory factors of, **1:**799–809
　　space travel and, **7:**146–47
　microscopic forms of, **1:**772
　mineralization, **1:**765, **3:**481
　modeling, **1:**812–13

Bone and bones *(Continued)*
　morphogenic proteins, structure and function, **1:**764
　nasal, of rabbit, **2:**707
　as on organ, **1:**783–84
　origins of, **1:**784
　osteoblasts, **1:**782–83. *see also* Osteoblasts
　osteoclasts, **1:**782. *see also* Osteoclasts
　osteocytes, **1:**782–83
　parathyroid hormone regulation
　　of calcium, **5:**661
　　of phosphate, **5:**662
　pelvic, **7:**38
　prehistoric and associated tools. *see* Taphonomy
　preosteoblasts, **1:**782–83
　process, **2:**825
　radiography, cephalometric, **2:**701–2
　regulatory factors, **1:**799–809
　　endogenous, **1:**805–7
　　endogenous, clinical aspects of, **1:**808
　repair, **2:**698
　　growth factors and, **7:**508
　shapes of, and joints, **1:**789
　shin, **7:**39
　skeletal system, **1:**784
　skull, **7:**34
　spongy, **2:**825
　stability, physiological, **2:**697
　strength, and bone density, **1:**774–75
　structural types of, **1:**772
　structure of, **1:**815
　sutural, **2:**697–98
　suture, **2:**704
　thigh, **7:**38
　thoracic, **7:**37
　as tissue, **1:**782–83
　tissue, characteristics of, **1:**771–73, 782–83
　trabecular, **2:**640
　　remodeling in, **1:**815–17
　turnover of its minerals, **1:**773–74
　of vertebral column, **7:**36–37
　vitamin D effect on, **7:**868
　zinc metabolism in, **7:**882
Bone cancer. *see* Bone neoplasms
Bone-derived transforming growth factor. *see* Transforming growth factor beta
Bone development, **1:**781–90, **2:**825
　bone-forming cells, **1:**785–86
　cell condensation, **1:**787
　cell migration, **1:**786–87
　developmental pathways, **1:**762
　direct osteogenesis as made of, **1:**784
　ectopic, **1:**785

Bone development *(Continued)*
    embryonic development, **1:**781–90
        cell condensation in, **1:**787
        cell migration in, **1:**786–87
        growth in, **1:**789–90
        shapes of bones and joints in, **1:**789
        signals for bone cell differentiation in, cell differen-
            tiation, **1:**787–89
    embyronic development, of face, **2:**690
    epithelial-mesenchymal interactions, **1:**788
    growth, **1:**789–90, **2:**215, 812
    growth factors, **1:**787–88
        function of, **1:**806–7
        synthesized by cells from adjoining tissues, **1:**807
    growth in adulthood, **4:**31
    hormonal factors, **1:**788–89
    indirect osteogenesis as mode of, **1:**784–85
    inductive interactions, **1:**788
    mechanical factors, **1:**788–89
    mesodermally derived skeleton, **1:**786
    modes of, **1:**784–85
    morphogenesis
        first-order, **1:**789
        second-order, **1:**789
    neural crest cells, **1:**786
    osteoblasts, **1:**785
    osteocytes, **1:**785
    osteogenesis
        direct, **1:**784
        indirect, **1:**784–85
    primordium, **1:**787
    signals for cell differentiation, **1:**787–89
    somites, **1:**786–87
Bone diseases
    in children, **5:**709
    vitamin D and, **7:**871
Bone fractures *see* Fractures
Bone loss, age-related. *see* Osteoporosis
Bone loss, osteoclastic. *see* Bone resorption
Bone marrow, **2:**753–55
    autologous, for transplantation, **1:**796
    and B-lymphocyte, **2:**754
    beta cell development in, **1:**566
    effect of radiation on, **6:**413
    macrophage production, **4:**854
    mitosis in, **5:**72
    noradrenergic innervation of, **5:**380
    normal, differentiation of, **1:**791–92
    and stem cell, **2:**754
Bone marrow transplantation, **1:**791–98
    for acute lymphoblastic leukemia, **1:**794
    for acute myelogenous leukemia, **1:**794–95

Bone marrow transplantation *(Continued)*
    for aplastic anemia, **1:**794
    autologous, **1:**796
    for carcinomas of breast, brain, and intestinal tract,
        1:796
    chemotherapy prior to, **1:**793
    for chronic granulomatous disease, **1:**794
    for chronic myelogenous leukemia, **1:**794–95
    clinical, **1:**792–93
    clinical procedures, **1:**792–93
    for congenital agranulocytosis, **1:**794
    cyclosporin a therapy, **4:**363
    for enzyme deficiencies, **1:**796
    for erythroid function disorders, **1:**796
    future prospects for, **1:**796–97
    for Gaucher's disease, **1:**796
    gene therapy, **1:**796–97
    for genetic diseases, **1:**796
    graft-versus-host disease, **1:**792–93
    for granulocyte actin deficiency, **1:**794
    haploidentical transplants, **1:**792
    histocompatibility, **1:**792, 794
    histoincompatible transplants, **1:**792
    for HIV-1, **1:**797
    HLA-identical siblings, **1:**792
    for Hunter's disease, **1:**796
    for Hurler's syndrome, **1:**796
    immunotoxin therapy, **4:**421
    for lymphoma, **1:**795
    matched unrelated donor transplants, **1:**792
    in pediatrics, **5:**704
    possible recipients for, **1:**793–96
    present status of, **1:**793–96
    for severe combined immunodeficiency, **1:**793–94, 797
    solid-organ transplants, **1:**797
    for Wiskott-Aldrich syndrome, **1:**794
Bone matrix, **1:**771, 783
Bone neoplasms, **7:**695–704
    classification of, **7:**696
    diagnosis of, **7:**696–98
    epidemiology of, **7:**695–98
    etiology of, **7:**695–98
    FBJ-MSV in, **3:**702
    fos gene and, **3:**705
    lymphoma, **7:**701
    metastases, hypercalcemia of malignant etiology,
        4:278–79
Bone plates
    cortical, **2:**825
    cribiform, **2:**825
Bone proteins
    collagens, **1:**763–64

Bone proteins *(Continued)*
noncollagenous, **1:**764
structure and functions, **1:**764–66
Bone remodeling
activation as cellular event in, **1:**820
and bone disease, **1:**823
bone-forming and bone-resorbing cells, **1:**18–819
and calcium homeostasis, **1:**813–14
cellular and biochemical events in, **1:**817–21
cement lines in, **1:**817
compact and trabecular, **1:**815–17
concept of, applied to bone loss, **1:**823
differences between remodeling and modeling, **1:**775
effect of local factors on, **1:**823
effect of systemic factors on, **1:**821
effects of mechanical loading on, **1:**821
formation as cellular event in, **1:**820
functions of, **1:**813
Haversian (internal), **1:**814
human, **1:**811–24, :811–24
mechanical advantages of, **1:**814–15
quantum concept of, **1:**815–17
regulation of, **1:**821–23
renewed modeling of, **1:**812–13
resorption and formation coupled in, **1:**821
resorption as cellular event in, **1:**820
reversal as cellular event in, **1:**820
sequence of cellular events in, **1:**819–21
Bone resorption, **1:**762, **4:**278
in bone remodeling, **1:**820, 821
calcitonin and, **3:**483
countermeasure to space travel, **7:**150–51
effect of vitamin D on, **1:**804
hydroxyproline and, **3:**481
IL-1 and, **4:**278, 474
microarchitecture of cancellous bone, **1:**777
microarchitecture of cortical, **1:**776–77
molecules stimulating, **1:**807
morphologic types of, **1:**778
osteoclast-mediated, **1:**762, 766–67
TNF and, **4:**474
Bone sialoprotein I, **1:**764
Bone sialoprotein II, **1:**764
Bone tumors. *see* Bone neoplasms
Bony pelvis of archaic *Homo sapiens*, **1:**825–31
Boom famine
defined, **3:**561
Sen on, **3:**562
Bordetella pertussis, **4:**460–61
Bordetella pertussis toxins. *see* Pertussis toxins
Boredom susceptibility, **6:**810

Boron
biological function of, **7:**606
deficiency of, **7:**608
toxicity of, **7:**608
Boron compounds, trigonal, **3:**398
Borrelia burgdorferi, Lyme disease and, **7:**894–95
Botts-Moralis scheme, **3:**429–30
Botulinum toxins, **2:**345, **3:**734
Botulism, **3:**666
Boveri, T., **3:**849
Bovine leukemia virus, **6:**646
Bowel, large. *see* Intestine, large
Bowel, small. *see* Intestine, small
Bowel bacterial overgrowth syndrome, **1:**244, 246
Bowels. *see* Intestines
Bowlby's model of attachment, **2:**857
Bowman's glands, **5:**288–89
Bowman's zone, **3:**536–37
Brachial plexus, **5:**354
Brachydactyly, **3:**850, 851
Braconid viruses. *see* Baculoviridae
Brady, **3:**777
Bradycardia, **2:**141, 145, 337, 340
Bradykinin, **2:**145
functions of, **5:**730
inflammation role, **4:**470
in postprandial intestinal hyperemia, **4:**555–56
receptors, **4:**596–97
structure, **4:**590
Brain, **2:**1–51, 241, 254, 431–36, 599. *see also* Central nervous system
absorbtion/secretion, intestinal, **2:**37
aging of, **3:**901
anatomy, **2:**2–3, **5:**348–49, 7–8, 333–37, 679
of adult, **5:**348–50
comparative, **2:**599–679
schema, **2:**679
asymmetry, **2:**7–8
anatomical, **2:**7–8
cerebral, **2:**7–8
functional, **2:**7
handedness, **2:**8
and sex, **2:**8
atrial natriuretic factor synthesis in, **1:**471
atrophy, and alcohol, **1:**171–712
base, photograph of, **2:**358
blood-brain barrier. *see* Blood-brain barrier
brain-gut axis, **2:**31
capillaries, **1:**716–17, **5:**32, 39
cells, number of, **2:**1
central gray area, **2:**333–40
and cholesterol, **2:**413

Brain (*Continued*)
  and chromatin structure, **2:**431–36
  cognitive aspects, **2:**547–58
    connectionism, **2:**550–52
    encodingism, **2:**553–55
    information processing, **2:**548–50
    interactivism, **2:**555–56
    intermodulation, **2:**557
    robotics, **2:**547
    studies, **2:**547
  crosstalk and, **5:**155–56
  cytoarchitechture, **2:**335
  development, **2:**861
    gangliosides in, **3:**B738
    omega-3 fatty acids in, **5:**537–38
    technology and, **3:**503
  disorders, and word meaning, **6:**485
  dysfunction, minimal. *see* Attention deficit disorder
      with hyperactivity
  dyslexia-dysgraphia and, **3:**216–17
  edematous, **5:**41
  equipotential theory of Flourens, **2:**860
  feeding behavior and, **3:**592
  fetal
    fatty acids in, **5:**539
    thyroid hormones and, **3:**340
  food intake and, **5:**519–20
  frontal lobe. *see* Frontal lobe
  functional areas, **4:**197–98
  functional changes in, lesion-induced, **6:**39
  functions
    basic, listed, **2:**1
    emotion, **2:**9–10
    higher, **2:**8–10
    intestinal, **2:**31–40
    language, **2:**9
    memory, **2:**8–9
    space, **2:**10
  G protein mRNA in, **4:**22–23
  Gall's theory, **2:**860
  ganglioside transport in, **3:**744
  in gastric circulation regulation, **3:**762
  gene expression, **2:**431–36
  glucose and, **3:**324
  gray periaqueductal, **2:**333
  heart-lung machine complications, **4:**95
  hemispheres. *see* Hemispheres
  higher function. *see* Higher brain function
  hindbrain. *see* Hindbrain
  influence on immune system. *see* Psychoneuroim-
      munology
  interactions, **2:**32

Brain (*Continued*)
  language development and, **3:**217
  lateralization, **2:**861
  learning and, **2:**599
  lobes, **2:**599, 719
  mammalian
    body weight and, **5:**306, 307
    glycolysis in, **3:**939
  mass action, **2:**860
  measurement, **2:**861
  meninges of, **5:**351
  messenger, **2:**11–29
  metabolic control of, **3:**380
  microtubule protein from tissue, **5:**894
  microtubule protein phosphorylation in rats, **5:**897–98
  midbrain. *see* Mesencephalon
  midsagittal section of, **6:**2
  modulation of ongoing activity in, **2:**549
  of monkey, **2:**335
  MT, **4:**200–201
  myelogenesis in, **5:**276
  natriuretic peptide, **6:**89
  neocortex, **2:**4
  nerve sympathetic, **2:**37–38
  neuroanatomy, and memory, **4:**674–75
  neurons. *see* Neurons
  neurotransmitters. *see* Neuroregulators
  nucleus, **2:**35
  organization, **2:**1–5, 7–8
    anatomical, **2:**2–3
    cellular, **2:**1–2
    cerebral, **2:**7–8
    chemical, **2:**5
    functional, **2:**5–7
    neurocortical, **2:**5–7
    physiological, **2:**3–5
  parietal lobe damage, effects on orienting, **1:**481
  pathway, **2:**31
    afferent, **2:**31
    efferent, **2:**31
  peptides. *see* Neuropeptides
  permeability, **1:**721
  posteriolateral view of, **7:**120
  receptors in, **5:**421–25
  as resource pool, **5:**156
  right hemisphere communication disorders, **1:**123
  sites and function, gastrointestinal, **2:**32
  size
    in fossil record, **3:**488–89
    reduced after damage, **2:**865
  spectrin, **2:**41–51
  subdivisions

Brain *(Continued)*
  dorsal, **2:**333-34
  dorsolateral, **2:**333-34
  medial, **2:**333-34
  ventrolateral, **2:**333-34
  substance uptake in
    effect of cerebral blood flow in, **1:**720-21
    effect of protein binding on, **1:**720
    model describing, **1:**719-20
    techniques for determining, **1:**722
  in sudden infant death syndrome, **7:**320
  of suicide victims, **2:**845
  switchboard model, outdated, **2:**549
  synaptic physiology of, **7:**383-93
  thalamus, in horizontal section of, **7:**439
  tissue studies, **2:**845
  vagus, **2:**37
  ventricles, anatomy of, **5:**350-51, 350-51
  versus mind in reasoning, **6:**494
  vessels supplying, **7:**299-301
  vocalization, **2:**339
  volume, **2:**599
  weight, **2:**1
  written language and, **3:**224-27
  zinc, **7:**883, 887
Brain-derived neurotrophic factor, **5:**335
Brain diseases
  CuZnSOD and, **3:**186
  dyslexia-dysgraphia and, **3:**217
  mercury-induced, **3:**387
Brain dysfunction, minimal. *see* Attention deficit disorder
    with hyperactivity
Brain electrical activity mapping. *see* Brain mapping
Brain injuries
  effects on orienting, **1:**481
  focal, deficits associated with, **5:**314
  inaccurate face representation and, **3:**226
  neocortex and, **5:**313
Brain mapping, **3:**216-17
Brain stem, **2:**2, 39, 673
  anatomy of, **5:**349-50
  ascending pathways of, **3:**239
  caudal
    amygdaloid projections to, **4:**721-23
    ben nucleus of the stria terminalis projections to,
      4:721-23
    hypothalamus projections to, **4:**719-21
    mesencephalon projections to, **4:**714-19
    prefrontal cortex projections to, **4:**723
  defined, **3:**553
  eye movement and, **3:**557-58
  longitudinal section through, **5:**349

Brain stem *(Continued)*
  in micturition control, **5:**44-45
  premotor mechanisms of, **3:**558
  and reflex, gastric, vagovagal, **2:**32
Brain tissue transplantation, in Parkinson's disease, **5:**671
Brain uptake index (BUI), **1:**721-22
Brainwashing. *see* Persuasive communication
Branchial arches, formation, **4:**658
Brandtzaeg, **5:**138
BrdU
  cell cycling and, **3:**624
  defined, **3:**621
Bread
  in "basic four" food groups plan, **3:**655-56
  nutrient composition of, **5:**495-96
  white, **2:**101
Breakfast foods, **3:**40
Breast
  alpha dispersion in, **3:**11, 12
  alveolar buds, **4:**604
  anatomy of, **6:**564-65
  development, **4:**603-4
  ducts, **4:**604
  and hormones, **2:**54, 57-58
  nutritional regulation of lipoprotein lipase in, **3:**587
  pineal gland and, **2:**495
  pituitary gland and, **2:**11-29, 143, 834
  preparation for lactation, **4:**604
  prolactin and, **6:**15
  and puberty, **2:**4-55
  and steroid receptors, **2:**57-58
Breast feeding. *see also* Milk, human
  effects on intestinal microflora, **1:**319
  incidence and advantages, **4:**609
  milk-ejection reflex, **4:**607-8
  neonate infections by, **4:**457
  protection against childhood lymphoma, **4:**342
Breast neoplasms, **2:**53-65, 116
  amplification of *neu* oncogene and, **5:**548-49
  androgen and, **2:**56-57
  antihormones and, **2:**54
  aromatase and, **2:**57
  biology, **2:**53-65
  cell heterogeneity, **2:**60-61
  cell lines, **2:**64
  chemically induced, **2:**55
  chromosome of abnormality, **2:**64
  dietary fat and, **2:**54-55
  dietary influence, **2:**55
  dietary lipids and, **5:**508-9
  endocrine-related, **2:**53-57
  environment-related, **2:**54-55

Breast neoplasms (*Continued*)
estrogen and, **2:**56–57, **7:**266, 744
blocking therapy, **2:**249
receptor-positive tumor, **2:**59
etiocholanolone and, **2:**56
genetic influence, **2:**55
genetic instability and, **2:**61
growth factors and, **2:**61–62
hormone therapy and, **2:**58
16 alpha-hydroxyestrone and, **2:**57
hypercalcemia of malignancy, **4:**278–79
lymph node status and, **2:**59
major disease of women, **2:**53
metastasis, **2:**59–60
neu oncogene amplification and, **5:**548–49
oral contraceptives and, **5:**605–6
progesterone receptor-positive tumor and, **2:**59
psychotherapy and, **5:**374
radiation, **6:**415
steroids and, **2:**56–57
Breath-hydrogen test, **4:**614, 620
Breathing
laryngeal functions in, **4:**660
mechanics of, **6:**600–602
and sleep, **7:**73–74
"Breathing" (enzymology), **3:**412
Breeding, **4:**28. *see also* Inbreeding
Brefeldin A, **4:**11–12
Breslow's microstaging, **4:**890–91
Bridge, cytoplasmic, **2:**273
and cell division, **2:**273
and signaling, chemical, **2:**273
Bright disease. *see* Glomerulonephritis
Brightness
defined, **2:**575
and luminance, **2:**579–80
Broca aphasia. *see* Aphasia, Broca
Broca-reference weight
defined, **3:**683
in West Germany, **3:**687
Broca's area. *see* Frontal lobe
Brochothrix thermosphacta, **3:**668
Brodman, **5:**307, 308, 309
Bromdeoxyuridine. *see* BrdU
Bromide, methyl, **2:**345
and fumigation of stored grain, **2:**345
and seizure, **2:**345
1-(p-Bromobenzoyl)-piperazine-2,3-dicarboxylic acid, **2:**524
Bromocriptine, **2:**23, 28–29, **4:**606
Bromodeoxyuridine, **5:**250

Bronchi
airway obstruction of, **6:**360–61
anatomy of, **6:**591–92, 592
diameter of, control of, **6:**362
smooth muscle of, pathologic stimuli of, **6:**362
structure of, **6:**359–60
Bronchioles. *see* Bronchi
Bronchiolitis, in infants and toddlers, **4:**461, **5:**706
Bronchitis, in infants and toddlers, **4:**460–61
Bronchoconstriction
in gas poisoning, **3:**387
laryngeal reflexes in, **4:**649
Bronchodilators, **1:**195
Bronchogenic carcinoma. *see* Carcinoma, bronchogenic
Bronchopulmonary dysplasia, **5:**705
Bronchus-associated lymphoreticular tissues, **5:**139
Broth dilution, in antimicrobial susceptibility testing, **1:**559
Brown, Jim, **3:**631
"Brown atrophy," **3:**901
Brown fat, **1:**61
definition of, **1:**57
features distinguishing brown and white, **1:**62
function of, alterations affecting, **1:**61
functions of, **3:**576
Brown-Sequard syndrome, **7:**231
Bruce effect, **5:**529
Bruch's membrane, **3:**538–39
Brugia malayi, **2:**388
Brush border. *see* Microvilli
Bruxism, **2:**798
Buffer strength, **5:**871
Buffers
bicarbonate, **5:**875–76
defined, **5:**871
intracellular acid load and, **5:**873
Bufuralol, **2:**746, 747
pharmokinetic properties of, **7:**111
polymorphism, **2:**746
Bulbourethral glands, **6:**561–62
Bulimia
characterization, **4:**997
defined, **3:**17
Bulimia nervosa, **2:**836
behavioral aspects of, **3:**592
course of, **3:**247
defined, **3:**243
as eating disorder, **1:**352
on restrained eating classification, **3:**689
Bumetanide, **3:**98
Bundle bone, **2:**777, 825
alpha-Bungarotoxin, **2:**522

Bundle bone *(Continued)*
    of His, **2:**183
Bupivacaine (marcaine), **2:**784
Bupropion, **1:**309
Burkitt, D., **3:**773
Burkitt-type acute lymphatic leukemia, **2:**452
Burkitt's lymphoma, **2:**64, 255, 450, 452, **4:**848
    characterization, **4:**848
    defined, **3:**770
    EBNA-1 and, **3:**470
    Epstein-Barr virus and, **3:**464
    in immunosuppressed patients, **4:**320
    oncogenic factors, **4:**343
    tumors in, **2:**255
Burns. *see* Thermal injuries
Burr, **3:**576
Bursa-dependent lymphocytes. *see* B lymphocytes
Bursae
    knee joint, **1:**376
    near synovial joints, **1:**373
Burst cells, in saccades, **3:**558
Buschke-Löwenstein tumors. *see* Condylomata, giant
Buserelin, **2:**300
Buthionine sulfoximine, **3:**917
Butler, E., **3:**778
Butler's field theory, **2:**790
Buturphanol, **2:**786
Butylated hydroxyanisole, **4:**223
Butylated hydroxytoluene
    activity of, **3:**681
    food analysis, **4:**223
Butyrlcholinesterase, **1:**441, **3:**102
Byssinosis, **7:**556

C

C. parvum, immunotherapy with, **4:**347
C-4-S. *see* Chondroitin-sulfate
C-jun/AP-1, defined, **3:**701
C protein. *see* Proteins, C
C-reactive protein, **1:**31
CA. *see* Cell-adhesion molecules
Ca^+, voltage-sensitive channels, **5:**421
Ca^{2+}
    angiotensin II and, **5:**730
    cholecystokinin and, **5:**718
    cholecystokinin receptors and, **5:**731
    G proteins and, **3:**946–47
    gangliosides and, **3:**733
    gonadoliberin and, **5:**732
    messenger system, **3:**331–33
    muscle contraction and, **5:**206, 225–26

Ca^{2+} *(Continued)*
    nerve growth factor and, **5:**336
    in olfactory receptor cells, **5:**526
    parathyroid hormone and, **3:**324
    in parenchyma cell walls, **3:**37
    in peptide binding, **5:**717
    pump, **4:**562–63
$C3_a$, inflammatory role, **4:**471
CAAT consensus sequence, defined, **5:**205
Cabbage fiber, **3:**44
Cabergoline, **2:**28
Cachectin. *see* Tumor necrosis factor
Cadmium, **2:**111
Caenorhabditis elegans, **2:**245
    and cell death studies, **2:**245
    longevity of, **3:**895
    muscles of, **5:**164
    mutations in, **5:**169
Caffeine, **2:**67–80
    addiction, **2:**78
    amount
        in beverages, **2:**69
        in drugs, **2:**69
        in foods, **2:**69
    and anorexia, **2:**79
    and anxiety, **2:**78–79
    assay techniques, **2:**71
    and athletes, **2:**73
    and bacteria, **2:**76
    biochemistry, **2:**69–71
    birth-related risks, **2:**75–76
    and birth-related risks, **2:**75–76
    and birth weight, **2:**76
    and blood pressure, **2:**70, 75
    in breast milk, **2:**75
    and cancer, **2:**76–77
    and cardiac output, **2:**70
    and cardiovascular risk, **2:**77–78
    chemistry, **2:**69
    and cholesterol, **2:**70
    clearance, **2:**71
    and cognitive performance, **2:**73–75
    and congenital malformation, **2:**76
    and coordination, muscular, impaired, **2:**73
    and depression, **2:**79
    and diazepam, **2:**79
    distribution in the body, **2:**69–70
    in double blind design, **2:**72
    effect of, **2:**70, 75
    ergogenic aid, **2:**73
    gastrointestinal risks, **2:**77
    and heart rate, **2:**70

Caffeine *(Continued)*
    impairment of accurate timing, **2:**73
    and increased mental acuity, **2:**75
    and increased mental arousal, **2:**73
    individual sensitivity to, **2:**70
    and lithium therapy, **2:**79
    and mental health, **2:**78–79
    and mutation in bacteria, **2:**76
    and nervous system, **2:**70
    and osteoporosis, **2:**77
    and panic disorders, **2:**78–79
    and performance, **2:**73–75
    and physical health, **2:**75–78
    and pregnancy, **2:**75–76
    problems, enumerated, **2:**67
    psychoactive, **2:**68
    reliability of self-reports, **2:**72
    research methodology, **2:**71–73
    and sleep disturbance, **2:**77
    and stomach acid, **2:**77
    suspected as mutagen, **2:**76
    tolerance to, **2:**70
    toxicity, **2:**71
    as 1,3,7-trimethylxanthine, **2:**69
    use of, **2:**68
    variables
        biological, **2:**71–72
        psychological, **2:**72
    volume of cup of coffee, **2:**71
    withdrawal symptoms, **2:**78
    xanthine, methylated, **2:**67
Calcarine sulcus, **2:**684
Calcification, physiologic, **1:**765, **3:**481
Calcisomes, in Ca^{2+} messenger system, **3:**331–33
Calcitonin, **2:**27, 35, 36, 38, 52
    in bone and calcium metabolism, **1:**800–801
    bone cell function and, **1:**765
    estrogen and, **3:**483
    fetal, **3:**346
    function of, **6:**89
    hypercalcemia of malignancy treatment, **4:**280–81
    hypersecretion of, **1:**808
    parathyroid hormone and, **5:**664
    perinatal, **3:**349
    regulation of bone and calcium metabolism, **1:**803
    regulation of bone resorption, **1:**762
    role in bone and calcium metabolism, **1:**800–801, 803
    synthesis and function, **7:**490–91
Calcitonin-gene related peptide, **2:**35, 36, 523, **6:**89
    in autonomic neurons, **1:**541
    functions of, **5:**424
    in gut-associated lymphoid tissue, **5:**381

Calcitonin-gene related peptide *(Continued)*
    as neuroregulator, **2:**365
    sensory neuron, visceral innervation and, **7:**795–96
    substance P and, **5:**381
Calcitonin receptors. *see* Receptors, calcitonin
Calcitriol
    ectopic, **3:**334
    estrogen replacement therapy and, **3:**483
    fetal, **3:**346
    fetal calcium absorption and, **3:**346
    interaction with parathyroid hormone, **5:**664
    perinatal, **3:**349
    synthesis, **3:**327, **5:**662
Calcium, **2:**46, 50, 89–99, 169, 175, 181–82, 308–16, 329, 330, 346
    absorption hypothesis, **4:**616
    in acrosome reaction pathways, **3:**604
    binding by alpha-actinin, **5:**223
    biochemistry, **2:**89–99
    body, as measure of bone mass, **1:**775–76
    and calmodulin kinase, **2:**96
    in central nervous system lesions, **7:**223
    colon carcinogenesis and, **3:**774
    concentration
        in bone, **1:**811–12
        intracellular, control of, **2:**95–98
    dietary fiber and, **3:**43
    discovery of, **2:**90
    distribution, in red cells, **6:**533–34
    elevated, effects on red cells, **6:**535–36
    extracellular, in space, **2:**91
    fetal, **3:**346
    function of, **2:**90, **5:**48–49
    in glycogen cascade, **3:**925–26
    heart contractility and, **3:**509
    homeostasis, and bone remodeling, **1:**813–14
    hormonal regulatory factors in bone metabolism, **1:**800–805
    and hormones, **2:**91
    in hypercalcemia of malignancy, **4:**279, 281
    indicator, fluorescent probes and, **5:**576–77
    intake during lactation, **4:**605
    ligation, **2:**90–91
    loss, space travel and, **7:**146–47
    measurements
        comparisons, **7:**542
        dual-photon absorptiometry for, **7:**541–42
        neutron activation analysis for total body, **7:**538
        total body, **7:**537–43
    metabolism
        effect of kidney dialysis on, **1:**409
        ion pumps, **4:**568–69

Calcium
  metabolism *(Continued)*
    perinatal, **3:**349
  in metaphase spindles, **5:**70
  in microtubule regulation, **5:**17
  muscle contraction and, **5:**233–36
  myocardial, **2:**125, 127, 129, 135
  in old rat cells, **5:**507
  parathyroid hormone regulation, **5:**660–61, 707–8
    of bone, **5:**661
    intestinal metabolism, **5:**661–62
    of kidney, **5:**661
  permeability, in red cells, **6:**533–34
  in phototransduction, **6:**619
  in platelet receptors, **6:**56
  pump, **2:**284–86
    and calmodulin, **2:**96
    function, **2:**284–86
    muscle relaxation and, **5:**236–37
    in muscle reticulum, sarcoplasmic, **2:**285
    of plasma membrane, **2:**96–97
    in red cells, **6:**534–35
  pump proteins, amino acid sequence, **4:**561
  red cell, **6:**533–36
  regulation
    in bone, **1:**774
    calcitonin and, **7:**491
    secretion of parathyroid hormone, **5:**662–63
  secretion, in adrenal medulla, **1:**651–53
  as signal transducer, **2:**90–92
  in sperm capacitation, **3:**602
  in sperm hyperactivation, **3:**604
  states of, **6:**533
  transfer to conceptus, **4:**26
  transport, **2:**310–11
    ATP-dependent, **2:**96
    in red cells, **6:**533–35
    requirements, **4:**560
    sarcolemma and, **5:**237
    sarcoplasmic reticulum and, **5:**237
    vitamin D and, **7:**867–68
  troponin and, **2:**172, **5:**175–76
  in vasomotor tone of lungs, **6:**347–48
Calcium, dietary, national recommendations on, **3:**27
Calcium-activated neutral proteases. *see* Calpain
Calcium-activated phospholipid-dependent kinase. *see* Protein kinase C
Calcium adenosine triphosphatase, **2:**96
Calcium antagonists. *see* Calcium channel blockers
Calcium-binding proteins, **2:**92–95, **3:**184, **4:**962
Calcium-calmodulin-dependent protein kinase. *see* Protein kinase C Type II

Calcium channel blockers, **2:**81–88, 96, 172, 176
  action of, **2:**82
  benzothiazepines, **2:**84–85
  binding site, **2:**83
  chemistry, **2:**83–85
  in Chinese medicine, **2:**81
  clinical relevance, **2:**87–88
  clinical uses, **2:**87
  defined, **2:**81
  dependence, **2:**86
  depression of heart pumping action, **2:**81
  dihydropyridine, **2:**84–85, 172
  in dilation of blood vessels of heart, **2:**81
  diltiazems, **2:**85
  for hypertension, **4:**294
  inorganic, **2:**83–84
  isradipine as, **2:**172
  list of compounds used as, **2:**84
  nifedipine as, **2:**172
  nitrendipine as, **2:**172
  organic, **2:**83–84
  phenylalkylamines, **2:**84–85
  piperazines as, **2:**84, 85
  stroke treatment with, **7:**305–6
  synthetic, **2:**81–88
  as therapeutic drug, **2:**81
  tissue selectivity, **2:**86
  verapamil as, **2:**81, 84–85
Calcium channels, **2:**81–83, 95–96, 172, 522
  cellular signal transduction and, **7:**19
  electrical activity, **2:**81–83
  function, **2:**95
  heterogeneity, **2:**83
  location, **2:**95
  in membrane, **2:**310
  role in neurotransmitter release, **7:**385
  types, **2:**83, 95–96
Calcium-mediated signaling, **6:**533
Calcium metabolism disorders, in children, **3:**353
Calcium phosphates, **2:**212
Calcium-phospholipid dependent protein kinase. *see* Protein kinase C
Calcospherite, **2:**775
Calculus, dental, **2:**805
California Psychological Inventory, **1:**610
Callosity, ischial, **2:**595
Calmidazolium, **2:**317
Calmodulin, **2:**42, 44, 46–48, 89, 92, 93, 96, 269, 271, 302
  as Ca^{2+} receptor protein, **3:**333
  cell receptors and, **6:**515
  defined, **2:**41
  in microtubule regulation, **5:**17

Calmodulin (*Continued*)
  phosphorylase kinase and, **3:**925
  Sertoli cell function and, **7:**426
  smooth muscle contraction and, **5:**234
Caloric expenditures, in work, **3:**507–8
Caloric intake
  cumulative, **3:**687
  in pregnancy, **3:**31
  psychogenetic model of, **3:**685
  space travel and, **7:**144
Caloric reduction, suggestions for, **3:**657
Caloric restriction
  immunity and, **5:**506–7
  and longevity, **4:**789
  longevity and, **3:**894–95
  protooncogenes and, **5:**515
  T-cells and, **5:**514–15
Caloric value. *see* Heat of combustion
Calories
  defined, **3:**243, **5:**471, 503
  popular misconceptions of, **3:**245
Calpain, **2:**316, 317, **5:**17
Calves (animals), twin, **3:**309
Campylobacter bowel infections, **4:**462
Campylobacter jejuni, **3:**665, **4:**450
Canal of Schlemm, **3:**537–38, 539
Cancellous bone. *see under* Bone and bones
Cancer. *see* Neoplasms; specific types of neoplasms; terms
    beginning with Tumor
Cancer genes. *see* Oncogenes
Candida albicans, **4:**458
Canine distemper vaccine, **5:**676
Canine X-linked muscular dystrophy, **5:**203
Cannabidiol, **4:**905–6
  alcohol and, **3:**212
  as anticonvulsant, **5:**444
  therapeutic use, **4:**910
Cannabinoids. *see* Marijuana and cannabinoids
Cannabinol, **4:**905–6
Cannabis. *see* Marijuana and cannabinoids
Cannabis delusional disorder. *see* Psychoses, substance-
    induced
Cannabis sativa, **5:**444
Cannibalism, hog, **7:**648
Cannon, Walter B., **3:**377
5' Cap, defined, **5:**205
Capacitance, electrical. *see* Electrical capacitance
Capacitation, sperm. *see* Sperm capacitation
Capacity
  analogies of, **5:**154
  defined, **5:**149
  performance and, **5:**150–51

Capillaria philippinensis, **2:**387
Capillaries, **2:**183–84
  anatomy, **5:**32–36
  carbonic anhydrase and, **5:**876
  defined, **3:**755
  fluid/solute exchange in, **5:**31–42
  in gastric blood flow regulation, **3:**760
  intestinal, **4:**549–50
  oxygen consumption by, **5:**622
Capillaries, lymphatic *see* Lymphatic system
Capillarization, exercise training and, **3:**520
Capillary absorptive pressure
  defined, **5:**31, 36
  edema and, **5:**40
  net capillary filtration, **5:**31
Capillary columns (gas chromatography), **3:**204
Capillary filtration pressure
  calculation of, **5:**37
  defined, **5:**31
Capillary permeability
  in psoriasis, **3:**451
  for selected tissues, **5:**38
Capillary pressure
  defined, **5:**36
  edema and, **5:**39–40
  histamine and, **5:**41
Caprine arthritis-encephalitis virus, **5:**677
Capsid, **4:**187
Capsomeres, **4:**187
Captopril, **2:**175–76
  angiotensin-converting enzyme and, **3:**396
Carabelli's trait, **2:**789, 792, 793
Carassius auratus. *see* Goldfish
Carbachol, **2:**493
Carbamate pesticides, **7:**560–61
Carbamezapine, **2:**843
Carbamoylphosphate, **5:**879
Carbanion, **2:**528–29
Carbinolamine, **2:**534
Carbocaine (mepivacaine), **2:**784
Carbocyanines, **5:**577
Carbohydrate hydrolases, ganglioside degradation and,
    3:732
Carbohydrate metabolism disorders
  in children, **3:**353
  ganglioside degradation and, **3:**731–32
Carbohydrates
  in antibodies, **3:**802, 805, 808–9
  and atherosclerosis, **2:**107
  and cardiovascular disease, **2:**107
  and caries. *see* Caries, dental
  complex carbohydrates, **2:**101, **3:**649

Carbohydrates *(Continued)*
  composition in foods, **2:**102-4
  consumer characteristics, **2:**105-6
  consumption pattern, **2:**101-2
  craving for, among dieters, **3:**689
  deficiency, in ketosis, **3:**575
  defined, **3:**71, **5:**471
  diet, effects of, **2:**106-7
  digestion of, **3:**79, **4:**612-13, 104
  and enzymes, **2:**104
  food analysis, **4:**217-18
  in foods, **2:**104
  ganglioside chains, **3:**728-29, 731
  and hyperactivity in children, **2:**107
  membrane, biogenesis, **4:**977
  metabolism
    oral contraceptives and, **5:**608
    pH and, **5:**821-28
    respiratory quotient of, **5:**475
  nutrition and, **2:**101-7, **5:**501
  satiating power of, **6:**729
  in sperm receptor function, **6:**510-11
  sweeteners, **7:**373-76
Carbomonoxyhemoglobin. *see* Carboxyhemoglobin
Carbon anhydride. *see* Carbon dioxide
Carbon dioxide, **2:**282, 365-66
  bicarbonate buffer pH and, **5:**876
  crop production and, **1:**143-44
  diffusion of, **5:**873
  in food preservation, **3:**668
  forms of, in blood, **6:**600
  hemoglobin transport, **4:**138
  importance of, in blood, **6:**600
  oxygen exchange and, **6:**370-74, 568, 598-99
  oxygen transport in blood, **5:**621
  pulmonary excretion of, **5:**877-78
Carbon dioxide laser vaporization, **3:**866
Carbon disulfide, **2:**346, 347
Carbon monoxide, **2:**345
  binding to hemoglobin, **6:**600
  inhaled, disposition and effect of, **7:**579
Carbon monoxide poisoning, **5:**669
Carbonic anhydrase
  in astrocytes, **1:**441
  pH_e and, **5:**876
  in saliva, **6:**694
Carbonic anhydrase inhibitors, **3:**96-97
Carbonyl phosphate, **2:**539-40
Carbonyl-terminal regions, **4:**183-84
Carboxyfluoresceine, **5:**873
Carboxyhemoglobin, **2:**345, **3:**390
Carboxyl esterases, **5:**286

Carboxyl methyltransferase, **3:**419
Carboxyl terminus, of histones, **3:**127
Carboxylates
  contact germicidal efficiency of, **7:**111
  germicidal efficiency of, **7:**111
Carboxylic acid, **3:**582
Carboxylic ara A. *see* Cyclaradine
Carboxylic ester hydrolases, **3:**50
Carboxymethyl-cellulose, **3:**37
Carboxypeptidase N, **2:**613
Carboxypeptidases, **7:**885
Carcinoembryonic antigen, **4:**341, **5:**319
Carcinogenesis, **2:**109, 116, 448-49
  cell aberration required, **2:**449
  colon, bile acids in, **1:**625
  endocrine pancreas and, **7:**573
  enzyme-altered organ foci in, **5:**318-19
  HeLa cells, **4:**108
  industrial cause, **2:**116
  initiator-promoter theory of, **7:**515
  malignant cell transformation, **4:**583-84
  multistage, oncogene amplification in, **5:**549-50
  patterns of, **7:**668
  physical carcinogenesis, **4:**344
  at point of contact, **2:**116
  promoters, in food, **3:**671, 680-81
  promoting agents, **3:**383
  transplacental, **7:**418
  tumor clonality and, **7:**692-93
Carcinogens, **2:**109
  cell transformation, **4:**583
  chemical carcinogens, **4:**343-44
  defined, **3:**383
  dietary, **3:**674-81
  effect on keratinocytes, **4:**586
  fibronectin and, **5:**905
  industrial processes as, **2:**112
  inhibitors, **3:**671, 680-81
  metals as, **7:**552-53
  pesticide residues in food and, **7:**562
  quantitative risk assessment, **3:**674
  trichothecenes as, **7:**665
Carcinogens, environmental, **2:**109-23, **4:**343-44
  genotoxic, **2:**120
  hazard identification of, **3:**388
  industrial, listed, **2:**112-13
  and leukemia, **4:**691
  mechanisms of, **2:**119-22
  nonindustrial listed, **2:**114-15
  toxicological science and, **3:**385
Carcinoid syndrome, **4:**600-601
Carcinoids, defined, **3:**769

Carcinoma. *see also* Neoplasms
  aerobic glycolysis in, **3:**940
  colonic, **3:**771
  effect on polarized epithelial cells, **6:**65–66
  endometrial, **3:**482
  gastric, **3:**772
  hepatocellular, **5:**607
  immunodiagnosis, **4:**346
  immunotherapy, **4:**346–49
  killing of cells by macrophages, **4:**858
  medullary, **7:**491
  nasopharyngeal, **3:**464–65
  of vulva, **3:**865
Carcinoma, bronchogenic, **3:**334
Carcinoma, small cell lung, **5:**547–48
Carcinoma, squamous cell, **4:**581
  c-erb B-1 oncogene amplification and, **5:**548
  of cervix
      oral contraceptives and, **5:**607
      papillomaviruses and, **3:**863
  ectopic hormone production and, **3:**335
  of genitals, papillomaviruses and, **3:**864
  of penis, papillomaviruses and, **3:**863
Cardia, def.ned, **3:**75
Cardiac beriberi. *see* Beriberi, cardiac
Cardiac contractility. *see* Myocardial contraction
Cardiac glycosides, hypokalemia and, **3:**99
Cardiac index, **2:**179
Cardiac intropism. *see* Myocardial contraction
Cardiac lesions, in carcinoid syndrome, **3:**769
Cardiac muscle. *see* Myocardium
Cardiac myosin light-chain isoforms, **5:**215
Cardiac output, **2:**186
  and caffeine, **2:**70
  control of, vascular capacitance in, **7:**759
  in dynamic exercise, **3:**518
  factors determining, **7:**759
  maximal, aerobic exercise and, **3:**508–9, 510, 512
  and stroke volume, **2:**186
  work intensity and, **3:**516
Cardiac transplantation. *see* Heart transplantation
Cardiac vesicles, **5:**162
Cardiolipins, **4:**970, **5:**884–85
Cardiology, pediatric, **5:**703–4
Cardiomyopathies, and selenium deficiency, **6:**796
Cardiotonic steroids. *see* Cardiac glycosides
Cardiovascular deconditioning syndrome, **2:**147, 149
Cardiovascular diseases
  animal models in human, **5:**678
  disease models, animal, **1:**256–57
  infectious agents and, **4:**451
  longevity and, **4:**785

Cardiovascular diseases *(Continued)*
  nitric oxide deficiency and, **3:**371, 376
  oral contraceptives and, **5:**600–604
  prevention of, **4:**788
  tobacco smoking and, **7:**517–18
Cardiovascular reflexes, laryngeal, **4:**650
Cardiovascular response, postflight, **2:**149
Cardiovascular system, **2:**167–77
  aerobic exercise and, **3:**508–10
  aging of, **3:**901
  alcohol effects, **1:**174–75
  antianginal drugs, **2:**171–72
  antiarrhythmic drugs, **2:**172–74
  anticoagulants, **2:**170–71
  antihypertensive drugs, **2:**176–77
  antiplatelet drugs, **2:**169–70
  birth control pills and, **1:**692–93
  chemistry, **2:**179–87
  cholesterol-lowering drugs, **2:**167–69
  exercise and, **3:**515–21
  fetal, at third month, **3:**611
  heart failure and, **2:**174–76
  hypothalamus role, **4:**310
  marijuana and cannabinoid effects, **4:**908
  pharmacology, **2:**167–77
  physiology, **2:**179–87
  psychophysiological disorders, **6:**336–37
  radiation effects, **6:**414
  as recipient of visceral afferent nerves, **7:**797
  serotonin effects, **6:**833
  space travel and, **7:**147–48, 151
  sudden infant death syndrome and, **7:**319
  thrombolytic drugs, **2:**170
  trichothecenes effects, **7:**663–64
Caries. *see* Dental caries
Cariogenicity, **2:**101
Carnitine esters, **2:**507
Carnitine shuttle, **3:**583
Carnivores, gut anatomy of, **3:**60
Carnivorousness, **2:**189–99
  among Eskimos, **2:**190
  by chimpanzee, **2:**194–95
  hunter-gatherer diet, protein in, **2:**191–92
  hunting hypothesis, **2:**189–91
  in prehistory, **2:**195–99
  and protein in nutrition human, **2:**193–94
beta-Carotene
  chemical structure of, **7:**846
  formation of, key steps in, **7:**847
  immunity and, **5:**512–13
Carotene, in nutrition education, **3:**25

Carotenoids
  in cancer regression, **3**:775
  food analysis, **4**:219
  as vitamin A precursors, **7**:846-47
Carrier females. *see* Heterozygote
Carrier males. *see* Nonpenetrant males
Carrier proteins
  hormone transport and, **3**:328
  in protein folding, **3**:419
Carriers, genetic. *see* Heterozygote
Cartesian philosophy, **3**:17
Cartilage, **2**:201-15, 572, 705-6
  articular, **1**:365-70, 372-73, 373, **2**:210, 213, 215
  in bone healing, **2**:202
  cell. *see* Chondrocyte
  composition, **2**:203-8
  deformation, **1**:368-69, **2**:210
  diseases affecting, selenium deficiency and, **6**:796-97
  elastic, **2**:201, 210, 212, 640
  fetal, **2**:202
  fibrous, **2**:201, 211-12
  formation, **2**:202-3
  glycosaminoglycan, **2**:206-8
  growth, **2**:202-3
  hyaline, **1**:373, **2**:201, 211, 213-15, 567, 640
  loaded, **2**:210
  macromolecules, **2**:203-4
  matrix, **1**:365-67, **2**:202-4, 206, 210, **3**:525
  maturation, **2**:202-3
  mineralization, **2**:202, 212
  and nutrition, **2**:210-11
  ossification, echondral, **2**:212
  permeability, **2**:201
  physial, **2**:214-15
  plasticity, **2**:203
  porosity, **2**:210
  proteoglycans of, **6**:259-61
  repair, **2**:212, **7**:507-8
  tissue fluid, **2**:203
  types, **2**:202, 211-12
  volume, **2**:202
  water content, **2**:210, 572
Casein, **4**:605
Casein kinase II, **5**:895-96
Castration and cell death, **2**:248
CAT scan, radionuclide. *see* Tomography, emission-
    computed
Catabolism pathways, **2**:503, 506
Catabolite repression, defined, **3**:940
Catagen, **4**:41
Catalase, **2**:655
  peroxide detoxification and, **3**:914

Catalysts, enzymes as, **3**:425
Catalytic site, enzyme. *see* Active site, enzyme
Cataracts, lactose absorption and, **4**:615
Catarrhines, evolution of, **3**:489
Catatonia, in schizophrenic disorders, **6**:771
Catechol-o-methyltransferase, **2**:14
Catecholamines, **2**:12-13, 22-25, 171, 173, 217-24, 513,
    652, 717, 831, 843. *see also* specific compounds
  and behavior, **2**:217-24
  biosynthesis of, **1**:643-45, **2**:218
  and criminality, **2**:717
  and depression, a hypothesis, **2**:843
  distribution in body, **2**:128
  effect on adrenal medulla, **1**:73, 76
  in fatty acid uptake, **3**:588
  in fetal adrenal medulla, **3**:347
  function of, **1**:641-42
  mediators of brain adoptive systems, **2**:221-23
  as neuroregulators, **2**:12-13, 217
  in pheochromocytoma, **3**:353
  thymocytes and, **5**:380
  transport, **1**:648
  in vitro studies of, **5**:382
Cation exchange capacity
  defined, **3**:35
  of dietary fiber, **3**:42-43
Cation pumps, membrane transport, **4**:974
Cats
  feline leukemia virus, **4**:342
  and toxoplasmosis, **6**:277-78
Cattle
  dicoumarol intoxication of, **5**:266-67
  seasonal intoxication of, **5**:263
Caudal pontine medial tegmentum, **4**:714-15
Caudal pontine reticular formation, **4**:712-14
Caudate nucleus, **2**:359
Caudate-putamen atrophy, in Huntington's disease, **4**:271
Cause of death
  immediate, **5**:99
  underlying, **5**:99
Cavernous sinus, **4**:984
Cavity, oral. *see* Mouth
CCAAT box, **4**:147, 256
5'-CCGG-3'
  inactivation of, **3**:157
  5-methyldeoxycytidine and, **3**:153
CCK$_B$ receptors. *see* Receptors, CCK$_B$
CD2, as T cell receptor, **7**:398
CD3, **4**:832-33, **7**:399
CD4
  T cell expression of, **7**:480-81
  as T cell receptor, **7**:399

CD4-positive T cells, **4**:361–62, 835
    AIDS and, **4**:345
    leishmaniasis and, **4**:388
    lymphocyte count, in HIV infection, **1**:153
CD5 beta cells, **1**:566
CD7, T cell expression of, **7**:480–81
CD8, **4**:361–62, 835
    T cell expression of, **7**:480–81
    as T cell receptor, **7**:399
CDC2 protein kinase. *see* Protein P34CDC2
CDP-diacylglycerol, **5**:886
CDR 3, H chain, **4**:316
Cebidae, Epstein-Barr virus in, **3**:464
Cecum, anatomy, **3**:65
Celery, psoralens in, **5**:266
Celiac axis, **3**:756
Cell adhesion, **2**:270
    alterations in tumor cell adhesion, **4**:1014
    laminin effects, **4**:626–28
    mechanisms during implantation, **4**:427–28
    molecule, **2**:330
    tumor cells, **4**:1012–13
Cell adhesion molecules
    functional properties, **4**:960–61
    melanoma-associated, **4**:960
Cell aggregation, during bone development, **1**:787
Cell aging. *see* Cell survival
Cell assembly abnormalities. *see* Cytoarchitecture
Cell biology, molecular, **2**:253–66
Cell body. *see* Perikaryon
Cell communication, **2**:273–78, 322–23, 330
    by bridges, cytoplasmic, **2**:273
    by gap junctions, **2**:274–78
    in hemopoietic system, **4**:172–73
    receptors and, **7**:15–21
    sensitivity of signaling, **2**:328–29
    signal transduction and messengers, **7**:20
    signals, **2**:299, 321–31
    by chemical signal, **2**:273
    specificity of, **7**:15–16
Cell compartmentation
    energy metabolism and, **3**:379
    flow cytometry of, **3**:624
Cell cycle, **2**:243–44, 253, 254, 255–57
    defined, **5**:65
    events of, **5**:71–72
    flow cytometry and, **3**:623–24
    mammalian, **5**:67
    regulation of, **7**:464
Cell death. *see* Cell survival
Cell differentiation. *see also* Histodifferentiation
    bone, growth factor signals, **1**:787–89

Cell differentiation (*Continued*)
    effect of transforming growth factor on, **7**:637
    extracellular matrix and, **3**:525
    gangliosides and, **3**:734
    isoenzyme analysis, **4**:578
    laminin effects, **4**:626–28
    in muscle, **5**:165
    sphingolipids and, **7**:183–84
Cell division, **2**:242–44, 243, 260–64, 253–56
    cell cleavage, defined, **5**:71
    defined, **5**:56
    DNA synthesis, **2**:263
    exogenous gangliosides and, **3**:741
    extracellular matrix and, **3**:525
    gangliosides and, **3**:734–35
    hormone regulation of, **3**:326
    parameters, listed, **2**:256
    pH_i rise and, **5**:874
Cell growth. *see* Cell division
Cell heterogeneity leukemias, **4**:693
Cell junction, **2**:267–78
    adhesive, **2**:267–70
    anchoring, **2**:268–70
    gap-. *see* Gap junction
Cell kinetics. *see* Cell cycle
Cell line
    characteristics peculiar to HeLa cells, **4**:107–11
    generation of, **3**:157
    gliogenesis and, **1**:434–35
    hamster, E2A promoter-CAT gene construct in, **3**:159
    HeLa, variants, **4**:110–11
    lymphoblastoid. *see* Lymphoblastoid cell lines
    mammalian. *see* Mammalian cell lines
    transforms, **2**:264–66
Cell-mediated cytotoxicity
    antibody-dependent, **5**:292
    basic model of, **5**:291
    killer cells and, **5**:291–92
    role of major histocompatibility complex in, **5**:292
Cell-mediated immunity
    and aging immune system, **6**:75–76
    effect of trichothecenes on, **7**:664–65
    endometriosis and, **3**:363
    to *Entamoeba histolytica*, **1**:228
    macrophages and, **4**:859, 861
    to malaria, **4**:879
    retinoid role, **4**:824–26
Cell membrane, **2**:130–32, 235, 236–39
    channels, taste cells and, **7**:532–33
    complex, hair, **4**:43
    dielectric properties of, **3**:10
    drug transport through, **5**:839

Cell membrane *(Continued)*
    effect of trichothecenes on, **7:**667
    exogenous gangliosides and, **3:**740
    fatty acids and, **3:**582
    fluid mosaic model for, **6:**542
    hormone receptor interaction at, **3:**330
    linoleic acid and, **3:**577
    phospholipid transfer in, **5:**889
    photoelectron microscopy and, **5:**905-7
    pituitary, **2:**300
    potential
        fluorescent indicators of, **5:**577
        pH$_i$ and, **5:**875
    protein translocation and, **6:**249-50
    receptor functions of, **5:**514
    receptors, for endogenous ligands, **6:**516-19
    structure, zinc and, **7:**883
    toxins damaging, **5:**3-4
    transport, **2:**238-39, 279-89
Cell membrane permeability
    determinants of, **5:**839
    pH and, **5:**822-23
    proton translocation and, **5:**873
Cell movement, **2:**241-43
    during bone development, **1:**786-87
    collagen and, **3:**530
    extracellular matrix and, **3:**524-25
    fibronectin and, **3:**526
    laminin effects, **3:**528, **4:**626-28
Cell nucleus, **2:**291-97
    DNA, **2:**291-95
    envelope, **2:**292-97
    lamina, **2:**292, 295-96
    microtubule, **2:**291-92
Cell populations
    cross-partition methed, **1:**362
    separation in aqueous two-phase systems, **1:**362
Cell receptors. *see* Receptors, endogenous substances
Cell senescence. *see* Cell survival
Cell surface
    signal-receiving systems at, **7:**16
    signal transduction at, **7:**15-21. *see also* Signal transduction
Cell surface glycoprotein. *see* T lymphocyte
Cell survival, **2:**249-51
    effect of radiation on, **6:**445-47
    laminin effects, **4:**626-28
    in normal aging, **3:**900
    survival curve for radiated mammalian, **6:**445
Cell transformation
    assays, for human papillomaviruses, **5:**638-39
    defined, **3:**701

Cell transformation *(Continued)*
    by human *fos* protein, **3:**702, 704-5
    transformed cells, **4:**1009
Cell viability. *see* Cell survival
Cell volume
    pH$_i$ and, **5:**875
    regulation, **4:**569-70
Cell wall
    bacterial
        antibiotic sensitivity of, **1:**276
        biosynthesis, antibiotic inhibitors of, **1:**271-85, 291
        gram stain response of, **1:**274-76
        structure of, **1:**274-76
    of dicotyledonous plants, **3:**37-39
    of monocotyledonous plants, **3:**39-40
    uronic acid in, **3:**42
Cells
    aberration, **2:**235-44
    aberration, acquired, **2:**449
    acinar, **6:**690
    aerobic, and radiation dose, **6:**447
    beta cell activation, immunology, **1:**563-72
    bipolar, of retina, **7:**837
    centriole, **2:**291, 292
    centrosome, **2:**291-92
    chromosomes in, **2:**292-95, 419
    condensation, during bone development, **1:**787
    control-cells, immunological, **2:**248-49
    dehydration of, as signal for thirst, **7:**474
    DNA, **2:**291-95
    doubling, Hayflick limit and, **3:**893
    electromagnetic radiation and, **6:**445-47
    enzyme control of chemical reactions, **3:**423-38
    epidermal, radiation effect, **6:**464
    ganglion, retinal, **7:**803-5, 837-38
    glial, of retina, **7:**838
    heat effects, **7:**450
    herpesviruses-infected, **4:**192-93
    heterogeneity in breast cancer, **2:**60
    hormone effects on, **3:**330
    in human development, **2:**245-52
        in adult tissues, **2:**247-48
        in embryo, **2:**246-47
        regulation, **2:**245-52
    hybrid, in studying tumor suppressor genes, **7:**708
    maintenance, **2:**235
    model of stress response, **7:**465
    morphology, and laminin effects, **4:**626-28
    mRNA, **2:**291, 294
    myocardial, **2:**126-32
    nerve, types of communication between, **7:**225
    nuclear envelope, **2:**291-97

Cells (*Continued*)
  organization, **2:**291-92
  P cells, and vision, **7:**804
  pH regulation, **4:**569
  physiology, and transforming growth factor, **7:**637-38
  postsynaptic, **7:**386-87
  proliferation, **7:**636
  pyramidal, **2:**677.686
  radiation dose rate and, **6:**446
  Renshaw, and spinal cord control, **7:**217
  specialization, metabolic, **2:**241
  sphingolipids, **7:**180-89
  staining of, microiontophoresis and, **5:**7-8
  suicide, **2:**245
Cells, artificial. *see* Artificial cells
Cells, cultured, **2:**235
  mitosis in, **5:**901
Cellular inclusions, defined, **5:**667
Cellular models. *see* Models, cellular
Cellular oncogenes. *see* Oncogenes, cellular
Cellular retinoic acid-binding protein, **7:**857
Cellular retinol-binding protein, **7:**857-58
Cellulose, **2:**101, 103
  alpha-cellulose residues, in parenchymatous plant
      tissues, **3:**37
  defined, **3:**35
  in plant cell walls, **3:**37
Cementoblast, **2:**823
Cementogenesis, **2:**817, 823-24
Cementum, **2:**769, 776, 817, 824
  classifications, **2:**776
  formation, **2:**823-24
  structure, **2:**824
  types, **2:**824
Centimorgan, defined, **3:**823
Centipedes, **7:**597
Central death rates. *see* Mortality
Central nervous system, **2:**3, 6, 31, 341-49, 875, 38-39
  acute alcohol effects, **1:**170
  aging of, **3:**900-901
  axonal regeneration in, **5:**338-39
  axonal transport in, **3:**743-45
  and behavior, **2:**6
  and caffeine, **2:**70
  cell types, **2:**343
  chronic alcohol effects, **1:**171-73
  crosstalk and, **5:**155-56
  divisions, listed, **2:**3
  exposure, **2:**342
  fluid compartments of, **1:**717
  functions, **2:**344-46, 875
  gangliosides and, **3:**739

Central nervous system (*Continued*)
  gastrointestinal tract and, **7:**286
  genesis of, **3:**286
  infections, effect of quinolones on, **6:**410
  integration, biological models of, **7:**223-25
  kinins and, **4:**595-96
  lead poisoning and, **3:**386-87
  lesions in
      effect of, **7:**223
      organic chemicals causing, **7:**549
  marijuana and cannabinoids, **4:**908
  meninges, anatomy of, **5:**350-51
  motor units and, **5:**176-77
  myelin, molecular organization of compact, **5:**278-79
  myelin proteins in, **5:**276-78
  myelinated internode of, **5:**274
  myelination, in peripheral nervous system, **5:**273-74
  nerve growth factor and, **5:**336-38
  nerve regeneration inhibition in, **5:**345
  neuronal cells, laminin effects, **4:**628-30
  pathology of, in Parkinson's disease, **5:**668-69
  persuasion routes of, **5:**808-10
  in puberty onset, **3:**355
  radiation effects, **6:**412-13
  regeneration in, **5:**342
  spinal cord dysfunction and, **7:**223-25
  spinal shock as dysfunction of, **7:**227-29
  structure, **2:**3, 343-46
  toxic agents, **2:**1-9, 346-48
  toxicology, **2:**1-9, 344-48
  transmitters. *see* Neuroregulators
  trial natriuretic factor and, **1:**475
  trichothecenes effects, **7:**663
  ventricles, anatomy of, **5:**350-51
Central nervous system depressants, alcohol and, **3:**212-13
Central nervous system stimulants. *see* Analeptics
Central precocious puberty, defined, **3:**358
Centration, and behavior, **1:**579-80
Centre d'Étude du Polymorphisme Humain, **3:**879
Centric fusions. *see* Robertsonian translocations
Centrioles, **2:**243, 244, 291, 292
  defined, **5:**11
  in prophase, **5:**68-69
Centromere, **2:**244, 444, 445, 470-73, 476, 477
  cytologically defined, **2:**473
  defined, **2:**465, **5:**65, 259
  function of, **3:**872
Centrosome, **2:**243, 291, 292
  in G$_2$ phase, **5:**68
  as microtubule organizing center, **5:**68, 70
Cephalexim, **2:**785
Cephalometry, roentgenographic, **2:**689

Ceramides
    defined, **3:**725
    formation, **4:**744
Cercopithecidae, fossils of, **3:**489
Cereals
    in "basic four" food groups plan, **3:**655–56
    dietary fiber in, **3:**39–40
    diseased, human intoxication by, **5:**263
    ergotized, **5:**263
    fermentability of, **3:**44
    handling of, **5:**268–69
    milling of, **3:**41
    moldy, toxicity of, **3:**675
    nutrient composition of, **5:**495–96
Cerebellar vermis, atrophied, **3:**711
Cerebellum, **2:**312
    anatomy, **2:**312, **5:**350
    in convulsive seizures, **6:**785
    in dysgraphia-dysarthria, **3:**218
    optokinetic reflex and, **3:**557
Cerebral amyloidosis, **1:**210, 215
Cerebral artery, posterior, **7:**301
Cerebral blood flow. *see under* Blood circulation
Cerebral cortex, **2:**335, 673–87
    anatomy, **2:**674–77, 679
    area, total, **2:**3, 675
    cell types, **2:**677
    changes in newborn, **6:**823–24
    chemical synapse of the, **7:**384
    defined, **2:**1
    divisions of, **2:**5
    dyslexia-dysgraphia and, **3:**219
    evolution, **2:**679
    functional areas, **4:**197–98
    histamine receptors in, **5:**423
    5-HT$_2$ receptors in, **5:**422
    human visual, stained section of, **7:**841
    inputs to hypothalamus, **4:**306
    maps, **5:**310–13
    network modeling of, **5:**388
    neurochemistry, **2:**685–87
    ontogeny, **2:**677–78
    parietal, **2:**10
    patterns, **2:**679
    piriform, **2:**679
    prefrontal, **2:**681–84
        disorders, **2:**681
        lesions, **2:**681–82
    quantification
        of surface area, **2:**675
        of volume, **2:**675

Cerebral cortex (*Continued*)
    and sexual response, **6:**867
    somatosensory
        lateral sulcus of, **7:**125
        pathways from receptors to, **7:**122–25
        subdivisions of, **7:**125
    spatial representational components of, **3:**220
    striate, and vision, **7:**806–13
    synapse modification in, **6:**39
    and vision processing, **7:**826–27
    visual areas of, **7:**813–14
        efferent projections from, **7:**843–44
        in visual system, **7:**840–43
Cerebral hemispheres. *see* Hemispheres
Cerebral ischemia, transient, **7:**303
Cerebral specialization, **2:**351–55, **4:**132
    and asymmetry
        anatomical, **2:**352
        functional, **2:**352–53
        in other species, **2:**354
    concepts, **2:**351, 352
    difference between
        handedness, **2:**354
        sexes, **2:**353
        songbirds, **2:**354
        species, **2:**354
    evolution, **2:**354–55
        language, gestural, **2:**355
        speech, **2:**355
    history, **2:**351–52
    testing techniques for, **2:**353
Cerebral specialization, complementary, **5:**313
Cerebrospinal fluid, **2:**674
    characteristics of, **5:**350
    chemical composition of, **1:**718
    control of, **1:**718–19
    detection of bacterial pathogens in, **1:**554–55
    of spinal cord, **7:**195–96
Cerebrotendinous xanthomatosis, **1:**626
Cerebrovascular disorders, **7:**299–308
    central nervous system transmitters in, **5:**413
    classification of syndromes, **7:**301
    death rate from, **5:**100, 101
    diagnosis of, **7:**304
    infarction and, **7:**302
    major syndromes, **7:**301–2
    mechanisms of, **7:**302–3
    nonhemorrhagic, oral contraceptives and, **5:**601
    oral contraceptives and, **5:**601–2
    risk factors for, **7:**303–4
    tempo of, **7:**303
    treatments, **7:**305–8

Cerebrovascular disorders *(Continued)*
  vascular abnormalities contributing to, **7**:303
Cerebrovascular system, **2**:357–70
  anastomosis, leptomeningeal, **2**:362–63
  anatomy, **2**:357–64
  arterial wall barrier, **2**:367
  artery, **2**:357–62
    carotid, **2**:357–59
    cerebellar, **2**:361–62
    cerebral, **2**:359–61, 367–69
  blood-brain barrier, **2**:366–67
  blood flow, regulation of, **2**:365–66
    autoregulation, **2**:365
    control, **2**:365–66
  carotoid system, **2**:357–59
  circle of Willis, **2**:362
  development stages, **2**:357
  in embryo, **2**:357
  innervation, **2**:364–65
  lesions, **2**:357
  neurotransmitters, **2**:365
  regulations, **2**:365–66
  sinuses, **2**:363–64
Cerebrum. *see* Brain
Ceruloplasmin, **1**:31, **2**:650–52, 659
Cervical cancer. *see* Cervix neoplasms
Cervical plexus, **5**:354
Cervical squamous dysplasia, **5**:607
Cervical vertebrae, **7**:767–68
  degenerative disease of, **4**:55
  spin-echo sequence, **4**:869–70
Cervicitis, **6**:899
Cervix, anatomy of, **6**:564
Cervix neoplasms
  cervical intraepithelial neoplasia
    lesions of, **3**:861
    papillomaviruses and, **3**:863, 864
    treatment of, **3**:866
  E6 HPV-16 proteins in, **5**:639–40
  E7 HPV-16 proteins in, **5**:640
  incidence of, **3**:859, 860
  integration and transcription in HPV-16 in, **5**:639
  oral contraceptives and, **5**:606–7
  papillomaviruses and, **3**:863
  radiation therapy, **6**:416
Cesarian section, **2**:596
Cestoda, **2**:389
5'-CG-3' dinucleotide combinations, **3**:155
5'-CG-3' dinucleotide sequences, **3**:152
5'-CG-3' nucleotide combinations
  DNA methylation and, **3**:160
  spreading DNA methylation and, **3**:159

5'-CG-3' nucleotide combinations *(Continued)*
  as "traffic lights," **3**:161
Chagas' disease. *see* Trypanosomiasis, American
Chagoma, **7**:685
Chan, **3**:213
Chancroid, **4**:450, **6**:409
Chang, M. C., **3**:599, 601
Charcoal, activated. *see* Activated charcoal
Charcot-Leyden crystals, **1**:189
Chartins, phosphorylation of, **5**:898–99
Chelating agents, ascorbic acid, **1**:431
Chemical carcinogenesis. *see* Carcinogens, environmental
Chemical mutagens. *see* Mutagens, chemical
Chemical reaction rates
  enzymes and, **3**:425
    multisubstrate, **3**:428–29
    with one substrate, **3**:425–28
    velocity of, **3**:427
  in transition state theory, **3**:424–25
Chemical reductants, ascorbic acid, **1**:431
Chemical speciation, defined, **3**:251
Chemical sympathectomy. *see* Sympathectomy, chemical
Chemically gated channels, **4**:973
Chemiluminescence, measurement of, **5**:580
Chemiluminescent assays, **4**:355
Chemiosmotic energy
  in ATP synthesis, **5**:56
  defined, **5**:55
Chemistry, clinical, **5**:684
Chemoreceptors, **2**:179
  pH homeostasis and, **5**:878
  ventilation control and, **5**:878
Chemotaxis, **2**:321
  defined, **2**:321
  as function of polymorphonuclear neutrophils, **5**:428–29
  inflammation and phagocyte, **5**:815
  macrophages, **4**:856
Chemotherapeutic agents, as skin radiosensitizers, **6**:464–65
Chemotherapy, **2**:371–79
  animal, **1**:294
  antagonism, **2**:371–79
  antihypertensive, **4**:293–94
  antimicrobial, guide to, **1**:552
  antiparasitic, **2**:381–94, **4**:121
  antiviral, **2**:395–402
    of AIDS, **2**:399–401
    broad-spectrum, **2**:399
    latency, **2**:401–2
    problems, **2**:401–2

Chemotherapy
  antiviral (*Continued*)
    resistance, **2:**401-2
    strategy, **2:**395-96
  in bone marrow transplantation, **1:**793
  combination-chemotherapy, **2:**371-74
  cytotoxic
    for leukemia, **4:**695-96
    lithium effects, **4:**751-52
  development of antibiotics and, **1:**271
  differential toxicity, **2:**381
  herpesviridae, **4:**193
  with hyperthermia, **4:**298, 299
  for lymphoma, **4:**849-50
  marijuana and, **5:**444
  in pediatrics, **5:**704
  personality disorders and, **5:**782
  pneumonia and, **5:**706
  potentiation, **2:**371-79
  prior to bone marrow transplantation, **1:**793
  solid tumor, lithium effects, **4:**752
  synergism, **2:**371-79
  viral, **7:**789-91
Chenodeoxycholic acids, **1:**624
Cherubism, **5:**615
Chest. *see* Thorax
Chewing. *see* Mastication
Chicken pox, **4:**189
Chief cells, parathyroid, **5:**658
Chilblains, **4:**798. *see also* Frostbite
Child. *see also* Prepuberty; Puberty
  assessment of pain in, **5:**631-32
  dreaming, **3:**194-95, 196
  hypothermia, **4:**803
  selective placement, in behavioral genetics research, **3:**837
  social relationships, **4:**545
Child abuse, **5:**711
Child development
  developmental aspects of hemispheric interactions, **4:**133-34
  growth, **4:**26-27
  visual perception, **5:**748
Child development disorders, **3:**217
Child language. *see* Language development; Language development disorders
Child nutrition
  diet for, **3:**659
  dietary fiber and, **3:**43
  in overfed communities, **3:**25
Child rearing, Japanese vs. American, **4:**703
Childbirth. *see* Labor

Chimera, of twins, **3:**312
Chimeric antibodies. *see* Antibodies, chimeric
Chimeric mice. *see under* Mice
Chimerism
  donor-specific tolerance, **4:**364-65
  lymphohematopoietic, **4:**364-65
Chimpanzees, **2:**189-90
  carnivorousness, **2:**194-95
  characteristics of, **6:**141
  in hominid evolution, **3:**495
  hunting by, **2:**195, 197
  mating, **4:**934
  omnivorous, **2:**195
  sex differences in, biocultural, **6:**847
Chinese hamster, **2:**457
Chiral chromatography, **4:**208
Chiral separations, **1:**364
Chlamydia, **4:**450
Chlamydia trachomatis, **4:**450, 458, 461
  detection of, in genital tract cultures, **1:**558
  effect of quinolones on, **6:**409
  infections
    trends in, **6:**898-99
    in women, **6:**898-99
  nongonoccoccal urethritis, **6:**899
Chlamydial cervicitis, **4:**450
Chloral hydrate, **2:**786
Chlorambucil, **2:**110-14
Chloramines, **7:**110
Chloramphenicol, **1:**290(t)
Chloramphenicol acetyltransferase, **3:**156
Chlordiazepoxide, **5:**443
Chlorella, **3:**160
Chlorhexidine, **2:**787
Chloride ion channel, **2:**316
Chlorides
  in cystic fibrosis diagnosis, **5:**706
  function, in body, **5:**49-50
  transport, in salivary gland cells, **6:**692
Chlornaphazine, **2:**110-14
Chloroform, **2:**111
Chloroimipramine, **2:**13
Chloromethyl methyl ether, **2:**110, 112
Chlorophyll, **6:**119
Chloroplast, **2:**238
Chloroquine, **2:**383-84, 391-92
  antimalarial, **4:**884
  in combination, **2:**384, 392
  in malaria, **2:**383
  resistance, **2:**392
Chlorosis. *see* Anemia, hypochromic
Chlorothiazide, **3:**97

Chlorpromazine, **2:**217, **4:**272
Choice behavior, in food acceptance, **3:**639–40
Cholangiopancreatography, endoscopic retrograde, **3:**5
Cholanoic acid, **1:**623–24
Cholecalciferols, **1:**765, **3:**323
    analogs of, **7:**863
    catabolism of, **7:**863
    classification of genes regulated by, **7:**865
    excretion of, **7:**863
    formation, regulation of, **7:**863
    hormonal action of, **7:**864–68
    liver metabolism of, **7:**862
    metabolic transformations of, **7:**864
    metabolism of, **7:**861
    model for mechanism of action of, **7:**866
    nonclassical target systems for, **7:**868
    receptors, tissue distribution of, **7:**867
    renal metabolism of, **7:**862
    skin synthesis of, **7:**861–62
    storage of, **7:**862
    transport of, **7:**862
Cholecystokinin, **2:**27, 33, 36, 338, 686
    antagonists for, **5:**723
    biliary tract diseases and, **5:**723
    in control of meal size and satiety, **1:**350
    functions of, **5:**730–31, **6:**89
    gastrin and, **5:**717–18
    in infants, **3:**51
    pancreatic secretion and, **3:**78
    satiety and, **4:**248
Cholelithiasis
    cholesterol gallstones, **1:**626
    in ultrasonography, **3:**7
Cholera, "pancreatic," symptoms of, **5:**722
Cholera B subunit, GM1 ganglioside and, **3:**741
Cholera infantum. *see* Diarrhea, "summer"
Cholera toxin
    in ADP ribosylation, **4:**14, 23
    gangliosides and, **3:**734
    intracellular activity of, **5:**2–3
Cholesterol, **2:**167–69, 411–18, 539, 546, 745
    abnormalities, and lipoprotein receptors, **6:**33
    absorption of, **2:**418, **3:**573
    and bile acid, **2:**168
    in brain, **2:**413
    in breast milk, **4:**605
    in caffeine, **2:**70
    within cells, **2:**413
    composition of myelin, **5:**275
    conversion
        into bile acids, **2:**414
        into steroid hormones, **2:**414
    customary diets and, **3:**18

Cholesterol (*Continued*)
    distribution, **2:**412–14
    drugs for lowering, **2:**167–69
    esterified, **2:**412
    in fatty streak, **2:**413
    in fetoplacental unit, **3:**345
    fish oils and, **3:**579
    formula, structural, **2:**412
    free, **2:**412
    functions of, **3:**568
    hepatic triacylglycerol lipase and, **3:**586–87
    in jejunum, **2:**418
    levels, and birth control hormones, **1:**693
    and lipoprotein, **2:**167
    in membrane, **2:**413
    membrane fluidity and, **4:**1003
    metabolism, **2:**412–13, **3:**774, **6:**32–33
    properties, **2:**412
    reverse transport, **2:**418
    in sperm plasma membrane, **3:**600, 602
    in steroid biosynthesis, **7:**258–62
    in steroid hormone synthesis, **3:**327, **7:**244
    structure, **2:**412
    synthesis, **2:**414–15, **3:**78–79, 330
    thiazide therapy and, **3:**98
    transport in reverse, **2:**418
    unesterified, **2:**662
    uptake by cells, **2:**415–17
    uptake in the arteries, **1:**453
    in whole body, **2:**412–13
Cholesterol-7-alpha-hydroxylase *P*-450, **2:**745
Cholesterol, familial hypercholesterolemia. *see* Hyper-
    cholesterolemia, familial
Cholesterol acyltransferase. *see* Acyl-coenzyme
    A-cholesterol acyltransferase
Cholesterol esters, **2:**662
    in gingiva, **4:**36
    lipoprotein lipase and, **3:**586
    storage of, **3:**583
Cholesteryl esters, **2:**412, 413, 414, 662
Cholestipol, atherosclerosis therapy with, **1:**454
Cholestyramine, **2:**168
    atherosclerosis therapy with, **1:**454
Cholic acid, structure of, **1:**624
Choline, **2:**13, 14
    astrocyte uptake of, **1:**439–40
    in phosphatidylcholine synthesis, **5:**887
Choline acetyltransferase, **2:**13, 14, 337
    depletion in Alzheimer's disease, **1:**214–15
    nerve growth factor and, **5:**337
Choline esterase, in muscle contraction, **7:**26
Cholinergic axons, **5:**339

Cholinergic dilator system, **3:**519
Cholinergic fibers, **5:**339
Cholinergic innervation, of lymphoid organs, **5:**378
Cholinergic neurons
    gangliosides and, **3:**739
    nerve growth factor and, **5:**336–37, 339
    pancreatic polypeptide and, **5:**719
Cholinergic receptors. *see* Receptors, cholinergic
Cholinesterase, **2:**347, 522
Cholinesterase inhibitors, **7:**107
    soft analog design of, **7:**108–9
    in treatment of myasthenia gravis, **5:**258
Chondroblastoma, **7:**699–700
Chondrocytes, **2:**202, 203, 207, 571–72, 646
    hypertrophy, **1:**785
Chondrodystrophy, defined, **6:**265
Chondrogenesis, **2:**202
Chondroitin sulfate, **2:**206, 207, 567, 573
    in gingiva, **4:**35–36
    structure of, **6:**258
Chondroitin sulfate proteoglycans
    biochemical and immunological characterization,
        4:958–59
    functional properties, **4:**959–60
Chondroma, **7:**699
Chondronectin, **2:**207, 212
Chondrosarcoma, **7:**700
CHOP antineoplastics, **4:**850
Chordata, elastin in, **3:**264
Choriocapillaris, **3:**538
Choriogonadotropic hormone, **1:**86
Chorion
    defined, **3:**293
    Drosophila, **3:**789
    insect, **3:**786
Chorionic gonadotropins. *see* Gonadotropins, chorionic
Chorionic sac
    defined, **3:**293
    in early abortion, **3:**294–95
    examination of, **3:**295–96
Chorionic somatotrophic hormone, **3:**338
Chorionic villi, in early abortion, **3:**294
Chorionic villi sampling, **1:**704
    defined, **3:**303
    in hemophilia diagnosis, **3:**147
Choroid, anatomy, **3:**538–40
Choroid plexus, **1:**718, **2:**359, 361, 362
Chou and Chou dose-effect analysis, **2:**378
Christmas disease. *see* Hemophilia B
Chromaffin cells
    and adrenal gland, **1:**72
    adrenal medullary, **1:**538

Chromaffin cells *(Continued)*
    secretion in, **1:**651–53
Chromaffin granules, CuZnSOD and, **3:**186
Chromaffin system
    at birth, **3:**350
    fetal, development of, **3:**346
    nerve growth factor and, **5:**334
Chromaffin vesicles, **1:**642–43
Chromagranins, in adrenal medulla, **1:**642
Chromatic aberration, as factor in visual illusions, **7:**831
Chromatids, **2:**243, **4:**952, 244, 954. *see also* Sister chromatid exchange
    in anaphase, **5:**71, 902
    defined, **5:**56
    in metaphase, **5:**70
Chromatin, **2:**291, 419–34
    condensed, **2:**433–34
    conformation of gene, **2:**431, 434
    defined, **2:**419, **3:**125, 465
    folding
        histones and, **4:**236
        into 30nm filaments, **2:**424–25
    fragments, **2:**419–30
    gene expression, **2:**431–36
    histone and, **2:**292, 426
    Histone H1 and, **3:**132
    isolation, **2:**421
    linkers, **2:**471
    in mammalian brain, **2:**431–36
    nuclease and, **2:**419, 431, 433
    nucleosome reconstitution and, **3:**130
    regulatory proteins and, **3:**131
    structure, **2:**431–36, **3:**125–26, **4:**236–37
Chromatographers, **3:**751
Chromatography
    defined, **3:**199
    in drug analysis, **3:**203–7
    retention factor, defined, **3:**199, 203
    retention time, **3:**199, 203
    reversed phase, **4:**218
Chromatography, adsorption, **3:**407, **4:**207
Chromatography, affinity, **3:**407–8, **4:**209
Chromatography, column, **3:**406–7
Chromatography, gas, **3:**749–54, 751–52
    column selection, **3:**751–52
    complex smell analysis using, **5:**760–61
    in drug analysis, **3:**203–5
    "megabore" column, **3:**752
    packed columns
        in drug analysis, **3:**203
        features of, **3:**752

Chromatography, gas (*Continued*)
  partition ratios
      defined, **3**:749–50
      sample residues and, **3**:753
  peak resolution, defined, **3**:750
  peak width, defined, **3**:750
  relative retention ratios
      in column selection, **3**:752
      defined, **3**:749
  retention time, defined, **3**:750
  stationary phase
      in drug analysis, **3**:203
      selection of, **3**:752–53
  wall-coated open tubular column, **3**:204
Chromatography, gel
  in drug analysis, **3**:205
  in enzyme isolation, **3**:408–9
Chromatography, high-performance liquid, **4**:205–13
  adsorption chromatography, **4**:207
  affinity chromatography, **4**:209
  band-broadening, **4**:206–7
  bonded phases, **4**:207–8
  capacity factor, **4**:205–6
  chiral chromatography, **4**:208
  chromatographic retention, **4**:205–6
  columns, **3**:205–6, **4**:210
  coupled to mass spectrometry, **4**:212
  derivatization, **4**:211
  detectors, **4**:210–12
  discriminating power, **4**:212
  distribution coefficient, **4**:205
  drug analysis using, **5**:837–38
  electrochemical detection, **4**:211
  flame ionization, **4**:211
  fluorescence detection, **4**:210–11
  food analysis. *see* Food analysis, HPLC
  gradient elution and automation, **4**:212–13
  infrared detection, **4**:211
  injection devices, **4**:209–10
  instrumentation, **4**:209–13
  ion-exchange chromatography, **4**:208
  kinins, **4**:593
  lipids, **4**:734–36
  mass detector, **4**:211–12
  modes, **4**:207–9
  partition chromatography, **4**:207–8
  plate number, **4**:207
  pumps, **4**:209
  refractive index detection, **4**:211
  resolution, **4**:206
  reversed phase, **4**:208
  size-exclusion chromatography, **4**:208

Chromatography, high-performance liquid (*Continued*)
  solvent reservoir, **4**:209
  stationary phase, **4**:208–9
  theory, **4**:205
  UV visible detection, **4**:210
Chromatography, high pressure liquid
  defined, **3**:671
  in drug analysis, **3**:205–6
  gas-liquid chromatography and, **3**:204
  in heterocyclic aromatic amine isolation, **3**:677
  microbiological assay and, **3**:209
Chromatography, ion-exchange, **3**:205
Chromatography, liquid, **3**:750
Chromatography, paper, **5**:837
Chromatography, partition, **4**:207–8
Chromatography, size-exclusion, **4**:208, 217
Chromatography, supercritical fluid
  in drug analysis, **3**:206–7
  gas chromatography and, **3**:750
Chromatography, thin layer, **3**:207
Chromatography, two-dimensional, **3**:172
Chromatophores, **5**:20
Chromatosomes, **2**:420–21
  structure of, **3**:127
Chromium, **2**:111–12
  biological function of, **7**:606
  deficiency, **7**:608
  oxidation states, **7**:552
  toxicity, **7**:688
Chromomycin A3, **2**:458
Chromophores
  absorption spectra of tissue, **5**:589
  light absorption in tissue by, **5**:588
Chromosome 3, **3**:177
Chromosome 6
  interferon gamma and, **3**:183–84
  map of, **7**:251
Chromosome 8, **3**:183
Chromosome 11, **1**:96–98
Chromosome 21
  band 21q22, **3**:176
  band 21q22.3, **3**:184–85
  Down's syndrome and, **3**:175, **4**:951–52
  genes assigned to, **3**:178
  mapping of, **3**:180–85
  messenger RNA and, **3**:177
Chromosome aberrations, **2**:447–50
  anomaly and consequences, pathogenic, **2**:437–44, 451–55
  defined, **3**:819
  in disease, **3**:821–22
  mitosis errors and, **5**:72

Chromosome aberrations *(Continued)*
    in primate evolution, **3:**884
    from psychedelic drugs, **5:**449
Chromosome abnormalities, **2:**441. *see also* Abnormalities
    acquired during prolonged cultivation, **4:**584
    birth defect-related, **1:**699-700
    as cause of spontaneous abortion, **1:**5-7
    and criminality, **2:**713
    in early spontaneous abortions, **3:**302, 303
    frequency of, **5:**247
    growth-disorganized embryos and, **3:**302
    and leukemia, **4:**692
    and lymphomas, **4:**845
    maternal age and, **3:**814
    oncogene amplification and, **5:**545-46
    and study of tumor clonality, **7:**690-91
    trisomy 13, **1:**699-700
    trisomy 18, **1:**699
    trisomy 21. *see* Down's syndrome
Chromosome banding, **2:**445, 448, 466, 467, **4:**253
    isochores and, **4:**254-55
    in metaphase, **3:**839
    metaphase bands, **3:**871
Chromosome deletion, **2:**439, 441, 445, **7:**708
Chromosome mapping, **2:**467-68, 482-87, 485, **3:**823-33
    of chromosome 21, **3:**180-85
    defined, **3:**869
    of Epstein-Barr virus, **3:**467, 470-71
    genetic, **2:**480-82
    in photoelectron microscopy, **5:**908
    physical, **2:**482-87
    principles of, **3:**849-50
    restriction fragment-length polymorphism probes and,
        3:138
    restriction map, **2:**482
    strategy, **2:**482
    techniques of, **3:**877-78
Chromosome probes. *see* DNA probes
Chromosome rearrangements
    defined, **5:**243
    rate of, **5:**247
Chromosome resistance to antibiotics, **1:**291-92
Chromosome walking, **2:**483, **3:**878
Chromosome X, **2:**437, 442, 443
    affective disorders related to, **1:**98
    in sex determination, **6:**283-84
Chromosome X and Y
    abnormal interchanges involving, **6:**287-88
    DNA sequence homologies between, **6:**292
    electron microscopy of, **6:**284
    meiotic pairing between, **6:**283-84
Chromosome Y, **2:**438, 442
    male specific gene on, **6:**838-39

Chromosome Y *(Continued)*
    in sex determination, **6:**283-84
Chromosomes, **2:**292-95, 465-77, **4:**252-53
    acrocentric, **2:**441
    alteration, **2:**439
    analysis, TV-based, **5:**566
    in anaphase, **5:**71
    aneuploidy, **2:**437
    arm, **2:**437, 441
    autosomal, **2:**445, **3:**135, 475, 476
    breakage, **2:**441
    and buffer, **2:**458
    centromere. *see* Centromere
    chromatin. *see* Chromatin
    crossing-over, **2:**244
    cytoplasmic, **2:**475
    defects, pregnancy failure, **4:**430
    defined, **3:**317, **5:**259, 791, 813, 869
    DNA, **2:**425, 456, 465, 470, 476, 477
    double-minute
        amplified genes on, **3:**788
        defined, **3:**785
        oncogene amplification and, **5:**545-46
    duplication, **2:**440
    dyes, fluorescent, **2:**458
    elements, functional, **2:**472-74
    end, sticky, **2:**473
    end, telomeric, **2:**484
    error, meiotic, **2:**439
    evolution rate of, **3:**886
    fiber 30nm, **2:**476
    flow cytometry, **2:**459-60, **3:**625
    flow histogram, **2:**458
    fluorescence distribution pattern of human, **5:**566
    folding, **2:**419-30
    function, **2:**472-74
    gene libraries, specific, **2:**455-64
    gene location on, **3:**796
    and genome, **2:**437, 448
    haploidy, **2:**437
    and histone, **2:**476
    homogeneously staining regions, oncogene amplifica-
        tion and, **5:**545-46
    human, listed, **2:**456
    identification, **2:**465-67
    idiotype, **2:**447
    insertion, **2:**441
    instability, cancer risk associated with, **5:**324-25
    inversion, **2:**440, 444, 445
    isochromosome, **2:**437
    isoenzyme locations, **4:**574
    isolation, **2:**457-59, 469

Chromosomes (*Continued*)
jumping, **2:**486
karyotype, **2:**437
Klinefelter's syndrome, **2:**713
lampbrush, **2:**292
landmark, **2:**445
lesions produced by electromagnetic radiation, **6:**444
linkage analysis, **2:**480–81
and locus, genetic, **2:**480
lyonization, **2:**443
maintenance of, **3:**872
major histocompatibility complex, **2:**477
and malignancy, **2:**441, 442
markers
HeLa cells, **4:**108
in studying tumor clonality, **7:**688–91
metaphase, **2:**468–69
methylation, **2:**476
microdissection, regional, **2:**469
micronucleus formation, **2:**468
in mitosis, **3:**849, 850
molecular genetics, **1:**705
nuclear, **2:**475
nucleosome, **2:**476
number, HeLa cells, **4:**108
pairing, **2:**470
pattern, **2:**445–54
in human cancer, **2:**445–54
in leukemia, **2:**445–54
Philadelphia, **2:**441, **7:**690
polymerase chain reaction, **2:**481
polymorphism, **2:**485
polytene-, **2:**292
in prometaphase, **5:**70
protein, **2:**470–72
purification by flow sorting, **2:**455
purity, measurement of, **2:**460
region, **2:**445, 447
and RNA, **2:**465
sequence, telomeric, **2:**484
sex, **2:**442–45. *see also* Sex chromosomes
pseudoautosomal region of, **6:**283–92
staining, **2:**457–59
structure of higher order, **2:**465–72
studies, molecular, **2:**475–88
subunit structure, **2:**419–30
Sutton and Boveri on, **3:**849
synapses at meiosis, **4:**951–52
telomere, **2:**477
translocation, **2:**440, 441, 447, **4:**951–52, **5:**326
and x-rays, **2:**467–68
XYY complement, **2:**713

Chromosomes, mice, pair 16, **3:**186–87
Chronic disease
antibody-antigen complexes in, **1:**300–301
inhaled toxicants and, **7:**583
Chronic fatigue syndrome. *see* Fatigue syndrome, chronic
Chronic obstructive pulmonary disease. *see* Lung diseases, obstructive
Chronis mononucleosis, diagnosis of, **3:**465
"Chunking" (cognition), **3:**644
"Chunking" (phonology), **3:**218
Chylomicrons, **1:**451, **2:**662, 663
in cholesterol absorption, **3:**573
in fat absorption, **3:**81, 573
as lipoproteins, **6:**27–28
remnants, as lipoproteins, **6:**28
Chyme
defined, **3:**47, 71
gastric acid secretion and, **3:**76
in small intestine, **3:**77
Chymotrypsin, **2:**42, 47
Cicatrix
astrocytic scars, **5:**345
hypertrophic, **2:**574
pelvic, **3:**364
Cigarette smoking. *see* Smoking
Ciguatoxin
poisoning, algal toxins and, **7:**591
sources of, **3:**673
Cilazapril, captopril and, **3:**396
Cilia, **2:**241, 242
defined, **5:**11
nose, functional role of, **6:**358
olfactory, G proteins and, **4:**17
in upper respiratory tract defense, **5:**691
Ciliary arteries, **3:**548
Ciliary body, **3:**539–40
Ciliary epithelium, **3:**539–40
Ciliary neuronotrophic factor, **5:**335
Cimetidine, **2:**77
Cinchona alkaloids, **1:**181–82
Cingulate cortex, **2:**5
Cingulin, **2:**272
Cingulum, **2:**789, 792
Cinoxacin, **6:**404
Ciprofloxacin, **6:**403–4
Circadian oscillator, **2:**491, 492, 495
Circadian rhythm. *see also* Periodicity
ACTH secretion and, **5:**398
daily cycles, **2:**490
desire for specific foods and, **4:**1006
endocrine function and, **5:**396
hypothalamus role, **4:**311

Circadian rhythm *(Continued)*
  reproductive functions, **4**:931
  sleep disorders of, **7**:80-81
  sleep research and, **7**:74-75
Circle of Willis, **2**:362
Circulation. *see* Blood circulation; Microcirculation
Circumcision
  cultural aspects of, **6**:884
  incidence, **4**:935
Circumduction, as joint movement, **5**:133
Cirrhosis. *see* Fibrosis
Cis-acting elements
  defined, **3**:107
  in gene transcription regulation, **3**:111-12
Cisterns
  lumbar, **4**:989
  subarachnoid, **4**:985
Citalopram, **2**:846
Citanest (prilocaine), **2**:784
Citrate cycle, **2**:503-10
  intermediates, formation of, **2**:505-7
  regulation, **2**:508-10
  schema of, **2**:504
Citrates, **2**:503-10
  lyase, **2**:506
  synthetase, **2**:506, 509, 510
Citreoviridin, **5**:266
Citric acid cycle. *see* Tricarboxylic acid cycle
L-citrulline, nitric oxide synthesis and, **3**:375
Cl⁻/HCO₃ antiporter, **4**:565-66
Clark's microstaging, **4**:890-91
Class II antigens. *see* Histocompatibility antigens class II
Class II genes. *see* Genes, immune response
Class II heparin-binding growth factor. *see* Fibroblast growth factor, basic
Classifications
  population, history of, **6**:100-101
  racial, **6**:101
  virus, **7**:772-73
Clathrim, **2**:239
Claviceps purpurea
  effect on humans, **1**:180-81
  ergotism and, **5**:263
Clavicles, **7**:27
Clavulanic acid, as enzyme inhibitor, **3**:400
Clay, cravings for, **3**:31-32
Cleavage, cell. *see* Cell division
Cleft lip, **5**:611
  multifactorial causes, **1**:700
  prevalence of, **1**:702(t)
  in stage 21 embryo, **3**:297

Cleft palate, **2**:251, **5**:611
  multifactorial causes, **1**:700
  prevalence of, **1**:702(t)
Cleidocranial dysplasia, **5**:615
Clindamycin, **2**:384
Clinical chemistry. *see* Chemistry, clinical
Clinical genetics. *see* Genetics, medical
Clinical markers. *see* Biological markers
Clinical pathology. *see* Pathology, clinical
Clinical trials
  overviews. *see* Meta-analysis
  randomized, for health care technology, **4**:62-63
Clitoris, **4**:936, **6**:564
Clodronate, **4**:280-81
Clomiphene citrate
  as regimen for ovulation induction, **7**:347
Clomiphine citrate
  in treatment of spontaneous abortion, **1**:5
Clonal selection theory, **4**:332-33
Clones, **2**:445
  jumping as mapping tool, **2**:485
  telomeric, **2**:484
  tumor
    spatial distribution of, **7**:691
    types of, **7**:687
  tumor analysis of, **7**:691
Clonic seizures, in epilepsy, **3**:456
Clonidine, **2**:39, 127, 325, 326, 834, 845, 848, 850
Cloning, **2**:455
  defined, **2**:455
  DNA encoding autoantigens, **1**:339
  of mammalian adrenergic and related receptors, **1**:84-85
Cloning, molecular. *see also* DNA, cloned; Genomic clones
  in antibody production, **3**:806-7
  defined, **3**:823, **5**:205
  Leishmania genes, **4**:683
  of myosin genes, **5**:211
  in *in vitro* DNA manipulation, **3**:876
Cloning, molecular of, DMD gene, **5**:193-94
Cloning vectors. *see* Genetic vectors
Clonorchiasis, **2**:389
Clonorchis sinesis, **2**:389
Clorgyline, **2**:493
Clormiphene, **2**:60
Closed-head injury in children, **2**:864-65
  and aphasia, **2**:864
  and hemispheric damage, **2**:864-65
Clostridial neurotoxin. *see* Tetanus toxin
Clostridium
  habitats, **1**:241

Clostridium (Continued)
morphology, **1:**239
resistance to antimicrobial agents, **1:**246
virulence, **1:**242, 246
Clostridium botulinum, **2:**345
in food, **3:**666
nitrites and, **3:**679
Clostridium botulinum toxins. *see* Botulinum toxins
Clostridium perfringens
in food, **3:**666
membrane damaging toxin of, **5:**3
in skin and wound infections, **1:**557
Clostridium tetani, **4:**445
Clostridium tetani toxin. *see* Tetanus toxin
Clotting cascade, initiation, **4:**470–71
Clubfoot, prevalence of, **1:**702(t)
Cluster designations groups, defined, **4:**693
Cluster headaches, **4:**53–54
CO_2, microvessel exchange of, **5:**34, 39
Co-morbidity, antidepressant drugs for, **1:**308
Coaccervation, defined, **3:**261
Coagulation, blood. *see* Blood coagulation
Coagulation factors. *see* specific factors
Coal mine dust, toxicity, **7:**555
Coal tar, **2:**112, 114
Coamplification
advertent, **3:**789
inadvertent, **3:**788
Coat laminar. *see* Sarcolemma
Cobalt
biological function of, **7:**606
deficiency, **7:**608
significance, in body, **5:**49
toxicity, **7:**608
Cobra venom factor, **4:**370–71
alpha-Cobratoxin, structure of, **7:**589
Coca-Cola, **2:**512
Coca plant, **2:**512
Cocaine, **2:**511–20, 843
abuse of, **5:**441
alcohol and, **3:**213
binges, **2:**517–19
and brain hemorrhage, **2:**518
and cardiovascular system, **2:**518
and cerebral aneurysm, **2:**518
and Coca-Cola, **2:**512
complications, **2:**517–19
medical, **2:**517–19
psychiatric, **2:**518–19
crack smoking, **2:**514, 515
crush phase, **2:**517
fatal, **2:**517

Cocaine (*Continued*)
and Freud, **2:**512
"grease slice" return, **2:**517–19
and heart attack, **2:**518
history, **2:**512–13
and hyperinsomnia, **2:**517
and hypertension, **2:**518
and hyperthermia, **2:**518
isolation in 1855, **2:**512
as local anesthetic, **2:**514
morbidity, **2:**517–19
mortality, **2:**517–19
pharmacology, **2:**513–15
anesthetic effect, local, **2:**514
application, medical, **2:**515
behavior stimulation, **2:**514–15
catecholamine effects, **2:**513–14
and seizure, **2:**518
sinus, defined, **2:**511
snorting, **2:**516
treatment for abuse of, **2:**519
U.S. epidemic of 1980's, **2:**515
Coccygeal vertebrae, **7:**768
Coccyx, **7:**768
Cochlea, **2:**521–25
anatomy, **3:**232
and neurotransmission, **2:**521–25
physiology, **3:**232–38
and synaptic vesicle, **2:**522
Cochlear diseases, **3:**241
Cochlear hair cells. *see* Hair cells
Cochlear microphonic potentials, **3:**238
Cochlear nucleus, in auditory information flow, **3:**239, 240
Code of hygienic practice, defined, **3:**661
Codeine, **1:**179, **2:**746, 747, 785, 786
Codominance, defined, **3:**854
Codon, defined, **5:**75
Codon coding
for *Escherichia coli*, **3:**166
in RNA sequence analysis, **3:**164
Coefficient de parenté, **4:**433
Coefficient of coancestry, **4:**433
Coefficient of consanguity, **4:**433
Coefficient of kinship, **4:**433
Coenzyme A
biochemistry, **2:**539–41
biosynthesis of, **6:**120
"carnitine shuttle" and, **3:**583
reactivity of, **3:**437
Coenzyme B_{12}, as catalyst, **3:**437
Coenzyme CoF420, biochemistry, **2:**543
Coenzyme M, biochemistry, **2:**543

Coenzyme Q, **1**:459–60, **2**:541
Coenzymes
    ascorbic acid, **1**:431
    biochemistry, **2**:527–45
    biochemistry, defined, **2**:527
    cellular chemical reactions and, **3**:423–38
    defined, **3**:393, 919
    function of, **3**:434–38
    high-energy phosphate transfer and, **3**:435–37
    in high-energy transition state stabilization, **3**:425
    oxidation-reduction and, **3**:437
    substrates of, **3**:428
Coercion. *see also* Violence
    persuasion and, **5**:799, 800
Coevolution, **2**:511, 744
Cofactor
    defined, **2**:527
    metabolism, in adrenal medulla, **1**:646
Coffee, **2**:68. *see also* Caffeine
Cognition, **2**:854
    affective influences on, **1**:108–11
    aging and, **3**:901
    in argument evaluation, **5**:803
    cognitive development, during adolescence, **1**:66–67
    cognitive processes, in media-related aggression, **1**:8–9, 1058
    cognitive representation in brain, **2**:547–58
    cognitive task, **2**:861
        language, **2**:861
        reading, **2**:861
    consistency, cognitive, **1**:487
    control, of attention, **1**:482–83
    defined, **3**:637
    emotion as arousal and, **1**:105
    in food acceptance, **3**:637–39, 645
    independence of affect and, **1**:111
    measurement, in psychobiology, **1**:609–12
    objective performance instruments, **1**:611–12
    observational rating scales, **1**:610–11
    positron emission tomographic studies of, **3**:224
    self-report tests, **1**:609–10
Cognition disorders
    dyslexia-dysgraphia and, **3**:217
    in fragile X syndrome, **3**:714–15
    in learned helplessness, **4**:668
    nerve growth factor and, **5**:337
    in older adults, **1**:123–24
Cognitive consistency theories, **4**:538–39
Cognitive dissonance theory, **4**:539
Cognitive map, hippocampal function, **4**:232
Cognitive mediation, of operantly conditioned autonomic
        responding, **1**:663

Cognitive neuroscience, **2**:547
Cognitive orientation, **7**:236–38
Cognitive psychology, **2**:547
    learning of violent scripts, **1**:1058:1:8–9
    priming of aggressive associations, **1**:1058:1:9
Cognitive structural theory, **1**:106–7
Cognitive theory of self-development, **2**:855–58
Cognitive therapy, personality disorders and, **5**:782
Cohen, **5**:374
Cohort life tables, defined, **5**:102
Coital headaches, **4**:58
Coitus, **6**:880–81
Col, **4**:33–34
Colcemid, **2**:457, 458
Colchicine, **2**:448
    characteristics of, **1**:181–82
    structure of, **7**:589
Cold erythema, **4**:797–98
Cold-induced vasoconstriction, **4**:796
Cold stress, **4**:792–93
    metabolism, **4**:797
    peripheral response, **4**:795–96
Cold urticaria, **4**:797
Coleonol. *see* Forskolin
Colestipol, **2**:168
Coley's toxin, **4**:346, 348
Colipase-dependent lipase
    bile salt-stimulated lipase and, **3**:50
    bile salts and, **3**:51
    gastric lipase and, **3**:52
    milk triacylgylcerol hydrolysis and, **3**:52
    properties of, **3**:48–49
    secretion of, **3**:51
    triacylglycerols and, **3**:51
Colitis
    acute amoebic, **1**:222, 229
    chronic nondysenteric, **1**:222
    invasive amoebic, **1**:222
Collaborative Group for the Study of Stroke in Young
        Women, **5**:601
Collaborative Study of the Psychobiology of Depression,
        1:612
Collagen, **2**:130, 561, 201, 253, 566, 567, 639–42, 652,
        658, 774, 777, 823–26, 829
    aggregation, **2**:564
    and aging, **2**:561, 570
    amino acid composition, **2**:565
    in autoimmunity, **1**:524–25
    in basement membranes, **4**:624
    biosynthesis, **2**:560–63
    bone, **1**:771–72
    bone, structure and function, **1**:763–64

Collagen (*Continued*)
  in cartilage, **2:**568
  in cartilaginous tissue, **1:**365–67
  collagenase, **2:**568
  in connective tissue, **3:**262
  crosslinking, **2:**569–71
  defined, **3:**649
  degradation products, **2:**569
  of dermis and subcutaneous tissue, **7:**47
  distribution in body, **2:**565
  elastin and, **3:**525
  in embryonic development, **2:**566, **3:**529–30
  excess synthesis, pathological, **2:**574
  fiber, **2:**559–61, 564, 777
  fibrillar, **2:**204, 205, 559, 564
  and fibrosis, **2:**574
  function, **2:**559–74
  gene expression, **2:**561–62
  gene map, **2:**562
  in gingiva, **4:**34–35
  and glycine, **2:**559, 561
  glycosylation, **2:**563
  of interstitium, **5:**35
  metabolism, **2:**568
  microfibril of, **2:**204
  mineralization, **2:**574
  molecule, structure of, **2:**559–60
  mRNA, **2:**562, 563
  in myogenesis, **5:**170
  as platelet receptor, **6:**54–55
  polymerization, endothermic, **2:**561
  polypeptide chains of, **3:**183
  posttranslational modification, ascorbic acid role, **1:**431
  procollagen, **2:**560–64, 641
  soluble, **2:**561
  structure of, **2:**559–74, **6:**221
  synthesis
    in soft tissue repair, **7:**501, 503–4
    in vitro, **2:**571–72
  translation, **2:**562–63
  triple helix of, **2:**561, 563
  Type I, **1:**764
  Type V, **1:**764
  types, listed, **2:**564–68, 640
  and wound healing, **2:**571
Collagen-proteoglycan interaction, **2:**573
Collagenase, **2:**568–69, 646
  in gingiva, **4:**36
  production, laminin effects, **4:**626–28
Colliculus, inferior
  in auditory information flow, **3:**240
  and convulsive seizure initiation, **6:**784

Colliculus, inferior (*Continued*)
  sound source and, **3:**240
Colliculus, retinal, **7:**843
Colliculus, superior, **5:**389
Colloid-osmotic lysis, **5:**4
Colloidal gold. *see* Gold colloid
Colon, **2:**33, 39
  anatomy, **3:**65, 66, 82
  carcinogenesis, **1:**625
  descending, **3:**82
  dietary fiber and, **3:**43–44
  motility in, **3:**82
Colonic diseases, in children, **5:**707
Colonic polyps, identification of, **3:**776
Colonization, hominid, **3:**504
Colony-stimulating factor, **1:**686–87, **2:**752, **4:**836
  in acute phase response, **1:**29
  as biological response modifiers, **1:**686–87
  clinical applications, **1:**687
  erythropoietin, **1:**687
  granulocyte, **1:**686–87
  granulocyte-macrophage, **1:**686–87
  hematopoietic, clinical applications, **1:**686–87
  hemopoietic cells and, **4:**173–75
  interleukin-3, **1:**686–87
  macrophage, **1:**686–87
  macrophage production and, **4:**854
  polymorphonuclear neutrophil maturation regulation by, **5:**427–28
  receptors for, **5:**816
Color, **2:**575–92
  anomaly, **2:**590
  complementarity, **2:**578
  constancy, **2:**582–83
    -opponent model, classical, **2:**586–87
    algorithms for, **2:**589–90
    alternative model, **2:**587–88
    in daylight, **2:**583
    in tungsten light, **2:**583
  protanomaly, **2:**590
  retinal sensitivity to, **7:**810–11
  simultaneous contrast of, **2:**582
  space, **2:**581–82
    alternative, **2:**582
    axis, **2:**581–82
  variation and vision, **2:**590–91
Color blindness. *see* Color vision defects
Color vision, **2:**575–90
  aspects of color
    perceptual, **2:**576–83
    physiological, **2:**583–90
  blackness, **2:**582

Color vision (*Continued*)
  color anomaly, **2**:590
  complementarity, **2**:578
  contrast effects, **2**:582
  darkness, **2**:582
  deficiency, **2**:590
  and deuteranopia, **2**:590
  human, **2**:548
  metamer, **2**:578-79
  metamerism, **2**:579
  and objects, **2**:582-83
  and protanopia, **2**:590
  three-dimensionality, **2**:578
  trichromacy, **2**:579
  tritanopia, **2**:590-91
  variations of, **2**:578, 590-91
  whiteness, **2**:578
Color vision defects, **2**:590
  gene, **2**:480
  as genetic marker for affective disorders, **1**:98-99
  red-green deficiency, **2**:590
Colorants, food analysis, **4**:223
Colorectal neoplasms
  etiology of, **3**:773-74
  genetic aspects of, **3**:771
  genotoxic factors, **3**:774
  oral contraceptives and, **5**:608
  prevention of, **3**:776
Colostrum, **4**:604-5
Colposcopy, in papillomavirus infection diagnosis, **3**:861-62
Column chromatography. *see* Chromatography, column
Columns, capillary. *see* Capillary columns (gas chromatography)
Columns, HPLC. *see under* Chromatography, high-performance liquid
Columns, packed. *see under* Chromatography, gas
Coma, **2**:345
Combination index of drugs, **2**:371, 373
Combined oral contraceptives. *see* Contraceptives, oral, combined
Combustion, heat of. *see* Heat of combustion
Commensalism, **1**:262
Commissural connections, **4**:231
Commissure, **2**:674
Commissurotomy, **2**:352-53
Commitment, developmental
  defined, **3**:281
  pattern of, **3**:283
Committee on Safety of Medicine in the United Kingdom, **5**:602
Communicable diseases. *see* Infectious diseases

Communication
  animal models, **1**:258
  nonverbal, in older adults, **1**:121
Communication disorders, **7**:155-62. *see also* Hearing disorders; Speech disorders
  of auditory processing, **7**:156-57
  hearing impairment as, **7**:155-57
  in older adults, **1**:121-24
  phonation and, **7**:157
  right hemisphere, in older adults, **1**:123
Community psychology, **4**:81-82
Comorbidity, in behavioral genetics twin studies, **3**:837
Compartment compression syndrome, **4**:806
Compartment syndromes, **5**:41
Compartmentalization, muscle, **3**:696
Competition
  aggression and, **1**:116
  in multiple-task interactions, **5**:150-54
Competitive enzyme inhibitors, **3**:395
Complement, **2**:601-17
  and anaphylatoxin, **2**:612-14
  cascade, **4**:471
  and cell death via transmembrane pores, **2**:612
  component, terminal, **2**:610-12
  components, **4**:384
  cytotoxicity, **2**:610-12
  decay-accelerating factor, **2**:604, 608
  deficiency of, **2**:616-17
  diseases related to, **2**:616
  function, inflammatory, **2**:612-14
  and hemolysis, **2**:612
  inflammation and, **2**:612-14, **4**:471
  and membrane attack, is complex, **2**:610-11
  opsonization, **2**:614-16
  protein, defined, **3**:801
  in regulation of beta cell response, **1**:569
  and spasmogenicity, **2**:612
Complement activation, **1**:298, **4**:408-9
Complement C1, **2**:603, 604
Complement C3, **1**:31, **2**:603-6
Complement C3/C5, **2**:603-4, 607-8, 610
Complement C4, **2**:603-4
Complement C5, **2**:610, 613-14
Complement C9, **2**:610-11
Complement-dependent cytotoxicity
  carbohydrates and, **3**:805
  defined, **3**:801
Complement membrane attack complex, **2**:601
Complement pathway, alternative, **2**:606-10
  convertase, **2**:607-8, 610
  decay-accelerating factor, **2**:608
  factors, **2**:607

Complement pathway, alternative *(Continued)*
  feedback, positive, **2:**609-10
  importance of C3, **2:**606-9
  nomenclature, **2:**606
  protein factors, five, **2:**606-7
Complement pathway, classical, **2:**602-4
  diagram, **2:**602
  nomenclature, **2:**603
  protein factors, five, **2:**602
Complementary-determining residues, **3:**809
Complementary DNA. *see* DNA
Complementary specialization, cerebral, **5:**313
Complementation (genetics), defined, **3:**791
Complex inheritance, principles of, **3:**855-57
Compliance, attitude change and, **1:**493
Composite x-ray laser microscope, **5:**583-84
  and phase fluorescence microscope, **5:**583
  schematic of, **5:**582
Compound action potential, defined, **3:**238
Compound proposition, **2:**620
Comprehension
  difficulty, in children, **2:**624
  reading, in older adults, **1:**120
  of sentences, by older adults, **1:**120
Comprehension stage, of attitude change, **1:**493
Computation, by neural networks, **5:**392
Computer reasoning. *see* Artificial intelligence
Computer vision systems. *see* Artificial intelligence
Computerized emission tomography. *see* Tomography,
    emission-computed
Computerized tomography. *see* Tomography,
    computerized
Computers
  flow cytometry and, **3:**625
  role in health care technology, **4:**66-67
Conception, interactive. *see* Representation
Conceptus, defined, **3:**293, 609
Concussion, **4:**56
Condensation, cell. *see under* Cells
Condensation (psychoanalysis), in dreams, **3:**192
Conditionals
  assertions in pre-school children, **2:**623-24
  comprehension of, **2:**622
  evaluation of a simple conditional, **2:**624
  and "if," **2:**622-23
  and language, early, **2:**623
  and modus ponens, **2:**619, 621
  and modus tollens, **2:**619, 621
  and proof, conditional, **2:**621
  quantified, **2:**622, 624
Conditioning, **2:**629-37, **4:**79
  and associationism, **2:**630

Conditioning *(Continued)*
  aversive, **2:**632
  and behavior, **2:**629-30
  and bond, **2:**630
  desensitization, systematic, **2:**632
  extinction, experimental, **2:**631
  and fears, irrational, **2:**632
  flooding technique, **2:**632
  instrumental, **2:**306, 629, 632-34
  versus learning, **2:**630, 636-37
  operant, **2:**629, 633-36
  salivation in dog, **2:**630
  stimulus and response, **2:**630
  token economy as, **2:**635
Conditioning, classical, **2:**305, 629-32
  biofeedback and, **1:**661-62
  extinction of, **2:**629, 631
  of immune responses, **5:**374
  in spinal networks, **5:**808
Conditioning, operant, biofeedback and, **1:**661-62
Condoms
  as birth control device, **1:**694
  for prevention of acquired immune deficiency
      syn-drome, **1:**155
Conductance, defined, **2:**305
Conductance, electrical. *see* Electrical conductance
Conduction
  as means of body heat exchange, **1:**753
  thermoregulation, **4:**795
Conduction dysphasia. *see* Dysphasia, conduction
Conduction velocity, of muscle fiber axons, **3:**721, 723
Condyle, **2:**705
Condyloid joints, **1:**374
Condylomata, giant, defined, **3:**863
Condylomata acuminata
  incidence of, **3:**862
  malignant conversion of, **3:**863
  papillomavirus infection and, **5:**638
  signs of, **3:**861
  treated grafts, **4:**586
Cone shells, toxins produced by, **7:**597
Cones
  photopigmented, **2:**575, 583-85, 591
  terminals, in retina, **6:**624
Confocal brightfield reflection microscopy, **6:**753
Confocal interference microscopy, **6:**752, 753-54
Confocal laser scanning microscope, **5:**578
Confocal microscopy, **6:**751-54
  effects of pinhole or slit size on, **6:**752-53
  imaging in, **6:**751-52
  with laser-beam path, **5:**573
  optical sectioning effect in, **6:**751-52

Confocal scanning microscopy, **5:**570-72
Confocal transmission microscopy, **6:**753
Conformational enzymes, **4:**580
Conformational epitopes, **4:**329
Conformers, **4:**580
Confusional states, acute, **1:**130
Congenital adrenal hyperplasia. *see* Adrenal hyperplasia,
    congenital
Congenital defects. *see* Abnormalities
Congestive heart failure. *see* Heart failure, congestive
Congression, defined, **5:**70
Coniine, **1:**178
Conjoined twins. *see* Twins, conjoined
Conjugase, **2:**536
Conjugate acid-base pairs. *see* Acid-base pairs, conjugate
Conjugate eye movements, defined, **3:**553
Conjugation, genetic, **3:**824
Conjunctiva
    anatomy, **3:**546-47
    effect of ultraviolet on, **5:**595
Conjunctival cul-de-sac. *see* Fornix conjunctivae
Conjunctivitis, neonatal, **4:**458
Connectin, **5:**168, 223-24
Connection, in conditioning, **2:**630
Connectionism, **2:**547, 550-52
    critique of, **2:**551-52
    defined, **5:**391
Connectionist model, **4:**199, 202
Connective tissue, **2:**639-48
    adaption to environment, **2:**648
    adult, **2:**247-48
    aging, **2:**645-47
    bone, **2:**640
    cartilage, **2:**640
    classification, **2:**640
    collagen in, **3:**262
    composition, **2:**640-43
    consistency, **2:**639
    definition, **2:**639
    elastin in, **3:**262
    fat, **2:**640
    functions, **2:**642-46
    gingival, **2:**826
    and hormone, **2:**646
    ligament, **2:**572, 640
    list of, **2:**643
    macromolecules, listed, **2:**641
    matrixextracellular, **2:**646
    and metabolite, **2:**646
    mineralization, **2:**644
    morphogenesis, **2:**645-46
    in muscle, **2:**640, **5:**181

Connective tissue *(Continued)*
    organogenesis, **2:**645
    and plasminogen activator (tPA), **2:**170
    properties, biochemical, **2:**572-73
    protein in, **2:**646
    regulation, **2:**645-46
    repair, **2:**645-46
    specificity, **2:**300
    stiffness, **2:**643
    storage function, **2:**644
    stress, **2:**643
    structure, **2:**640-43
    tendon, **2:**572, 640
    transduction, **2:**646
    types, **2:**640
    viscoelasticity, **2:**644
Connective tissue diseases, **2:**646-48
    autoimmunity and, **1:**332-35
    generalized, **2:**647
    primary, **2:**647
    secondary, **2:**647
    sports injury on impact, **2:**648
Connective tissue tumors. *see* Neoplasms, connective
    tissue
Connexin, **2:**132, 275, 276
Connexon, **2:**132, 275
Consanguinity, **3:**851, **4:**431
Conscience, **2:**711
    and anxiety, **2:**711
    formation, **2:**711
Consciousness, **2:**853
    epileptic seizures and, **3:**456
    splitting, in marijuana use, **5:**444
Consensus conferences, for health care technology,
    4:64-65
Consensus patterns
    defined, **3:**163
    in sequence analysis, **3:**164
Consensus sequences
    base pair sequences and, **3:**122
    defined, **3:**107
    for prokaryotic promoter sites, **3:**166
Conservation, evolutionary. *see* Evolutionary conservation
Conserved sequence. *see* Base sequence
Constant regions, antibody
    defined, **3:**801
    in tumor control, **3:**810
Constitutive secretory products, **4:**10
Constructional apraxia. *see* Apraxia, constructional
Constructional dyspraxia. *see* Dyspraxia, constructional
Contiguity principle, **2:**630
Contiguous gene syndrome, **2:**440

Continuing Survey of Food Intakes by Individuals, **3:**650

Continuous capillaries, defined, **5:**32

Continuum, evolutionary, animal-human, **2:**869

Contraception, **1:**691–96

    benefits of, **1:**695

    complications of, **1:**692–95

        cardiovascular, **1:**692

        depression as, **1:**694–95

        lipid alterations as, **1:**693

        thromboembolism as, **1:**693–94

    effective and safe, **1:**692

    history of, **1:**691–92

    maternal mortality and, **4:**919

Contraceptive agents, **1:**692, 693, 695, **2:**54, **5:**600

Contraceptive devices, **1:**694

Contraceptives, oral, **1:**692

    disease risk and, **5:**599–609

    hypertension associated with, **4:**289–90

Contraceptives, oral, combined

    constituents of, **5:**599

    defined, **5:**599

    side effects of, **5:**600

Contraceptives, oral, multiphasic, **5:**600

Contraceptives, oral, sequential, **5:**600

Contractile component, muscle, **5:**184

Contractile proteins. *see under* Proteins

Contractile ring, cellular, **5:**71

Contrast media

    in angiography, **3:**3–4

    defined, **3:**1

    water-soluble iodine-containing, **3:**4

Conus eleasticus, **4:**656

Convection

    versus dialysis, in artificial kidney, **1:**404–5

    as means of body heat exchange, **1:**752–53

    thermoregulation, **4:**794

Convention Concerning Safety in the Use of Asbestos, **1:**425

Conversion headaches, **4:**55

Convertase C3/C5, **2:**610

Convulsions

    and brain gene expression, **6:**786–87

    brain propagation and control of, **6:**782–83

    conditions provoking, **6:**780

    forebrain and initiation of, **6:**781–82

    general description of, **6:**780–81

    hindbrain initiation of, **6:**784

    initiation of, **6:**781–82

    metabolic and electrographic mapping studies of, **6:**785–86

    neurotransmitter mechanisms for initiation of, **6:**787–88

Convulsions *(Continued)*

    pentylenetetrazol-induced, **6:**783–84

    susceptibility and hindbrain, **6:**784

Convulsions, febrile, **3:**457

Convulsive ergotism, **5:**263

Cookery

    amino acid pyrolysis products in meat, **3:**677

    gastrointestinal cancer and, **3:**766

    heterocyclic aromatic amines and meat, **3:**677

Cooking and eating utensils, **3:**595

Coomassie blue stains, **6:**193

Coomb's test, **1:**515

Coordination number, **2:**89

Coping behavior. *see* Adaptation

Copper, **2:**649–60

    biological function of, **7:**606

    in body, **2:**653

    deficiency, **7:**609

    deficiency manifestations, **2:**650

        cardiovascular, **2:**652

        in infant, **2:**650–51

    excess in Wilson's disease, **2:**651

    excretion, **2:**653

    in hemocyanin, **2:**650

    hypocrupemia, **2:**651

    and hypoferremia, **2:**650

    imbalance, immunity and, **5:**514

    interaction with zinc, **7:**889

    intoxication from kidney dialysis, **1:**409

    and Menke's disease, **2:**650–51

    metabolism, human, **2:**649–60

    in oxidation reduction reactions, **5:**52

    poisoning, **2:**651

    significance of, in body, **5:**49

    toxicity, **7:**609

Copper, dietary. *see* Dietary copper

Copper-zinc superoxide dismutase

    brain damage and, **3:**186

    Down's syndrome and, **3:**177

    genetic locus of, **3:**184

Copper-zinc superoxide simutase, transgenic mice and, **3:**187

Corals, toxins produced by, **7:**596

Core histones. *see* Histones, core

Cori, Carl, **3:**923

Cori, Gerty, **3:**923

Corn, **2:**104

Corn starch, **2:**101

Corn sweetener, **2:**101

Cornea

    anatomy, **3:**535, 536–37

    and collagen, **2:**566–67

Cornea *(Continued)*
  effect of ultraviolet on, **5:**595
  and Kayser-Fleischer ring, **2:**651
Corneal epithelium, **3:**536, 537
Corneal stroma, **3:**536–37
Corneoscleral limbus, **3:**537–38
Corona penetrating enzyme, **3:**604–5
Coronary artery bypass, **4:**97
Coronary disease
  death rate from, **5:**100, 101
  DNA marker diagnosis of, **3:**148
  oral contraceptives and, **5:**602–4
  as psychophysiological disorder, **6:**337
  type A personality, **4:**75
Coronary vessels, **2:**182
Corpus, stomach, **3:**755
Corpus callosum, **2:**8, **4:**132–33, **5:**348, 353, 674, 871
  dyslexia and, **3:**224
  functions of, **3:**223
  neocortex and, **5:**309
Corpus luteum
  defects, ovulation induction and, **7:**348
  defined, **3:**627
  formation of, **3:**634–35
  implantation, **4:**429
Corpus striatum, **2:**217
Corpus striatum, bird, **5:**306
Corpuscles, Meissner, **7:**121
Corpuscles, Pacinian, **7:**121
Corpuscles, Ruffini, **7:**121
Corpuscles, simple, as hair receptors, **7:**676
Correlation coefficient, **3:**856
Correspondent interference theory, **5:**790–91
Cortex, cerebral. *see* Cerebral cortex
Cortical column, defined, **5:**305
Cortical filters, for vision measurement, **7:**819
Cortical granules, **3:**605, 606
Cortical plate, **2:**678
Cortical response, of sensation seekers, **6:**812–13
Cortical sulcus, **2:**674
Cortical zone, tertiary, **3:**221
Corticosteroids. *see* Adrenal cortex hormones
Corticosterone, **2:**26
  aging and, **3:**896
  G protein mRNA and, **4:**21
Corticotrophs, of anterior pituitary, **6:**16–17
Corticotropin. *see* Adrenocorticotropic hormone
Corticotropin-releasing hormone, **2:**35–38, 834, 851
  and corticotroph activity, **6:**16–17
  cortisol and, **5:**396
  defined, **5:**393
  fetal, **3:**342

Corticotropin-releasing hormone *(Continued)*
  function of, **6:**89, **7:**286
  regulation of, **5:**398
Corticotropinlike intermediate lobe peptide, **3:**343
Corti's organ. *see* Organ of Corti
Cortisol. *see* Hydrocortisone
Corynebacterium diphtheriae, **1:**556, **4:**445, 460
Corynebacterium parvum, **1:**676, 678
Cosmetics, and hair proteins, **4:**47
Cosmic radiation, **6:**427–33
  average dose of, in United States, **6:**442
  cosmic rays as source of, **6:**430–31
  radiation belts in space, **6:**428–29
  solar flares as source of, **6:**431–32
  as source of radiation, **6:**430–31
  as source of space radiation, **6:**430–31
  sources of, **6:**428–32
  South Atlantic anomaly as source of, **6:**429–30
  and space missions, **6:**432–33
  and space travel, **6:**432–33
Cosmids
  cloning, **2:**462–63
  defined, **2:**462
Cost-effectiveness analyses, for health care technology, **4:**65–66
Cotransporters, **2:**286–87, **4:**558
  defined, **2:**279
  sodium-glucose, **2:**286–87
Cotrimoxazole, uses of, **1:**329
Coughing
  laryngeal functions in, **4:**660
  laryngeal reflexes in, **4:**648–49
Coumarin toxicity, **5:**266–67
Counseling, **4:**80. *see also* Genetic counseling
Counter-regulatory hormones. *see under* Hormones
Counterion condensation theory, **2:**424
Countertransport, in renal function, **3:**96
Coupled transport, membranes, **4:**974
Coupling
  defined, **5:**149
  force, **5:**159–60
  homologous, **5:**159
  in mechanical crosstalk analogue, **5:**154–55
  phase, **5:**158–59
  space-related, **5:**159
  temporal, **5:**157–58
Coupling sites, respiratory chain
  in ATP synthesis, **5:**56
  defined, **5:**55
Covalence, enzyme thermoinactivation and, **3:**413, 414, 415–16, 421
Covariation principle, **5:**789

Cowper's glands. *see* Bulbourethral glands
Cowpox virus, **4**:464
CR3, **4**:475–76
Crabtree effect, defined, **3**:940
"Crack" cocaine. *see* Cocaine
Cranial nerves
    anatomy, **5**:352–53
    first. *see* Olfactory nerve
Craniofacial complex, **2**:689–90
    defined, **2**:689
    embryonic, **2**:689–90
Craniofacial growth, **2**:689–96
    analysis, **2**:692–93
    bony landmarks, **2**:692
    factors, **2**:690, 691
    genetics, **2**:693–95
    heritability of animal growth rates, **2**:694–95
    inheritance of trait, **2**:691
    of nasal spine, **2**:692–93
    prenatal, **2**:689–90
    sella, **2**:692–93
    twin studies, **2**:693–94
    variance, genetic, **2**:694
    x-ray analysis, **2**:692
Cravings, dietary. *see* Dietary cravings
Crayfish axon, **2**:274
Creatine, **2**:182
    amino acid pyrolysis products and, **3**:677
    phosphotransferase, **2**:182
Creatine kinase, **4**:575–77
    in cancer tissue, **4**:578
    in muscular dystrophy diagnosis, **5**:191, 202
    organ specificity, **4**:578
Creatinine, heterocyclic aromatic amines and, **3**:677
Creativity, scientific thinking and, 471–72
"Credit-assignment" problem, back-propagation and, **5**:391
Cretinism, thyroxin deficiency and, **5**:707
Creutzfeldt-Jakob disease, **1**:210, **2**:764, 214
Crevicular fluid, **2**:778
Cri-du-chat disease, **2**:439
    DNA marker diagnosis of, **3**:149
Crib death. *see* Sudden infant death syndrome
Cribiform plate, **2**:779, 825
Crick, Francis, **3**:869
Cricoarytenoid muscle, lateral, **4**:656
Cricoarytenoid muscle, posterior, **4**:656
Cricoid cartilage, laryngeal, **4**:655
Cricothyroid muscle, **4**:657
Criminality, **2**:709–19
    adoption studies, **2**:712–13
    and age, **2**:710

Criminality (*Continued*)
    and androgen, **2**:716
    appearance, **2**:713–14
    and arousal, **2**:714
    and automatism, **2**:718
    and brain pathology, **2**:717–19
    and catecholamines, **2**:717
    and chromosome abnormality, **2**:713
    and correlates, biochemical, **2**:716
    and electroencephalogram, **2**:714
    and epilepsy, **2**:717–18
    and ethnicity, **2**:709
    family studies, **2**:712
    genetic aspects of, **3**:844–45
    and genetics, **2**:711–14
    and mental illness classification, **3**:838
    neurochemistry, **2**:714–17
    and physique, **2**:713–14
    and premenstrual tension, **2**:716–17
    and psychopathy, **2**:711
    and psychophysiology, **2**:714–17
    and sex chromosome abnormalities, **3**:821
    and tendency to break rules, **2**:710
    and testosterone, **2**:716
    theory of, **2**:710–11
    twin studies, **2**:712
Criminology, **2**:709–11
Critical micellar concentration, **4**:729–30
Cro protein, recognition sequences for, **3**:122
Cromolyn, **1**:195
Crop protection, antibiotics for, **1**:295
Crop science. *see also* Agronomy
    disciplines within
        crop ecology, **1**:145
        crop modeling, **1**:145
        crop production and management, **1**:144
        molecular biology and genetic engineering, **1**:144
        plant breeding and genetics, **1**:144
    effects of environment on production, **1**:143–44
    food, feed, fiber and natural product production
        beverages, **1**:142–43
        fiber and forest crop production, **1**:142
        food from animal products, **1**:142
        leading world food crops, **1**:141
        medicinals and related derivatives, **1**:143
        other natural products, **1**:143
        spices, flavorings, perfumes, and other essential oils, **1**:143
        vegetable oils, fats, and waxes, **1**:143
Crops
    ecology, crop sciences and, **1**:145
    management, crop sciences and, **1**:144

Crops *(Continued)*
  modeling, crop sciences and, **1:**144–45
  production, crop sciences and, **1:**141–45
Cross-bridges, muscle
  defined, **5:**211
  in stretching, **5:**185
  in walk-along contraction theory, **5:**226
Cross-cultural comparison
  of customary diets, **3:**18–19
  health psychology, **4:**77
"Cross fostering" studies, in behavioral genetics research,
    3:842
Cross-partition method, for proteins, **1:**362
Cross-reactivity tumor, **4:**339
Cross-tolerance
  barbiturate/alcoholic, **5:**443
  cocaine/amphetamine, **5:**441
  defined, **5:**439
Cross-transfusion, intrauterine. *see* Fetofetal transfusion
Crossing-over (genetics)
  defined, **3:**849
  genomic mutations and, **3:**883
  illustrated, **3:**850
  linkage maps and, **3:**877
  mutation and, **3:**873
Crosslinking of collagen, **2:**569–671
  and aging, **2:**570
  inhibition, **2:**571
Crossmatching, tissue. *see* Histocompatibility testing
Crosstalk
  defined, **5:**149
  functions of, **5:**155–56
  mechanical analogue of, **5:**154–55
  "mental," **5:**160
  motor, **5:**155, 157–60
  perceptual, **5:**156–57
  between signaling pathways, **2:**329–30
Croup, **4:**460
Cruciate ligaments, **1:**376
Cruciferous vegetables, **3:**649
Cryogenics, **4:**808
Cryoglobulinemia, **4:**798
Cryosectioning, **2:**724
Cryotechniques, **2:**721–30
Cryptosporidiosis, **2:**382–83, **6:**279
  and spiramycin, **2:**382
Cryptosporodium, **4:**463
Crystallins, longevity of, **3:**419
Crystallography, X-ray, **3:**928, 929
CTF/NF-1, DNA-binding domains and, **3:**113
CTP-phosphocholine cytidyltransferase, **5:**887
Cues, peripheral, , **5:**804, 807

Cultural characteristics
  affective response to food, **1:**501
  of feeding behavior, **3:**591, 596
Cultural conformity. *see* Social conformity
Culture, as measure of population differentiation, **6:**103–4
Culture, tissue. *see* Tissue culture
Culture media
  differential, **1:**553
  enriched, **1:**553
  selective, **1:**553
Cumulus cell, oocyte, **5:**553–54
Curare, **2:**522
Curettage, in nonviable pregnancies, **1:**4
Cushing's syndrome, **1:**77, **2:**28–29, 500, **7:**280
  ACTH hypersecretion in, **5:**398
  and cortisol level, **2:**501
  defined, **5:**393
  steroids and, **7:**268
Cusp, **2:**769, 789, 792
Cutaneous sensory receptors, **7:**675
Cutin, dietary fiber and, **3:**39
Cutis laxa, elastin mRNA and, **3:**269
CuZnSOD. *see* Copper-zinc superoxide dismutase
6-Cyano-2, 3-dihydroxy-7-nitroquinoxaline, **2:**524
Cyanocobalamine, **2:**537
Cyanophyta. *see* Algae, blue-green
Cybernetics, theory of self-regulation, **1:**107
Cycasin, **2:**111
Cyclaradine, **2:**397
Cyclic AMP. *see* Adenosine cyclic monophosphate
Cyclic AMP-dependent protein kinase. *see* Protein kinases
Cyclic GMP. *see* Guanosine cyclic monophosphate
3',5'-Cyclic GMP phosphodiesterase, **5:**417
2'3'-Cyclic nucleotide-3'-phospho-hydrolase
  myelin-associated, **5:**278
Cyclin, **2:**243, 264, **5:**66, 67
Cyclomania, **4:**997
Cyclooxygenase, **2:**169, 170, **6:**351
Cyclophosphamide, **2:**114
  in bone marrow transplantation, **1:**795
  immunosuppressive/immunopotentiating effects,
    1:685–86
Cycloserine
  effect on peptidoglycan biosynthesis, **1:**278
  mode of action, **1:**290(t)
  structure of, **1:**273
Cyclosporins
  immunouppressive effects, **1:**685
  and induction of glucose intolerance, **7:**572–73
  preventing graft rejection, **4:**363
  psoriasis and, **3:**449
Cyclotrons, as source of neutrons, **7:**538

Cyproheptadine, **2:**29, **7:**572
Cystathionine beta-synthase, genetic locus of, **3:**182
Cystatins, in saliva, **6:**694
Cysteamine, **2:**539
Cysteine, **2:**541
    in glutathione synthesis, **3:**911, 917
    in microfibrillar proteins, **3:**262
Cysteine desulfhydrase, **2:**534
Cysteinylglycine
    in gamma-glutamyl cycle, **3:**913
    hydrolysis of, **3:**908, 912
Cystic fibrosis, **2:**463, 731–36
    advances in treatment of, **5:**706–7
    and chloride ion, abnormal, **2:**732
    chromosomal mapping of, **3:**138
    and chromosome-7, **2:**732
    diagnosis, genetic, **2:**735
    and DNA analysis, **2:**732–33
    DNA marker diagnosis of, **3:**143–44
    frequency of occurrence, **2:**731
    gene, **2:**487
        identification of, **6:**384–85
        isolated, **2:**731–33
        mapping, **2:**732–33
    genetic aspects of, **3:**831–32
    genotype correlated with phenotype, **2:**734–35
    inbreeding and, **4:**440
    inheritance, autosomal recessive, **2:**732
    ion transport, abnormal, **2:**732
    and mucus accumulation, **2:**731
    mutation, **2:**734
    newborn screening test for, **5:**436–37
    pedigree of, **3:**830
    and salivary glands, **6:**697
    sweat and electrolytes in diagnosis of, **2:**731
    symptoms, clinical, **2:**731
    treatment, **2:**735
    triacylglycerol absorption in, **3:**51
Cysticercosis, **2:**389
Cysticercus, **2:**389
Cystine, in irreversible enzyme thermoinactivation, **3:**417, 418
Cystinosis, glutathione and, **3:**908
Cystinyl-*bis*-glycine, hydrolysis of, **3:**912
Cysts, **2:**381, 382
    Entamoeba histolytica, **1:**217–22
    Giardia lambia, **1:**266
    oral, **5:**614
    Toxoplasma gondii, **1:**268
Cytidine, DNA demethylation and, **3:**156
Cytoarchitecture, **2:**333, 335, 673, 676
    abnormalities and dyslexia, **3:**216

Cytoarchitecture (*Continued*)
    defined, **5:**305
    of neocortex, **5:**310
Cytochalasin B, **2:**272
Cytochalasin D, **2:**272
Cytochalasins, **2:**272
Cytochrome b$_{561}$, **1:**651
Cytochrome C
    in electron transport chain, **1:**459–60
    mutants of, **3:**421
Cytochrome monooxygenases
    in cancer, **7:**516
    and drug transformation, **7:**104–5
Cytochrome oxidase, **2:**650, 652, 655
    composition of, **5:**59
    in electron transfer, **5:**62
    encoded subunits of, **5:**60
    functions of, **5:**61
    isoforms of, **5:**61
    mitochondrial diseases and, **5:**62
    oxygen tensions and, **5:**622
Cytochrome P-448, **2:**118
Cytochrome P-450, **2:**655, 737–49
    absorption of light by cysteine residue at 450nm, **2:**739
    age 400 million years, **2:**744
    amino acid sequences, **2:**743
    and arachidonic acid metabolism, **2:**738, 745
    biochemistry, **2:**738–41
    and breast cancer, **2:**744
    and cancer, **2:**741, 744, 745
    catalytic cycle, **2:**739–40
    and cholesterol degradation, **2:**745
    and codeine, **2:**746
    deficiency, **2:**738
    defined, **2:**737, **3:**671
    distribution in nasal mucosa, **5:**286
    drug metabolism and, **5:**832–33
    and drug oxidation, **2:**746
    enzymes, and nasal cavity xenobiotics metabolism, **5:**285–86
    enzymology, **2:**739–41
    evolution, **2:**743–44
    and fatty acid metabolism, **2:**745
    functions, **2:**738
    genes, **2:**743
    as hemoprotein, **2:**737
    list of human, **2:**742
    microsomal electron transport to, **7:**250
    mitochondrial electron transport to, **7:**247
    N-hydroxylation and, **3:**677
    nitrosamines and, **3:**679
    nomenclature, **2:**741–43

Cytochrome P-450 *(Continued)*
    and oxygen, **2**:738
    and palmitic acid metabolism, **2**:745
    pathway, in lung, **6**:351
    physical properties, **2**:738–39
    polymorphism and drug oxidation, **2**:746
    and prostatic cancer, **2**:745
    reactions catalyzed by, **2**:740
    regulation, **2**:748
    in steroid hormone synthesis, **7**:244–46
    and steroidogenesis, **2**:741, 744–45
    substrates for, **2**:742
    and xenobiotic metabolism, **2**:745–48
Cytochrome P-450b, **5**:515
Cytochrome P450 reductase, **7**:249
Cytochrome P450aro, **7**:252–53
Cytochrome P450c11, **7**:251–52
Cytochrome P450c17, **7**:248–49
Cytochrome P450c21
    genetic determinants of, **7**:249–50
    molecular genetics of, **7**:250–51
    in steroid hormone synthesis, **7**:249–50
Cytochrome P450scc, **7**:246–48
Cytofluorometers, **5**:574
Cytofluorometry, flow. *see* Flow cytometry
Cytogenetic maps, in genome study, **3**:877–78
Cytogenetics, scope of, **5**:684
Cytokeratins, **5**:224
Cytokines, **1**:680–84, **2**:751–62, **4**:835–36. *see also* Tumor
        necrosis factor
    in acute phase response, **1**:26–29, 34
    antiradiation effects of, **6**:422–23
    bone cell function-related, **1**:765
    and cellular reactions, **1**:34
    characterization, **1**:677, **4**:823
    cloning of, **5**:734
    cutaneous, **3**:449–52
    defined, **3**:445, **5**:373, 503
    effects on hemopoietic cells, **4**:173–75
    enhancement of immune response to tumor vaccines,
        1:678
    functions, **1**:29, **2**:751, **4**:823, **5**:731
    immune cells and, **5**:504
    immune functions of, **5**:511
    in immune inflammation, **4**:473–75
    in immune response, **2**:751–62, **4**:393, 471
    interferon-alpha, **1**:680–81
    interferon-beta, **1**:680–81
    interferon-gamma, **1**:680–81
    interleukin-1, **1**:682–83
    interleukin-2, **1**:683
    interleukin-3, **1**:683

Cytokines *(Continued)*
    interleukin-4, **1**:683
    interleukin-5, **1**:683
    interleukin-6, **1**:684
    interleukins, **7**:481–82
    macrophage-derived, **5**:816
    neurons and, **5**:385–86
    nutrition and, **5**:510–12
    properties, **2**:755
    tumor necrosis factor, **1**:681–82
Cytokinesis. *see* Cell division
Cytokinetics. *see* Cell cycle
Cytology
    analytical, **3**:621
    defined, **5**:681
    surgical pathology and, **5**:683
Cytolysis, by *Entamoeba histolytica*, **1**:225–26
Cytolytic reactions, immune system, **4**:408–9
Cytomegaloviruses, **2**:398, **4**:187, 189–90
    as cause of spontaneous abortion, **1**:8
    in day-care centers, **4**:463
    fetal infections, **4**:456–57
    and gancyclovir, **2**:398
    late E2A promoter and, **3**:158
    neonate infections, **4**:457, 458
    resistance, **2**:398
    tests for, **1**:739
Cytophotometry, **5**:570
Cytoplasmic actin, defined, **5**:216
Cytoplasmic filaments. *see* Cytoskeleton
Cytoplasmic granules
    defined, **5**:725
    formation, and Golgi apparatus, **4**:11
Cytoplasmic inheritance. *see* Maternal inheritance
Cytosine, **2**:117
Cytoskeleton, **2**:130, 325
    cytocortical, polarized organization of, **6**:62
    cytoplasmic, polarized organization of, **6**:62
    defined, **3**:523, **5**:65
    in development of polarized epithelial cell, **6**:60–62
    fibers of, **5**:66
    filaments, in myofibril organization, **5**:169
    integrins and, **3**:531
    MAP-2 phosphorylation and, **5**:898
    membrane connections, **4**:975
    microtubule components of, **5**:18
    in mitosis, **5**:69
    photoelectron microscopy and, **5**:907
Cytosol, **2**:507, 509
    communication with mitochondrion, **4**:1004
    defined, **3**:317
Cytosol pH, in fructolysis, **5**:827

Cytotoxic cells, natural. *see* Killer cells, natural
Cytotoxic drugs. *see* Antineoplastic agents
Cytotoxic reactions, immune system, **4:**408–9
Cytotoxic T lymphocytes. *see* T lymphocytes, cytotoxic
Cytotoxicity, **2:**751
    cellular, antibody-dependent, **2:**233
    graft rejection, **4:**362
    lymphocytes, **4:**835
    natural killer cells, **4:**339–40
Cytotoxins
    Entamoeba histolytica, **1:**224–25
    heterodimeric, **4:**417
Cytotrophoblast, **4:**425
Cytoxicity, antibody-dependent cell. *see* Antibody-dependent cell cytoxicity
Cytoxicity, complement-dependent. *see* Complement-dependent cytotoxicity

D

D cells, and somatostatin, **5:**720
D-Penicillamine, **2:**52, 652
D-sleep. *see* Sleep, REM
d-Tubocurarine, **1:**662–63
Dahlberg's tooth districts, **2:**790
Daimox. *see* Acetazolamide
Dairy products
    in "basic four" food group plan, **3:**654
    nutrient composition of, **5:**496–97
Dalton, defined, **5:**205
Danazol, for endometriosis, **3:**368
Dane particle. *see* Hepatitis B virus
Dapsone, specificity of action, **1:**327
Daptomycin, **1:**278
Darwin III (automaton), **5:**390
Daspus. see Armadillos
Data pooling. *see* Meta-analysis
Daunorubicin, **4:**370
Davis, Bernard, **3:**933
Day-care centers, and infectious diseases, **4:**463
Daydreaming, **7:**471
DDE, peregrine falcon and, **3:**254
DDT
    for *Anopheles* mosquitoes, **4:**883
    glutathione and, **3:**915
    peregrine falcon and, **3:**254
    toxicity
        in fish, **7:**565
        in foods, **7:**561
De Groat, **5:**44
De novo methylation, **3:**155–56
De Rerum Naturae (Lucretius), **3:**195

Dead time (gas chromatography). *see* Gas hold-up time
Deamidation
    defined, **3:**411
    in enzyme thermoinactivation, **3:**416–17
    rate of, **3:**419
Deamination
    as cause of DNA damage, **6:**548
    of 5-methyldeoxycytidine, **3:**154–55
Death, **5:**99–103, 489. *see also* Mortality
Death certificates, standards for, **5:**99–100
Debrisoquine, **2:**747
Decamethonium, **2:**522
Decapacitation factor, **3:**601
Decay-accelerating factor, **2:**601
Decibels, **3:**229
Decidua, **3:**293
Decidual cell reaction. *see* Ovum implantation
Decision analyses, for health care technology, **4:**65–66
Decision-making
    in food acceptance, **3:**639–40
    mood effects, **1:**110
    thinking and, **7:**472
Declaration of Helsinki, **1:**658
Deconditioning, cardiovascular, syndrome in astronaut, **2:**147
Decorin
    defined, **6:**261
    structure and function, **1:**764
Deductive reasoning, **6:**495–98
Deer breeding, **4:**931
Defecation, high-fiber diets and, **3:**767
Defensive behavior, **1:**114–15
Deficiency triene, in total parenteral feeding, **3:**577
Deformities. *see* Abnormalities
Degeneration, hepato-lenticular. *see* Wilson's disease
Degenerative diseases, **2:**251
    and cell death, **2:**251
        in Huntington's chorea, **2:**251
        in muscular dystrophy, **2:**251
    of the cervical spine, headaches caused by, **4:**55
    customary diets and, **3:**18
    Western diets and, **3:**27
Deglutition
    defined, **3:**74
    laryngeal functions in, **4:**660–61
    laryngeal reflexes, **4:**649–50
    monitoring of, **3:**595
    neural basis of, **5:**371
Degranulation, as function of polymorphonuclear neutrophils, **5:**428–29
Dehancers. *see* Silencers (genetics)

Dehydration
    in aldosterone deficiency, **3**:353
    diarrhea and, **5**:707
    hypothermia, **4**:801
Dehydroalanine, in beta-elimination, **3**:417
Dehydroascorbic acid, **1**:430
Dehydroemetine, **2**:389
Dehydroepiandrosterone sulfate
    action of, **1**:72–74
    in adrenal gland development, **1**:71
    in fetoplacental unit, **3**:345
Dehydrogenases, NAD- and flavoprotein-linked,
    1:459–60
Dehydrolysinonorleucine, **3**:266, 267
3-Dehydrosphinganine, **2**:534
Delinquency. *see* Criminality; Juvenile delinquency
Delirium
    dementia and, **1**:130
    toxic psychosis and, **5**:445
Delirium tremens. *see* Alcohol withdrawal delirium
Delta rhythm
    defined, **3**:191
    in normal sleep, **3**:193
Delta rule, for error size, **5**:391
Deltoid muscle, recruitment rate of, **5**:180
Delusional disorder, **4**:997
Delusional headaches, **4**:55
Dementia
    defined, **5**:667
    infectious agents in, **4**:451
    Parkinson's disease and, **5**:667
    primary, **1**:210, 210(t)
    pseudodementia, **1**:210
    secondary, **1**:209–10
Dementia, presenile, **1**:210
Dementia, senile, **1**:130, 209, 210, **2**:763–68
    and alcoholism, **2**:764
    and Alzheimer's disease, **2**:764
    and attention decrease, **2**:765
    and brain changes, **2**:764
    clinical features, **2**:765–67
    defined, **2**:763
    definition, **1**:130
    and depression, **2**:764
    diagnosis, **1**:131–32, **2**:767
    and drug intoxication, **2**:764
    effect on communication, **1**:123–24
    etiology, **1**:130–31, **2**:763–64
    incidence (U.S.A.), **2**:763
    incurability when degenerative, **2**:768
    management, **2**:767
    nerve growth factor and, **5**:337

Dementia, senile *(Continued)*
    as neurobehavioral syndorme, **2**:765
    neuroimaging techniques useful, **2**:767
    normal aging and, **3**:900
    pathology, **2**:764–65
    reversibility, **2**:768
Dementia praecox. *see* Schizophrenia
Demethylation
    defined, **3**:151
    mechanisms of, **3**:156
Demodex folliculorum, **2**:390
Demography, of older adults, **1**:129
Demyelinating diseases, **5**:271–83
Denaturation
    defined, **3**:403
    of DNA, **3**:108
Denaturation, enzyme. *see* Enzyme denaturation
Dendrites, **2**:1, 2, 341, 343, 344, **4**:198–99
    defined, **5**:341
    motor neuron organization, **7**:215–17
Dendrochiotoxicosis, **5**:263
Dendrocytes, dermal, in psoriasis, **3**:451
Dengue fever, **7**:522
Densitometry, **1**:673
Dental biology, **2**:769–81, 783–88, 817–29
    anatomy, **2**:769–81
    dentine. *see* Dentine
    embryology, **2**:817–29
    enamel, **2**:773–74, 817
    histology, **2**:817–29
    lamina, defined, **2**:817
    organ, defined, **2**:817
    pain, **2**:825
    pulp, **2**:775–76, 817
    teeth, **2**:770–73
    tissues, **2**:773–78, 825
Dental caries, **2**:101, 107, 805–15
    and age distribution, **2**:814
    in antiquity, **2**:810, 811
    apatite dissolution by bacteria, **2**:805
    and autoimmunity theory, **2**:805
    and carbohydrates, **2**:812
    of dentine, **2**:807
    in diabetes, **2**:813
    diagnosis, **2**:805
    and diet, **2**:812
    diet-related, **5**:488
    dimorphism, sexual, **2**:813
    and disease, systemic, **2**:813
    in Down's syndrome, **2**:811
    and economic status, **2**:814
    of enamel, **2**:807

Dental caries (*Continued*)
 factors listed in schema, **2:**812
 in forensic medicine, **2:**809
 in fossil animals, **2:**809
 and genetics, **2:**810
 heritability, **2:**810–11
 in hypothyroidism, **2:**813
 indices listed, **2:**806
 and lactobacilli, **2:**811
 multifactorial, **2:**809
 pathology of, **5:**612
 prevalence, **2:**810–11
 prevalence in females, **2:**813, 814
 of pulp, **2:**807
 radiography, **2:**808–9
 in "Rhodesian man," **2:**810
 scoring technique, **2:**805–8
 sequelae to, **5:**612
 in Sjögren's disease, **2:**813
 and *Streptococcus mutans*, **2:**811
 in syphilis, **2:**813
 theories on origin of, **2:**805
  acidogenic, **2:**805
  autoimmunity, **2:**805
  chemicoparasitic, **2:**805
  proteolysis chelation, **2:**805
  proteolytic, **2:**805
 in twins, **2:**810
 WHO recommendations, **2:**807
Dental enamel, **2:**773–74, 817, 819, 821–23
 caries of, **5:**612
 diagram, **2:**773
 hypoplasia of, **5:**611
Dental pulp, **2:**817, 824–25, 775–76, 819, 824
Dentate gyrus, **4:**227–28
 connections, **4:**230–31
Denticulate ligaments, **4:**990
Dentigerous cyst, **5:**614
Dentine, **2:**774–75, 817, 819, 821–23
Dentinogenesis, **2:**817, 822–23
Dentinogenesis imperfecta, **5:**612
Dentistry
 patient, **2:**783
 patients, **2:**783
 pharmacology, **2:**783–88
  analgesics, **2:**786–87
  anesthetics, local, **2:**783–84
  antibiotics, **2:**784–85
  antiplaque agents, **2:**787
  preventatives, **2:**787
  sedatives, **2:**786–87
Dentition, **2:**769, 770, 789–90

5'-Deoxyadenosylcobalamine, **2:**536
Deoxycholic acid, **1:**624
2'-Deoxycoformycin, **2:**397
2-Deoxyglucose
 and convulsive seizure studies, **6:**785–86
 staining, in olfactory bulb study, **5:**526–27
Deoxynevalenol
 animal feeding disorders and, **5:**264
 of *Fusarium roseum*, **5:**263
Deoxynucleoside triphosphates, in enzymatic repair,
  3:172
Deoxyribonuclease I
 DNA and, **3:**128
 protection assay, of DNA-protein interactions, **3:**158
Deoxyribonucleases, **2:**1, 431, 433, 436, 479
Deoxyribonucleic acid. *see* DNA
Deoxyribonucleotides
 biosynthesis pathways, **5:**462–63
 concentrations, **5:**469
Deoxyribose moiety, in oligonucleotide synthesis,
  3:169–70
Deoxytransferase, **1:**565–66
Dephospho-coenzyme A, **2:**541
Dephosphokinase, **2:**541
Depolarization, **2:**134–35, 305, 137, 183, 217, 222, 305,
  521–23
Depomedroxyprogesterone acetate, **5:**600
Deprenalin. *see* Selegiline
Deprenyl. *see* Selegiline
Depression, **2:**219, 841–51. *see also* Antidepressive agents
 ACTH suppression in, **5:**398
 and adrenal corticosteroids, **2:**834
 after childbirth, **2:**833
 and aggression, **2:**839
 animal models for, **2:**838–39
 antecedents, evolutionary, **2:**838–39
 antidepressant drugs for. *see* Antidepressive agents
 and anxiety, **2:**832, 835
 appetite loss, **2:**841
 and attribution, **2:**856
 atypical, **5:**93
 Beck's cognitive model of, **2:**855, 858
 Beck's self-rating scale, **2:**842
 Beck's self-schema, **2:**855–56
 biological basis of, **2:**851
 biological features, **2:**831, 833–36
 and biological rhythm, **2:**500
 and caffeine, **2:**79
 chronic, **2:**832
 classification, **2:**842
 clinical features, **2:**841–43
 as complication of birth control pill, **1:**694

Depression *(Continued)*
 and cortisol excretion, **2:**851
 delusional, **5:**93
 in dementia, **2:**764, 766
 diagnosis, **2:**831–32
 diagnostic criteria for, **1:**94
 and dopamine, **2:**835, 837
 double, **5:**93
 electroconvulsive therapy, **2:**838, 843
 and electroencephalogram, **2:**835
 endocrine effects, **2:**834
 endogenous, **2:**832, 842
 energy loss, **2:**841
 epidemiology, **2:**842
 in families, **2:**833
 frequency, **2:**842
 genetic aspects of, **3:**843
 and glandular fever, **2:**833
 and growth hormone, **2:**834
 Hamilton's rating scale, **2:**842
 and homovanillic acid, **2:**835
 and 5-hydroxyindoleacetic acid, **2:**836, 837
 and hyperactivity of the hypothalamic-pituitary-adrenal
   axis, **2:**851
 and hypercortisolemia, **2:**851
 and hypomania, **2:**832
 and hypothyroidism, **2:**832
 and imipramine, **2:**836
 immune system and, **6:**320–21
 and increased right frontal lobe activity, **2:**835
 and insomnia, **2:**835
 and irritability, **2:**835
 learned helplessness and, **4:**668, 671
 and lithium treatment, **2:**838
 loss of interest, **2:**841
 mania, **2:**832, 833, 842
 marijuana and, **5:**446
 masked, **1:**132
 measurement, **2:**842
 and melatonin, **2:**833
 and monoamine oxidase inhibitor, **2:**842
 and monoamine transmitter theory, **2:**836
 and mood, **2:**831
 in mothers of fragile X children, **3:**714
 and neuroleptics, **2:**832
 neurotic, **5:**93
 neurotransmitters and, **1:**132, **2:**836, 844–45, **5:**410
 and noradrenaline, **2:**837
 and norepinephrine, **2:**9
 in older adults, **1:**132–33
 and Parkinson's disease, **2:**833
 pathogenesis, **2:**832–33

Depression *(Continued)*
 and personal loss, **2:**832
 and plasma cortisol, **2:**851
 pleasure loss, **2:**841
 as predictor of suicide, **7:**331
 prevalence in the U.S.A., **2:**842
 psychobiology, **2:**836–37
 and psychosurgery, **2:**842
 and rapid eye movement, **2:**835
 reactive, **2:**832
 and reserpine, **2:**832
 seasonal, **2:**833
 seasonal type of, **5:**94
 secondary, antidepressants for, **1:**308
 self-critical, **2:**856
 serotonin and, **2:**9, **6:**832
 and sex difference, **2:**833
 sleep loss, **2:**841
 and sleep pattern, **2:**835
 and stress, **2:**832
 suicidal concerns and, **5:**94
 symptoms, **2:**831, 841–42, 841–42
 theory of, **2:**9, 843–44
 thermal injuries and, **7:**456–57
 and thyroid hormone treatment, **2:**838
 and thyroid-stimulating hormone, **2:**834
 treatment, **2:**837–38, 842, **5:**95–97
 in twins, **2:**833
 weight loss, **2:**841
 in winter, **2:**833
 and X-chromosome, **2:**833
Depression, bipolar. *see* Bipolar disorder
Depressive disorder, **2:**832, 842
 brain monoamine neurotransmitter systems and,
   1:612–13
 characterization, **4:**997
 cognitive attitudes about life and, **1:**613
 effects of tricyclic drugs, **1:**613–14
 major depressive disorder, **1:**94, 308, **5:**92
Deprinil. *see* Selegiline
Depth perception
 and binocular vision, **1:**638
 characteristics and measurements of, **7:**824–26
 network modeling of, **5:**389
Depurination, as cause of DNA damage, **6:**548
Derivatization
 for gas-liquid chromatography, **3:**205
 for high-performance liquid chromatography, **3:**206
Dermal dendrocytes. *see* Dendrocytes, dermal
Dermal papilla, **4:**40–41
Dermal ridges
 counting, among relatives, **3:**856

Dermal ridges *(Continued)*
    counting of, **3**:87–88
    development of, **3**:85–86
Dermatan sulfate, **2**:206, 573
    in gingiva, **4**:35–36
    structure of, **6**:258
Dermatitis, **4**:463–64, **7**:47
Dermatoglyphic core, defined, **3**:86
Dermatoglyphics, **2**:593, **3**:85–92, 856
Dermatomyositis
    antigens associated with, **1**:336
    autoantibodies associated with, **1**:336
    characteristics of, **1**:335
Dermis. *see* Skin
Descemet's membrane, **3**:537
Desensitization, receptors, **4**:975
Desensitization, systematic, **2**:632
    and extinction of fear, **2**:632
    and stimulus gradient, **2**:632
Desferrioxamine, **2**:655
Desipramine, **2**:392
Desmethylimipramine, **2**:747, 834, 846, 849
    6-OHDA and, **5**:383
Desmin, **2**:130, **5**:224, 132
Desmocalmin, **2**:269
Desmoglein, **2**:269
Desmoplakin, **2**:269
Desmosine, **2**:206, **3**:266
Desmosome, **2**:132, 267, 268
Desmoyokin, **2**:269
Desquamation, defined, **7**:43
Desynchronized sleep. *see* Sleep, REM
Detectors (high-performance liquid chromatography),
    3:205, 206, **4**:210–12
Detergents, solubilized proteins, **4**:971–73
Determination, genetic. *see* Eugenics
Detoxification
    nasal cavity, **5**:285–89
    in treatment of addiction, **1**:49
Detrusor muscle, innervation of, **5**:43
Deuteranopia, **2**:590
Developed countries, diets of, **3**:26–28
Developing countries
    children of, **5**:701
    diets of, **3**:26
    liver disease in, **5**:607
Development. *see* Child development; Child development
    disorders; Growth; Language development; Per-
    sonality development
Development, behavioral. *see* Personality development
Developmental Gerstmann's syndrome. *see* Gerstmann's
    syndrome

Dexamethasone, **2**:26, 27, 492
    ACTH suppression and, **5**:398
    depression test, **2**:851
    elastin and, **3**:265
    suppression test, **1**:132, **2**:834
Dexterity. *see* Motor skills
Dextran, **2**:140
Dextran, alpha-Dextrin, **2**:104
Dextran-polyethylene glycol systems, **1**:357
    composition and concentration, **1**:357–58
    isolation of membranes and organelles, **1**:361
    manipulation of phase properties, **1**:358
    phase diagram, **1**:357
    protein extractions, **1**:363–64
Dextrin, limit, **3**:923
Dextroamphetamine, **1**:310
Dextromethorphan, **1**:179, **2**:746, 747
Dextrose. *see* Glucose
Diabetes insipidus
    vasopressin deficiency in, **5**:397
    vasopressin in, **6**:20
Diabetes mellitus, **2**:106, 329, 648
    antibodies in, **4**:408
    autoantibodies to receptors and, **1**:515
    as autoimmune disease, **1**:528
    and dental caries, **2**:813
    disease models, animal, **1**:257
    drug-induced, **7**:571–72
    and ethnicity, **2**:106
    G protein defects in, **4**:23–24
    infectious agents in, **4**:451
    kinins and, **4**:601
    longevity and, **4**:789
    oral contraceptives and, **5**:608
Diabetes mellitus, adult onset. *see* Diabetes mellitus, non-
    insulin-dependent
Diabetes mellitus, insulin-dependent, **2**:106, **3**:353, **5**:708
Diabetes mellitus, juvenile onset. *see* Diabetes mellitus,
    insulin-dependent
Diabetes mellitus, non-insulin-dependent, **3**:335, 896
Diabetes mellitus, type II. *see* Diabetes mellitus, non-
    insulin-dependent
Diacylglycerol. *see* Diglycerides
Diagnosis
    Alzheimer's disease, **1**:213
    DNA markers in, **3**:135–50
    in genetic counseling, **3**:815
Diagnosis, prenatal. *see* Prenatal diagnosis
Diagnostic pathology. *see* Pathology, diagnostic
Diagnostic radiology. *see* Radiology, diagnostic
Dialects, language development and, **4**:637

Dialysis. *see also* Artificial kidney; Hemodialysis; Peritoneal dialysis
  versus convection, in artificial kidney, **1**:404
  economic impact of, **1**:408
  equilibrium, for determining concentration of drugs in blood, **1**:708
  hypercalcemia of malignancy treatment, **4**:281
  research opened by, **1**:410
Dialysis, extracorporeal. *see* Hemodialysis
Diamond, Louis, **5**:704
Diapedesis, **5**:428–29
Diaphragma sellae, **4**:981–82
Diarrhea, **2**:38, 104–5, 650
  acid hypersection and, **5**:721
  bacterial etiology, **1**:557
  in children, in developing countries, **5**:707
  and copper deficiency, **2**:650
  Howland and Marriott on, **5**:701
  infectious, **4**:449–50, 463
  infective, in early industrial cities, **3**:21
  mechanism for, **2**:38
  opioids and, **5**:719
  in "pancreatic cholera," **5**:722
  secondary lactase deficiency, **4**:618
  secretory, **4**:462
  "summer," **5**:700
  and sympathectomy, **2**:38
  vasoactive intestinal polypeptide and, **5**:718
Diarrhea, infantile, **4**:462
  secondary lactase deficiency, **4**:618
Diarthroses, **1**:371, 373
Diaschisis, **7**:225
Diastolic blood pressure, **4**:285
Diazepam, **2**:493, 787
  anticonvulsant action of, **6**:787
  and caffeine, **2**:79
  in treatment of drug addiction, **1**:49
Dibenz (a,b) anthracene, **2**:110
  6,7-Dichloro-3-hydroxy-2-quinoxaline carboxylic acid, **2**:524
Dichorial placentas, **3**:308
Dichorial twins. *see under* Twins, monozygotic
Dichroism, defined, **5**:567
Dichromacy, red-green, **2**:590
Dicotyledonous plants
  cell walls of, **3**:37–39
  seeds of, **3**:39
Dicoumarol. *see* Bishydroxycoumarin
Dictyostelium, **5**:165
Dictyostelium discoideum, **5**:217
Dideoxycytidine
  in treatment of AIDS, **1**:152

2'3'-Dideoxycytidine, **2**:400
2'3'-Dideoxyinosine, **2**:400
3',4'-Dideoxynorlaudanosoline-1-carboxylic acid (DNLCA), **1**:179–80
Didhydrochalcones, as sweeteners, **7**:377
Dieldrin, peregrine falcon and, **3**:254
Dielectric constant. *see* Relative permittivity
Dielectric loss, defined, **3**:9
Dielectric properties of tissue, **3**:9–15
Dielectric relaxation, **3**:9. *see also* Relaxation time
Diencephalic amnesia, **4**:677
Diencephalon, **2**:3, **5**:349, 673, 678, 682
Dientamoeba fragilis, **6**:269
Diet, **2**:191–93, 798–99, **3**:17–30. *see also* Food groups; Nutrition
  American, problems with, **5**:485–89
  of American ethnic minorities, **3**:650–51
  assessment, in the elderly, **4**:899
  British, **3**:36, 44
  bulk in, **5**:51
  and calories, **2**:192
  for children, **5**:708
  Chinese, **3**:21
  colorectal cancer and, **3**:774
  customary
    evolutionary perspectives on, **3**:19–23
    food selection in, **3**:17–19
  diet-related diseases, **5**:485–88
  digestive system gross anatomy and, **3**:60–61
  diseases related to, **5**:485–88
  in early industrial cities, **3**:21–22
  eating behavior, in Western countries, **3**:591, 596–97
  food acceptance and, **3**:639
  guidelines for Americans, **5**:487
  of hominids, **3**:500–501
  immune system aging and, **6**:77
  immune system and, **5**:503–5
  importance of, **5**:483–84
  improvement of, **3**:656–58
  during lactation, **4**:605
  and meat, **2**:191–92
  minerals in, **5**:486–87
  in modern cities, **3**:22–23
  nutritional quality of foods, **5**:483–502
  and pathology, **2**:799
  problem nutrients in, **5**:488–89
  and prostate cancer, **6**:187–88
  recommendations
    of Committee on Diet and Health, Food and Nutrition Board, Commission on Life Sciences, **5**:488
    national, United States, **3**:651–52
    Surgeon General's report on, **5**:488

Diet *(Continued)*
  restrictions in kidney dialysis, **1**:409
  selenium in, **6**:790–91
  special, **3**:658–59
  and teeth, **2**:798
  in traditional cities, **3**:21
  vitamins in, **5**:486–87
Diet, Nutrition, and Cancer Report, **3**:651
Diet, reducing, **3**:25, 593, 683–90
Diet, sodium-restricted, in pregnancy, **3**:349
Diet, synthetic. *see* Food, formulated
Diet and Health, **3**:655
Diet surveys. *see* Nutrition surveys
Dietary anticarcinogens. *see under* Anticarcinogens
Dietary aversions, in pregnancy, **3**:31–33
Dietary calcium. *see* Calcium, dietary
Dietary carcinogens. *see under* Carcinogens
Dietary copper, defective cross-linking and, **3**:267
Dietary cravings
  in pregnancy, **3**:31–33
  sensory influences on, **5**:753–54
Dietary deficiencies, interferons and, **4**:525
Dietary fats, **2**:192–93, **3**:567–80
  in "basic four" food group plan, **3**:653
  defined, **3**:649
  in developed countries, **3**:29
  immunity and lipids, **5**:507–10
  national recommendations on, **3**:27
  in nutrition education, **3**:25
  spoilage of, **3**:663
  in Western diets, **3**:80
Dietary fiber, **2**:101, **3**:35–45
  defined, **3**:649, **5**:501, 765
  health food movement and, **3**:22
  mineral absorption and, **3**:42–43
  mutagen inhibition and, **3**:678
  satiating power of, **6**:729
Dietary Goals for the United States, **3**:651
Dietary Guidelines, **3**:651
Dietary guidelines, nutritional quality and, **5**:485
Dietary iron
  for children, **5**:708
  hookworm infection and, **4**:115
  phytates and, **3**:43
  recommended dietary allowances of, **5**:484
  vegetable origin, **4**:115
  in Western diet, **2**:656
Dietary legislation. *see* Legislation, food
Dietary modification
  assessment in, **3**:23–24
  nonnutritional, **3**:25

Dietary proteins, **2**:191–93, **6**:229–35. *see also* Protein-calorie malnutrition
  balanced composition of, **6**:231
  chemical composition of, **6**:229–30
  in diet of hunter-gatherer, **2**:192–93
  digestible energy from, **5**:476
  digestion of, **3**:80, **6**:232
  distribution of, **6**:232
  effect of age on utilization of, **6**:233–34
  elimination of excess, **6**:232–33
  immunity and, **5**:505–6
  intake, limits of, **2**:193
  intake during lactation, **4**:605
  and Kalahari San people, **2**:194
  nutritional quality of, **5**:499
  protein destruction and nutritional quality of food, **5**:493–94
  quality and quantity of, **6**:230–32
  requirements, **6**:233, 234–35
  sources of, **6**:230–32
  synthesis and breakdown of, **6**:233
  and undernutrition, **6**:235
  utilization of, **6**:233–35
Dietary restraint
  defined, **3**:589
  in Western countries, **3**:591
Dietary salt. *see* Sodium, dietary
N,N-Diethyl-m-tobramide (Deet), **2**:390
Diethylcarbamazine, **2**:388
Diethyldithiocarbamate, **1**:684
Diethylstilbesterol, **1**:8
Differential phase contrast, **6**:750–51
Differentiation. *see also* Histodifferentiation
  defined, **3**:609
  transforming growth factor-beta and, **7**:637
Differentiation, cell. *see* Cell differentiation
Differentiation of environment, **2**:555
Differentiation of functions (behavioral), **1**:582
Diffuse pulmonary fibrosis, **7**:583
Diffuse-type gastric carcinomas, **3**:769
Diffusion, facilitated, **2**:279
Diffusion, passive. *see* Biological transport
Diflunisal, **2**:785
  alpha-Difluoromethylornithine, **2**:392
gamma-Difluoromethylornithine. *see* Eflornithine
DiGeorge anomaly, **1**:698
Digestible energy
  defined, **5**:471
  from protein, **5**:476
Digestion
  of DNA, **3**:128
  of fat, **3**:572–73

Digestion (*Continued*)
  hormonal control of, **5:**715–16
  of human milk lipids, **3:**47–56
Digestive system, **3:**71–83
  afferent innervation of. *see* Viscera, afferent systems of
    anatomy, **3:57–69**
      gastrointestinal cancer and, **3:**767–68
    epithelium lining, types of, **6:**470–71
    fetal
      from fourth month, **3:**618
      at third month, **3:**611
Digital dermatoglyphics. *see* Dermatoglyphics
Digital subtraction angiography. *see* Angiography, digital
    subtraction
Digitalis, **2:**173–75
Digitonin, **3:**12
Digits. *see* Fingers
Diglycerides, **2:**19, 304, 308
  behavior and distribution, **4:**743
  in Ca^{2+} messenger system, **3:**333
  as cellular messenger, **7:**20
  characterization, **4:**742
  in gingiva, **4:**36
  metabolism, **4:**742–43
Digoxin, **2:**175, 176
  monitoring of, **3:**208
Dihydrobiopterin, **2:**535
Dihydrofolate. *see* Tetrahydrofolate
Dihydromorphone, **2:**786
Dihydropteroate synthetase, inhibition of, **1:**327
1,4-Dihydropyridines, **2:**84–86
  as vasodilators, **2:**87
Dihydroquinone, **2:**544
Dihydrotestosterone. *see* Stanolone
2,3-Dihydroxy-6,7-dinitroquinoxaline, **2:**524
9-(1,3-Dihydroxy-2-propoxymethyl)-guanine. *see* Ganci-
    clovir
Dihydroxyacetone phosphate, in second-stage glycolysis,
    3:936
1,25-Dihydroxycholecalciferol. *see* Calcitriol
1,25-Dihydroxyvitamin D$_3$. *see* Calcitriol
Diiodotyrosine receptors. *see* Receptors, thyroid hormone
Dilation, **2:**403, 404–8
Dilipoyl dehydrogenase, **2:**540, 542
Diltiazem, **2:**85–87, 96, 172, 176
2,3-Dimethoxy-5-methylbenzoquinones, **2:**541
p-Dimethylaminoazobenzene, **2:**110
7,12-Dimethylbenz(a)anthracene, **2:**55, 110, 121
Dimethylnitrosamine, **2:**110, 121
  carcinogenicity of, **3:**679
  chemistry of, **3:**678
5,5-Dimethyloxazolidine-2,4-dione, **5:**872

Dimethylphthalate, **2:**390
Diniconazole, **2:**121
1,4-Dinitrobenzene, **2:**118
Dinitrophenols, Pasteur effect and, **3:**940
Dinitropyrene, **2:**121
Dinoflagellida, **3:**673, **7:**590–91
  algal toxins and, **7:**590–91
Diopter, **3:**533
Dioptrics, **3:**533
Dipalmitoyl-palmitic acid, **5:**886
Dipalmityllecithin, **5:**618
Dipetalonema streptocerca, **2:**388
Diphenhydramine, **2:**787
Diphosphatidylglycerols. *see* Cardiolipins
2,3-Diphosphoglycerate, **4:**137, **6:**600
Diphtheria, in infants and toddlers, **4:**460
Diphtheria, toxin, **5:**1–3
Diphtheria-Tetanus-Pertussis vaccine, **4:**466
Diphtheria toxin, **4:**418
  conjugates with, **4:**368–69
Diphtheria vaccines, **4:**445
Diphyllobothrium latum, **2:**389
Diplacusis, **5:**740
Diploidy, **2:**291, **5:**201
Dipoles, molecular. *see* Molecular dipoles
Dipylidium caninum, **2:**389
Dipyridamole, **2:**169–70
Direct repeats. *see* Repetitive sequences, nucleic acid
Directionality perception, **3:**223
Disaccharides, **2:**103
Disaggregated cultures, **4:**582–83
Disclosure, truth. *see* Truth disclosure
Disconjugate eye movements. *see* Vergence eye move-
    ments
"Disconnection syndromes," **3:**223
Discontinous capillaries, **5:**33
Discrimination, conditioned, **2:**631
Disease models, animal, **1:**251–59
  aging, **1:**258
  AIDS, **1:**256
  alcohol addiction, **1:**258
  atherosclerosis, **1:**450–51
  birth defects, **1:**704
  cancer, **1:**256
  cardiovascular disease, **1:**256–57
  diabetes mellitus, **1:**257
  experimental metastasis, **4:**1008
  hepatitis B, **1:**255
  Huntington's disease, **4:**272
  hyaline membrane disease, **1:**258
  infectious diseases, **1:**255–56
  leprosy, **1:**255–56

Disease models, animal *(Continued)*
    malaria, **1:**255
    memory loss, **1:**258
    metabolic disorders, **1:**257
    non-A, non-B hepatitis, **1:**255
    nutritional deficiency, **1:**257
    polio, **1:**255
    reproductive disorders, **1:**257-58
    surgical procedures, **1:**258-59
    transplantation procedures, **1:**258-59
    treatment of animals, **5:**678-79
Disease reservoirs, **7:**892
Disease resistance. *see* Immunity, natural
Disease vectors, **2:**455
    competence, and transmission of infectious agents, **7:**892
    defined, **2:**455
    retroviruses as, **6:**653-59
    role in zoonoses, **7:**893-96
Diseases. *see* specific diseases
Disinhibition. *see also* Inhibition (psychology)
    defined, **3:**591
    as sensation-seeking trait, **6:**810
Disk, intercalated, **2:**132
    intervertebral, **2:**566, 640
Disomy, uniparental, **2:**438-39
Disopyramide, **2:**173
Dispermic mosaicism. *see under* Mosaicism
Displacement (psychology), **3:**192
Disseminated intravascular coagulation, **7:**498
Dissociative disorders, **4:**998
Distributed representation, **5:**387
Distribution constant (gas chromatography), **3:**750
Distribution equation, **2:**375
Distributive justice, **4:**539
Disulfides
    beta-elimination and, **3:**417
    crosslinks, **2:**540, 563
    defined, **3:**907
    immunoglobulins bonds, **4:**832
    intracellular glutathione and, **3:**912
    in potentially reversible enzyme thermoinactivation, **3:**415
    protein thermostability and, **3:**420
Diuresis, **2:**143
    and antidiuretic hormone, **2:**143
    and atrial natriuretic factor, **1:**473
    and hypothermia, **4:**801
    and sodium excretion, **2:**143
    in weightlessness (spaceflight), **2:**151
Diuretics, **2:**176, **3:**93-100,
    for hypertension, **4:**294

Diuretics, loop
    characteristics of, **3:**98-99
    hypercalcemia of malignancy treatment, **4:**281
Diuretics, osmotic, **3:**99-100
Diuretics, potassium-sparing, **3:**99
Diuretics, thiazide
    characteristics of, **3:**97-98
    loop agents and, **3:**98
Diurnal rhythm. *see* Circadian rhythm
Diver's bronchorrhea, **4:**797
Divers' hypothermia, **4:**802
Divistyramine, **2:**168
DNA, **2:**1-5, 41, 235, 291-95, 341, 419-30, 469-70, 826, 829
    alterations, and cancer, **6:**415-16
    B-form, **2:**292
    cell aging and, **3:**900
    and centromeres, **2:**472, 473
    chain elongation, proteins in, **5:**467-68
    and chromatin, **2:**476
    chromosomal, **2:**419-30
    chromosomal band, **4:**253
    compaction of, in eukaryotes, **3:**125
    complementary. *see* DNA, complementary
    consensus, sequence, defined, **2:**475
    content
        of cell, **2:**254
        of chromorones, listed, **2:**456
    copy, **2:**431
    defined, **1:**139, **2:**53, 439, 737, 791, 869, **3:**107, **5:**681
    denatured, in hybridization, **3:**888
    digestion of, **3:**128
    electromagnetic radiation and, **6:**444
    of eukaryotic genes, **3:**793
    fluorescence, **2:**459
    fragments, **4:**253
    fragments, cloned, **2:**455
    function of, **3:**109
    gene library. *see* Gene library
    in genetic evolution, **3:**493
    as genetic material, **6:**683-85
    grooves of
        deformed, **3:**129
        DNase I and, **3:**128
    hybridization, **2:**475, 731
    intrachromosomal amplified, **3:**102
    inversion, defined, **3:**101
    joining diversity, **4:**327
    jumping clone, **2:**485
    junk-, **2:**478
    length, **2:**432
    metabolism, **5:**463-64

DNA *(Continued)*
  methylation, **2:**476, **3:**151–62
  non-coding, **3:**873–74
  nontumor-, **2:**53, 293
  and nucleosome monomer, **2:**471
  in nucleosomes, **3:**125–33
  and open reading frame, **2:**475
  organization, **2:**292
  parasitic, **2:**478
  perturbations, in nucleosome, **3:**129
  photochemistry, **7:**66–67
  photoelectron imaging of, **5:**907
  polymorphism, megabase, **2:**480, 485
  properties of, **3:**108
  repair nuclease and, **2:**293
  replication fork, defined, **3:**441
  replicon, **2:**293, 294
  restriction mapping. *see* Restriction mapping
  and RNA hybridization, **2:**432
  separation of. *see* Electrophoresis, pulsed field gel
  sequence analysis, **3:**172
  skin aging and, **3:**903
  staining, in flow cytometry, **3:**623
  structure, **2:**469, **3:**870, **6:**682
  supercoiled. *see* DNA, superhelical
  synthesis. *see* DNA replication
  as target of carcinogen, **2:**118
  and telomere, **2:**472
  thermal untwisting. *see* DNA replication
  transcription. *see* DNA replication
  transfer, **3:**101
  types of, **3:**107
  of uterine cells, **3:**624
  variant structures of, **3:**119
  variation, and population genetics, **6:**107–9
  vectors, **2:**462–63
DNA, adenovirus
  de novo methylation and, **3:**155
  transactivator of, **3:**158
DNA, amplified. *see* Gene amplification
DNA, cloned, **2:**432, 460–63, **3:**175
DNA, complementary, **2:**41, 53–55, 395, 419–22, 473, 737
  defined, **4:**13, **5:**205
  of DMD gene, **5:**194
  for glucocerebrosidase, **3:**778
  of pertussis toxin substrates, **4:**18
DNA, double-minute
  defined, **3:**101
  in neuroblastoma cell, **3:**103
  in tumors, **3:**102
DNA, gyrase, **3:**129, **6:**402–3
DNA, linker, **2:**421, 422, **3:**128

DNA, mitochondrial
  in evolution research, **3:**502
  human origins and, **3:**888–89
  in inheritance, **3:**858
  as measure of population differentiation, **6:**103
  respiratory chain and, **5:**59–60
DNA, neoplasms, **2:**53
DNA, papillomavirus, **3:**864–65
DNA, plant
  DNA methylation and, **3:**161
  5-methyldeoxycytidine in, **3:**156
DNA, recombinant
  in antibody molecule production, **3:**803
  defined, **3:**101
  and disease diagnosis, **6:**523–31
  Down's syndrome and, **3:**185–86
  immunoconjugate synthesis with, **4:**376
  immunotoxin design, **4:**419
  peptide sequence analysis and, **5:**714
  technology, **2:**455–56
  transgenic mice and, **3:**187
DNA, satellite, **2:**477
DNA, superhelical
  in nucleosome, **3:**129
  transcription and, **5:**469
DNA, viral
  as activated oncogene, **4:**342
  herpesviridae, **4:**192
  sequences in genomes, **4:**190–91
DNA amplification, **3:**101–5
DNA-based polymorphisms (RFLPs), **1:**96
DNA-binding proteins, **2:**431, 435, 436, **3:**119–20
  in embryonic body pattern formation, **3:**291
  of Epstein-Barr virus, **3:**471, 472
  methyl groups and, **3:**152
  muscle-specific genes and, **5:**168
DNA binding sites, **3:**117–24
  electrostatic interactions, **3:**118
  nonspecific binding, **3:**120
  steroid-receptor complex binding to, **3:**331
  of transacting factors, **3:**112
DNA-binding sites
  categories of, **3:**112–13
  defined, **3:**107
DNA damage
  aging and, **3:**895
  alkylation as cause of, **6:**548–49
  bulky adducts as cause of, **6:**549
  deamination as cause of, **6:**548
  depurination as cause of, **6:**548
  endogenous, **6:**547–48
  exogenous, **6:**548–49

DNA damage *(Continued)*
  glycosylases for repair of, **6:**550–51
  ionizing radiation as cause of, **6:**548
  leukemia-related, **4:**691–92
  metabolic processes causing, **6:**548
  mismatched bases as cause of, **6:**548
  mispairing errors, repair of, **6:**553–54
  oxygen radicals as cause of, **6:**548
  repair of, **6:**547–54
  ultraviolet irradiation as cause of, **6:**548
DNA fingerprinting
  defined, **5:**686
  methods, **2:**24, 483
  minisatellite probes and, **3:**137
  overlapping clone maps and, **3:**878
  restriction endonuclease, **4:**191
  VNTR probes and, **3:**137
DNA gel retardation assay, **2:**435
DNA helix destabilizing proteins. *see* DNA-binding
    proteins
DNA index, **2:**54–55
  cell kinetics and, **3:**623
  defined, **3:**621
DNA insertion elements, **4:**255
DNA library. *see* Gene library
DNA ligases, **2:**293, **3:**172, 441, 462
DNA markers. *see* Genetic markers
DNA methyltransferases
  action of, **3:**155
  defined, **3:**151
  inhibition of, **3:**158
  methylation control and, **3:**152
DNA motifs, defined, **3:**701
DNA mutations, **2:**118
  and analysis of genetic disease, **6:**524
  and carcinogenesis, **2:**118
  diagnosis of cancer-associated, **6:**528
  direct detection of, **6:**526–27
  ribonuclease A cleavage in detection of, **6:**527
DNA polymerase delta, **3:**442, 443
DNA polymerase I, **3:**809
DNA polymerase II, **3:**443
DNA polymerase III, **3:**443
DNA polymerases, **2:**293, 395, 473
  alpha primase, **2:**263, 264
  chain reaction, **2:**11
  defined, **3:**439
  in DNA replication, **3:**441–43
  DNA replication and, **5:**467–68, **6:**684
  of Epstein-Barr virus, **3:**471
  3'Ä5' exonuclease and, **3:**441
  in polymerase chain reaction, **3:**173

DNA polymorphism. *see* Polymorphism (genetics)
DNA probes
  for chromosome 21, **3:**180
  defined, **3:**135
  in detection of bacterial pathogens, **1:**553
  and diagnosis of genetic diseases, **6:**524–25
  fluorescent, **2:**455, 481
  in food monitoring, **3:**669
  multilocus, **3:**138
  oligonucleotides as, **3:**173
  pseudogenes and, **3:**792
  types of, **3:**137–38
DNA-protein interactions
  defined, **3:**151
  sequence-specific methylation and, **3:**158–59
DNA rearrangement. *see* Gene rearrangement
DNA repair
  base-specific responses in, **6:**550–51
  cancer risk associated with disorders of, **5:**324–25
  direct removal mechanisms of, **6:**549–50
  of free-radical damage, **3:**894
  and hyperthermia, **4:**301
  longevity and, **3:**895, **4:**785
  mechanism of, **2:**293
  mutation and, **5:**250, 253–54
  nucleotide excision as mechanism of, **6:**551–53
  photoreactivation as mechanism of, **6:**549
  SOS regulation of, **6:**553
DNA repetitious regions. *see* Repetitive sequences, nucleic
    acid
DNA replication, **2:**255, 293, 294, 395, **3:**169–73, **4:**251,
    6:683–85
  antibiotic inhibitors, **1:**291
  bidirectional, **3:**104
  enzyme isolation and, **3:**404
  enzymology of, **3:**439–44
  fork, **2:**293, 294
  gene amplification from unscheduled, **5:**546
  gene transcription and, **3:**107–16
  histone protein synthesis relationship, **4:**237–38
  histones and, **4:**237–38
  in interphase, **5:**66
  leading and lagging strands in, **6:**684
  maintenance methylation inhibition and, **3:**158
  nucleosomes and, **3:**133
  oocyte growth and, **5:**558
  and origin of replication, **2:**62
  principles of, **3:**870
  proteins associated with, **5:**467–68
  semiconservative
    defined, **3:**439
    mechanics of, **3:**440

DNA replication *(Continued)*
  synthesizing machinery, **2:**253
  telomeres and, **7:**401
  thermal untwisting of, **3:**128
  and thymidine, **2:**255
  ultracentrifugation in determining mechanics of, **7:**720
  unscheduled, **5:**250
DNA restriction enzymes, **2:**483–84, 661, 669–71, **5:**463–64
  activation of, **3:**153
  allele-specific oligonucleotides and, **3:**139–40
  binding sites and, **3:**122
  cleavage sites, **4:**191
  defined, **3:**135, 151, 169, 823, **5:**205
  EcoR1, **2:**461, 668
  and endonuclease, **2:**419
  in gene technology, **3:**108, 129
  genetic disease diagnosis using, **6:**526
  Hind III, **2:**463
  inhibition of, **3:**152
  in *in vitro* DNA manipulation, **3:**876
  list of, **2:**484
  methylation and, **3:**152
  nucleosome and, **3:**128
  in pulsed field gel electrophoresis, **6:**383
  recognition sequence, **2:**484
  sequence-specific methylation and, **3:**153
  in sickle cell anemia diagnosis, **3:**145
  site, **2:**475
DNA topoisomerases, **3:**441
DNA tumor viruses, **3:**124
DNAse. *see* Deoxyribonuclease
DNase I. *see* Deoxyribonuclease I
Docosahexaenoic acid, **2:**170
  as omega-3 fatty acid, **5:**533, 534
  sources of, **5:**533–34
  of tapeworm, **2:**389
Dolichol, **2:**415
Domestic animals. *see* Animals, domestic
Dominance, cerebral, **2:**351–52, **5:**313
Dominance, manual, **5:**314
Dominance (genetics)
  defined, **3:**135, 847
  molecular basis of, **3:**854
Dominant alleles, **3:**823
Donathlandsteiner antibody, **1:**514
levo-DOPA. *see* Levodopa
Dopa decarboxylase, **5:**672
Dopamine, **2:**5, 9, 11, 12, 16, 18, 22–24, 29, 217–29, 330, 337, 523, 686, 831, 835, 850
  in autonomic neurons, **1:**540
  blockers, for Huntington's disease, **4:**272

Dopamine *(Continued)*
  characteristics of, **1:**178
  depleters, for Huntington's disease, **4:**272
  dopa decarboxylase and, **5:**672
  drug interference with, **5:**397
  and emotion, **2:**219–20
  levels in Huntington's disease, **4:**271
  maintenance of lactation, **4:**606
  metabolites, relationship to anxiety and depression, **1:**612–13
  as neuromodulator in retina, **6:**629
  Parkinson's disease and, **2:**219, **5:**668–69
  prolactin and, **3:**339
  prolactin regulation and, **5:**399
  in prolactin release, **6:**15
  in schizophrenia, **6:**778
  schizophrenia hypothesis and, **2:**9
  secretion, in adrenal medulla, **1:**76
  in sensation seeking, **6:**816
  synthesis, in adrenal medulla, **1:**644
Dopamine beta-hydroxylase, **2:**541
Dopamine beta-monooxygenase, in adrenal medulla, **1:**642–44
Dopamine receptors. *see* Receptors, dopamine
Dopaminergic neurons, **3:**739
Dopaminergic system, **2:**218
L-Dopaquinone, **2:**541
Doppler apparatus, in ultrasonography, **3:**7
Dorsal column-trigeminal complex, **7:**122
Dorsal horn, of spinal cord, **7:**202–3
Dorsal interosseous muscle, first. *see* First dorsal interosseous muscle
Dorsal nucleus of Clarke, **7:**204
Dorsal root ganglia. *see* Ganglia, spinal
Dorsal telencephalon, bird, **5:**306
Dorsal ventricular ridge, reptile, **5:**306
Dorsolateral pontine tegmental field, **4:**714
Dosage compensation (genetics)
  defined, **3:**709
  DNA methylation and, **3:**159
  methylation and, **3:**872
  in muscular dystrophy, **5:**192
  principles of, **3:**855
Dose-response relationship, toxicological
  estimation of, **3:**388
  foods and, **3:**673
Double-contrast examinations (radiology), **3:**4
Double gastrulation. *see under* Gastrulation
Double ligation model. *see* Models, double ligation
Double-minute chromosomes. *see under* Chromosomes
Double-minute DNA. *see* DNA, double-minute
Double mosaicism. *see under* Mosaicism

Double-nucleation model, 7:10–11
Double ovulation. see Ovulation, double
Doubling, cell. see under Cells
Down, John Langdon, 3:176
Down regulation, hormonal, 3:330
Down's syndrome, 1:698, 2:437–39, 441, 442, 859,
    3:175–89, 863
    amyloid fibrils, 1:215
    fragile X syndrome and, 3:717
    frequency, 2:438
    genetic aspects of, 3:839–40
    heterokaryotes and, 3:306
    inherited forms of, 1:699
    maternal age association, 1:699, 2:439
    molecular genetics of, 3:175–89
    physical characteristics, 1:699
    prevalence of, 1:702(t)
    risk factors, 1:699, 2:439
    trisomy 21 and, 2:481, 3:839–40, 4:951–52
Doxorubicin
    chemoimmunoconjugates with, 4:370
    immunosuppressive/immunopotentiating effects, 1:686
Dracunculus medinensis, 2:388
Dream anxiety attacks. see Nightmares
Dreaming, 3:191–97
Drift, ocular, 3:557
Drinking
    abnormalities, and autism, 1:506
    neural basis of, 5:371–72
Drinking, alcohol. see Alcohol; Alcoholism
Drip infusions. see Infusions, intravenous
Droplets, inhaled
    clearance and retention of, 7:580–82
    mechanism of deposition of, 7:579–80
Dropsy. see Edema
Drosophila
    cAMP, 2:317
    chemical mutations in, 5:169
    DNA, 2:472
    embryonic body pattern formation in, 3:284
    eye, 2:242
    mutants, 2:317
Drosophila melanogaster
    DNA amplification in, 3:101
    genetic map of, 3:826
    genetic recombination in, 3:825
Drug abuse. see Substance abuse
Drug addiction. see Substance dependence
Drug carriers, artificial cells as, 1:380
Drug combinations
    automation analysis computerized, 2:377–78
    Chou and Talaly combination index method, 2:377

Drug combinations (Continued)
    dose effect, 2:375
    dose reduction index, 2:371
    fractional product method of Webb, 2:377
    index, 2:371, 376
    intoxication and dementia, 2:764
    isobologram equation of Loewe, 2:376–77
    mass action law, 2:372, 375
    median-effect principle, 2:377
    therapy, 2:371–75
    Webb's fractional product method, 2:377
Drug design
    activated soft compounds and, 7:109–10
    active metabolite principle and, 7:110–12
    design of natural steroids as, 7:112–13
    enzyme isolation and, 3:404
    inactive metabolite approach in, 7:113–16
    parasite infections and, 5:654
    soft analogs and, 7:106–9
    target enzymes in, 3:393–94, 401
    therapeutic index in, 7:102
Drug industry, fermentation in, 3:938
Drug resistance, 2:514
    defined, 3:801
    gene amplification and, 3:102, 786, 787–88
    genetics of, 5:650–51
    HeLa cell lines, 4:111
    during lymphoma therapy, 4:850
    parasite, 5:650–51
    in tumors, glutathione and, 3:917
Drug resistance, cancer chemotherapy, 4:921–28
    change in susceptibility, 4:921
    combination of noncross-resistant agents, 4:925–26
    Luria-Delbruck model, 4:922–24
    model for single class of drug-resistant cells, 4:924–27
    multidrug resistance, 4:926
    necrotic regions, 4:924
    passaged experimental tumor systems, 4:927
    predicted cure rates, 4:925–26
    Skipper-Schabel model, 4:922–23
    stem cell division rate, 4:924
    Zero resistant cell probability, 4:924–25
Drug resistance, microbial, 1:246, 248–49, 328
    acquired resistance, 1:292
    chromosomal resistance, 1:291–92
    in developing nations, 1:293
    natural resistance, 1:291
    plasmid-mediated, 1:293
    resistance plasmids, 1:291
    susceptibility testing, 1:559
Drug synergism, 2:1, 374–75
Drug therapy. see Chemotherapy

Drugs. *see also* specific drugs
    abolition of pain by, **5**:413
    absorption, **5**:841–42
    acidic, protein interactions of, **1**:710
    activated soft, **7**:109–10
    analysis, **3**:199–209
    antimony-containing, for snail fever, **1**:263
    biliary excretion of, **5**:846–47
    bioavailability, **5**:842
    biodisposition, **5**:841–49
    biotransformation, **5**:845–46
    blood binding of, **1**:707–13
    blood concentration of, **1**:708
    blood protein binding of, **5**:843
    brain response to, animal models, **1**:258
    defined, **3**:199
    distribution in blood components, **1**:709
    dosage
        naive averaging of data to determine, **5**:852
        nonlinear mixed effect model for determining, **5**:852
        population kinetics and, **5**:852
        standard two-stage method for determining, **5**:852
    effect on behavior and emotions, **1**:613–14
    electrochemical detection, **3**:206
    electron-capture detection, **3**:205
    exaggerated response to, **5**:830–32
    excretion, lactation and, **5**:847
    extraction, **5**:837
    half-life, **5**:844
    hard
        defined, **7**:104
        and risk of sudden infant death syndrome, **7**:320–21
    implications in, **1**:707
    intermediate free hypothesis, **1**:711–12
    ionization, **5**:840
    laws governing, **1**:45–46
    levels and activity
        hysteresis and, **5**:856
        proteresis and, **5**:857
        receptor binding as factor in, **5**:854–55
        summary of kinetic-dynamic models to measure, **5**:855
    lipid solubility of, **5**:840
    lipoprotein binding of, **5**:840
    metabolism
        basic steps in, **5**:831
        defined, **3**:211
        liver enzymes important in, **5**:832
        metabolic pathways associated with, **5**:849
        phases in, **5**:845–46
        pulmonary, **6**:350

Drugs
    metabolism (*Continued*)
        stereoselective kinetics and, **5**:848–49
        and toxicity, **7**:103–6
    metabolism enzymes, **5**:833
    microiontophoresis and pressure injection of, **5**:5–9
    molecular weight, bile excretion and, **5**:846
    monitoring
        in pathological states, **1**:712–13
        therapeutic, **5**:857–58
    multiple binders, in blood, **1**:708
    neuroactive, **2**:11
    nonnarcotic, **5**:439–49
    oral, used in dentistry, **2**:785
    pharmacokinetics, **5**:835–61, 849–51
        analysis, **5**:849–51
        analytical methods in, **5**:836–39
        chimeric antibodies in, **3**:810
        clearance, **5**:843–44
        consequences of, **1**:709–11
        of diffusion, **5**:840
        direct approach to, **5**:849
        distribution, **5**:842–43
        effect of age and disease on, **5**:853–54
        elimination, **5**:843–47
        empirical approach to, **5**:849–50
        formulation and, **5**:841
        in health and disease, **5**:853
        history, **5**:836
        kinetic-dynamic relationships of, **5**:854–57
        mean residence times, **5**:845
        membrane transport of, **5**:839
        pharmacogenetics as factor in, **5**:847
        physiochemical properties of, **5**:840–41
        physiological approach to, **5**:851
        quantitative structure, **5**:840–41
        renal excretion, **5**:845
        stereoselective kinetics, **5**:848
        study of major body fluids used for, **5**:837
        symbols and derivatives to explain, **5**:859–60
        systems dynamics approach to, **5**:851
        theoretical approach to, **5**:850–51
    in pituitary hormone secretion disruption, **5**:396–97
    plasma concentration, and intravenous administration, **5**:844
    polarity, bile excretion and, **5**:846–47
    protein complexes in blood, **1**:707–8
    protein interactions of, **1**:710
    repeat-dose kinetics of, **5**:847
    soft, **7**:101–17
        analog design of, **7**:106–9
        analogs of, **7**:106–9

Drugs
  soft (*Continued*)
    defined, **7:**105
    design process of, **7:**102–3
  solubility, absorption and, **5:**841
  sulfites as ingredients in, **7:**341
  tachyphylaxis, **2:**514
  targeting, **1:**713
  therapeutic window of, **5:**835
  toxicity
    and dementia, **2:**342, 748
    and drug design, **7:**102–3
  transport
    across blood-brain barrier, **1:**723, 727
    and blood binding, **1:**711–12
    pH partition hypothesis of, **5:**840–41
  volume of distribution of, **5:**843
Drugs of abuse. *see* Street drugs
DSM-III-R, **4:**653, 668, 981, 995
Dual-task performance, **2:**862
DuBois, Eugene, **3:**492
Dubos, Renee, **3:**932
Duchenne muscular dystrophy, **2:**464, 479
  newborn screening test for, **5:**437
  pulsed field gel electrophoresis and, **6:**385
Ductus deferens, anatomy of, **6:**560–61
Duffy factor, **4:**882
Duke's classification, **3:**775
Duncan's syndrome. *see* Lymphoproliferative disorders
Duodenum
  anatomy, **3:**62–63, 64–65, 76
  ascorbic acid absorption, **1:**430
  bicarbonate secretion in, **7:**288
  Dyad symetry, defined, **2:**419
  fat digestion in, **3:**81
Dura mater, **5:**351
  cranial, **4:**980
  spinal, **4:**988, **7:**195
Dural septae, **4:**981
Dural venous sinuses, **4:**982–9734
Duration, exercise, **3:**515
Dust, toxicity, **7:**553–57
  coal mine, **7:**555
  organic, **7:**556–57
Dye marking, of cells, **5:**7–8
Dyes
  food analysis, **4:**223
  as pH indicators, **5:**577
  synthetic, in food analysis, **4:**223
Dynein, **2:**7, 8, 241, 242, **5:**20
Dynorphin, **2:**34
  functions, **5:**731–32, **6:**89

Dynorphin (*Continued*)
  levels in Huntington's disease, **4:**271
Dysarthria, **2:**649
  defined, **2:**763, **7:**158
  hypothermia, **4:**800
  lesioned cerebellum and, **3:**218
  in older adults, **1:**122–23
Dyscontrol syndrome, **2:**718
Dysentery, amebic, **1:**222, 266–67. *see also Entamoeba histolytica*
Dysgraphia, **3:**215–27
Dysgraphia, efferent, **3:**218
Dyskeratocytes, defined, **3:**861
Dyslexia, **3:**215–27. *see also* Dyslexia-dysgraphia
  beginning reading and, **6:**491–92
  hemispheric interactions, **4:**134
  occipital pole and, **3:**220–21
  perception and, **5:**743
Dyslexia, acquired, **2:**9, **3:**223, **4:**202
Dyslexia, ecological
  brain dysfunction and, **3:**221–22
  defined, **3:**215
  primate behavioral, **6:**127–33
Dyslexia-dysgraphia
  and child development disorders, **3:**217
  and occipital-parietal area, **3:**221
  and occipital-parietal cortex, **3:**221
  and occipito-parietal-(temporal) system, right, **3:**222–23
Dyslipoproteinemia, **2:**661, 668, 671
Dysmenorrhea, **3:**361
Dysomia, defined, **5:**529
Dysorthographia
  basal ganglia lesions and, **3:**224
  dysarthria and, **3:**218
  left prefrontal convexity impairment and, **3:**218
Dyspareunia
  defined, **3:**361
  in endometriosis, **3:**365
Dysphonia, **7:**157
Dysphoria, defined, **2:**511
Dysplasia, defined, **3:**765
Dyspnea, defined, **5:**259
Dyspraxia
  defined, **7:**158
  parieto-temporal-occipital lesions and, **3:**221
Dysthymia
  antidepressant drugs for, **1:**308
  characterization, **4:**997
  as form of depression, **5:**93
Dystrophin
  alpha-actinin and, **5:**222, 223

Dystrophin *(Continued)*
    analysis of, **5:**202
    defined, **5:**191
    in DNA marker diagnosis, **3:**147
    evolutionary conservation of, **5:**198
    gene, **2:**479, **2:**44
    isoforms of, **5:**199
    localization of, **5:**196–98
    mutation and, **5:**245
    structure of, **5:**196,197
    tissue specificity of, **5:**198–99

E

E-cadherin, **2:**272
E2A gene, inactivation of, **3:**157
E2A promoter-CAT gene construct
    in cell line generation, **3:**157
    inactivation of, **3:**157
    inhibition of, **3:**158
Ear, **3:**229–41
    in fragile X syndrome, **3:**711
    muscles, **2:**599
    organs in, and space travel, **7:**149–50
    sensitivity range of, **5:**739
    sound transduction in, perception and, **5:**739
Ear, external, **3:**229–30
Ear, middle, **3:**230–32
Early promoters. *see* Promoter regions (genetics)
Easter Island, **3:**91
Eastern equine encephalitis virus. *see* Encephalitis virus,
    Eastern equine
Eating. *see also* Satiation
    abnormalities, **3:**591–92
    autism and, **1:**506
    brain centers associated with, **5:**519–20
    cardiovascular response, **4:**553–54
    feeding behavior and, **3:**589–92
    food acceptance and, **3:**638
    homeostatic regulation of, **5:**485
    and hunger, **6:**724–25
    influence of smell on, **5:**753
    influence of taste and/or smell disorders on, **5:**754
    influence of taste on, **5:**753
    innate sensory preferences and, **5:**754
    meal size
        gastric factors as physiological control of, **1:**349–50
        and satiety, **1:**349–51
    measurement of, **3:**594–95
    mechanisms by which sensory factors influence, **5:**754
    motivational parameters in, **6:**726–27

Eating *(Continued)*
    nutrient-related health disorders and, **5:**753
    oral stimulation and, **5:**754
    sensory influences on, **5:**753
    system for regulation of, **5:**519
Eating behavior. *see* Feeding behavior
Eating disorders, **3:**243–50
    anorexia nervosa as, **1:**352
    bulimia nervosa as, **1:**352
    characterization, **4:**997
    in children, **5:**709
    obesity as, **1:**351
    restrained eating as, **1:**351
    serotonin and, **6:**832
Eating utensils. *see* Cooking and eating utensils
EBNA-1, **3:**470
EBNA, varieties of, **3:**467–69
EBV-determined lymphocyte-detected membrane antigen,
        3:470
Eccrine glands, **7:**44–45
Echinococcus granulosus, **2:**390
Echinosphaerium nucleofilium, **5:**19
Echography. *see* Ultrasonography
Echotomography, computer. *see* Ultrasonography
ECO DNA topoisomerase II. *see* DNA, gyrase
Ecocycle, **4:**700
Ecology, gastrointestinal cancer and, **3:**765–67
Ecosystem, defined, **3:**251
Ecotoxicology, **3:**251–59, 391
Ectoderm, monoamniotic twinning and, **3:**312
Ectoderm, primitive. *see* Blastoderm
Ectodermal dysplasia, **2:**697
Ectoparasites, **2:**390
Ectopic hormone syndromes, **4:**275
Ectopic hormones. *see under* Hormones
Eczema, **5:**710, **7:**47
EDAX. *see* Energy-dispersive analysis of X-rays
Edelman's classification for development, **2:**873
Edema, **2:**145, 175
    defined, **3:**93
    and diuretics, **2:**175
    facial, in spaceflight, **2:**149, 151
    hereditary angioneurotic, **4:**384
    kwashiorkor-related, **4:**894
    pulmonary, extravascular water and, **6:**376
    thiazides and, **3:**97
    types of, **5:**40–41
Edema, interstitial
    consequences of, **5:**35
    defined, **5:**31
    safety factor, **5:**31, 39–40
    tissue fluid pressure, defined, **5:**36

Edkins, **5:**713

Edogram, in feeding behavior research, **3:**595

Education

　body weight and, **3:**684

　dyslexia-dysgraphia and, **3:**226

　in ergonomics, **3:**478

　health psychology, **4:**80–81

　for prevention of birth defects, **1:**703

　programs in treatment of autism, **1:**511

Education, nutrition. *see* Nutrition education

Education, special, **3:**716–17

Edward syndrome. *see* Trisomy-18

Edwards, R. G., **3:**600

EF-hand, **2:**89, 91, 93, 94

　model, **2:**94

　proteins, listed, **2:**94

　tree, evolutionary, **2:**94

Effector, gastrointestinal, **2:**33

Effector cells

　access to, **4:**394

　access to regionally located antigens, **4:**394

　response amplification, **4:**394

　selection and immune deficiencies, **4:**397–98

　types generated, **4:**399–400

　unique spectra, **4:**397

Effector organ, **2:**137

Effector pathway, **4:**362

Effector system, **2:**326

Efferent, **2:**333, 337

Efferent fiber, **2:**682, 685

Effort. *see* Exertion

Eflornithine, **2:**386

　as inhibitor of ornithine decarboxylase, **6:**70

　therapy for protozoal infections, **6:**72

EGF receptor. *see* Receptors, epidermal growth factor-
　　urogastrone

Egg-binding protein, **6:**511

Egg shell, of peregrine falcon, **3:**254

Egg yolk, fatty acid levels in, **5:**542

Eggs

　in "basic four" food groups plan, **3:**654

　Salmonella contamination of, **3:**664

Eggs, amphibian, **3:**284

Eggs, sea urchin, **5:**902

Ego, **2:**853, 854. *see also* Personality development; Self

　cognitive-developmental, **2:**856–58

　cultural construction, **4:**702–4

　defense, **2:**854

　defined, **6:**308

　identity and aging, **4:**705–6

　LSD use and, **5:**448

　natural opposition within, **4:**704–5

Ego (*Continued*)

　notion of self, **4:**698

　and personality disorder, **2:**856

　psychology of Heinz Hartmann, **2:**854

　repressed, in psychoanalysis, **6:**308

　sublimation, **2:**954

　Sullivan's interpersonal theory, **2:**853–54

Egocentrism

　during adolescence, **1:**66

　and behavior, **1:**579–80

　distinguished from altruism, **1:**198–99

　helping, **1:**200

　helping, motives for, **1:**200–202

Ehlers-Danlos syndrome

　defined, **3:**261

　fragile X syndrome and, **3:**713

Ehrlich, Paul, **1:**271–72, **5:**137

Eicosanoids

　characterization, **4:**748

　defined, **3:**567

　diet lipid-mediated, **5:**508

　functions of, **3:**579

　in gastric mucosal blood flow regulation, **3:**760

　in inflammation, **1:**313, 314

　as second messengers, **5:**417

　synthesis, **3:**579–80

Eicosapentaenoic acid, **2:**170

　as omega-3 fatty acid, **5:**533, 534

　sources of, **5:**533–34

Einstein, Albert, **5:**904

Ejaculation, **7:**174

Ejaculatory duct, **6:**560–61

Ejection fraction, ventricular, **2:**179

Elastase, acyl enzymes and, **3:**397

Elastic energy, **5:**185

Elastic fiber, **2:**639, 642

Elastic fibers

　components of, **3:**262

　formation of, **3:**265–67

　tropoelastin and, **3:**264

Elastin, **2:**201, 215–16, 571, 639, 640, 652, **3:**261–69

　collagens and, **3:**525

　of dermis and subcutaneous tissue, **7:**47

　diseases related to, **3:**268–69

　organogenesis and, **3:**267–68

Elastolysis, alpha$_1$-antiproteinase and, **3:**268

Elbow joint, angular velocity in, **3:**698

Elder abuse, **1:**135–36

Elderly persons. *see* Aged

Electrical activity

　in fetal nervous system, **3:**618–19

　myocardial, **2:**134–36

Electrical activity (Continued)
    of neocortex, 5:314
    olfactory, 5:526
Electrical activity mapping, brain, 3:216-17
Electrical capacitance, defined, 3:9
Electrical conductance, of biological tissues 3:14
    of cellular structures, 3:11
    defined, 3:9
    dielectric polarization and, 3:11
Electrical impedance. see also Tetrapolar bioelectrical im-
        pedance analysis
    defined, 3:9
    of erythrocyte suspensions, 3:10
Electrical properties
    astrocytes, 1:438-39
    passive, 3:9
Electrical responses
    in auditory nerve, 3:238
    of cochlea, 3:237-38
Electrically gated channels, 4:973
Electrochemical gradient, 4:973
Electroconvulsive therapy, 2:842, 5:96-97
    for depression, 2:838
    side effects, 2:838
        amnesia, 2:838
        confusion, 2:838
Electrodermal response, in sensation seekers, 6:812
Electrodes, miniaturized. see Microelectrodes
Electroencephalography, 2:709, 714, 715
    in benign partial epilepsies, 3:458, 459
    defined, 3:191, 455
    in depression, 2:835
    in dream research, 3:193-94
    in epileptic absence seizures, 3:456
    in febrile convulsions, 3:457
    in infantile spasms, 3:458
    of olfactory bulb activity, 5:527
Electrofocusing. see Isoelectric focusing
Electrolytes
    absorption of, 3:81
    changes, countermeasures to space travel, 7:151
    defined, 3:71
    dielectric characteristics of tissues and, 3:13
    replacement, thermal injury and, 7:452-53
Electromagnetic radiation
    characteristics of, 7:65-66
    properties of, 5:587
Electromagnetic spectrum, 5:587-88
Electromorphic protein variants. see Proteins, electromor-
        phic
Electromyographic activity
    defined, 3:691

Electromyographic activity (Continued)
    in m. biceps caput longum, 3:695
    in movement coordination, 3:691-92
Electromyography, 3:699
    defined, 5:173
    electromechanical delay in, 5:184
    in muscular dystrophy, 5:191-92
Electron energy loss spectroscopy, 3:276
Electron microscopy. see Microscopy, electron
Electron probe x-ray analysis, 5:584
Electron transport
    ATP synthesis and, 5:61-62
    between redox centers, 5:61
    of respiratory chain complex subunits, 5:60-61
    via the mitochondrial electron transport chain,
        1:459-60
Electron transport chain, 2:238, 239
Electron tunneling, defined, 5:55
Electrons
    interaction with body tissue, 6:439
    as source of space radiation, 6:428
Electrophiles, 2:118
    defined, 3:671, 907
    detoxification of, 3:910, 915
Electrophoresis
    conventional, effect of alternating fields on, 6:380
    in enzyme isolation, 3:408
    isoenzymes, 4:577
Electrophoresis, acrylamide gel, 3:877
Electrophoresis, agar gel, 6:380
    for allele detection, 3:140
    in restriction fragment-length polymorphism detection,
        3:137
Electrophoresis, gel, 2:475
    defined, 3:125
    denaturing gradient gel, 6:527
    in DNA-sequencing analysis, 3:173
    in in vitro DNA manipulation, 3:876
    two-dimensional, trisomic cells and, 3:177
Electrophoresis, polyacrylamide gel, in oligonucleotide
        purification, 3:172
Electrophoresis, pulse field gel, 2:475, 6:379-86
    alternating fields and, 6:380-81
    applications of, 6:383-86
    choice of restriction enzymes for, 6:383
    defined, 2:475, 6:379
    in Duchenne muscular dystrophy defect, 6:385
    equipment used for, 6:381
    field switching in, 6:381-82
    in genetic mapping, 3:832
    identifying genes in, 6:383-84
    interpretation of data produced by, 6:385-86

Electrophoresis, pulse field gel *(Continued)*
  mapping of cystic fibrosis gene, **6**:384–85
  in muscular dystrophy diagnosis, **3**:147
  need for, **6**:379–80
  sample preparation and handling in, **6**:382
  size markers in, **6**:382–83
  switch interval and, **6**:382
Electroporation, **3**:805
Electrotherapy, in treatment of mania and depression, **1**:95
Element method, finite, **2**:689
Elements, **2**:813
  anti-carcinogenic, **2**:813
  essential for life, **5**:48
  favoring caries, **2**:81
  heavier, body requirement for, **5**:51
  light, in body, **5**:50
  trace. *see* Trace elements
Elimination
  animal, **3**:256
  cellular, **3**:900
ELISA. *see* Enzyme-linked immunosorbent assay
Ellipsoidal joints, **1**:374
Elongation factor 2, **4**:418
EMBL data library, **3**:163
Embolism
  heart-lung machine, **4**:95
  tumor cell, **4**:1012–13
Embryo, **2**:870, **4**:423–30
  aborted fetuses and, **3**:294
  body pattern formation, **3**:281–91
  congenital defects, **7**:412–13, 414
  defined, **3**:293, 599, 609
  development during implantation, **4**:428–29
  developmental stages in, **3**:298, 299–300
  endometrium preparation for blastocyst implantation, **4**:429
  epithelial penetration stage, **4**:424–25
  examination of, **3**:296–98
  growth, **4**:26
  growth disorganization
    abnormal chorionic sac and, **3**:297
    chromosomal numerical abnormality and, **3**:302
    defined, **3**:293
  hemoglobin, **4**:143
  implantation site, **4**:427
  macerated, **3**:298, 301
  melanoma, **4**:887
  previllous stages, **4**:427
  primitive streak, **4**:429
  time sequence, **4**:423–24
  transition from free to attached blastocyst, **4**:424
  trophoblast differentiation, **4**:426–27

Embryo *(Continued)*
  in vitro fertilization, **4**:430
  in vitro models, **4**:430
Embryo, preimplantation. *see* Blastocyst
Embryofetopathology, **3**:293–304
Embryogenesis, **2**:245–47
  and cell death, **2**:245–47
  and limb bud formation, **2**:246
  and sex determination, **2**:246
  of twins, **3**:301–15
Embryonic induction
  primary, **3**:286–87
  secondary, **3**:286
Embryonic neural crest cells. *see* Neural crest cells
Embryotoxins. *see* Teratogens
Emerogenes. *see* Genes, suppressor, tumor
Emesis, **5**:259
Emetine, **1**:267
Emigration and immigration, Upper Paleolithic, **3**:502
Eminence, median, **2**:11
Emmetropia, **3**:533
Emotional lability, theory of, **1**:105
Emotions, **2**:9–10, 217. *see also* Affect; Externality, concept of
  appraisal theories, **1**:107–8
  as arousal plus cognition, **1**:105
  attraction as, **4**:537
  attributional theory, **1**:106
  in close relationships, **1**:105–6
  cognitive structural theory, **1**:106–7
  control of, **2**:9
  dimensional structure, **1**:101–2
  drug effects on, **1**:613–14
  effects of alternative future worlds, **1**:107
  excitation transfer theory, **1**:104–5
  expressing, **5**:786–89
  facial expression and, **5**:807
  facial feedback theory, **1**:103–4
  falsification of, **5**:787–88
  in feeding behavior research, **3**:596
  headache associated with, **4**:55
  hippocampal functions, **4**:232
  hypothalamus role, **4**:311
  information complexity effects, **1**:106
  as interruptions, **1**:107
  James-Lange theory of, **1**:102–3
  measurement, in psychobiology, **1**:609–12
  muscular theory of, **1**:104
  neurotransmitter systems and, **1**:612–13
  norm theory, **1**:106
  objective performance instruments, **1**:611–12
  priorities and, **1**:107

Emotions *(Continued)*
    prototypical, **1:**102
    schema congruity theory, **1:**105
    schema-triggered affect, **1:**106
    self-report tests, **1:**609-10
    social constructivist view of, **1:**102
    in starvation process, **3:**563
    vascular theory of, **1:**104
Empathy-altruism hypothesis, **1:**205-6
Emphysema
    alveolar elastin and, **3:**268
    enzyme inhibitors and, **3:**397
    inhaled toxicants as cause of, **7:**583
    as type of airway obstruction, **6:**362-63
Emulsions
    defined, **3:**47
    in digestion, **3:**80
En cascade, defined, **5:**667
Enalapril, **2:**175, 176, **3:**396
Enamel. *see* Dental enamel
Enamel organ, **2:**818-21
Enamelin, **2:**773, 822
Encainide, **2:**173.746, 747
Encephalitis
    in infants and toddlers, **4:**461-62
    mosquito-borne, **7:**522-23
    Parkinson's disease and, **5:**669
    tick-borne, **7:**524
Encephalitis virus, Eastern equine, **7:**895
Encephalization
    defined, **3:**495
    of earliest hominids, **3:**496
Encephalomyelitis, allergic, **1:**526-27
    experimental
        as multiple sclerosis model, **5:**146-47
        T-cell receptor in, **5:**147-48
    nonadrenergic nerves and, **5:**383
Encephalomyelitis, myalgic. *see* Fatigue syndrome,
        chronic
Encephalomyocarditis virus, **4:**524
Encephalomyopathy, familial mitochondrial. *see* Mito-
        chondrial encephalomyopathy, familial
Encephalopathy, **2:**341, 345, 346
    Reye syndrome-related, **4:**465
Encodingism, **2:**547, 548, 553-55, 557
Endemic disease, defined, **7:**640
Endemic nephropathy, Balkan. *see* Balkan nephropathy
Endocardium, **2:**157
Endocrine action, defined, **5:**713
Endocrine agents, defined, **3:**318

Endocrine cells
    of oxyntic mucosa, **3:**768
    of small intestine, **3:**768
Endocrine diseases
    of children, **3:**352-53, **5:**707-8
    classes of, **3:**333-35
    depression-related, in older adults, **1:**132
Endocrine gland neoplasms
    octreotide for, **5:**723
    secretin and, **5:**722-23
Endocrine glands
    aging of, **3:**902-3
    classical, **3:**319-23
    defined, **3:**71, 243, 318-19
    effect of atrial natriuretic factor on, **1:**473-74
    fetal, **3:**611, 619
Endocrine hormones
    changes in, space travel and, **7:**144-45
    defined, **5:**394
    sleep and secretion of, **7:**75
Endocrine neoplasms
    of gastrointestinal tract, **3:**769
    multiple, peptide hormone disorders and, **5:**722
Endocrine syndromes, **4:**275
Endocrine system, **2:**322, **3:**317-35
    development of, **3:**337-59
    and hormones, **2:**322
    hypothalamic control, **4:**311-12
    zinc and, **7:**886-87
Endocrine tissues, nonclassic, **3:**323
Endocytosis, **2:**238, **4:**975, 239, 299, 302. *see also* Pha-
        gocytosis
    receptor-mediated, and LDL lipoproteins, **1:**452
    and secretion of proteins, **6:**241
Endoderm
    in body pattern formation, **3:**286
    during embryo development, **4:**428-29
    parietal, **4:**428-29
Endolymph, electrochemistry of, **3:**232
Endolysosome, **2:**238
Endometrial hyperplasia, **5:**604
Endometrial neoplasms, **5:**604-5
    characteristics of, **3:**364
    endometriomas, defined, **3:**361
Endometriosis, **3:**361-69
Endometrium, **2:**248
    decidualization, **4:**426-27
    defined, **3:**361, 609
    endometriotic implants, **3:**365
    beta-endorphin, **2:**5, 18, 323, 338, 339
    morphology of, **7:**738
    preparation for blastocyst implantation, **4:**429

Endomyosis, defined, **3**:361
Endomysium, defined, **7**:23
Endonucleases, in DNA repair, **6**:550–51
Endonucleases, restriction. *see* DNA restriction enzymes
Endoperoxides, **3**:578
Endoplasmic reticulum
    Ca^{2+} pump, **4**:563
    defined, **5**:205
    Golgi apparatus and, **4**:3–4
    phospholipid transfer from, **5**:889
    in protein sorting, **6**:239–40
    as routing signal receptor, **6**:247–48
    sorting and transport of proteins in, **6**:252–55
Endoribonucleases
    activation by (2',5')-oligoadenylate synthetase, **6**:664
    defined, **6**:661
Endorphins, **2**:26, 35–39
    abolition of pain by, **5**:413
    in disease therapy, **5**:723
    functions of, **5**:719, **6**:88, 731–32
    immunity and, **5**:375
    listed, **2**:17
    types of, **5**:727
Endoscopic retrograde cholangiopancreatography. *see* Cholangiopancreatography, endoscopic retrograde
Endosomes, **2**:238, **5**:875
Endospermous tissues, **3**:40
Endosteum, **2**:697
Endothelial cells
    antithrombogenic function of, **6**:350
    psoriasis and, **3**:451–52
    and radiation injury, **6**:424
Endothelial cells, pulmonary
    barrier function of, **6**:349
    homeostasis and, **6**:350
    immune responses of, **6**:350
    and lung injury, **6**:351–53
    metabolic activity of, **6**:350
    modulation of vascular tone by, **6**:349
    pharmacology of, **6**:348–49
    respiratory function of, **6**:350
Endothelial-leukocyte adhesion molecule 1, **4**:476
Endothelial-leukocyte cell interactions, **4**:475–76
Endothelins, **2**:366, **6**:349, 368
Endothelium, **2**:157, 183, 366
    function of, **6**:91
    postcapillary venules, **4**:396–97
    tissue-specific cell-surface molecules, **4**:1013
Endothelium, vascular
    effect of radiation on, **6**:424
    endothelium-derived relaxing factor and, **3**:371
Endothelium-dependent contracting factor, **6**:349

Endothelium-derived relaxing factor, **3**:371–76
Endotoxins
    bacterial, **7**:589–90
    immunotherapy with, **4**:348
    lung injury from, **6**:352
    salmonella, structure of, **6**:704
Energy
    adenosine triphosphate and, **1**:53–55
    body reserves, **4**:1005
    high, theoretical approach to concept of, **1**:54–55
    metabolism, space travel and, **7**:143–45
    for phagocytosis, **4**:857
    transduction, **2**:235
        events in, **1**:53–54
        proteins and, **6**:225–26
Energy, digestible. *see* Digestible energy
Energy, metabolizable
    defined, **5**:471
    of protein, **5**:476
Energy-dispersive analysis of X rays, **1**:423, **3**:274
Energy expenditure. *see* Energy metabolism
Energy intake, defined, **3**:589
Energy metabolism, **3**:377–81
    adrenal medulla and, **1**:641–53
    body fat in, **3**:576
    defined, **5**:56
    fats in, **3**:575–76
    hormone regulation of, **3**:326–27
Energy-reducing devices. *see* Labor-saving devices
Engineering psychology. *see* Human engineering
Enhanced velocity, muscle, defined, **5**:187
Enhancer elements (genetics)
    in antibody production, **3**:806
    defined, **3**:151, **5**:205, 794
    DNA binding sites and, **3**:124
    in gene transcription regulation, **3**:112
D-Ala-met-enkephalinamide, **2**:340
Enkephalins, **2**:18, 34, 323, 338
    in autonomic neurons, **1**:541
    functions of, **5**:731–32, **6**:89
    immunity and, **5**:375
    levels in Huntington's disease, **4**:271
    and pain control, **5**:367
Enolase. *see* Phosphopyruvate hydratase
Enovid pill, **5**:599
Entactin, in basement membranes, **4**:624
Entamoeba histolytica, **4**:450. *see also* Dysentery, amebic
    adherence of trophozoites to target cells, **1**:223–24
    bowel infections, **4**:462
    cysts, **1**:218, 220–21
    cytolytic activity, **1**:225–26
    cytotoxins, **1**:224–25

Entamoeba histolytica (Continued)
    diseases associated with, **1**:222–23, 222(t)
    encystation, **1**:218, 266
    epidemiology, **1**:218–19
    excystation, **1**:218
    host immune response to, **1**:226–28
    infection by, characteristics of, **6**:268
    infective dose, **1**:218
    intestinal colonization and adherence, **1**:223–24, 266
    life cycle, **1**:218–19, 266
    metabolism, **1**:219–22
    metacystic development, **1**:218
    phylogeny, **1**:219
    prevention of infection, **1**:229–30
    secretory enzymes, **1**:224–25, 266
    strain virulence, **1**:219
    structure, **1**:219–22
    trasmission of amoebic infection, **1**:218–19
    trophozoites, **1**:218–21
        complement-mediated resistance, **1**:227–28
    zymodemes, **1**:219
Entamoeba species, **6**:268
Enteric fever. *see* Typhoid
Enteric nervous system, **2**:31–33, 38, **4**:555
    of gastrointestinal tract, **3**:73
    small intestine motility and, **3**:77–78
Enteritis, **4**:449–50
    Campylobacter jejuni-induced, **3**:665
Enterobiasis, **2**:387
Enterobius vermicularis, **2**:387
Enteroclysis, in small bowel double-contrast examination,
    3:4
Enterococcus, **1**:319
Enterocyte migration, **4**:617
Enterogastrone, **5**:716
Enteroglucagon, **5**:718
Enterooxyntin, **5**:716
Enteroreceptor, **2**:33
Enterotoxins
    bacterial, **7**:590
    staphylococcal, **5**:4
Entitlement failure
    defined, **3**:561
    famine and, **3**:562
Entopeduncular nucleus, **6**:783
Entorhinal cortex, **4**:227, 229–30
    connections, **4**:230–31
    sensitivity to disease states, **4**:232
Enuresis. *see* Bedwetting
env genes. *see* Genes, *env*
env viral proteins. *see* Gene products, env
Envelope, nuclear. *see* Nuclear membrane

Environment
    aging and, **3**:891
    crop production and, **1**:143–44
    defined, **3**:251
    gene amplification and, **3**:786
    language performance in older adults and, **1**:126–27
    metabolic regulation role, **4**:1006
    performance and, **5**:122
Environmental exposure, birth defects related to,
    1:700–701
Environmental pollution
    ecotoxicology and, **3**:251
    effects of, **3**:253–55
Environmental specimen banks, **3**:259
Environmental toxicology. *see* Hazardous substances
Environmentalism, extreme. *see* Behaviorism
Enzymatic product inhibition
    defined, **3**:931
    of hexokinase, **3**:936
Enzyme conformation
    defined, **3**:411
    native, **3**:412
    structural elements of, **3**:431–33
    thermostability and, **3**:411–12
Enzyme deficiencies
    bone marrow transplantation for, **1**:796
    prenatal, diagnosis of, **3**:820
Enzyme denaturation, **3**:412–13, 414, 415
    alpha-amylases and, **3**:415
    solvating medium changes and, **3**:433
    temperature of, **3**:420
Enzyme inhibitors, **3**:393–401
    classes of, **3**:429
    irreversible, characteristics of, **3**:398–400
    noncompetitive, characteristics of, **3**:395
    patterns of, **3**:430
    reversible, characteristics of, **3**:395–98
Enzyme isolation, **3**:403–9
Enzyme-linked immunosorbent assay, **3**:468
    basic principles, **4**:354
    customary, **4**:352
    double sandwich, **4**:354
    drug analysis using, **5**:838
    enzymatic systems in, **4**:353
    kinins, **4**:594
    parasite detection, **4**:388
    single sandwich, **4**:352–53
    for thyrotrophin receptor antibodies, **4**:354–55
Enzyme precursors
    activation, **4**:180–81
    amino-terminal regions, **4**:184
    blood clotting cascade, **4**:178–79

Enzyme precursors *(Continued)*
functional properties, **4**:182–85
proteinase domains, **4**:183–84
structure, **4**:182–83
Enzyme reactions, classification of, **3**:436
Enzyme repair, **3**:172
Enzyme-substrate complexes, **3**:433–34
Enzyme tests
defined, **3**:403
lactose malabsorption diagnosis, **4**:619
methods of, **3**:405–6
Enzyme thermoinactivation
defined, **3**:411
irreversible, **3**:413–18
reversible, **3**:412–13
steps in, **3**:412
in vivo, **3**:418–19
Enzyme thermostability, **3:411–21**
Enzymes. *see also* Isoenzymes
antibody-combining sites and, **3**:809
antibody conjugates with, **4**:372–73
in artificial cells, **1**:382
binding to immunoglobulins, **4**:580
cellular chemical reactions and, **3**:423–38
chemical catalysis by, **6**:224–26
cooperative, kinetics of, **3**:428–29
defined, **3**:71, 317, 393, 439
denaturation, **3**:405
in DNA replication, **3**:439–44
extracellular matrix degradation, **4**:1014
functions, **3**:177–79, 423–38, **6**:224–25
genetic diseases, **4**:1001–3
in gingiva, **4**:36
in glycogen degradation, **3**:923–24
in glycogen formation, **3**:921–24
in Golgi apparatus, **4**:7–9
induction, and pesticide residues in food on, **7**:562
in lipoprotein metabolism, **6**:34
lytic, in soy sauce production, **3**:667
mammalian, **3**:405
mechanism-based enzyme inactivators, **3**:399–400
mechanochemical, **5**:19–20
membranes, **2**:237–38
metabolic, as substrates for phosphorylation, **6**:210–11
of microtubule protein phosphorylation, **5**:894
multifunctional, in nucleotide synthesis, **5**:465
muscle-specific isoforms and, **5**:163
nonisoenzymic molecular forms, **4**:580
of osteoclasts, **1**:766–67
oxygen transport in tissue, **5**:622
P-450, drug metabolism by, **5**:831
phenotypes in HeLa cells, **4**:108

Enzymes. *(Continued)*
protein-based, **3**:431–33
rate-limiting, **4**:1000–1001
regulation of, **6**:225
relaxed state, defined, **3**:919
renaturation, **3**:413, 415
respiratory, classification of, **5**:622–23
role in bacterial virulence, **1**:242
secretory, **1**:224–25
sperm, **7**:166
structure, ultracentrifugation in study of, **7**:716
taut state, **3**:919
thermophilic, **3**:419–20
variants, frequencies of, **6**:108
Enzymes, soluble
metabolic organization and, **5**:462
transferase, drug metabolism and, **5**:831
xenobiotic metabolizing, in nasal cavity, **5**:285–87
Enzymoelectrophoresis, **4**:577
Eocene primates, **6**:136–38
Eosin thiocyanate, **5**:580
Eosinophil cationic protein, **6**:663
Eosinophil-derived neurotoxin, **6**:663
Eosinophils
in allergic response, **1**:189
characteristics of, **5**:812
characterization, **4**:166
effect of platelet-activating factor on, **6**:47–48
functions, **4**:166
in inflammation, **1**:313
production in helminth parasitic diseases, **4**:387
respiratory burst in, **6**:583
Ependyma, **2**:319, 678
Ependymoglial cells, **1**:437–38
Ephedrine, **2**:512
characteristics of, **1**:178
Epicardium. *see* Pericardium
Epidemic disease, defined, **7**:640
Epidemiological Catchment Area, **3**:843
Epidermal cells, grafts of, **4**:583
Epidermal coverings, plant, **3**:39
Epidermal growth factor, **2**:61, 234
in bone cell differentiation, **1**:788
DNA synthesis and, **3**:173
effects on bone cell function, **1**:765
extracellular matrix molecules and, **3**:446–47
gangliosides and, **3**:734
in glioblastomas and squamous cell carcinomas, **5**:548
receptor, **2**:61
and transforming growth factor, comparative biological
activities of, **7**:628–29

Epidermal growth factor receptors. *see* Receptors, epidermal growth factor-urogastrone

Epidermal growth factor-related transforming growth factor. *see* Transforming growth factor alpha

Epidermal hyperplasia, psoriatic. *see* Psoriasis

Epidermal prickle cell layer, **3**:864

Epidermal proliferation, **3**:445-53

Epidermal ridges. *see* Dermal ridges

Epidermis
    development of, **7**:41-42
    layers of, **7**:42
    light penetration by, **5**:593
    sensory receptors of, **7**:677
    structure of, **7**:43-46

Epidermodysplasia verruciformis, **5**:637-38

Epidermolysis bullosa, **5**:615
    in children, **5**:710

Epididymis
    anatomy of, **6**:560
    defined, **3**:599
    sperm maturation in, **3**:600
    sperm transport in, **7**:173

Epidural hematoma, **4**:988

Epidural space
    cranial, **4**:980
    spinal, **4**:988

Epigenesis, **2**:53, 869

Epiglottis
    anatomy, **4**:655
    function of, **6**:589
    swallowing and, **3**:74

Epiglottitis, **4**:460

Epilepsy, **2**:717-18, **3**:455-61, 876
    atonic seizures, **3**:456
    benign partial epilepsies, in children, **3**:458-59
    central nervous system transmitters in, role of, **5**:412
    in children, **2**:864
    and criminality, **2**:717-18
    defined, **6**:780
    enzyme inhibitors and, **3**:399-400
    generalized seizures in, **3**:456
    hippocampal formation and, **4**:233
    hippocampal inhibition loss and, **5**:388
    partial seizures, **3**:456
    temporal lobe, **4**:233
    treatment with cannabidiol, **4**:910

Epilepsy, grand mal, **5**:443

Epilepsy, myoclonus, **5**:62

Epimastigote, defined, **7**:681

Epimysium, defined, **7**:23

Epinephrine, **2**:11, 12, 12, 91, 136, 140, 153, 171, 186, 217, 218, 221, 238, 241, 325, 330, 337, 513, 534, 717, 784, 787
    in autonomic neurons, **1**:540
    effect on adrenal gland, **1**:73
    in gastric circulation regulation, **3**:762
    glycogen breakdown and, **3**:925
    lipoprotein lipase and, **3**:587
    perinatal, **3**:350
    phagocytosis and, **5**:382
    plaque-forming cell response and, **5**:382
    as platelet receptor, **6**:55-56
    synthesis, in adrenal medulla, **1**:644-45
    in treatment of allergy, **1**:195
    in treatment of sulfite sensitivity, **7**:340

Epiphyses
    cold-induced necrosis, **4**:807
    ossification, **4**:28-29
    union to long bones, **4**:29

Episiotomy, defined, **3**:361

Episodic memory, **4**:673, 676-77

Episomal persistence, defined, **3**:463

Episomal state, **4**:189

Episomes. *see* Plasmids

Epithelial-mesenchymal interactions, **1**:788

Epithelial proton transport, **4**:569

Epithelium, **2**:271-73, 186, 217, 218, 221, 238, 241, 325, 337, 513, 534, 731, 777, 784, 787
    basement membranes and, **3**:524
    defined, **3**:71
    differentiation of, **5**:636-37
    Epstein-Barr virus and, **3**:470
    gastrointestinal, radiation effects on, **6**:469-70
    gamma-glutamyl transpeptidase in, **3**:912
    nose, functional role of, **6**:358
    oral, **2**:780
    penetration stage, **4**:424-25
    polarized, **6**:59-66
        diseases affecting, **6**:65-66
        distribution and general organization of, **6**:59-60
        embryonic development of, **6**:63-64
        functional surface domains in, **6**:61
        maintenance of, **6**:64-65
        as passive and active barriers, **6**:63
        protein sorting in, **6**:241
        protein transporting pathways in, **6**:63
        tissue distribution of, **6**:60
    tumor cell properties, **4**:584
    visceral, defined, **3**:93

Epitope, conformational, **4**:329

Epitopes, **1**:565, **4**:314
    in autoimmune disease, **1**:340

Epitopes *(Continued)*
    defined, **4:**402
Epoxide hydratase, **2:**118
Epoxide hydrolases, **5:**286
Epstein, **3:**777
Epstein-Barr virus, **2:**849, **3:**463-72, **4:**187-88, 190
    and B-lymphocyte transformation, **2:**230, 457
        correlation with nasopharyngeal carcinoma and
            Burkitt's lymphoma, **4:**343
        genome
            structure of, **3:**466
            transcription of, **3:**466
        latent, **4:**193
        mononucleus, **4:**190
        viral enzymes of, **3:**471
        viral proteins, **3:**470-72
        virion envelope proteins, **3:**472
Equal environments hypothesis, **3:**837
Equilibrium
    control in locomotion, **4:**771
    fusimotor system and, **3:**724
    maintenance during locomotion, **4:**777-78
    space travel effects on, **7:**149-50
Equilibrium constant, defined, **3:**425
Equine infectious anemia virus, **5:**677
Equine infectious anemia virus vaccine, **6:**646-47
Equipotentiality, **2:**859, 872
Equity theories of interpersonal attraction, **4:**539-40
Erectile tissues, genital. *see* Genital tissues, erectile
Ergoline, **2:**29
Ergonomics. *see* Human engineering
Ergonomics Society, **3:**475, 478
Ergosterol, **2:**745
Ergot, **2:**28, **7:**592, 29
Ergot alkaloids, **1:**180-81
    functions of, **5:**263
Ergotism, **5:**263
Erionite, **2:**114
Erogenic aid, **2:**67
Erogenous zones, **4:**935
Error signals, in back-propagation, **5:**391
Errors in randomized trials, **4:**62
Erythema
    cold, **4:**797-98
    ultraviolet radiation-induced, **7:**66-67
Erythorbic acid, **3:**679
Erythro-9-(2-hydroxy-3-nonyl)adenine, **2:**397
Erythroblastosis, **4:**409
Erythroblastosis fetalis, **4:**381
Erythroblastosis virus, avian, **3:**183
Erythrocyte membrane, **2:**42, **6:**537-46
    abnormalities, **6:**545

Erythrocyte membrane *(Continued)*
    acylation and lipid renewal in, **6:**539
    as bilayer couple, hypothetical treatment of, **6:**541
    composition, current concepts of, **6:**543
    functions of, **6:**538
    lipid asymmetry of, **6:**541-42
    lipid composition of, **6:**538-42
    lipid disposition, **6:**539-41
    organization, current concepts, **6:**543
    proteins
        and cell shape, **6:**542
        relative positions of, **6:**542-43
Erythrocyte volume
    detection method, **4:**124
    loss in spaceflight, **4:**123-24
    magnitude of loss in spaceflight, **4:**124-25
    pattern of loss in spaceflight, **4:**125-26
    recovery after spaceflight, **4:**127-28
Erythrocytes
    anion transporter, **4:**974
    artificial, **1:**382
    autoantibodies to, **1:**514
    calcium and, **6:**533-36
        distribution, **6:**533-34
        elevated, **6:**535-36
        permeability, **6:**533-34
        in sickle cell anemia, **6:**535-36
    calcium transport, **6:**533-35
        calcium pump, **6:**534-35
    characterization, **4:**167
    destruction by macrophages, **4:**862
    electrical impedance of, **3:**10
    envelopes, *Plasmodium* proteins in, **4:**875-76
    functions, **4:**167
    heart-lung machine complications, **4:**96
    life span of, **6:**119-20
    lipid, metabolic pathways for, **6:**541
    lipid exchange, pathways for, **6:**541
    parasite modification of, **5:**648
    Plasmodium-infected, **4:**875
    sickled, **4:**881-82
    somatic mutation and, **5:**249
    space-born, functional capacity, **4:**128
Erythrocytes, chicken, **5:**19
Erythrocytes, sheep, norepinephrine and, **5:**382
Erythroid function disorders, **1:**796
Erythromycin, **2:**785
    mode of action, **1:**290(t)
    motilin and, **5:**723
Erythropoiesis, **2:**649
Erythropoietin, **2:**259, 654
    red blood cell production and, **4:**173

Erythropoietin (Continued)
    uses for, 1:687
Erythroxylon coca (coca plant), 2:512, 5:441
Escherichia coli
    bowel infections, 4:462
    cell wall of, 1:274-77
    codon bias plot for, 3:166
    diarrhea caused by, 4:449-50
    DNA content of, 3:125
    DNA polymerases of, 3:443
    in DNA replication study, 3:440-41
    in DNA-sequencing analysis, 3:173
    in DNA synthesis, 3:172
    in food infections, 3:665-66
    genetic mapping of, 3:824, 825
    infant intestinal tract, 1:319
    lipoprotein, as adjuvant for subunit vaccines, 7:753-54
    neonatal infections, 4:458
    phospholipid synthesis in, 5:883-85
    in protein synthesis, 3:805
    toxins, intracellular activity of, 5:1-3
Eskimos
    dermatoglyphics of, 3:89
    as meat-eaters, 2:190
    origin, based on teeth, 2:797
    prolonged bleeding of, 3:579
    tooth as tool, 2:800-803
Esophageal neoplasms, 2:116
Esophageal sphincter, upper, 3:73
Esophagus
    alcohol consumption and, 1:175
    core temperature measurement, 4:793
    in digestive process, 3:74
    motility in, stress and, 7:289
Essential nutrients. see under Nutrients
Esterases
    organophosphorate insecticides and, 3:252
    phenotype in HeLa cells, 4:107-8
Esters
    characterization, 4:741-42
    hydrolysis of, 3:53
Estradiol, 2:56, 57, 247, 248, 493
    for bone loss, 3:482
    calcitonin and, 3:483
    corpus luteum and, 3:634
    in estrous cycle, 3:897
    fertilization and, 1:1-4
    1-alpha-hydroxylase and, 3:483
    hypothalamopituitary axis and, 3:632
    LH receptors and, 3:633
    in oocyte maturation, 5:555

Estradiol (Continued)
    in pubescent girls, 3:355
    sex differentiation and, 4:243
Estradiol-17B, synthesis, 3:629
    two-cell theory of, 3:630
Estradiolvalerate, 3:482
Estriol, 2:56, 3:345
Estrogen-inducible mRNA. see under RNA, messenger
Estrogen replacement therapy
    for bone loss, 3:482-83, 484
    menopause, 4:708
Estrogenic substances, conjugated, 3:482
Estrogens, 2:54, 56-57, 114, 403, 744, 833
    biosynthesis of, 7:263-64
    in birth control, 1:691-92, 693
    in bone and calcium metabolism, 1:804
    in bone loss, 1:776
    in bone remodeling, 1:822
    and breast cancer, 2:59-60, 7:266
    cognitive/motor tasks and, 4:248
    deficiency, and menopause, 4:708-9
    endometrial disorders and, 5:604
    in estrous cycle, 7:739-40
    in female reproduction, 1:743
    female sexual behavior and, 4:244-45
    fetal, 3:342
    follicular, and oocyte maturation, 5:555-56
    folliculogenesis, 7:346-47
    functions of, 7:274-75
    mammogenesis, 4:604
    in menstruation, 7:739-40
    metabolic transformation of, 7:279-80
    osteoporosis and, 3:481-86
    ovarian synthesis, 7:252-53
    placental synthesis, 7:253-54
    in pregnancy
        biosynthesis of, 7:275
        prolactin and, 3:339
    for prostate cancer, 3:369
    in sequential oral contraceptives, 5:600
    sperm capacitation and, 3:601
    in steroid contraceptives, 5:599
    uterine response to, 7:737-44
    venous thromboembolism and, 5:604
    zearalenone and, 5:267
Estrone, 2:56, 3:483
Estrus, 4:931-32
    aging and, 3:897
    endocrine regulation of, 7:739-40
Estuaries, pollution of, 3:253
Ethacrynic acid, as loop agent, 3:98

Ethambutol
    mechanism of action, **1**:327
    medical uses of, **1**:329
Ethanol. *see* Alcohol, ethyl
Ethers, **4**:741–42
Ethics, medical. *see also* Bioethics
    Declaration of Helsinki, **1**:658
    Hippocratic Oath, **1**:657
    human experimentation, **1**:657–58
    Nuremburg Code, **1**:657–58
    self-determination rights of the elderly, **1**:136
Ethidium, in DNA staining, **3**:623
Ethinamate, **2**:787
Ethinyl estradiol, **2**:111
    for bone loss, **3**:482
    in combined oral contraceptives, **5**:600
    norethindrone acetate and, **5**:603
    stroke risk of, **5**:602
Ethinyl estradiol 3-methyl ether. *see* Mestranol
Ethinyl nortestosterone. *see* Norethindrone
Ethionamide, mechanism of action, **1**:327
Ethiopia, famine in (1984-1985), **3**:562
Ethmoid bone, **5**:523
Ethology, **2**:870
Ethyl alcohol. *see* Alcohol, ethyl
Ethylene dibromide, **7**:565
Ethylene oxide, **2**:110
N-ethylmaleimide, dynein and, **5**:20
Ethynyl estradiol. *see* Ethinyl estradiol
Etidocaine (Duranest), **2**:784
Etidronate, **4**:280–81
Etiologic science, **2**:67
Etofibrate, **2**:169
Eubacteria, **1**:274
Euchromatin, defined, **3**:797
Eugenics, **2**:870
Eukaryotic cells, **2**:236
    defined, **2**:465, **3**:791, 823
    DNA binding sites in, **3**:118
    DNA in, **3**:125
    DNA polymerases of, **3**:443
    gene transcription in, **3**:110–11
    genetic mapping of, **3**:824–26
    genomes, **4**:252
    promoter regions of, **3**:123–24
    ribosomal protein genes in, organization of, **6**:675–77
    ribosomal proteins of, evolution of, **6**:674–75
    transfection of, **3**:175
Eukaryotic genes, **3**:793–96
Eumelanins, in hair, **4**:45
Eunuchs, sexuality of, **6**:883
European Molecular Biology Laboratory 1. *see* EMBL

Euteleogenesis. *see* Insemination, artificial
Euthanasia, **1**:657
Evaporative heat loss, **1**:753, **4**:795
Event-related potentials (ERPs), **1**:173
Evolution, **2**:593, 737, 860–62, **3**:487–93. *see also* Micro-
    evolution
    anatomical, **3**:883
    body pattern formation in, **3**:283–84
    of customary diets, **3**:19–23
    defined, **3**:881
    of dermatoglyphic features, **3**:86
    of genes, **3**:797–98
    germ-line mutations and, **3**:873
    glycolysis and, **3**:932
    of hominid strategies, **3**:495–504
    human
        chronology of, **3**:498
        and *Plasmodium* evolution, **4**:880–81
    language, **4**:633–39
    nucleotide pairs in, **3**:154
    primate
        at genome level, **3**:884–86
        at molecular level, **3**:886–87
        rate of, **3**:886–87
    and sexuality, **6**:879
Evolutionary conservation
    of actin genes, **5**:217
    of DMD gene, **5**:196
    of dystrophin, **5**:198
Ewing's sarcoma, **2**:451, **7**:700–701
Exanthem subitum, **4**:462
Exanthems, in preschool children, **4**:463–64
Exchange protein, **2**:92
Excitation-contraction coupling, **5**:164
Excitation transfer theory (emotions), **1**:104–5
Excitatory amino acid receptors. *see* Receptors, excitatory
    amino acid
Excitatory postsynaptic potentials, **7**:386
Excitatory potential, postsynaptic, depolarizing (EPSP),
    2:310
Excitotoxins, and Huntington's disease, **4**:272–73
Exercise, **3**:505–13
    aerobic
        adaptations to, **3**:512–13
        cardiovascular system and, **3**:508–10
        "fight or flight" syndrome and, **3**:510
        heart work fitness and, **3**:506
        intensity of, **3**:505–6
        sympathetic nervous system in, **3**:510–11
    anaerobic, pH and, **5**:824
    bone cell function and, **1**:765
    cardiac output in the aged, **3**:901

Eye *(Continued)*
    tunic, anatomy, **3:**536-40
    vitamin A uptake by, **7:**856
    zinc metabolism in, **7:**883
Eye movements, **3:**553-59. *see also* Sleep, REM
    muscles for, **4:**261
    reading perception and, **5:**743
    reading process and, **6:**482
    smooth pursuit
        brain control of, **3:**559
        characteristics of, **3:**556
        defined, **3:**553
    speed reading and, **6:**491
Eyebrows, **4:**40
Eyelids, 545-46
Eysenck's theory
    of criminality, **2:**711
    of personality, **5:**773

F

F-actin. *see* Actins
F factor, **3:**824
Fab region, of immunoglobulin E, **1:**184
Face, **2:**599. *see also* Neolithic face pattern
    bone growth, embryonic, **2:**690
    dyslexia-dysgraphia and representation of, **3:**222
    genetic factors, intrinsic, **2:**690
    and jaw, **2:**707
    and oral function, neural basis of, **5:**361-72
    and tooth. *see* Tooth
Facial expression
    muscles of, **4:**261
    persuasion and, **5:**807
Facial feedback theory, **1:**103-4
Facilitated diffusion
    defined, **3:**581
    as mechanism of transport, in blood-brain barrier, **1:**724
Factor, tissue. *see* Tissue factor
Factor Va, cofactor activity, **4:**185
Factor VII, **4:**180
Factor VIII, **4:**181
    deficiency, **4:**153-54
    deficiency, in hemophilia, **2:**464, **3:**147-48
    gene defects in hemophilia A, **4:**155-57
Factor VIIIa, cofactor activity, **4:**185
Factor IX
    deficiency, **4:**154, 181. *see also* Hemophilia B
    gene mutations causing hemophilia B, **4:**157-59
    as genetic marker for affective disorders, **1:**98-99

Factor X
    extrinsic activator, **4:**180
    intrinsic activator, **4:**181
Factor Xa, binding to cell membrane, **4:**180-81
Factor XI
    deficiency, **4:**155
    gene mutations causing hemophilia C, **4:**161-62
Factor XII
    domains, **4:**184-85
    kringle domains, **4:**184
Factor XIII, **4:**180
Falco peregrinus. see Peregrine falcon
Falk, **3:**630
Fallopian tubes, defined, **3:**599
Falx cerebelli, **4:**981
Falx cerebri, **4:**981
Familial Alzheimer's disease. *see* Alzheimer's disease
Familial hypercholesterolemia. *see* Hypercholesterolemia, familial
Familial mitochondrial encephalomyopathy. *see* Mitochondrial encephalomyopathy, familial
Familial polyposis syndrome. *see* Polyposis syndrome, familial
Familiarity, and interpersonal attraction, **4:**542
Family
    correlations, in complex trait analysis, **3:**856
    and criminality, **2:**712
    food technology and, **3:**21
    health psychology, **4:**81
Family characteristics
    of the eating-disordered, **3:**245
    genetic segregation and, **3:**853
Family history. *see* Pedigree analysis
Family planning. *see* Contraception
Famine, **3:**561-66
    early warning systems, **3:**561, 565-66
    relief work for, **3:**564-65
Farabee, W. C., **3:**850
Farber, Sidney, **5:**704
Farm animals *see* Animals, domestic
Farmer's lung, **7:**556
Farnesyl pyrophosphate, **2:**415
Fascia, **2:**640
Fasciculation, **2:**347
Fasciculus
    arcuate, lesions of, **3:**223
    inferior longitudinal, **3:**223
Fasciola hepatica, **2:**389
Fascioliasis, **2:**389
Fasciolopsis buski, **2:**389
Fast-fiber skeletal muscles. *see* Muscle, skeletal, fast-fiber
Fast twitch motor unit, defined, **5:**178

Exercise *(Continued)*
 cardiovascular function and, **3:**515–21
 decreasing risk of sex hormone-sensitive cancers and, **1:**747
 delayed menarche and amenorrhea, **1:**746
 dynamic
  cardiovascular adjustments to, **3:**515–19
  defined, **3:**515
  peripheral blood flow control in, **3:**519–20
 glycogen breakdown and, **3:**926
 headaches associated with, **4:**58
 hepatic lactate disposal and, **5:**826
 induced acidosis and, **5:**825
 isometric. *see* Isometric contraction
 isotonic. *see* Isotonic contraction
 lactic acidosis and, **5:**877
 longevity and, **4:**784, 789
 male reproduction and, **1:**748–49
 muscle fibers and physical training, **5:**179
 physical training, defined, **3:**505
 plasma bicarbonate level and, **5:**878
 skeletal muscle and physical training, **5:**189–90
Exercise-induced hypothermia, **4:**802
Exercise therapy, for the aged, **3:**904
Exertion, **2:**106
 dietary protein needs and, **6:**234
 generalized central capacity and, **5:**154
 glucose and, **2:**106
 glycogen and, **2:**106
Exertion, mental, **5:**803
Exocrine glands
 defined, **3:**71
 in digestive process, **3:**78–79
 secretion, defined, **5:**394
Exocytosis, **2:**238, **4:**975, 239
 in adrenal medulla, **1:**651
 defined, **3:**327
 G proteins and, **4:**17
Exons, **2:**294, 295, 479. *see also* Introns
 defined, **3:**791, **4:**13, **5:**205, **6:**685
 of DMD gene, **5:**194–96, 199
 in gene evolution, **3:**797
 shuffling, **2:**480
3'→5' Exonuclease, **3:**441, 442
Exoribonuclease, defined, **6:**661
Exotoxin A, **4:**418
Exotoxins
 diversity of, **5:**1
 intracellular site of action of, **5:**1–3
Expansive gastric malignant tumors. *see* Intestinal-type gastric malignant tumors
Experimental pathology, defined, **5:**681

Expert testimony, **5:**801, 804
Explant cultures, **4:**582
Explosives, from fermentation processes, **3:**938
Exposure, environmental. *see* Environmental exposure
Extension, as joint movement, **5:**133
External plexiform layer, olfactory, **5:**524
Externality, concept of
 defined, **3:**683
 eating behavior and, **3:**685
 obesity and, **3:**685
Extracellular matrix, **2:**639, **3:**523–31
 altered tumor cell adhesive properties, **4:**1014
 defined, **5:**341
 degradation, **4:**1014
 effects of transforming growth factor-beta on, **7:**636–37
 epidermal growth factor and, **3:**446–47
 epidermal growth regulation and, **3:**449
 formation, **1:**783
 Golgi apparatus role in formation, **4:**10
 membrane connections, **4:**975–76
 mineralization, **1:**785
 nerve regeneration and, **5:**345
 tumor cell effects, **4:**1013
Extracellular matrix proteins, **5:**171
Extrachromosomal amplified DNA, **3:**102–3
Extracorporal membrane oxygenation, **4:**97
Extracorporeal dialysis. *see* Hemodialysis
Extraembryonic membranes, defined, **3:**609
Extrafusal muscle fibers
 defined, 719
 spindles and, **3:**719, 720
Extraocular muscles, anatomy, **3:**547–48
Extrapyramidal tracts, defined, **5:**667
Extraversion, **2:**711
 biological basis of, **5:**774
 as personality description, **5:**770
Extremities
 dorsal root ganglia and, **5:**334
 plain film radiography of, **3:**3
Extrinsic feedback. *see* Feedback, extrinsic
Eye, **2:**599, **3:**534–45. *see also* Optic system
 anatomy, **3:**533–51
 autoimmunity and, **1:**528
 autonomic innervation of, **1:**542
 bovine lense fibers, **3:**12
 characteristics of, **7:**835–36
 different spectral regions through, **5:**595
 fetal, **3:**611
 horizontal section of, **7:**836
 optical radiation through, **5:**594–95
 river blindness, **1:**265
 spectral absorption, **5:**595

Fasting
  gallbladder and, **3:**78
  gastric emptying and, **3:**75
  glycogen level in, **3:**921
  hepatic fatty acid esterification and, **3:**586
  hunger pains and, **3:**75
  immunity and, **5:**506
  lipoprotein kinase activity and, **3:**587
  small intestine motility and, **3:**77
Fat cells. *see* Adipose tissue
Fat-soluble vitamins food analysis, **4:**219–20
Fathead minnows, **3:**257
Fatigue
  maximal aerobic fitness and, **3:**506–7
  motor skills acquisition and, **5:**128
  in motor units, **5:**178
  pH and, **5:**825
Fatigue syndrome, chronic
  Chronis mononucleosis and, **3:**465
  role of infectious agents in, **4:**451
Fats
  absorption of, **3:**81
  characterization, **4:**741–42
  components of, **6:**437
  digestion, in infants, **3:**47
  digestion of, **3:**80
  elemental composition of, **6:**437
  energy storage in, **5:**474–75
  fatty acid and cholesterol content of, **5:**501
  mass density of, **6:**437
  metabolism, respiratory quotient of, **5:**474
  nutrient composition of, **5:**497
  production, crop sciences and, **1:**143
  role of bile acids in digestion, **1:**5625
  role of bile acids in emulsification, **1:**625
  satiating power of of, **6:**729
  saturated, in food, **3:**571
Fats, animal, **2:**192–93, 195, 240, 241, 640
Fatty acid binding proteins, **3:**587–88
Fatty acids, **2:**170, 241, 507
  biosynthesis, in *Escherichia coli*, **5:**883–85
  cellular uptake of, **3:**581–88
  chemistry of, **3:**568
  common, **3:**570
  composition of myelin, **5:**275
  defined, **3:**317
  dietary recommendations of, **5:**542–43
  distribution of among triglycerides, **3:**570
  in egg yolks, **5:**542
  esterification of, **3:**583
  facilitated diffusion of, **3:**587–88
  food analysis, **4:**222

Fatty acids (*Continued*)
  in foods, **5:**500
  G proteins and, **4:**17
  gangliosides and, **3:**731
  hormone regulation of, **3:**324
  of human milk, **5:**539–40
  metabolic fate of, **3:**574
  metabolism, **4:**737–40
  in milk formulas, **5:**540
  monoaclyglycerols and, **3:**55
  names, formulas, and properties, **4:**738
  naturally occurring, **3:**570
  nomenclature of, **3:**568–69
  in ocean fish, **2:**170
  oxidation of, **3:**575, 583
  physical chemistry of, **3:**581–82
  plant content, **5:**533, 535
  remodeling, trypanosome, **5:**653
Fatty acids, essential
  animal studies and, **5:**537–39
  defined, **3:**567
  human studies and, **5:**539–41
  sources of, **3:**576–80
  utilization of, **3:**50
Fatty acids, free. *see* Fatty acids, nonesterified
Fatty acids, nonesterified, **3:**584–86, **4:**36
  in fetal nutrition, **3:**48
  gastric lipase and, **3:**52
  triacylglycerols and, **3:**51
Fatty acids, omega-3, **3:**579–80
  aging and, **5:**538–39
  animal studies and, **5:**537–39
  desaturation of, **5:**534–35
  dietary recommendations, **5:**541–43
  elongation of, **5:**534–35
  in fetal growth, **5:**539
  in fish, relative content of, **5:**536
  functions of, **3:**579–80, **5:**537–39
  in growth and development, **5:**533–43
  human studies and, **5:**539–41
  mammary tumors and, **5:**509
  metabolism of, **5:**535–37
  in pregnancy, **5:**539
  rheumatoid arthritis and, **5:**510
  sources of, **5:**533–34
Fatty acids, omega-6
  dietary balance with Omega-3 fatty acids, **5:**541–42
  dietary recommendations of, **5:**541–43
Fatty acids, short-chain, **3:**43
Fatty acids, unsaturated
  common, **3:**569
  immune dysfunction and, **5:**506–7

Fatty acids, unsaturated (*Continued*)
  products arising from free-radical oxidation of, 7:876
Fatty acids, volatile, 3:586
Fatty alcohols, 4:737–40
Fatty liver
  in alcohol consumption, 1:173
  lipoprotein accumulation and, 3:585
Fatty streak, arterial. *see* Atherosclerosis
Faunivory consumption
  defined, 3:57
  gut anatomy and, 3:60
Fc (fragment crystallizable)
  chimeric antibodies and, 3:809
  defined, 3:801
  region of immunoglobulin E, 1:184
Fc receptors. *see* Receptors, FC
Fc signaling
  antigen-presenting cells and, 1:303–4
  mechanisms, 1:301
  prevention by T cells, 1:303
Fear behavior, 2:340, 632
Febrifugine, antimalarial, 4:884
Febrile convulsions. *see* Convulsions, febrile
Fecapentenes, colorectal cancer and, 3:774
Feces
  anaerobic bacteria isolated from, 1:241
  colonic blood in, 3:776
  pH and colon cancer, 3:774
  production of, 3:82
  and trichinosis transmission by, 7:648–49
  weight of, and dietary fiber, 3:43–44
Fecundity. *see* Fertility
Federal Republic of Germany. *see* Germany, West
Feed additives
  antibiotics in, 1:294
  antimicrobial resistance, 1:294
  regulatory agencies, 1:295
  synthetic antimicrobials in, 1:329–30
Feedback
  hormonal, 5:376
  in metabolic control, 3:380
  muscle/nerve, 3:721
  in neural-immune interactions, 5:373
  in primary hormonal hypersecretion, 3:334
Feedback, artificial *see* Feedback, extrinsic
Feedback, augmented *see* Feedback, extrinsic
Feedback, extrinsic, 5:126–27
Feedback, information
  defined, 5:121
  for skill learning, 5:126–28
Feedback, intrinsic, 5:126–27
Feedback, kinematic/kinetic, 5:128

Feedback, negative
  in adrenocortical axis, 3:896
  in ATP production/utilization, 3:377
  in endocrine systems, 3:324–25, 355
  in estrous cycle, 3:897
  in gonadotrophin release, 3:634
  in gonadotropin regulation, 5:398
  in homeostasis, 5:396
  in hypergonadotropic hypogonadism, 3:358
  in neuroendocrine system, 3:897
Feedback, positive, 3:632
Feedback, response-produced. *see* Feedback, intrinsic
Feedback inhibition, 2:241, 3:919
Feeding behavior, 3:589–97, 683–90. *see also* Eating disorders; Food
  cardiovascular response, 4:553–54
  in decerebrate organisms, 5:809
  fixed-point scales in research, 3:596
  hormonal influences, 4:248–49
  hypothalamus role, 4:309–10
  neural basis of, 5:371–72
Feelings. *see* Emotions
Feline immunodeficiency virus, 2:401
Feline leukemia virus, 4:342
Feline retroviruses, 6:645–46
Felodipine, 2:96
Felten, D., 5:380
Felten, S., 5:380
Felty's syndrome, 4:751
Females. *see also* Sexes; Women
  delayed puberty in, 6:343
  maturation, 4:27
  prevalence of female babies, 2:875
  puberty in, sequence of, 6:340
  stages critical to development of sex, 6:837–39
Feminization, testicular, 2:30, 301
Femur, 7:38
Fenestrated capillaries, defined, 5:32–33
Fenfluramine, 1:510, 2:13, 846, 847
Fenformin. *see* Phenformin
Fengabine, 2:850
Fenofibrate, 2:169
Fermentation
  ATP synthesis and, 3:932
  defined, 3:931
  of dietary fiber, 3:44
  of food, gut anatomy and, 3:60–61
  in food processing, 3:666–67
  practical applications of, 3:938
Ferricytochrome c. *see* Cytochrome c
Ferritin, 2:653, 655, 4:126
Ferrochelatase, 2:655

Fertility, **4:**429–30. *see also* Menarche
　animal studies of, **1:**747–48
　body fat, menarche and, **1:**741–50
　DZ twinning and, **3:**307
　evolutionary rate and, **3:**886
　female, biological determinants of, **1:**748
　global, effect of nutrition and physical exercise on, **1:**749–50
　lactation and, **4:**607–8
Fertility factor. *see* F factor
Fertilization, **2:**875, **3:**599–607
　internal, **4:**929–30
　receptor molecules, **6:**505–11
　sperm receptors and, **6:**506
　in vitro
　　animal models, **1:**257
　　exogenous gonadotrophins in, **3:**633
　　history of, **3:**600
　　LH surge and, **3:**633, 634
　　twinning and, **3:**307
Fertilization pathways, **6:**506–9
　features common to mammalian and nonmammalian, **6:**508–9
　in mammals, **6:**507–8
　in nonmammals, **6:**506–7
Fetal alcohol syndrome, **1:**701, 175–76, **2:**346, **7:**416–17
Fetal blood, of monochorial twins, **3:**311
Fetal blood sampling, **1:**729–35
　fetal anoxemia as indication of, **1:**732–33
　fetal infection as indication of, **1:**731–32
　fetal therapy as indication of, **1:**734–35
　indications of, **1:**731
　nonspecific biological indices of infection in, **1:**732
　prenatal diagnosis as indication of, **1:**731
　prenatal pharmacology as indication of, **1:**734–35
　techniques and results of, **1:**729–30
Fetal circulation, **3:**616–18
　defined, **3:**609
　in monochorial twins, **3:**310
　shunts, defined, **3:**609
Fetal death, **7:**406
　twin transfusion and, **3:**311
Fetal development, **3:**609–19
　animal models, **1:**257
　collagen in, **3:**529–30
　in development week 16, **3:**296
　energy demand in, **3:**48
　ganglioside expression in, **3:**730
　IGF effects, **4:**508–10
　index of, maternal meat aversion and, **3:**33
　omega-3 fatty acids and, **5:**539
　of twins, **3:**314–15

Fetal diseases, infectious, **4:**456–57
Fetal distress, vasopressin and, **3:**347
Fetal growth retardation
　catecholamine count in, **3:**347
　postnatal catch-up and, **3:**347
Fetal hemoglobin, **4:**143
Fetal tissues
　biosynthesis of IGFs in, **4:**504–6
　insulinlike growth factor receptors, **4:**508
Fetofetal transfusion, **3:**311
Fetoplacental unit
　defined, **3:**344
　physiology of, **3:**345
Fetotoxins. *see* Teratogens
Fetus, **1:**733, **2:**346, **3:**609–19
　aborted embryos and, **3:**294
　and caffeine, **2:**75
　calcium transfer to, **4:**26
　circulating IGFs, **4:**507
　examination of, **3:**298–302
　immune system, **4:**331
　laryngeal reflexes, **4:**651–52
　locomotion in, **4:**779
　macerated, with midline cleft, **3:**303
　maternal alcohol consumption and, **1:**175–76
　maternal smoking and, **7:**518
　maturation of different systems in, **1:**733
　oxygen transport in, **5:**623
　previable, defined, **3:**293
　radiation effects, **6:**414
　regulation of IGF biosynthesis in, **4:**507–8
　teratogenic effects. *see* Abnormalities
　thrombocytopenia, **1:**733
　transplacental passage of molecules in, **1:**733
　V-regions, **4:**331
　zinc metabolism in, **7:**888
Fetus, acardiac. *see* Acardiac fetus
Fetus papyraceus, cause of, **3:**311
Fever, **2:**753. *see also* Hyperthermia
　in acute phase response, **1:**30–31
　regulation of body temperature and, **1:**759–60
Fiber optics
　in laser Doppler technique, **3:**759
　in reflectance spectrophotometry, **3:**759
Fiber tracts (white matter). *see* White matter
Fibers
　as airborne toxicants, **7:**577
　inhalation of, **7:**579–82
　production of, **1:**142
　and zinc, **7:**890
Fibers, asbestos. *see* Asbestos
Fibrae zonulares. *see* Zonular fibers

Fibrates, for atherosclerosis therapy, **1**:454
Fibric acid, **2**:168-69
Fibril, contractile, **2**:126
Fibrillin, **2**:639, 641
Fibrillogenesis of collagen, **2**:564
Fibrin, **2**:170, 171, 270, 668
    plasminogen activators and, **3**:810
    polymerization, **4**:179-80
    in thermal injury healing, **7**:455
    in wound healing, **3**:525
Fibrinogen, **2**:170
    abnormal, **4**:162
    in acute phase response, **1**:31
    domains, **4**:179-80
    in platelet aggregation, **6**:56
    thrombin action on, **4**:180, 182
Fibrinolysin, **2**:170
Fibrinolysis, **4**:182
Fibroblast growth factor, **2**:62
    keratinocyte growth and, **3**:447
    muscle differentiation and, **5**:168
    role in bone cell differentiation, **1**:788
Fibroblasts
    in aging immune system, **6**:76
    disaggregated cultures of keratinocytes, **4**:582
    fluorescence micrograph of, **5**:575
    functions, **4**:402-3
    growth of, gangliosides and, **3**:734
    human, **2**:258-59, 264, 456, 572, 645, 646, 753
        fluorescence scan of, **5**:569
    as mesenchymal cells, **5**:162
    microtubules in, **5**:13
    migration, in soft tissue repair, **7**:501
    murine, **2**:258, 261
        microscopic images of, **5**:906
    paracrine keratinocyte growth factors and, **3**:451
    transformation, **4**:583-84
    tropomyosins in, **5**:218
    tumor cell effects, **4**:1014
Fibrocartilage, **2**:640
Fibrocystic disease of pancreas. *see* Cystic fibrosis
Fibrocytes, defined, **7**:47
Fibroids, uterine. *see* Uterine neoplasms
Fibronectin, **2**:91, 207, 566, 614, 639, 641, 642, 645
    in cell migration, **3**:524, 530
    in embryonic development of bone, **1**:786
    functions of, **3**:526
    in gingiva, **4**:36
    laminin and, **3**:528
    membrane connections, **4**:975-76
    photoelectron microscopy and, **5**:906
    salivary, **6**:695

Fibronectin (*Continued*)
    synthesis in tumor cells, **4**:1014
    in wound healing, **3**:525
Fibrosarcoma, **7**:701-2
Fibrosis, **2**:655
    alcoholic, **1**:173-74, **2**:655
    dermal, **2**:574
    primary biliary, **1**:528
    pulmonary, **2**:574
Fibrous astrocytes, **1**:435
Fibrous dysplasia, **5**:615
Fibrous joints, **1**:371, 372
Fibula, **7**:39
Fibular collateral ligament, **1**:376
Fick equation, **3**:516
Fick principle, **6**:597
"Fight-or-flight" state
    aerobic exercise and, **3**:510
    gastric circulation and, **3**:762
    hormonal responses in, **5**:396
    Filament 30mn. *see* Nucleosome
Filariasis
    lymphatic, **2**:388
    treatments for, **4**:448
Filopodia, defined, **5**:343
Fimbria, **4**:227-28, 231
Finger domains, **4**:184-85
Finger Tapping Test, **1**:611
Fingerprints. *see* Dermatoglyphics
Fingers
    dermatoglyphics of, **3**:89-90
    hyperextensibility of joints, in fragile X syndrome,
        3:711
Firing, neural. *see* Neural firing
First cranial nerve. *see* Olfactory nerve
First dorsal interosseous muscle, **5**:180
Fish
    in "basic four" food group plan, **3**:654
    cholesterol content in, **5**:536
    fertilization, **4**:929
    nutrient composition of, **5**:497
    omega-3 fatty acid content of, **5**:534
    poisonous, **7**:598
Fish-eye disease, **2**:667
Fish oils
    autoimmune disease and, **5**:510
    cholesterol and, **3**:579
    mammary tumors and, **5**:509
    prolonged bleeding and, **3**:579, 580
Fish tapeworm, **2**:389
Fish venoms, **3**:673

Fishbein-Ajzen model
    in measurement of attitudes to food, **1**:499
    in prediction of food consumption, **1**:503
Fission
    binary, asexual reproduction, **4**:929
    membrane bilayers, **4**:971
Fissure, longitudinal, **2**:674
FITC
    defined, **3**:621
    in flow cytometry, **3**:623
Fitness
    defined, **3**:505
    elements of, **3**:510
Fitness, aerobic
    limits on, **3**:506
    plasticity and, **3**:512
    significance of, **3**:506-7
Fixation, ocular, **3**:557
Flagellum, **2**:241, 242
Flame ionization detection, **3**:205
Flashbacks, drug
    in LSD use, **5**:448
    in marijuana use, **5**:446
Flashbulb memory, **4**:661, 676
Flatfoot, in fragile X syndrome, **3**:711
Flatulence, **2**:104
Flavin adenine dinucleotide (FADH), **2**:505-8
    pyrophosphatase, **2**:530
    synthetase, **2**:529, 530
Flavin-containing monooxygenase activity in nasal cavity,
        5:286
Flavin mononucleotide (FMN)
    phosphatase, **2**:530
    as riboflavin 5'-phosphate, **2**:524
Flavivirus vaccines, **7**:525
Flaviviruses, **7**:5231-25. *see also* Togaviridae
    control measures for, **7**:525
    diseases associated with, **7**:522-24
    epidemiology of, **7**:524
    molecular biology of, **7**:524-25
Flavocoenzymes, **2**:529-30
    functions, **2**:529-30
    and hydroquinone, **2**:530
    metabolism, **2**:529
Flavohydroquinone, **2**:531
Flavokinase, **2**:529, 530
Flavonoids, plant. *see* Plant flavonoids
Flavoprotein-linked dehydrogenases, electron transport
        chain, **1**:459-60
Flavoquinone, **2**:531
Flavor, and food preferences, **1**:345
Flavosemiquinone, **2**:531

Flax plant, **3**:786
Fleas, **2**:390, **7**:896
Flecanide, **2**:173
Flexion, as joint movement, **5**:133
Flexion torque, intramuscular EMG activity and, **3**:694
Flies, **2**:390, **7**:895-96
Flippase, **5**:889
Flooding technique of conditioning, **2**:632
Floppy baby syndrome, **5**:62
Flouren's theory of brain functioning, **2**:860
Flourescence spectometry. *see* Spectometry, fluorescence
Flow cytometry, **2**:455, **3**:621-26, **5**:574
    in diagnostic pathology, **5**:686
    sorting of chromosomes, **2**:455, 457
Flow cytometry, laser, **2**:459-60, **3**:623
    and fluorescence, **2**:459
Flowmeters, **2**:459-60
    basic design of, **5**:575
Flubendazole, **2**:390
Fluid bilayer, **4**:970-71
Fluid exchange, microvascular, **5**:31-42
Fluid therapy
    defined, **5**:699
    for diarrhea, **5**:707
Fluids
    absorption of, **3**:81
    retention in essential hypertension, **4**:291-92
    total body, regulation, **4**:570-71
Fluke diseases, **2**:386, 389
Fluoresceins, **2**:468
    as fluorescent probe, **5**:577, 578
    microfluorimetry studies using, **5**:566
Fluorescence
    defined, **3**:621
    of DNA, **2**:459
Fluorescence-activated cell sorting. *see* Flow cytometry
Fluorescence anisotropy decay, **5**:579
Fluorescence lifetimes
    microspectrofluorometer and, **5**:569-70
    pixel-by-pixel determination of, **5**:578-80
Fluorescence microscopy. *see* Microscopy, fluorescence
Fluorescence polarization assays, **4**:356
Fluorescence polarization immunoassay, **3**:208
Fluorescence resonance energy transfer, **5**:581-82
Fluorescence tests, **4**:355-56
    in drug analysis, **3**:206
    HPLC, **4**:210-11
Fluorescent antibody technique
    indirect, **2**:41, 42
    test for antinuclear antibodies, **1**:516
Fluorescent dyes, **5**:574-78, 872-73
    aequorin, **5**:577

Fluorescent dyes *(Continued)*
   calcium, **5:**576-77
   calcium indicator and, **5:**576
   in flow cytometry, **3:**623
   fluo-3, **5:**576
   fluorochromes, types of, **5:**575-76
   photoactivation, **5:**581
   photobleaching recovery, **5:**580-81
   probes, defined, **5:**561
   in protein detection, **6:**194
   quin-2, fura-2, **5:**576
   as vital stains, **5:**576
Fluorescent probes. *see* Fluorescent dyes
Fluoridation of drinking water, **2:**787, 813
Fluorides
   cariostatic, **2:**813
   in dentistry, **2:**787
Fluorimetry, **3:**202
Fluorine
   biological function of, **7:**606
   toxicity, **7:**609
Fluorochromes. *see* Fluorescent dyes
5-Fluorocytosine, antifungal activity, **1:**327
6-^{18}F-fluorodopa, in PET studies of Parkinson's disease, **5:**670
Fluorography, to detect radioactive proteins, **6:**197-98
Fluoroisothiocyanate. *see* FITC
Fluoroscopy
   biplane, in diagnosis, **3:**2
   in diagnosis, **3:**1
   in gastrointestinal radiology, **3:**4
5-Fluorouracil-leucovorin, **2:**372
Fluoxetine, **2:**844
   antidepressant effects, **1:**309
Fluvoxamine, **2:**13
Foam cells, **1:**449, **2:**411, 417
Focal adhesions, membranes, **4:**975-76
Focal brain damage. *see* Brain damage, focal
Fodrin, **2:**42
   defined, **2:**41
   role of, in polarized epithelial cells, **6:**62
Folates, **2:**535, 536
   food analysis, **4:**221
Folic acid
   bacterial synthesis of, **1:**326
   food analysis, **4:**221
   in fragile X demonstration, **3:**710
Folic acid deficiency, **5:**513
Folin method, for protein detection, **6:**192
Folivory consumption
   defined, **3:**57
   gut anatomy and, **3:**60

Follicle-associated epithelial cells. *see* M cells
Follicle-stimulating hormone, **2:**20, 27, 247, 248, 498
   antral follicle growth and, **3:**631
   fertilization and, **1:**1-4
   fetal, **3:**342
   and folliculogenesis, **7:**346-47
   function of, **6:**88
   hypothalmic dysfunction and secretion, **1:**743
   inhibin and, **3:**632
   and menstrual cycle, **6:**556-57, **7:**739
   in oocyte maturation, **5:**555, 557
   ovulation induction using, **7:**347-48
   perinatal, **3:**348
   for polycystic ovarian disease, **3:**630
   production and activity of, **6:**17-18
   Sertoli cell function and, **7:**425-26
Follicle-stimulating hormone receptors. *see* Receptors, FSH
Follicles
   antral
      development of, **3:**631-32
      oocyte, **5:**553
      steroid synthesis in, **3:**632
   dominant, **3:**632-33
   Graafian. *see* Graafian follicles
   growth of, **3:**627-35
   insect, gene amplification in, **3:**786
   as oocyte maturation inhibitor, **5:**556-57
   ovulation. *see also* Ovulation induction
      effect of various gonadotropins on, **7:**350
   primordial, **3:**628
   rupture of, **3:**634
Folliculogenesis, **7:**346-47
Folliculostatin. *see* Inhibin
Follitropin receptors. *see* Receptors, FSH
Folyl polyglutamate, **2:**535
Folyl polyglutamate synthetase, **2:**536
Fontanel, **2:**403
Food, **2:**34, 36, 194, 342. *see also* Diet; Eating; Nutrients; Nutrition
   acceptance, **3:**637-47
   affective responses to, **1:**499-501
      exposure as, **1:**500
      food ingredients as, **1:**500-501
      genetic disposition as, **1:**500
      liking and aversions as, **1:**499-500
      meal components as, **1:**501
      shifts in, **1:**501
      social and cultural context as, **1:**501
   ascorbic acid in, **1:**429
   attitudes toward, measurement of, **1:**497-99

Food (*Continued*)
    attributes
        sensory tests for, **5**:752
        trained panelists to determine, **5**:752
    aversions, sensory influences on, **5**:753-54
    and body temperature, **1**:756-57
    carrying of, by early hominids, **3**:496
    consumption
        attitudes and determinants of, **1**:497-504
        attitudes and their measurement in, **1**:497-99
        empirical models evaluating, **1**:503-4
        factors affecting, **1**:497, 501-3
        Fishbein-Ajzen model in prediction of, **1**:503-4
        space travel and, **7**:144
    defined, **3**:637
    energy content of, **5**:471-76
    essential nutrients, defined, **3**:576, 683
    fat and oil content of, **3**:571-72
    fatty acids found in, **5**:500
    fermentation, **3**:667
    gathering of
        by hominids, **3**:500-501
        social intellect and, **3**:504
    induced hyperemia, **4**:556-57
    ingredients, as effective response to, **1**:500-501
    innate mechanisms of, **1**:345-46
    intake, and ovulation, **1**:748
    learned, **1**:346-47
    nutritional quality of diets, **5**:483-502
    palatability, **5**:752
        fat and, **3**:576
        food acceptance and, **3**:640, 645
        satiety and, **3**:641
    pesticide residues in. *see also* Pesticides
        toxicology of, **7**:559-66
    pollutants in, **3**:256
    protein
        amino acid, digestibility of, **6**:231-32
        content and quality of, **6**:230-32
    restriction, and rodent aging process, **5**:479-80
    satiating power, **6**:723-33
        attitudes and, **6**:730-31
        caloric value of, **6**:728-29
        characteristics of, **6**:728-31
        eating patterns and, **6**:723-24
        habits and, **6**:729
        and hunger, **6**:724-25
        macronutrient content and, **6**:729
        measurements of, **6**:726-28
        palatability and, **6**:730
        processes controlling, **6**:725-26
        sweetness and, **6**:730

Food (*Continued*)
    selenium in, **6**:789-90
    sensory properties of, **5**:751-53
    snack food, acceptance of, **3**:641
    as source of salmonella, **6**:711
    in stomach, **3**:75
    sulfites, **7**:337-43. *see also* Sulfites
        detection of, **7**:342
        fate of, **7**:341
        as ingredients in, **7**:340-41
        likelihood of reactions to, **7**:341-42
        regulatory restrictions concerning, **7**:342
    swallowing of, **3**:74
    sweet, in gatherer-hunters' diets, **3**:20
Food, artificial. *see* Food, formulated
Food, formulated
    grazing and, **3**:596-97
    industrialization and, **3**:25
    in nineteenth century, **3**:21
    packaging of, **3**:668
    in twentieth century, **3**:22
Food acidulants, **3**:668
    food analysis, **4**:223
    as preservatives, **3**:668
Food additives
    food analysis, **4**:223-24
    polysaccharide, **3**:36, 40-41
    in processed foods, **3**:22
    as toxicants, **3**:672
Food allergies. *see* Food hypersensitivity
Food analysis, HPLC, **4**:215-26
    amino acids, **4**:221-22
    carbohydrates, **4**:217-18
    characteristics, **4**:215-17
    clean-up of complex samples, **4**:217
    contaminant determination, **4**:217
    contaminants, **4**:224-26
    fat-soluble vitamins, **4**:219-20
    food additives, **4**:223-24
    lipids, **4**:222-23
    sugars, **4**:217-18
    vitamins, **4**:218-21
    water-soluble vitamins, **4**:220-21
Food and Drug Act (1906), **3**:672
Food and Drug Administration, U.S.
    on food labeling, **3**:656-58
    mycotoxins and, **5**:261-62
Food aversions. *see* Dietary cravings
Food-borne pathogens
    defined, **3**:661
    species of, **3**:664-66
    spoilage associations of, **3**:662

Food-borne pathogens (*Continued*)
    temperature requirements of, **3**:663
Food choice. *see* Feeding behavior; Food preferences
Food composition. *see* Nutrients
Food contamination
    antibiotic feed additives, **1**:294
    Balkan nephropathy and, **5**:265
    factors in, **3**:661–63
    by mycotoxins, **5**:260, 261–62
    in preindustrial societies, **3**:21
    tapeworms, **1**:264–65
    Toxoplasma gondii, **1**:268
    trichina roundworms, **1**:265
Food cravings. *see* Dietary cravings
Food deprivation. *see* Famine; Hunger; Starvation
Food dispensers, automatic, **3**:595
Food groups, **3**:649–59
    in nutrition education, **3**:
    table of foods in, **5**:492
Food gums, dietary fiber of, **3**:40–41
Food handling
    economic aspects, **3**:668
    food poisoning and, **3**:666
    hygiene, **3**:661–70
    methods of, **5**:490
    nutritional quality and, **5**:490
    quality assurance in, **3**:668–69
Food hypersensitivity
    gastrointestinal allergy, **4**:410–11
    symptoms of, **3**:674
    types of, **3**:673–74
    Western diets and, **3**:26
Food intake. *see* Eating
Food irradiation
    acceptance of, **3**:668
    defined, **3**:661
Food labeling
    FDA requirements for, **3**:656–58
    food acceptance and, **3**:646
Food laws. *see* Legislation, food
Food microbiology, **3**:661–70
Food packaging, aseptic, **3**:668
Food poisoning, **3**:671–81
    prevention of, **3**:666
    sources of toxicants, **3**:672
    Western diets and, **3**:26–27
Food preferences
    and appetite, **1**:345
    in customary diets, **3**:18–19
    differences in sensory function and, **5**:752
    in feeding behavior, **3**:590
    food acceptance and, **3**:639–40, 642

Food preferences (*Continued*)
    and nutrition, **5**:751–57
    nutritional quality and, **5**:489–90
    preference and, **5**:752–53
    suggested pattern for daily, **5**:493
    United States, **3**:650–51
Food preservation
    factors in spoilage, **3**:661–63
    new technologies in, **3**:668
    nutritional quality and, **5**:490–91
    in traditional cities, **3**:21
Food production
    crop sciences and, **1**:141–44
    in world, **1**:142
Food technology, twentieth century, **3**:22
Foods, sweet, nutrient composition of, **5**:497
Foot, **2**:594
    muscles of, **4**:266–67
Foraging. *see* Hunter-gatherers
Foramina, scleral. *see* Scleral foramina
Force coupling. *see* Coupling, force
Force-velocity curve, muscle. *see under* Muscle, skeletal
Force-velocity relation, **3**:691
Forearm
    coordination of movements, **3**:691–99
    movement coordination in, **3**:691–99
    muscles of, **4**:264
Forebrain, **2**:3, 32
    anatomical areas, **2**:3
    and convulsive seizures, **6**:781–83
    nucleus in, **2**:32
    REM sleep and, **3**:197
    topographical regions, **4**:130–31
Foregut
    anatomy, **3**:61–64
    defined, **3**:57
    fetal, development of, **3**:59
Forensic medicine, **2**:809
    and dental caries, **2**:809
    DNA markers and, **3**:149
    identification, **4**:578
    scope of, **5**:685
Form change, **2**:689
Formal operational thought, **1**:66
Formaldehyde, **2**:111
    exposure routes of, **3**:385
    inhaled, disposition and effect of, **7**:579
    metabolism, in nasal cavity, **5**:288
Formaldehyde dehydrogenase, glutathione and, **3**:915
Formalin. *see* Formaldehyde
Formant frequencies
    defined, **4**:633

Formant frequencies (*Continued*)
　encoding, **4:**634–35
　perception of, **4:**635
Formate, **2:**535, 536
Formyl-methionyl-leucyl-phenylalanine, **4:**371
Fornix, **4:**228, 231
Fornix conjunctivae, **3:**533
Forskolin, **5:**899
Fos gene. *see* Genes, *fos*
Fosfomycin
　effect on peptidoglycan synthesis, **1:**278
　mode of action, **1:**290(t)
　structure of, **1:**273
Fossa, **2:**769
Fossils
　as clue to origin of *Homo sapiens*, **1:**827
　pelvic, of archaic *Homo sapiens*, **1:**827–30
Founder and drift hypothesis, **5:**868
Fourier analysis
　analytical gas chromatography and, **3:**754
　magnetic resonance image, **4:**866–67
Fovea
　defined, **6:**482
　in retina, **7:**836–37
Fovea centralis retinae, defined, **3:**533
Foveation, **1:**480–81
Foveolar cells, defined, **3:**765
Foxes, and rabies transmission, **7:**896–97
Fractures
　bone, factors leading to, **1:**778–80
　factors leading to, **1:**778–80
　hip, and elderly, **1:**779–80
　muscle atrophy in, **5:**189–90
　nontraumatic, fragile bones and, **1:**778–80
　vertebral, pathogenesis of, **1:**779
　wrist, **1:**779
Fragile sites, isochores and, **4:**255–56
Fragile X syndrome, **3:**709–17
　genetic aspects of, **3:**840–41
　inheritance pattern, **3:**715
　sex differences involving, **6:**861
Frame-shift hypothesis, **5:**199–201
Frank-Starling law, **2:**186
Frank-Starling mechanism, **2:**186
　defined, **3:**517
Free fatty acids. *see* Fatty acids, nonesterified
Free proton concentration. *see* Hydrogen ion concentration
Free radicals
　in aging, **3:**893–94
　cell aging and, **3:**900
　defined, **3:**371, **5:**667, 907
　glutathione and detoxification of, **3:**910, 914–15

Free radicals (*Continued*)
　vitamin E and, **5:**512
Free-running state, defined, **5:**396
Freebase, **2:**511. *see also* Cocaine
Freemartin cattle, **4:**357–58, 930
Freeze substitution technique, **2:**725, 727
Frequency, sound. *see* Sound frequency
Frequency-threshold curve. *see* Tuning curve
Fresh frozen plasma. *see* Plasma
Freud, Sigmund
　and cocaine use, **2:**512
　dream research of, **3:**192–93
　frustration, **2:**854
　model, **2:**853
　theory of self, **2:**853
　thought, **2:**853
　types of suicide developed by, **7:**329–30
Freund's adjuvant
　in experimental autoimmune disease, **1:**526
　for synthetic vaccines, **7:**752–53
Friction skin, defined, **3:**85
Frogs
　olfaction of, **5:**528
　toxins produced by, **7:**598–99
Frontal cortex, **5:**309–10
Frontal dysgraphia, **3:**218
Frontal lobe, **2:**8, 351, 9, 598, 682, 683, 766–67, 871
　damage to, **5:**313
　language and, **3:**218
　language processing and, **3:**225
　signs in dementia, **2:**766–67
Frontal operculum, **3:**218
Frostbite, **4:**806–7
　pathophysiological changes, **4:**807
　rewarming, **4:**808
Fructose, **2:**101–5
Fructose 1,6-biphosphate
　in first-stage glycolysis, **3:**933
　glycolysis and, **3:**940
　in intravenous feeding solutions, **5:**827–28
　in second-stage glycolysis, **3:**936
Fructosephosphates, **5:**821
Frugivory consumption
　defined, **3:**57
　gut anatomy and, **3:**60–61
Fruit
　ascorbic acid occurrence in, **1:**429
　in "basic four" food groups plan, **3:**655
　nutrient composition of, **5:**496
Fruit fly. *see* Drosophila
Frusemid. *see* Furosemide
Frustration, Freudian, **2:**854

Frustration-aggression hypothesis, **7**:234–35
FSH. *see* Follicle-stimulating hormone
FSH receptors. *see* Receptors, FSH
FSH-releasing hormone. *see* Gonadorelin
Fugu toxin. *see* Tetrodotoxin
Fulminant colitis, **1**:222
Fumarate, **2**:508
Functional age
    chronological age and, **3**:900
    defined, **3**:899
Fundus, **2**:35, 36
    defined, **2**:403
    of stomach, **3**:75
Fungal toxins. *see* Mycotoxins
Fungi
    antibiotics produced by, **7**:593–94
    polyamine metabolism in, **6**:73
    toxins produced by, **7**:591–93
Furcation of tooth, **2**:824
Furchgott, R. F., **3**:372
Furosemide
    hypercalcemia of malignancy treatment, **4**:281
    as loop agent, **3**:98
Fursemide. *see* Furosemide
Fusarinon-X, **5**:264
Fusarium, **3**:675, **7**:662
Fusarium nivale, **5**:263
Fusarium poae, **5**:263
Fusarium roseum
    deoxynevalenol of, **5**:263
    zearalenone and, **5**:267
Fusarium solanae, **5**:263
Fusarium sporotrichiodes, **5**:263
Fusarium tricinctum, **5**:263
Fused tetanus (tetanic contraction), defined, **5**:178
Fusidic acid, **1**:290(t)
Fusimotor system, **3**:719–24, **6**:171
Fusobacterium
    habitats, **1**:240
    infections of, **1**:246
    morphology, **1**:238–39
    resistance to antimicrobial agents, **1**:246
    virulence, **1**:241–42

G

G-actin. *see* Actins
G cells, **2**:254–55, 260–64
G$_1$ period, **5**:65, 66
G$_2$ period, **5**:65, 66
G-proteins. *see* Guanine nucleotide regulatory protein

GABA, **2**:5, 12, 13, 18, 23–25, 29, 34, 316, 330, 338, 340,
    493, 494, 522, 523, 534, 686, 850
    astrocyte uptake of, **1**:440
    in autonomic neurons, **1**:540
    bicuculline and, **2**:850
    in convulsive seizures, **6**:782–83
    enzyme inhibitors and, **3**:399
    epilepsy and, **3**:459–60
    as inhibitory transmitter in brain, **5**:411
    and initiation of convulsive seizures, **6**:787
    levels in Huntington's disease, **4**:271
    as neuroactive substance in retina, **6**:629
    as neuroregulator, **2**:13–14, 316, 522
    production during alcohol consumption, **1**:172
GABA receptors. *see* Receptors, GABA-benzodiazepine
GABA transaminase, inhibition of, **3**:399
Gadolinium-DPTA, **4**:870
gag genes. *see* Genes, *gag*
gag-pol fusion proteins, HIV-1, **1**:13
gag viral proteins. *see* Gene products, *gag*
Gagarin, Yuri, **2**:147
Galactomannans, **3**:39
Galactophores, **4**:607
Galactorrhea, **2**:29, **5**:393
Galactose, **2**:103–5
    levels, lactose malabsorption diagnosis, **4**:619–20
Galactosemia, **5**:436
Galactosylhydroxylysine, **2**:569
Galactosyltransferase, **2**:563
    in egg binding, **3**:605
    in epididymal fluid, **3**:600
    Golgi apparatus purification monitoring, **4**:6
Galacturonic acid, **3**:38
Galanin, **2**:27
    in autonomic neurons, **1**:541
    function of, **5**:721, **6**:89
Gallbladder
    anatomy, **3**:63
    cholecystokinin 33 and, **5**:730
    cholecystokinin and, **3**:78
    motility, stress and, **7**:290
Gallbladder diseases
    oral contraceptives and, **5**:608
    somatostatin and, **5**:722
Gallium nitrate, **4**:281
Gall's theory of brain organs, **2**:860
Gallstones. *see* Cholelithiasis
Galton, Francis, **3**:86, 837, 855, 856
Galton's minutiae. *see* Ridge details
Galvanic skin response, **5**:551–53
Gambierdiscus toxicus, ciguatera syndrome and, **7**:591

Gambling
pathological, **4**:998
societal aspects, **4**:84
Gamete intrafallopian tube transfer, **3**:603
Gametes, **2**:235
recognition, and fertilization, **6**:506
Gametogenesis, **2**:291, **4**:949-51, **5**:551-53
gamma-Aminobutyric acid. *see* GABA
Gamma-chains, **4**:146
Gamma efferent system
activation of, **3**:722
evolutionary advantage of, **3**:721
in movement, **3**:722
"reflex reinforcement" and, **3**:723
Gamma globulin vaccines, **4**:466
Gamma-glutamyl-GSH, **3**:912
Gamma-glutamylamino acids, **3**:913
Gamma-glutamylcyclotransferase, in 5-oxo-L-proline
synthesis, **3**:913
Gamma-glutamylcystein synthetase
deficiency of, **3**:916
glutathione synthesis and, **3**:910-11
Gamma-glutamyltranspeptidase, **2**:367, 535
deficiency of, **3**:916
in glutathione degradation, **3**:912
in mercapturic acid synthesis, **3**:915
tissue concentrations of, **3**:908
Gamma herpesvirinae, **4**:187-88
Gamma radiation
and chemical reactions, **6**:443
effect on chemical reactions, **6**:443
emission, after neutron activation, **7**:539
in food preservation, **3**:668
Ganciclovir (DHPG), **2**:398
and cytamegalovirus treatment, **2**:398
resistance to, **2**:398
Ganglia, **2**:33
basal, **2**:3
celiac, **2**:33
cultures, nerve growth factor and, **5**:334
mesenteric, **2**:33
and response to light, **6**:627
Ganglia, nodose. *see* Nodose ganglion
Ganglia, parasympathetic, **1**:538-39
ciliary, **1**:539
otic, **1**:539
spenopalatine, **1**:539
submandibular, **1**:539
Ganglia, retinal, **5**:389, **7**:803-5
Ganglia, spinal, **7**:208-9
neuronal death in, **5**:334
tachykinins and, **5**:720

Ganglia, sympathetic, **1**:537-38
nerve growth factor and, **5**:336
Ganglio series, defined, **3**:737, 743
Ganglion root, dorsal, **2**:33
Ganglioside antigens
biochemical and immunological characterization,
4:962-64
functional properties, **4**:964-65
Gangliosides, **3**:725-35
formation, **4**:744
neuronal differentiation and, **3**:737-42
structures, **4**:962-63
transport of, **3**:743-48
Gangliosidosis, pathogenesis of, **3**:732
Gangliotetraose structure, **3**:737, 743
Gangrenous ergotism, **5**:263
Ganja. *see* Marijuana and cannabinoids
Gap-1. *see* G_1 period
Gap-2. *see* G_1 period
Gap junctions, **2**:132, 274-78
channel, **2**:277, 278
conductance, electrical, **2**:276, 277
function, **2**:277-78
in membranes, **4**:973
protein, **2**:276
regulation, **2**:276-77
structure, **2**:275-76
Garden warbler, **2**:421
Gas chromatography. *see* Chromatography, gas
Gas hold-up time, defined, **3**:750
Gas poisoning, **3**:387-88
Gases
as airborne toxicants, **7**:577
convection
mechanisms of, **6**:568-70
in vertebrates, **6**:569-70
diffusion and convection, in respiratory system, **6**:568
inhaled, disposition and effect of, **7**:579
transport in lungs, **6**:574-75
Gastric acid, **2**:34, 35, 39
blood alkilinity and, **5**:877
blood circulation and, **3**:756, 761
bulimia nervosa and, **3**:247
cephalic secretion phase of, **5**:715
emptying, **2**:36
regulation control of, **2**:36
gastrin and, **5**:715
hydrochloric acid in, **3**:76
inhibition of, **5**:716
by peptide YY, **5**:719
nitrosation and, **3**:679
nutrient utilization and stimulation of, **5**:755-56

Gastric acid (*Continued*)
   secretion, **2:**34–35
      autonomic control of, **1:**547
      stress and, **7:**287–88
      vagal control of, **1:**547–48
Gastric antrum. *see* Pyloric antrum
Gastric arteries, **3:**756
Gastric cancer. *see* Stomach neoplasms
Gastric circulation, **3:**755–63
Gastric dysplasia, **3:**770
Gastric emptying, **2:**36, 39
   and cental regulation, **2:**36
   in digestive process, **3:**75
   and neuological disorder, **2:**36
Gastric inhibitory peptide, **6:**89
Gastric juice, **3:**76
Gastric lipase
   lipolysis and, **3:**53
   milk triacylglycerol hydrolysis and, **3:**52
   in preterm newborns, **3:**51
   properties of, **3:**49
Gastric motility, **2:**35–36
Gastric mucosa, **3:**61
   ascorbic acid absorption, **1:**430
   blood flow in, **3:**760–61
   cancer in, **3:**768
   cholecystokinin and, **5:**731
   gastrin and, **5:**717
   hyperplasia, in Zollinger-Ellison syndrome, **5:**721
   injury to, **3:**763
   physiology, **3:**75
Gastric parietal cells. *see* Parietal cells, gastric
Gastrin-releasing peptide, **1:**541, **2:**34, 38
Gastrins, **2:**34
   acid hypersection and, **5:**721
   detection, by radioimmunoassay, **6:**459–60
   discovery of, **5:**713
   function of, **3:**75, **5:**715, **6:**89
   gastric acid secretion and, **3:**76
   release of, **5:**716
   in vagus nerve, **5:**715–16
Gastritis, atrophic, **3:**772–73
Gastrocnemius muscle
   cat, fiber angulation in, **5:**183
   defined, **3:**691
   soleus and, **5:**188
   in vertical jump, **5:**187
Gastroenteritis
   in infants and toddlers, **4:**462
   salmonella-induced, **3:**664–65, **6:**710–11
Gastroepiploic arteries, **3:**756

Gastrointestinal diseases
   in children, **5:**707
   oral contraceptives and, **5:**608
   peptide hormones and, **5:**721–23
Gastrointestinal function, **2:**31–40
   and brain regulation of, **2:**31–40
   mammalian, **2:**31–40, 32–38
   stress-induced alteration of, **7:**285–92
Gastrointestinal hormones, **1:**350, **3:**73–74, **4:**555
Gastrointestinal motility
   in digestive process, **3:**75
   galanin and, **5:**721
   motilin and, **5:**720
   opioids and, **5:**719
   stress and, **7:**289
   tachykinins and, **5:**720
Gastrointestinal neoplasms, **3:**765–76
Gastrointestinal peptide hormones, **5:**713–23
Gastrointestinal stress responses, **7:**291–92
Gastrointestinal system
   aging of, **3:**902
   alcohol consumption and, **1:**175
   autonomic control of secretion, **1:**547
   autonomic innervation of, **1:**543
   capillaries of, **5:**32, 38–39
   changes in drug kinetics in, **5:**854
   diagnostic radiology of, **3:**4–5
   endometriosis in, **3:**369
   extrinsic innervation of, **7:**286
   histology of, **3:**62
   infections, effect of quinolones on, **6:**408
   kinins and, **4:**594
   mammalian, **3:**59, 60
   mucosal immunology of, **5:**139
   precancerous conditions, **3:**770–71
   radiation injury of, **6:**478–79
   radiation syndrome of, **6:**471–72
   stress
      and motor function in, **7:**289–90
      and secretion, **7:**286–89
      and water absorption, **7:**288
   trichothecenes and, **7:**663
Gastrointestinal wall, **2:**33
Gastrula, **4:**429
Gastrulation
   body pattern formation and, **3:**284–86, 288, 289
   commitment and, **3:**283
   defined, **3:**305
   double, **3:**312
   fibronectin in, **3:**524
   primary induction and, **3:**286
   in vertebrates, **3:**283, 285–86

Gate control system, of pain, **5:**627-28
Gating mechanism, **2:**280
Gaucher, Phillippe, **3:**777
Gaucher's disease
    bone marrow transplantation for, **1:**796
    molecular genetics of, **3:**777-83
Gautier, Marthe, **3:**176
Gel electrophoresis. *see* Electrophoresis, gel
Gel filtration. *see* Chromatography, gel
Gel retardation assay, **2:**435
Gel shift assay, **3:**158
Gelatin, **2:**568
Gels, defined, **3:**403
Gemfibrozil, **2:**169
    atherosclerosis therapy with, **1:**454
GenBank, services of, **3:**163
Gender identity, **3:**340
Gender identity disorders, **4:**997
Gene action regulation. *see* Gene expression regulation
Gene activation. *see* Gene expression regulation
Gene amplification, **3:**785-89
    chromosomal rearrangements and, **5:**546
    cytotoxic agents enhancing, **5:**546-47
    mechanisms of, **5:**546
    parasite drug resistance and, **5:**651
Gene clusters. *see* Genes, reiterated
Gene coding sequences, **5:**216
Gene conversion, **2:**737, **4:**953-54
Gene dosage effect
    secondary, **3:**177
    theory of, **3:**177-80
Gene expression
    in development of recombinant subunit vaccines,
        7:749-51
    DNA amplification and, **3:**103-4
    hepadnavirus effects, **4:**765
    immunity and, **5:**514-15
    interleukin-2, **4:**531-32
    in mammalian brains, **2:**318-19, 431-36
    methylation and, **3:**872
    nutrition and, **5:**514-15
    stages of, **3:**871-72
Gene expression regulation
    activated by inteferons, **4:**515
    combinatorial, **5:**167
    control of, **5:**171
    DNA methylation and, **3:**151-62
    in embryo cell positional memory, **3:**290
    of *fos* gene, **3:**705-6
    of G proteins, **4:**21
    of gangliosides, **3:**732
    in muscle, **5:**165, 166-67

Gene expression regulation *(Continued)*
    in muscular dystrophy, **5:**199
    of prepubescent growth, **3:**351
    sequence-specific methylations and, **3:**160
    trisomy 21 and, **3:**179-80
    of tubulin, **5:**17
    viral, in carcinogenesis, **3:**864-65
Gene families, **3:**874
Gene fusion. *see* Cloning, molecular
Gene imprinting. *see* Imprinting (genetics)
Gene inactivation
    methylation and, **3:**160
    sequence-specific promoter methylations and,
        3:156-59
Gene library, **2:**253, 257, 455, 462-64
    chromosome-specific, **2:**455-64
    construction in bacteriophage lambda, **2:**461
    defined, **2:**455
    human DNA for the first library (1978), **2:**455
    uses, **2:**463-64
Gene mapping. *see* Chromosome mapping
Gene pool
    defined, **3:**251
    pollutants and, **3:**254-55
Gene probes. *see* DNA probes
Gene products, *env*, **1:**12
Gene products, expressed by osteoclasts, **1:**766-70
Gene products, *gag*, **1:**12-13
Gene products, *nef*, **1:**14
Gene products, *pol*, **1:**13
Gene products, *rev*, **1:**13-14
Gene products, *tat*, **1:**13
Gene products, *vif*, **1:**14
Gene products, *vpu*, **1:**14
Gene rearrangement
    in antibodies, **3:**802
    in eukaryotes, **3:**794-95
    in lymphocytes, **3:**798
Gene-regulatory proteins. *see* Regulatory proteins
Gene targeting, for Down's syndrome, **3:**189
Gene therapy
    defined, **5:**699
    for Down's syndrome, **3:**189
    for Gaucher disease, **3:**782-83
    retroviral vectors in, **6:**659
    in treatment of genetic disease, **1:**796-97
Gene transcription, **2:**431
    convulsive seizures and brain, **6:**786-87
    histones and, **4:**238
Gene transfer. *see* Transfection
Gene transposition, and cancer, **5:**326-27
Genealogy, **4:**434

Generalizability
    defined, **5:**121
    as learning criterion, **5:**122
Generalized central capacity
    effort and, **5:**154
    models of, **5:**151–52
Generalized motor programs
    defined, **5:**121
    learning and, **5:**126
Generally Recognized as Safe (food additives), **3:**671
Generator current, olfactory, **5:**526
Genes, **3:**791–99
    actin, properties of, **5:**217
    of chromosome 21, **3:**178, 182–85
    construction, **2:**435
    defective, Parkinson's disease and, **5:**669–70
    defined, **3:**317, 813, 869, **5:**205
    desmin, vimentin genes and, **5:**224–25
    distribution, **4:**254
        race and, **6:**102
    diversity, in populations and races, **6:**109
    eukaryotic, **2:**478–80
    frequency in inbreeding, **4:**435
    fusion, of glucocerebrosidase, **3:**780, 781
    G protein, structure of, **4:**21
    genome and, **3:**870
    glucocerebrosidase
        characteristics of, **3:**778–79
        mutations in, **3:**779–80
    growth-regulated, listed, **2:**262
    heat-shock, **4:**100–101
    identical by descent and identical in state, **4:**433
    Ig heavy-chain structural, **4:**833–34
    insulin-like growth factors, **4:**504–6
    interferons, **4:**514–15
    interleukin-2, **4:**531–32
    isochores and, **4:**254
    isoenzymes as markers, **4:**577–78
    isolation of, and pulse field gel electrophoresis, **6:**379
    length of, **2:**479
    location of, **3:**878
    migration matrices, **4:**435
    orthologous, **2:**737
    osteoblast, regulation of, **1:**766
    osteoblast-specific, **1:**763–66
    polymorphism of, **6:**79–85
    pseudogenes, **2:**480
    regulation methods, listed, **2:**431
    ribosome biogenesis and ribosomal protein, **6:**675–77
    search, in disease, **2:**487
        in cystic fibrosis, **2:**487
    STE6 gene, alpha2 protein and, **3:**132

Genes *(Continued)*
    supergenes, **2:**480
    tropomyosin, properties of, **5:**218–19
    variation in, and population genetics, **6:**107–9
    vimentin, **5:**224
Genes, adenoviral
    DNA methylation and, **3:**153
    expression of, **3:**156
    late E2A promoter and, **3:**157
    spreading methylation and, **3:**159
Genes, class II. *see* Genes, immune response
Genes, DMD
    cloning of, **5:**193–94
    defined, **5:**191
    locus of, **5:**192–93
    mutations in, **5:**199–201
    regulation of, **5:**199
Genes, dominant. *see* Dominance (genetics)
Genes, *env*, **1:**12
Genes, *fos*, **3:**701–7
Genes, *gag*, **1:**13
Genes, homeo box, **3:**313
Genes, HPRT
    in Lesch-Nyhan syndrome, **5:**245
    mutations of, **5:**249
Genes, immune response, **4:**834
    DNA rearrangement in, **3:**795
Genes, mitochondrial
    eukaryote genes and, **3:**794
    structure of, **3:**857, 858
Genes, *myc*, in genital carcinogenesis, **3:**865
Genes, myosin
    cloning of, **5:**211
    heavy-chain, **5:**211–12
    light-chain, **5:**212
    protein isoform generation by, **5:**211
Genes, *nef*, HIV-1, **1:**14
Genes, *pol*, HIV-1, **1:**13
Genes, *ras*, **3:**865
Genes, reiterated
    in chromosome 21, **3:**184–85
    alpha gene cluster, **4:**146–47
    beta gene cluster, **4:**147
    theta gene cluster, **4:**146–47
    zeta gene cluster, **4:**146–47
Genes, reporter, **2:**435
    defined, **3:**701
Genes, retinoblastoma, **2:**64
    effect on retinoblastoma, **6:**636–37
    germ line mutations involving, **6:**637–38
    as tumor suppressor, **2:**451–54
    tumorigenic mutations and, **6:**637

Genes, *rev*, HIV-1, **1**:14
Genes, spliced. *see* DNA, recombinant
Genes, suppressor, tumor, **2**:53, 60–61, 451, **7**:705–9
  in colorectal precancerous process, **3**:771
  concept of, **2**:265
  defined, **2**:53
  early discovery of, **7**:705–6
  hybrid cells in studying, **7**:708
  and retinoblastoma, **6**:635–36
  tumor development and, **5**:330
  in "two-hit" theory of cancer genetics, **3**:822
Genes, *tat*, HIV-1, **1**:13
Genes, Tn, **5**:206
Genes, Tn-I, **5**:221
Genes, Tn-T, **5**:221, 222
Genes, transforming. *see* Oncogenes
Genes, *vif*, HIV-1, **1**:14
Genes, viral, **7**:772
  DNA methylation and, **3**:160
  in lytic cycle, **3**:471
Genes, *vpu*, **1**:14
Genes probes, DNA. *see* DNA probes
Genetic carriers. *see* Heterozygote
Genetic code, **6**:217
  recognition of transfer RNA and, **5**:76
Genetic counseling, **3**:813–17
  after miscarriage, **3**:302
  forensic pathology and, **5**:685
  in fragile X syndrome, **3**:717
  for hemophilia, **4**:163
  for prevention of birth defects, **1**:703
Genetic counselors, **4**:80
Genetic diseases. *see* Hereditary diseases
Genetic distance, **4**:952–53
Genetic drift
  DNA divergence and, **3**:883
  and variation, **6**:111–12
Genetic engineering
  bioethical concerns, **1**:658–59
  in crop sciences, **1**:144
  DNA restriction enzymes and, **3**:108
  nerve regeneration and, **5**:345
  transacting factors and, **3**:112
Genetic induction. *see* Gene expression regulation
Genetic locus, **2**:480
  defined, **3**:135, 819, **5**:243
  identification, **2**:480
Genetic markers, **2**:731
  in affective disorders, **1**:93–100
  in chromosome mapping, **3**:824, 828
  in diagnosis of Gaucher disease, **3**:782
  as diagnostic tools, **2**:735, **3**:135–50

Genetic markers *(Continued)*
  in muscular dystrophy diagnosis, **5**:202
  polymorphism, **2**:668–71
  of twins, **3**:306
  for zygosity, **3**:307
Genetic models. *see* Models, genetic
Genetic reaction range, defined, **3**:841
Genetic recombination. *see* Recombination, genetic
Genetic screening
  bioethical concerns, **1**:658–59
  scope of, **5**:685
Genetic segregation
  defined, **3**:847
  in families, **3**:852–53
  in peas, **3**:848
Genetic variation. *see* Mutation
Genetic vectors
  defined, **3**:135, 169
  DNA
    defined, **3**:801
    in transfection, **3**:805–6
Genetics, **3**:847–58
  absence of enzyme activity, **4**:1001–3
  in autoimmune disease, **1**:531–32
  biochemical, of Down's syndrome, **3**:175–89
  as cause of autism, **1**:507–8
  and criminality, **2**:711–14
  dental, **2**:793–94
  evolutionary, **3**:488
  as factor in adaptational physiology, **1**:40–41
  germ-line, **4**:833–34
  growth and development determinants, **4**:28
  histocompatibility antigens, **4**:358–59
  and hypertension, **4**:290
  isoenzyme patterns, **4**:575
  and leukemia, **4**:691
  longevity and, **4**:783–84
  Mendelian heredity, **2**:871
    defined, **3**:85, 791
    principles of, **3**:848–50, 854–55
  pediatrics and, **5**:703
  population, **6**:107–15
  population, polymorphism and, **6**:81–83
  reverse. *see* "Reverse" genetics
  shared, in behavioral development, **1**:582
Genetics, behavioral, **3**:835–45
Genetics, quantitative. *see* Quantitative genetics
Geniculate bodies, **3**:550–51
Geniculate nucleus, lateral, **2**:494, **5**:311, **7**:838–40
Geniculohypothalmic tract, **2**:494
Genital ducts. *see* Müllerian ducts; Wolffian ducts

Genital neoplasms, papillomaviruses and, **3:**859–67, **5:**638

Genital warts. *see* Condylomata acuminata; Papillomaviruses

Genitalia, female

   anatomy of, **6:**562–65

   development and puberty, **6:**341–42

   effect of radiation on, **6:**413

   sperm survival in, **7:**175

Genitalia, male

   anatomy of, **6:**559–62

   development and puberty, **6:**341–42

   effect of radiation on, **6:**413

   fetal, development of, **3:**341–42

   sperm maturation in, **3:**600

   sperm survival in, **7:**174–75

   spinal cord injury and, **5:**808

Genitourinary system. *see* Urogenital system

Gennari's stripe, **2:**684

Genome, human, **2:**437, 455

   clone, **2:**483

   composition, **2:**475–80

   and DNA, **2:**9

      repetitive elements, listed, **2:**477

   fingerprinting, **2:**483

   mitochondrial, **2:**480

   physical map, high-resolution, **2:**482

   polymorphism, megabase, **2:**485

Genome project, human. *see* Human genome project

Genomic clones, defined, **3:**175

Genomic library, **3:**869–79, **4:**251–58

   chromosomal band, **4:**253

   chromosomes, **4:**252–53

   coding sequences, **4:**252

   components, **4:**251

   compositional patterns, **4:**256–57

   defined, **3:**823

   families of repeated sequences, **4:**252

   hamster, late E2A promoter-CAT gene construct and, **3:**157

   isochores. *see* Isochores

   Leishmania, **4:**683

   neogenome, **4:**258

   organization, **4:**253–55

   overlapping, **2:**482–83

   paleogenome, **4:**258

   primate relations and, **3:**881–89

   sequencing, **4:**239

   size, **4:**252

Genomic mutations. *see* Mutations, genomic

Genomic sequencing, **3:**153

Genotoxic carcinogens

   biological markers of, **3:**390

   defined, **3:**383

   dose response estimation of, **3:**389

Genotype, **2:**689

   defined, **2:**689, **3:**135, 823, 847, 848, **5:**243

   variation in, population genetics and, **6:**107–9

Geography

   longevity and, **4:**784–85

   as measure of population differentiation, **6:**103–4

Geometric forms, **5:**157

Geometric isomerism, defined, **3:**569

Geophagia, **2:**649, **3:**31

Geriatrics, **3:**891–97, 899–905

   service delivery, **1:**135

Germ cells

   female, prenatal, **3:**628

   generation of diversity, **4:**327

   inheritance, **4:**833–34

   testicular, development of, **7:**427–28

Germ layers

   in body pattern formation, **3:**286

   defined, **3:**281

German measles. *see* Rubella

Germany, West

   nutritional disorders in, **3:**683

   obesity in, **3:**684

Germinal mutations. *see* Mutations, germinal

Germinal vesicle breakdown, **5:**554

Gerontogenes, in "failed" aging, **3:**893

Gerontology. *see* Geriatrics

Geropsychiatry, **1:**129–37

   active therapeutic intervention, **1:**136

   acute confusional states, **1:**130

   anxiety, **1:**134–35

   delirium, **1:**130

   dementia, **1:**130–31

   depressive neuroses, **1:**132–33

   elder abuse, **1:**135–36

   ethics, **1:**136

   Informed Consent Doctrine, **1:**136

   mood disorders, **1:**132–33

   organic mental disorders, **1:**130–32

   paranoia, **1:**133

   pseudodementia, **1:**132

   senile dementia of the Alzheimer's type, **1:**131–32

   sexuality, **1:**136

   sleep disorders, **1:**133–34

Gerstmann-Strüssler syndrome, **1:**214

Gerstmann's syndrome, **3:**221

Geschwind, **3:**223

Gestational age, defined, **3:**293

Gestrinone, **3:**368

GH-releasing hormone. *see* Somatotropin-releasing hormone

Giant cell tumors, **7:**702–3

Giant neurites. *see* Meganeurites

Giardia lamblia, **4:**450, **6:**274–75, 463

Giardiasis, **2:**382, **6:**274–75, 383
    drugs for, **2:**382
    secondary lactase deficiency, **4:**618

Gibbs, Willard, **3:**424

Gilles de la Tourette syndrome, **2:**217, 222, **4:**997
    haloperidol therapy, **2:**222
    obscenities, **2:**222

Gingiva, **2:**817, **4:**33–38
    biochemical composition, **4:**35–36
    blood and nerve supply, **4:**35
    collagen, **4:**34–35
    coloration, **4:**35
    enzymes, **4:**36
    glucose metabolism, **4:**37
    histology, **4:**34–35
    lipids, **4:**36
    metabolism, **4:**36–37
    respiratory rate, **4:**36–37
    turnover rates, **4:**37

Gingival diseases, **4:**33–34, 37

Gingival groove, **4:**34

Gingival sulcus, **2:**777

Gingivitis, **2:**805, **4:**37
    chronic, **5:**612

Gingivitis, necrotizing, acute, **5:**613

Ginglymus joints, **1:**374

Gla domains, **4:**184

Gla-protein, matrix, **1:**764

Glacial age. *see* Ice Age

Glandular fever. *see* Infectious mononucleosis

Glaucoma
    carbonic anydrase inhibitors and, **3:**97
    defined, **3:**533
    marijuana for, **5:**444

Glial cells, **2:**673
    defined, **5:**341
    nerve growth factor and, **5:**336
    number of, **2:**675
    scar tissue and, **5:**345
    Schwann cells and, **5:**342

Glial fibrillary acidic protein, **1:**434

Glial hyperactivity, **3:**897

Glicentin, **5:**718

Gliding joints, **1:**374

Glioblastoma, **5:**548

Glioblastoma-derived T-cell suppressor factor. *see* Transforming growth factor beta

Gliogenesis, **1:**434–35

Gliosis, **1:**444–45

Global Information and Early Warning System on Food and Agriculture, **3:**565

Global Ward Behavior Scale, **1:**611

Globin, **2:**455, 479
    gamma-Globulin, **2:**659
    beta-globin chain synthesis, **4:**148
    gamma-globin chain synthesis, **4:**148

Globin gene
    allelic diversity in, **3:**854
    clusters, **4:**146–47
    enhancer sequences, **4:**148
    expression and regulation, **4:**148
    as gene family, **3:**797–98
    beta-globin gene cluster, in primate evolution, **3:**887
    methylation state, **4:**148

Globin gene promoter, **3:**155

Glomerular filtration
    bicarbonates in, **5:**878
    defined, **3:**95

Glomerular filtration rate, **3:**902

Glomerular layer, olfactory
    anatomy, **5:**524
    mitral cell order and, **5:**530

Glomerulonephritis
    characterization, **4:**410
    in children, **5:**708

Glomerulus, **3:**94–95

Glottis, **4:**654, **6:**590

Glucagon, **2:**105, **4:**493–501
    actions in vivo, **4:**494, 496–98
    fetal, **3:**346
    function of, **5:**718–19, **6:**89, 732–33
    intestinal smooth muscle and, **5:**723
    lipoprotein lipase and, **3:**587
    mechanism of action, **4:**500–501
    perinatal, **3:**350
    satiety and, **1:**350, **4:**248
    structure, **4:**494

Glucocerebrosidase
    cDNA for, **3:**778
    defined, **3:**777

Glucocorticoid cascade hypothesis of aging, **3:**896–97

Glucocorticoid pathway, **7:**260

Glucocorticoid receptors. *see* Receptors, glucocorticoid

Glucocorticoid response element, **3:**132

Glucocorticoids, **2:**247, 754
    antiinflammatory action of, **7:**269–70
    in bone and calcium metabolism, **1:**804

Glucocorticoids (*Continued*)
  bone cell function and, **1:**765
  cardiovascular system and, **7:**270
  defined, **7:**269
  function of, **7:**267–70
  G proteins and, **4:**21
  in host defense, **1:**311–12
  hypercalcemia of malignancy treatment, **4:**280–81
  immunity and, **5:**374, 376, 385
  immunosuppression and, **7:**269–70
  inhibition of tissue repair and, **7:**505
  initiation of lactation, **4:**605–6
  interaction with parathyroid hormone, **5:**664
  metabolic effects on tissue, **7:**268
  nervous system and, **7:**270
  prepubescent growth and, **3:**351
  therapy, **2:**247
Glucokinase, **4:**1000
Gluconeogenesis
  defined, **5:**821
  inhibition of, by acidosis, **5:**826
  2-oxoglutarate and, **5:**828
  pathways, **4:**1002
  pH and, **5:**822
Glucooligosaccharides, **3:**645–46
Glucophores
  defined, **7:**361
  sweet, stereochemical requirements of, **7:**365–70
Glucose, **2:**102–6, 281–82, 286–87, 324, 504, 508, 787
  absorption of, **3:**79
  analogues, **2:**39
  in biological energy flow, **3:**932
  in glycogen chemistry, **3:**920
  intestinal hyperemia and, **4:**556
  intolerance, **2:**106
    induction of, **7:**572–73
  metabolism, **4:**37
  pH and, **5:**824
  reduced bonds of, **5:**474
  as signal molecule, **2:**324
  topography, diagram of, **2:**282
  transport, across blood-brain barrier, **1:**725
Glucose, blood. *see* Blood glucose
Glucose-6-phosphatase
  in glycogen degradation, **3:**923–24
  in liver, **3:**939–40
Glucose-1-phosphate
  in glycogen degradation, **3:**923, 924
  in glycogen formation, **3:**921
Glucose-6-phosphate
  in first-stage glycolysis, **3:**933
  in glycogen degradation, **3:**923, 924, 929

Glucose-6-phosphate (*Continued*)
  in glycogen formation, **3:**921
Glucose-6-phosphate dehydrogenase, **2:**481, **4:**1001
  allelic diversity in, **3:**854
  as genetic marker for affective disorders, **1:**98–99
  markers, in discrimination of monoclonal and polyclonal tumors, **7:**689
  NADPH and, **3:**913
  parathyroid hormone assay of, **5:**660
  phenotype in HeLa cells, **4:**107–8
  in uterus, estrogen-induced, **7:**742–43
  variations in genes coding for, **4:**882–83
Glucose-6-phosphate dehydrogenase deficiency, **4:**578–79
Glucose oxidase, antibody conjugates with, **4:**372–73
Glucose-phosphate isomerase, **3:**933
Glucose-sodium cotransporter. *see* Cotransporter
Glucose tests, **4:**619–20
Glucosyl phosphatidyl inositols, **4:**748
Glucosylgalactosylhydroxylysine, **2:**569
Glucosyltransferase, **2:**563
Glucuronidation
  in formations of primary bile acids, **1:**624
  in nasal cavity, **5:**286
Glucuronyltransferases, **5:**286
L-glutamate, as neuroactive substance in retina, **6:**629
Glutamate decarboxylase, **2:**16, 851
Glutamate dehydrogenase, **2:**533
Glutamate receptors. *see* Receptors, glutamate
Glutamates, **2:**5, 13, 312, 313, 316, 337, 339, 521–25, 536, 686
  epilepsy and, **3:**460
  and neurotransmission, **2:**312, 521–25
  as neurotransmitter in brain, **7:**384–85
  receptor subtypes, three, **2:**313
Glutamic acid
  in enzyme catalysis, **3:**434
  alpha-tubulin detyrosination and, **5:**18
  valine and, **3:**144
L-glutamic acid, epilepsy and, **3:**460
Glutaminase, **2:**14
Glutamine, **2:**506
  astrocyte uptake of, **1:**440
  metabolism, 2-oxoglutarate dehydrogenase and, **5:**822
  urea synthesis and, **5:**879
Glutamine synthase, **1:**440–41
Glutaraldehyde, **3:**278
Glutathione, **2:**655, **3:**907–18
  glycolysis and, **3:**932
Glutathione ethyl ester, **3:**917

Glutathione peroxidase, **2:**652
    peroxide detoxification and, **3:**914
    as secondary gene dosage effect, **3:**177
    selenium-dependent, function of, **6:**793-94
Glutathione reductase, **2:**533
    deficiency of, **3:**916
    intracellular thiols and, **3:**913-14
Glutathione synthetase deficiency, systemic, **3:**916
Glutathione transferases, **2:**118
    detoxication activities in nasal cavity, **5:**287
    electrophile/glutathione reactions and, **3:**915
Glutethimide, **2:**787
Gluteus maximus, **2:**595
Glycan chains, **4:**4
Glycemic index, **2:**101
Glyceraldehyde-3-phosphate
    in second stage glycolysis, **3:**936
    in third-stage glycolysis, **3:**936-37
Glyceraldehyde-3-phosphate dehydrogenase, heat shock-
        stimulated, **4:**103
Glycerides
    behavior and distribution, **4:**743
    defined, **3:**47
    fatty acids and, **3:**55
    free fatty acids and, **3:**52
    hydrolysis of, **3:**53
    metabolism, **4:**742-43
    triacylglycerols and, **3:**51
Glycerin, **2:**241
    defined, **3:**317
    from monoacylglycerol hydrolysis, **3:**53
    as osmotic diuretic, **3:**100
Glycerol. *see* Glycerin
Glycerolipids, **4:**970
Glycerolphosphate acyl transferase, **3:**583
Glycerophopholipids, **4:**743-44
Glyceryltrinitrate, **2:**172
Glycine, **2:**5, 323, 338, 522, 559, 561, 568
    in elastin, **3:**264
Glycine receptors. *see* Receptors, glycine
Glycocalyx, **2:**130
Glycogen, **2:**103-5, 238, 240, 241, 508, 534, 645,
        3:919-29
    breakdown of, hormone regulation in, **3:**324
    defined, **3:**317
    energy storage in, **5:**474-75
    in exercise, **3:**513
    metabolism, **7:**20-21
    phosphorolysis, **2:**534
Glycogen phosphorylase deficiency, **5:**227-28
Glycogen storage disease, **3:**924
Glycogen storage disease type V, **5:**227-28

Glycogen storage disease type VII, **5:**227
Glycogen synthetase, **3:**331
    in glycogen formation, **3:**921
    regulation of, by phosphorylation, **6:**210
Glycogenosis 5. *see* Glycogen storage disease type V
Glycogenosis 7. *see* Glycogen storage disease type VII
Glycolipids
    defined, **3:**725
    formation in Golgi apparatus, **4:**8
    in neuronal development, **3:**745
Glycolysis, **2:**240, **3:**931-41
    aerobic, in carcinoma, **3:**940
    defined, **3:**377, **5:**821
    pathways, **4:**1002
    phosphofructokinase and, **3:**184
    protons in, **5:**824
    rate-limiting enzymes, **4:**1000-1001
    respiratory alkalosis and, **5:**825
    in skeletal muscle, **5:**825
Glycoprotein, membrane. *see* Membrane glycoproteins
Glycoproteins, **2:**201, 206, 225, 270, 639, 734
    cell substrate-interacting, **4:**958
    in elastin, **3:**262
    enzymatic modifications in Golgi apparatus, **4:**4
    herpesviridae, **4:**191
    in milk fat globule membrane, **3:**48
    in nerve endings, **3:**744
    in neurons, **3:**747
    nuclear pores and, **5:**454
    in parenchymatous plant tissues, **3:**38
    radioactivity of, **3:**746
    of zona pellucida, **3:**605
Glycosaminoglycanase, **2:**646
Glycosaminoglycans, **2:**201, 639, 641, 643, 645
    gingiva, **4:**35-36
    structure of, **6:**258-60
    types of, **6:**257-58
Glycosidases, **2:**568
    of cortical granules, **3:**605
    intestinal, **4:**612-13
Glycosides
    as plant toxins, **7:**595
    terpenoid, as sweeteners, **7:**375-76
Glycosides, cardiac. *see* Cardiac glycosides
Glycosis, **3:**931-41
Glycosphingolipids
    behavior and occurrences, **4:**745-47
    classification of, **7:**180
    metabolism, **4:**745
    neutral, structure of, **3:**726, 728
Glycosylation, **2:**240
    bacterial antibody synthesis and, **3:**804

Glycosylation *(Continued)*
  chimeric antibodies and, **3:**808
  of collagen, **2:**563
  defined, **3:**801
  in enzyme isolation, **3:**409
  of G-protein-linked receptors, **5:**417
  O-linked, in Golgi apparatus, **4:**8
  transacting factor modification and, **3:**115
Glycosylation-enhancing factor (GEF), **1:**185
Glycosylation-inhibiting factor (GIF), **1:**185
Glycosyltransferases, **3:**729, 730
Glycyrrhizin, **7:**376
Glyoxylase I, **3:**915
Glyoxylase II, **3:**915
Glyoxylate cycle, **2:**507–8
GM1 ganglioside
  functions of, **3:**741
  oligosialogangliosides and, **3:**746
Gnathostoma, **2:**389
Goblet cells
  of conjunctiva, **3:**546
  defined, **3:**765
Goggle headaches, **4:**58
Goitrogens, **7:**486–87
Gold colloid, **2:**467
  for gene mapping, **2:**467
  in immunocytochemistry, **3:**277
  in nuclear transport, **5:**456, 458
  in photoelectron microscopy, **5:**906
Golden hamster, **2:**490, 492, 499
Goldfish, **5:**342
Golgi apparatus, **2:**11, **4:**1–12, 16, 19, 130, 238–40, 563
  in cellular constituent processing, **4:**10–11
  composition, **4:**6–9
  constitutive secretory products processing role, **4:**10
  drug effects on function, **4:**11–12
  enzymes in, **4:**7–9
  extracellular matrix formation, **4:**10
  function, **4:**3–4
  functional polarity and dynamics, **4:**9–10
  ganglioside carbohydrate chains and, **3:**728–29
  glycolipid formation, **4:**8
  isolation, **4:**5–6
  lipids in, **4:**6–7, 9
  lysosome formation role, **4:**4, 11
  in maintaining epithelial cell polarity, **6:**64–65
  membrane proteins, **4:**976
  N-asparagine-linked glycoprotein processing, **4:**7
  O-linked glycoprotein processing, **4:**7–8
  plasma membrane transport role, **4:**11
  polarity, **4:**1, 5
  in protein sorting, **6:**237–38

Golgi apparatus *(Continued)*
  proteins in, **4:**7
  proteoglycan sugar chain formation, **4:**8
  in secretory granule formation, **4:**11
  in secretory pathway, **6:**240–41
  sorting and transport of proteins in, **6:**252–55
  sphingomyelin formation, **4:**8
  structure and cytochemistry, **4:**1–5
  sulfation, **4:**4
  sulfation modification of secretion products, **4:**8–9
  topology, **4:**9
Golgi method, **2:**676
Golgi tendon organs
  muscle activation and, **3:**699
  role in movement, **6:**171–72
Gompertz curve, **5:**100
Gomphoses, **1:**372
Gonadal differentiation, **3:**340–41
Gonadal hormones, **6:**816
Gonadal steroids
  and gonadotrophin secretion, **6:**18
  perinatal, **3:**348–49
  role in bone and calcium metabolism, **1:**804
Gonadarche, defined, **3:**350
Gonadoliberin. *see* Gonadorelin
Gonadorelin, **6:**89
  amino acid sequence for, **7:**349
  deficiency of, **5:**395
  defined, **5:**393
  in fetal hypothalamus, **3:**342
  functions of, **3:**629, **5:**732
  for hypothalamic/pituitary disorders, **5:**399
  oocyte maturation induction and, **5:**558
  in ovulation induction, **7:**348–49
  prolactin and, **3:**339
  in regulating female reproduction, **1:**743
Gonadorelin analogues, **3:**368
Gonadostat, defined, **6:**343
"Gonadostat" theory, refuted, **3:**355
Gonadotrophs, **6:**17–18
Gonadotropin-releasing hormone. *see* Gonadorelin
Gonadotropins, **2:**24, 26, 323, 414, 498
  corpus luteum and, **3:**634, 635
  exogenous, follicle selection and, **3:**633
  hypothalmus dysfunction, weight loss and secretion, **1:**743
  in oocyte maturation, **5:**555, 557
  ovary and, **3:**629–30
  ovulation induction with, **7:**347–48
  production and activity of, **6:**17–18
  in puberty, **3:**355
  regulation of, **5:**398–99

Gonadotropins *(Continued)*
    twinning and, **3:**307
    in two-cell theory of follicular steroidogenesis, **3:**630
Gonadotropins, chorionic
    corpus luteum and, **3:**635
    fetal adrenal cortex and, **3:**344
    fetal testosterone secretion and, **3:**340, 342
    function of, **6:**91
    measurements of, in spontaneous abortion **1:**4
    and morning sickness, correlation between, **6:**125
    perinatal, **3:**348-49
Gonadotropins, pituitary, **5:**606
Gonads
    aging of, **3:**903
    hormone production in, **5:**708
Gonorrhea, **4:**450. *see also Neisseria gonorrhoeae*
    antibiotic resistance and, **6:**898
    infections, **6:**897-98
Gonyaulax. see Dinoflagellida
Goodman, Morris, **3:**488, 881
Goose pimples, **4:**800
Gorging. *see* Binge/gorge
Gorillas
    characteristics of, **6:**141
    mating, **4:**934
Gottesman, I. I., **3:**841
Gottschalk, A., **3:**726
Gout
    hyperuricemia and, **6:**392-93
    hypoxanthine phosphoribosyltransferase deficiency
        and, **3:**855
    purine metabolism and, **6:**391-92
Government agencies, famine and, **3:**565-66
Gowans, **5:**139
Graafian follicles
    defined, **3:**627
    development of, **3:**630-31, 633
    ripening, **4:**932
Gradient elution, HPLC, **4:**212-13
Graft, condylomata acuminata-treated, **4:**586
Graft rejection, **4:**360-62
    prevention, **4:**362-65
Graft-versus-host disease, **1:**525, 792-93
    in bone marrow transplantation, **1:**792-93
    cyclosporin a therapy, **4:**363
    immunotoxin therapy, **4:**421
Gram-negative bacteria, transmembrane channels, **4:**973
Gram-negative cell. *see* Cell wall, bacterial
Gram-positive cell wall. *see* Cell wall, bacterial
Gram stain
    for anaerobic bacteria, **1:**238
    in detecting bacterial pathogens, **1:**552-53

Grand mal epilepsy. *see* Epilepsy, grand mal
Grandmother neuron, infamous, **2:**549
Grandparents
    in genetic mapping, **3:**829
    of muscular dystrophy patients, **5:**201
Grant, James, **5:**701
Granulocyte-macrophage colony-stimulating factor. *see*
        Colony-stimulating factor
Granulocytes
    characterization, **4:**166
    and complement C5, **2:**613
    in electron microscopic imagery, **3:**277
    functions, **4:**166-67
    5-methyldeoxycytidine residues of, **3:**154
    responses to lithium, **4:**750-51
    types of, **4:**166
    and zinc, **2:**659
Granulocytosis, **4:**750-51
Granulomatous disease, chronic
    bone marrow transplantation for, **1:**794
    ectopic 1,25-dihydroxyvitamin D_3 production and,
        3:334
    as model for studying respiratory burst, **6:**582
    polymorphonuclear neutrophil activity in, **5:**430-31
Granulomatous reactions, **4:**413
Granulosa cells
    in corpus luteum formation, **3:**634
    defined, **3:**627
    LH surge and, **3:**633
    receptors on, **3:**633
Granulosis viruses. *see* Baculoviridae
Graphemes
    defined, **3:**215
    dyslexia-dysgraphia and, **3:**217
Grating microspectrofluorometer, **5:**567
Graves-Basedow's disease. *see* Hyperthyroidism
Grave's disease
    and autoimmunity, **1:**530
    thyroid-stimulating hormone in, **5:**400
Gravida, defined, **2:**403
Gravity, and body biochemistry. *see under* Space flight
Gray matter, cerebral, **1:**213
Gray matter, laminae, **7:**199-200
Gray matter, spinal cord, **7:**196-205
    afferent systems to, **7:**207-11
    chemoarchitecture of, **7:**201-5
    cytoarchitecture of, **7:**197-200
    dendroarchitecture of, **7:**200-201
    efferent pathways in, **7:**211-13
    laminal organization of, **7:**197-98
    propriospinal pathways to, **7:**205-7
Gray's theory of anxiety, **5:**774-75

Great apes. *see* Apes
Greater omentum. *see* Omentum
Green algae. *see* Chlorella
Green Revolution, in India, **3:**564
Greenhouse effect, **1:**144
Gregory, **5:**716
Grief. *see* Depression
Griseofulvin, **1:**292
Gronblad-Strandberg syndrome. *see* Pseudoxanthoma elasticum
Gross, Robert, **5:**703
Ground squirrel, **2:**498
    hibernation rhythm, **2:**491
Growth, **2:**201, 202, 214, 253, 254, 264, 870, **4:**25–31
    in adolescence, **4:**29–30
    in adulthood, **4:**30–31
    of bone
        during development, **1:**789–90
        effect of growth hormone on, **6:**13–14
    cellular, growth factors and, **4:**503–4
    in childhood, **4:**26–27
    congenital malformation syndromes, **4:**27
    craniofacial, **2:**697–708
        bone growth measurement, **2:**698–702
        normal, **2:**702–8
        postnatal, **2:**697–708
    defined, **2:**689, 872
    dental development, **4:**29
    developmental age, defined, **3:**293
    effects of protein-energy malnutrition, **4:**894–95
    embryo, **4:**26
    evolution, **2:**870–72
    fetal, IGF effects, **4:**508–10
    first year, **4:**26
    genetic determinants, **4:**28
    HeLa cells, **4:**110
    helminthiasis and, **4:**116–18, 120
    laryngeal reflexes, **4:**651–52
    larynx, **4:**657–58
    leukemic cells, **4:**692–93
    locomotion, **4:**779–81
    major events in, **2:**875–78
        birth, **2:**876
        feeding of infant, **2:**877
        fertilization, **2:**875
        fetal, **2:**876
        germ cell, **2:**875
        infant, **2:**875
        mother/child relationship, **2:**877
        play activity, **2:**878
        prenatal behavior, **2:**876
        separation of child from mother, **2:**878

Growth
    major events in (*Continued*)
        sleep/wake state, **2:**878
        social relationships, **2:**878
        sperm cell, **2:**875
        suckling, **2:**877
        weaning of child, **2:**877
    nutritional modification of, **4:**28
    oral, **2:**783–88
    ossification of round bones and epiphyses, **4:**28–29
    prepubescent, **3:**350–52
    principles, **2:**872–75
    processes, **2:**875–78
    pubescent, **3:**354
    secular trend in size and maturational timing, **4:**30
    sex differences, **4:**27–28
Growth-associated proteins. *see under* Proteins
Growth cones, neural
    gangliosides and, **3:**745
    in neuronal morphogenesis, **5:**899
Growth cones, neurite
    defined, **5:**341
    nerve growth factor and, **5:**334
    nerve regeneration and, **5:**343
Growth disorders, **4:**996–97
Growth factors. *see* Growth substances
Growth hormone, **2:**21–28, 54, 259, 834, 845, 846, 849
    in bone and calcium metabolism, **1:**805
    in bone cell differentiation, **1:**789
    levels in protein-energy malnutrition, **4:**894
    maintenance of lactation, **4:**606
    production, action, and control of release, **6:**13–14
    regulation of, **5:**399–400
    secretion
        and adenoma, **2:**21
        and sleep, **7:**75
Growth hormone-releasing hormone. *see* Somatotropin-releasing hormone
Growth retardation, **7:**406–8
Growth retardation, intrauterine. *see* Fetal growth retardation
Growth substance receptors. *see* Receptors, growth substances
Growth substances, **2:**61–62, 89, 255, 256, 258–60. *see also* Epidermal growth factor; Neuronotrophic factors; Transforming growth factor
    amines as, **3:**322
    as biological response modifiers, **1:**686–87
    in bone, **1:**763
    bone, synthesized by skeletal cells, **1:**806
    bone cell function-related, **1:**765
    from bone matrix, **1:**806–7

Growth substances (*Continued*)
  and breast cancer, **2:**61-62
  for cell culture, **2:**258-60
  and cellular growth, **4:**503-4
  for central nervous system, **5:**336
  defined, **2:**53, 321, **3:**445, **5:**333
  as differentiation signals in osteogenesis, **1:**787-89
  domains, **4:**184
  embryonic body pattern diversity and, **3:**291
  epidermal, **2:**61, 89, 259, 260
  extracellular matrix and, **3:**449
  fetal, **3:**338-39
  hormones as, **3:**326
  immunity and, **5:**375
  insulin-like, **2:**62, 259
  insulinlike growth factor-I and, **3:**484, 485
  and interleukins, **2:**255, 259
  mammary gland development, **4:**604
  nonhematopoietic, **1:**687
  in oocyte maturation induction, **5:**558
  peptides as, **3:**322
  perinatal, **3:**347
  platelet-derived, **2:**89, 258, 259
  platelet-derived, simian sarcoma virus and, **5:**874
  polypeptide, wound healing and, **7:**499-500
  pubescent, **3:**355
  in soft tissue repair, **7:**501-4
  targeting moiety of immunoconjugates, **4:**374-75
  tissue repair and, **7:**499-509
  trophic growth effect, defined, **5:**725
  tumor cell requirements, **4:**584
  uterine production of, estrogen-induced, **7:**743
  wound extracellular matrix and, **7:**503
  wound healing and, **7:**499-500
GSSG
  physical properties of, **3:**908
  reduction of, **3:**913
  transport of, **3:**911
GTP phosphohydrolase
  G protein region 47-53 and, **4:**20
  G protein region 223-227 and, **4:**20
GTPase. *see* GTP phosphohydrolase
Guanabenz, **2:**177
Guanfacine, **2:**177
Guanidine hydrochloride, **3:**420
Guanine, **2:**117
Guanine nucleotide regulatory protein, **2:**14, 89, 92, 238, 302, 304, 325, **3:**943-54, **4:**975
  and cellular signal transduction, **7:**17-19
  defined, **5:**415
  effector systems and, **5:**417, 420

Guanine nucleotide regulatory protein (*Continued*)
  and guanosine triphosphate, **2:**318
  special sense receptors and, **5:**416
  synapse neuromodulation and, **7:**388
Guanine nucleotide regulatory protein receptors. *see* Receptors, guanine nucleotide regulatory protein
Guanine nucleotides
  G protein region 223-227 and, **4:**20
  G protein region 292-296 and, **4:**20
Guanosine cyclic monophosphate
  defined, **3:**371
  endothelium-derived relaxing factor and, **3:**371
  functions of, **3:**331
  nitric oxide and, **3:**374
  in protein phosphorylation, **6:**202-3
  secondary messenger role of, **3:**375
Guanosine diphosphate, **4:**975
Guanosine monophosphate, **2:**313, **6:**619
Guanosine phosphates. *see* Guanine nucleotides
Guanosine triphosphate, **2:**318
  analogs of, **4:**14
  hydrolysis of, **4:**14
  signal transduction, **4:**975
  tubulin and, **5:**12
Guanosine triphosphate phosphohydrolase. *see* GTP phosphohydrolase
Guanosinetriphosphatase. *see* GTP phosphohydrolase
Guanoxan, **2:**746, 747
Guanyl cyclase, **2:**313
  in cyclic nucleotide regulation, **5:**417
  defined, **3:**371
  effect of atrial natriuretic factor on, **1:**467, 473, 475-76
  endothelium-derived relaxing factor and, **3:**371
  nitric oxide and, **3:**374
Guidance feedback hypothesis, **5:**127-28
Guide for the Care and Use of Laboratory Animals, **1:**255
Guillain-Barré Syndrome, **1:**527
  as myelin disorder, **5:**282
  as myeline disorder, **5:**282
Guinea worm, **2:**388
Gum. *see* Gingiva
Gut. *see* Intestines
Gut-associated lymphoid tissues
  defined, **3:**769
  innervation of, **5:**381
  plant toxins and, **5:**137
Gut glucagon. *see* Enteroglucagon
Gut tube, fetal, development of, **3:**57-60
Guthric test, for phenylketonuria, **5:**435
Gymnodinium. see Dinoflagellida
Gynecomastia, defined, **3:**357

Gyrus
    angular, dyslexia-dysgraphia and, **3**:222
    dentate, **2**:680
    left angular, Wernicke's dysphasia and, **3**:220
    postcentral, **2**:682
    precentral, **2**:682
    supramarginal, lesions of, **3**:223
Gyrus cinguli, **2**:680-82

H

H chains. *see* Heavy chains
H^+/O ratio, during oxidative phosphorylation, **1**:461
H-reflex, **6**:162-63
H^+ transporting ATPase, **1**:462, **4**:563-64
    K^+ pump, **4**:562
H-Y antigen, **3**:340
Haarscheiben, sensory receptors of, **7**:677
Habit, and memory, **4**:676
Habitat
    defined, **3**:251
    pollutant alteration of, **3**:252-53
Habituation, **2**:305
Hadar site, fossils of, **3**:490
Haeckel's biogenetic law, **2**:870
Haemophilus influenzae, **4**:450
    in infants and toddlers, **4**:459
    meningitis and, **4**:461
    neonatal infections, **4**:458
    Haemophilus influenzae, type B, **4**:446
Hageman factor, **4**:471
Hair, **2**:599-600, **4**:39-48. *see also* Kinky hair syndrome
    abnormalities and disorders, **4**:47-48
    bulb, **4**:40
    chemistry of, **4**:44-47
    classes of sensory receptors in, **7**:675
    cortex, **4**:43
    cuticle, **4**:43
    development, **4**:39-40
    evolution of bodily hair, **3**:496-97
    fibers, **4**:40, 41
    function, **4**:40
    growth, **3**:448, **4**:40-42
    parasites and, **4**:47-48
    radiobiology of, **6**:466
    secondary terminal, **4**:40
    as sensory receptors, **4**:40
    skin, **7**:45
    structure, **4**:42-44
    types of, **4**:39-40
    weathering of, **4**:47

Hair cells, **2**:521-25
    anatomy, **3**:232-34
    mechanotransduction in, **3**:234-35
Hair cells, outer
    cochlear microphonic and, **3**:238
    motility generation of, **3**:237
    stereocilia deflection and, **3**:236
    in two-tone suppression, **3**:239
Hair loss, **4**:41-42
Hairy cell leukemia, **4**:751
Hairy leukoplakia, **5**:613
Hakomori, S., **3**:732
Halcion. *see* Triazolam
Haldane, J. B. S., **3**:828, 829
Haldane effect, **6**:600
Half-field technique, visual, **2**:353
    with tachistoscope, **2**:353
Half-life, chemical persistence, **3**:256
Hallucinations, **2**:346, 348, 766
    and atropine, **2**:348
    cold stress and, **4**:797
    in LSD use, **5**:448
    in schizophrenic disorders, **6**:771
Hallucinogens
    abuse of, **5**:447-49
    marijuana and, **5**:444
    serotonin and abuse of, **6**:832
Halo effect, **5**:786
Haloperidol, **2**:217, 223, 843
    and Gilles de la Tourette syndrome, **2**:222
    for Huntington's disease, **4**:272
Halophiles, defined, **3**:662
Halstead-Reitan Battery, **1**:611, **2**:865-66
Hamartoma, defined, **3**:765
Hamartomatous polyps, **3**:771-72
Hamilton Depression Rating Scale, **1**:611, **2**:842
Hamster. *see* Golden hamster
Hamstring, **2**:595
Hand
    dermatoglyphics of, **3**:90
    movements, crosstalk and, **5**:156, 158, 159
    muscle spindles, **3**:720
    muscles of, **4**:264-65
    primate, **2**:598
    response to acute low temperature, **4**:795-96
    skin receptors in, **7**:120-22
    tactile receptors of, **7**:52-53
Hand muscles. *see* Muscle, hand
Handedness, **2**:862
Handicapped child
    external family resources as factor in coping with, **7**:296
    factors affecting parental response to, **7**:294-97

Handicapped child *(Continued)*
   families with, stress in, **7:**293–98
   family crisis-meeting resources and, **7:**295–96
   family resources as factor in coping with, **7:**295–96
   parent-professional interaction and coping with, **7:**296
   parental coping with, **7:**293–94
   parental expectations and perceptions of, **7:**296–97
   parental stages of stress associated with, **7:**297
   parental stress and nature of, **7:**294–95
   perceptions of
      cultural background and, **7:**296
      economic status and, **7:**296
      religious affiliation and, **7:**297
Handwriting. *see also* Dysgraphia
   brain zones for, **3:**217–18
HANES. *see* Health and Nutrition Examination Survey
Hanson, **5:**137
Haploidy, **2:**437
   in female, post-sperm penetration, **2:**437
   in sperm cell, male, **2:**437
Haplotypes, defined, **3:**135, 835
Haptic system, **7:**60–62
Haptoglobin, **1:**31, **2:**659
Hardy-Weinberg law, **4:**435
Harmaline, **1:**180
Harman, Denham, **3:**893
Harmine, **1:**180
Harriet Lane Home, **5:**701
Harris, Walter, **5:**699
Harrison Narcotic Act of 1914, **2:**512
Hartmann's theory of ego psychology, **2:**854
Hashimoto's disease. *see* Thyroiditis, autoimmune
Hashish. *see* Marijuana and cannabinoids
Hawthorne effect
   defined, **3:**473
   discovered, **3:**475
Hay, contaminated, **5:**263, 266
Hayflick limit, **3:**893
Hazard analysis critical control points, defined, **3:**661
Hazard identification, in toxicology, **3:**388
Hazardous substances, **3:**383–91
Head, **2:**599, 689–90
   axial spin-echo images, **4:**868, 870
   cutaneous fields of, **7:**672
   fetal, **2:**405–8
   movements, persuasion and, **5:**807–8
   muscles of, **4:**261
   myogenesis and, **5:**162
   T1-weighted image, **4:**869–70
   trigeminal innervation of skin, **7:**672–73
Headache, **4:**49–59
   anatomy of, **4:**50

Headache *(Continued)*
   classification of, **4:**50–51
   drug-associated, **4:**56–57
   exercise-associated, **4:**58
   infection-associated, **4:**57–58
   inflammatory, **4:**54–55
   ingested substance-associated, **4:**57
   inhaled substance-associated, **4:**57
   metabolic disorder-associated, **4:**58
   muscle contraction, **4:**54
   neuralgia, **4:**55–56
   postconcussion/post-traumatic, **4:**56
   psychologic/psychogenic, **4:**55
   tension, **4:**54
   traction, **4:**54–55
   vascular, **4:**51–54
Healing, society's response to, **7:**324–25
Healing, suffering, and the human condition, **7:**323–25
Health
   genetic aspects of, **3:**892–93
   relation to behavior, **4:**72–73
Health and Nutrition Examination Survey, **3:**650
Health care, **4:**72
   decision-making, **4:**75–76
   interpersonal issues, **4:**76–77
   utilization, **4:**82
Health care organizations, **4:**81–82
Health care rationing, bioethical concerns, **1:**659
Health care systems, in space. *see* Space stations
Health care technology, **4:**61–67
   defined, **4:**61
   directions for, **4:**66–67
   evaluation methods, **4:**62–66
   evaluation systems, **4:**66
   problems with, **4:**61–62
Health food movement
   advent of, **3:**22
   influence of, **3:**25
Health maintenance organizations (HMOs), **4:**66
Health Psychology, **4:**69–88
   addictions, **4:**83–87
   antecedents, **4:**69–71
   assessment, **4:**77–79
   behavior change applications, **4:**79–83
   behavior modification, **4:**79–80
   biofeedback, **4:**79
   counseling, **4:**80
   cross-cultural issues, **4:**77
   decisions in health care, **4:**75–76
   definition of, **4:**69
   education and persuasion, **4:**80–81
   health status assessment, **4:**77–78

Health Psychology (*Continued*)
    informed consent, **4:**76–77
    injury and illness prevention, **4:**74
    major research issues, **4:**72–74
    origins, **4:**71
    pain measurement, **4:**78
    patient assignment to treatments, **4:**78
    personality and illness, **4:**75
Health psychology. *see also* Psychology
    provider-patient interaction, **4:**76
    psychoimmunoneurology, **4:**74–75
    psychological risk factors, **4:**78
    public health interventions, **4:**82–83
    quality of life assessment, **4:**78–79
    risk-taking behavior, **4:**73
    scope of, **4:**71–72
    social support, **4:**76
    systems and community approach, **4:**81–82
Health Research Extension Act of 1985 (PL99-158), **1:**253
Health services for the aged, **1:**135
Health status assessment, **4:**77–78
Healthy People, **3:**651
Heape, Walter, **3:**600, 601
Hearing, **3:**229–41
    auditory attributes of, **5:**739
    perception and recognition, **5:**739–40
    threshold, **5:**739
    transduction, **5:**739
Hearing disorders, in older adults, **1:**121–22
Hearing loss
    as communication disorder, **7:**155–57
    middle ear muscles and, **3:**232
    in older adults, **1:**120
    olivocochlear bundle and, **3:**237
    perception and, **5:**740
    relationship of autism to, **1:**507
Heart, **2:**179–83
    alcohol consumption and, **1:**174–75
    anatomy, **2:**157–65
    artificial. *see* Artificial heart
    atrial natriuretic factor and, **1:**473–74
    atrial natriuretic factor synthesis in, **1:**470–71
    autonomic innervation, **1:**542, 545
    and autonomic nervous system, **2:**183
    birth control pills and, **1:**692–93
    blood supply, intrinsic, **2:**160
    cells, **2:**172
    chambers, **2:**158–60
    core temperature measurement, **4:**793
    dysfunction, and sudden infant death syndrome, **7:**319
    edema safety factors in, **5:**40
    function, **2:**183

Heart (*Continued*)
    innervation of, **5:**240
    metabolic control of, **3:**380
    muscle. *see* Myocardium
    nerve, afferent, **2:**138–40
    plain film radiography of, **3:**2
    platelet-activating factor and, **6:**50
    rabbit, myosin in, **5:**165
    radiation effects, **6:**414
    reflex bradycardia, **2:**141
    response of, **2:**140–41
    size, **2:**145
    vascular capacitance in control of output, **7:**759–60
    wall, **2:**157–58
Heart, mechanical, **4:**89–97. *see also* Artificial heart
    ancillary equipment, **4:**92–93
    applications, **4:**96
    artificial heart components, **4:**90
    artificial lung, **4:**90–91
    blood pumps, **4:**90
    blood temperature, **4:**91
    bubble oxygenator, **4:**91
    bypass flow rate, **4:**92
    cardiotomy suction, **4:**92
    cell saver, **4:**93
    centrifugal pump, **4:**90, 92
    complications, **4:**94–96
    components, **4:**89–93
    filters, **4:**91–92
    finger pump, **4:**90
    function, **4:**89–90
    heart and lung failure, **4:**96–97
    hemodynamics, **4:**93–94
    hypothermic-cardioplegic arrest, **4:**92
    metabolism during bypass, **4:**94
    in pediatrics, **5:**704
    ultra filtration device, **4:**93
Heart-assist devices, **4:**90–91
    left
        extracorporeal, **1:**387
        implantable types of, **1:**388–89
        as support for failing circulation, **1:**386–87
        as total artificial heart, **1:**387–88
    right, as support for failing circulation, **1:**386–87
Heart atrium, **2:**137, 140, 158–61
    auricles, **2:**159
    baroreceptor, **2:**145
    foramen ovale, **2:**159
    fossa ovalis, **2:**159
    left, **2:**159
    musculi pectinati, **2:**159
    myocyte, **2:**143

Heart atrium (*Continued*)
   nerve structures in, **2:**138–39
   receptor, **2:**145
   receptor discharge in cat, **2:**140
      and adrenalin, **2:**140
      and dextran, **2:**140
   receptor stimulation, **2:**140–43
      methods, **2:**140–43
      and urinary flow, **2:**143
   reflexes arising in, **2:**140–44
   right, **2:**158–59
   stimulation, **2:**140–43
Heart attack. *see* Myocardial Infarction
Heart beat, **2:**172
   initiation of, **2:**132
   regulation of, **5:**240
Heart contractility. *see* Myocardial contraction
Heart defects, congenital, **1:**702(t)
Heart diseases, **2:**87–88, 167–68, 661–72
   angina, **2:**146, 160
   and autonomic nervous system, **2:**183
   changes in drug kinetics in, **5:**854
   coronary, **2:**661–72
   genetics, molecular, **2:**661–72
   ischemia, **2:**167
   and sympathetic nervous system, **2:**183
   and parasympathetic nervous system, **2:**283
   and selenium deficiency, **6:**796
Heart enlargement, **3:**518–19
Heart failure, congestive, **2:**145, 174–76
   assistance of heart-lung machine, **4:**96–97
   hypokalemia and, **3:**99
   lung congestion, **2:**174
   mechanisms, diagram of, **2:**175
   pathophysiology, diagram, **2:**175
Heart-lung machine. *see* Heart, mechanical
Heart rate, **2:**8
   and caffeine, **2:**70
   increase, **2:**145
   larynx stimulation effects, **4:**650
   nervous system control of, **3:**516
   in static exercise, **3:**520
Heart rate, maximal
   measurement of, **3:**508
   myocardium and, **3:**517
Heart septum, **2:**158
Heart surgery
   centrifugal extracorporeal pumps, **1:**388
   heart-lung machine, **4:**96
Heart transplantation, **3:**517
Heart valves, **2:**158, 180, 182

Heart ventricle, **2:**158–61, 179, 180, 183, 184
   left, **2:**159–60
   myosin expression in, **5:**171
   preload, **2:**179, **3:**517
   right, **2:**159
   septal defect, prevalence of, **1:**702(t)
   wall, **2:**132
Heat
   balance, in regulation of body temperature, **1:**754–57
   biological effects on tissues, **5:**596
   exchange
      convention as means of, **1:**752–53
      equation expressing, **1:**754
      evaporation as means of, **1:**753
      insulation of clothes in, **1:**753–54
      radiation as mechanism of, **1:**753
   exercise and, **3:**511–12
   radiation of skin and, **6:**465
   skeletal muscle exercise and, **7:**30
   tolerance, in mammalian development, **7:**459–66
   transfer
      conduction, **4:**795
      convection, **4:**794
      evaporative heat loss, **4:**795
      radiation, **4:**794
Heat loss. *see* Body temperature regulation
Heat of combustion
   calculation of, **5:**475–76
   defined, **5:**471
   of macronutrients, **5:**474
   molecular structure and, **5:**472–74
Heat production. *see* Body temperature regulation
Heat shock, **4:**99–106, 104, 105
Heat-shock genes, **4:**100–101
   activation and development of, **7:**464
   identification and characterization of, **7:**462
   rat development and activation of, **7:**465
   thermolerance and induction of, **7:**461–62
Heat-shock proteins, **4:**101–4
   cell function of, **7:**463–64
   developmental activation of, **7:**464
   functions, **4:**101–4
   genetic locus of, **3:**183
   heat shock response and, **7:**460–61
   HSP15-30 protein family, **4:**103
   HSP70 protein family, **4:**101–2
   HSP90 protein family, 103
   identification and characterization of, **7:**462
   and immune response, **4:**105–6
   induction of, **7:**463–64
   molecular control of expression of, **7:**464–66

Heat-shock proteins *(Continued)*
nonheat shock inducers of, **4:**104–5
ubiquitin system, **4:**103
Heat-shock response
genetic nature of, **7:**460–61
thermotolerance and, **7:**461
Heat treatment. *see* Hyperthermia
Heavy chain disease, **4:**849
Heavy chain disease, IGA type. *see* Immunoproliferative
small intestinal disease
Heavy chains, **4:**316
of antibody molecule, **1:**564
CDR, **4:**3
Hebb rule
defined, **5:**387
learning and, **5:**390
Hedonic inverted V. *see* Acceptance triangle
Hedonic responses, **1:**499
Heimes' Driver, **1:**389
HeLa cells, **4:**107–11
storage, **4:**110
toxin resistance, **4:**111
variants, **4:**110–11
virus susceptibility, **4:**108, 111
Helicity, **2:**465
defined, **2:**465
triple, of collagen, **2:**561, 563
Helicobacter pylori, **3:**772
Helix, amphipathic
-loop model, **2:**93
defined, **2:**279
alpha-Helix, as transactivating domain, **3:**115
Helix, DNA, conformations of, **3:**107–8
Helix, double
defined, **3:**439
stability of, **3:**440
Helix-loop-helix DNA-binding domain, **3:**113–14
Helminthiasis
eosinophilia and IgE production, **4:**387
intensity of, **4:**114–15
nutrition effects, **4:**113–22
ascariasis, **4:**117–18, 121
strongyloidiasis, **4:**118–19, 121
study design, **4:**115
trichuriasis, **4:**119–20
prevalence, **4:**114
as scalar-borne zoonoses, **7:**898
transmission, **4:**114
Helminths
anthelminthic drugs, for ascariasis, **1:**270

Helminths *(Continued)*
cell invasion by
acetylcholinesterase and, **5:**648
superoxide dismutase and, **5:**648
muscle cell modification by, **5:**648
penetration of host tissues by, **5:**647
Helper cells, **1:**677, **2:**225, 649, 752, **4:**333–34. *see also*
Interleukin-2
in beta cell memory responses, **1:**570
characteristics of, **4:**811, **7:**481
functions, **4:**402–3, 823
inducer T-lymphocyte, **2:**225
induction of donor-specific tolerance, **4:**364–65
radiation effects, **6:**422
Helping
altruistic motives for, **1:**202–3
egoistic motives for, **1:**200–202
relation of altruism and egoism to, **1:**200
Helplessness, learned, **4:**665–71
in cats, **4:**667
defined, **4:**665
in dogs, **4:**665–67
in fish, **4:**667
in humans, **4:**667–68
in mice, **4:**667
model of depression, **4:**668, 671
neurochemistry of, **4:**668–69
and physical illness, **4:**669
in rats, **4:**667
theory of, **4:**666–67, 669–70
Hemagglutinins, influenza viruses, **4:**482–83
Hemato-encephalic barrier. *see* Blood-brain barrier
Hematocrit, **4:**123–24
Hematologic diseases
granulopoietic effects of lithiumj, **4:**750–51
hypercalcemia of malignancy and, **4:**279
malaria, **1:**267
Hematologic tests, **4:**619–20
Hematology, **5:**683
Hematoma
epidural, **4:**988
stroke caused by, **7:**302–3
subdural, **4:**987–88
Hematopoiesis
cyclic, lithium effects, **4:**751
in leukemia, **4:**690–91
normal, **4:**689–90
Hematopoietic cell growth factors, **4:**348
Hematopoietic cell growth receptors, **5:**816
Hematopoietic cells, **1:**762, 791–92, **4:**171–72
Hematopoietic stem cells, pluripotent, **4:**167–68

Hematopoietic system, **4**:165–75, 393, **7**:663
    antigens expressed by osteoclasts, **1**:769
    cell-cell interactions, **4**:172
    cell-matrix interactions, **4**:172–73
    effect of radiation of, **6**:413, 449
    soluble mediators, **4**:173–75
Hematoporphyrins, **4**:373
Hematoside, named, **3**:726
Heme, **2**:506, 655
    and iron, **2**:655
    stereochemistry and oxygen binding, **4**:142
    structure of, **6**:119
    synthesis by *Leishmania*, **4**:682
Heme-heme interaction, **4**:137
Heme oxygenase, **2**:656
Hemeralopia. *see* Vision disorders
Hemicellulose, **2**:101
    defined, **3**:35
    in endospermous tissues, **3**:40
    in lignified plant tissues, **3**:39
    in wheat grain, **3**:40
Hemicrania. *see* Migraine
Hemidecortication, **2**:859
    unilateral, **2**:859, 875
Hemidesmosome, **2**:270, 271, 778, 826
Hemispheres, **2**:5, 7–8, 351, 673–75, 681, 719, 861, 873
    arterial supply in, **7**:300
    dysfunction, **2**:861
    dyslexia-dysgraphia and, **3**:216
    generalized central capacity model and, **5**:152
    information organization in, **3**:218
    interaction, **4**:129–34
        developmental aspects, **4**:133–34
        facilitation, **4**:133
        integration, **4**:130–32
        interhemispheric connections, **4**:129–30
        mediation through forebrain commissures, **4**:130
        reciprocal inhibition, **4**:132–33
    language, **2**:861
    left, visuo-spatial processing in, **3**:221
    in multiple-resource models, **5**:153
    plasticity, **2**:861
    right, isolation of, **3**:223
    specialization, **2**:861
Hemispheric lobe, **2**:351
Hemochromatosis
    as complication from kidney dialysis, **1**:410
    idiopathic, **2**:654
    and iron excess, **2**:654
Hemocyanin, **2**:650
Hemodialysis. *see also* Artificial kidney
    blood conductivity and, **3**:15

Hemodialysis *(Continued)*
    complications from, **1**:408–10
    dialysate used in, **1**:402
    dialysis versus convection in, **1**:404–5
    historical background of, **1**:402
    membranes used in, **1**:404
    principles of, **1**:402
    shorter dialysis time, **1**:405
    uremic toxin associated with, **1**:405
Hemodynamic equilibrium, fetal. *see* Fetal circulation
Hemodynamics, **2**:185–86, **4**:93–94
Hemoflagellate infection, **7**:895–96
Hemoglobin, sickle, **4**:149, **7**:1–14
    amino acid sequence of, **3**:792
    as electromorphic variant, **5**:246
Hemoglobin F, **4**:146
Hemoglobin S
    chemical studies of copolymerization of, **7**:3
    crystal and fiber structure of, **7**:4
    dissociation curve of oxy-, **7**:7
    electron microscopy and fiber structure of, **7**:4
    gel structure, excluded volumes of fibers effect on, **7**:8
    gelation
        effect of nongelling hemoglobins on, **7**:8
        entrophy-driven processes implicated in, **7**:7
        kinetics of, **7**:8–9
        physiological molecules that affect, **7**:7
        progress curve of, **7**:10
        stochastic properties of, **7**:10
        temperature and pressure dependences of, **7**:6
    gels
        direct microscopic observations of, **7**:13–14
        rheology of, **7**:11–13
        shear rates and, **7**:12–13
    higher organization of, **7**:4–6
    kinetics
        delay time and, **7**:9
        relationship between equilibria and, **7**:9–10
    oxygen binding and polymerization of, thermodynamic linkage between, **7**:7
    polymerization
        double-nucleation model for, **7**:10–11
        mechanism of, **7**:10–11
    polymers, structure of, **7**:2–16
    thermodynamics of gelation of, **7**:6–8
    x-ray crystallography of, **7**:3–4
Hemoglobins, **4**:135–44, 145–49
    abnormality, in sickle cell anemia, **6**:535–36
    allosteric proteins, **4**:138–39
    alpha-chain, **3**:165
    amino acid sequence, **4**:135

Hemoglobins *(Continued)*
  binding of carbon monoxide to, **6:**598
  catabolism of, **6:**120–21
  cooperative effects, **4:**137–38
  developmental changes in globin gene expression,
      4:148
  diseases, **4:**143–44
  embryonic, **4:**146
  embryonic and fetal, **4:**143
  fetal, and malaria, **4:**882
  gamma-chains, **4:**146
  gas transport by, respiratory adaptation and, **6:**574–75
  genetic disorders, **4:**149
  globin gene clusters, **4:**146–47
  globin gene expression and regulation, **4:**148
  and iron, **2:**653, 655, 738
  levels and hookworm egg counts, **4:**115, 120
  light absorption by, **5:**588–89
  microspectrography of, **5:**563
  molecular architecture of, **7:**2–6
  molecular pathology, **4:**148–49
  noncoding region structure, **4:**147
  optical properties of, **5:**595
  oxygen affinity in, **4:**137–38, 145, **6:**224, 599
  oxygen binding to, **5:**620–22
  oxygen equilibrium, **4:**136–37
  polypeptide chains, **4:**135–36
  protein conformation of, **6:**221
  reflectance spectrophotometry measurement of, **3:**759
  salt bridges in, **4:**141–42
  structural variants, **4:**149
  structure and heterogeneity, **4:**145–26
  tetramer assembly, **4:**140–41
  three-dimensional structure, **4:**138–41
  transition from deoxy to oxy structure, **4:**141–42
Hemoglobinuria, paroxysmal, **2:**601, **4:**797, 655
  defined, **2:**601, 655
Hemolytic disorders, **6:**535–36
Hemoperfusion, **1:**378, 411
Hemopexin, **1:**31
Hemophilia, **4:**151–63
  clinical severity of, **4:**154
  DNA marker diagnosis of, **3:**147–48
  in European royal families, **3:**853
  genetic aspects of, **3:**852
  genetic counseling for, **4:**163
  mutations causing, **4:**155–62
  tissue factor and, **7:**498
  treatment of, **4:**162–63
Hemophilia A
  characterization, **4:**153–54
  factor VIII, **2:**464

Hemophilia A *(Continued)*
  factor VIII gene defects in, **4:**155–57
  gene for factor VIII, **2:**480
Hemophilia B, **3:**148
  characterization, **4:**154
  factor IX gene mutations and, **4:**157–59
Hemophilia C
  characterization, **4:**155
  factor XI gene mutations and, **4:**161–62
Hemoprotein. *see* Cytochrome P-450
Hemorrhage, **2:**146
  fish oils and, **3:**579, 580
  postpartum, **4:**916, 919
  stroke caused by, **5:**602, **7:**302
  subarachnoid, **5:**601
  sympathetic nervous system and, **3:**510
Hemorrhagins, snake venom, **7:**600
Hemosiderin, **2:**655, 656
Hemosiderosis, pulmonary, idiopathic, **2:**655
Hemosorption. *see* Hemoperfusion
Hemostasis and blood coagulation, **4:**177–85
  blood clotting cascade, **4:**178–79
  contact phase, **4:**181–82
  extrinsic activator of Factor X, **4:**180
  fibrin polymerization, **4:**179–80
  fibrinolysis, **4:**182
  finger domains, **4:**184–85
  functional properties of proenzymes, **4:**182–85
  Gla domains, **4:**182–84
  growth factor domains, **4:**184
  initiation of clotting, **4:**180
  intrinsic activator of Factor X, **4:**181
  physiology, **4:**178–79
  proenzyme structure, **4:**182–84
  protein cofactor structural and functional domains,
      4:185
  protein structural domains, **4:**182–85
  prothrombin activator, **4:**180–81
  regulation of reactions, **4:**182
  tandem repeat domains, **4:**185
  termination of clotting cascade, **4:**182
"Hemp insanity." *see under* Psychoses
Henderson-Hasselbalch equation, **2:**375
  on bicarbonate buffers, **5:**876
  on pH_e balance, **5:**880
Hendrick, Hal W., **3:**479
Henry's Law, **5:**617
Hensen's mode, **5:**161
Hepadnavirus infection, **4:**761–62
Hepadnaviruses
  characterization, **4:**756–60
  presence in hepatocellular carcinoma, **4:**762–63

Hepadnaviruses *(Continued)*
    viral oncogenic mechanisms, **4:**763-67
Heparan sulfate, **2:**645
Heparin, **2:**171, 573
    growth factor regulation and, **3:**449
    in lipolysis, **3:**574-75
    structure of, **6:**258
    termination of clotting cascade, **4:**182
Heparin-binding growth factor. *see* Fibroblast growth factor
Heparin sulfate proteoglycan, **4:**624
Hepatic portal circulation. *see under* Blood circulation
Hepatic triacylglycerol lipase. *see* Lipase, hepatic
Hepatitis, alcoholic, **1:**173-74
Hepatitis, chronic active, **1:**528
Hepatitis, toxic, **5:**260-61
Hepatitis, viral, **1:**738
Hepatitis, viral, non-A, non-B, **1:**255
Hepatitis A
    immune globulin for, **4:**447
    transmission in day-care centers, **4:**463
Hepatitis B
    blood tests for, **1:**738-39
    disease models, animal, **1:**255
    neonatal infections, **4:**458
    sexual behavior and, **6:**895-96
    trends in infections by, **6:**895-96
Hepatitis B virus, **4:**755-68. *see also* Hepadnaviruses
    characterization, **4:**756-60
    hepatocellular carcinoma and, **5:**607
    liver cancer and, **3:**676, **4:**342
    vaccines for, **4:**446, 767
Hepatitis C, **1:**739, 740
Hepato-lenticular degeneration. *see* Wilson's disease
Hepatocytes, **3:**78
Hepatotoxicity, of organic chemicals, **7:**550
Herd immunity, **4:**444
Herd medication, **1:**294
Hereditary angioedema. *see* Angioedema, hereditary
Hereditary diseases
    absence of enzyme activity, **4:**1001-3
    bone marrow transplantation for, **1:**796
    deficient DNA repair in, **5:**254
    denaturing gradient gel electrophoresis in detection of, **6:**527
    diagnosis of, **3:**140-48
        allele-specific oligodeoxynucleotides in, **6:**526-27
        polymerase chain reaction and, **6:**526
        recombinant DNA and, **6:**523-28
        ribonuclease A cleavage in, **6:**527
    direct diagnostic analysis of, **6:**526-28
    DNA methylation and, **3:**160-61

Hereditary diseases *(Continued)*
    dyslexia-dysgraphia as, **3:**216
    frequency of, **6:**525
    genetic mapping and, **3:**828, 878
    genetic variation and, **3:**799
    glutathione metabolism and, **3:**908
    indirect diagnostic analysis of, **6:**524-26
    known mutations in, direct detection of, **6:**526-27
    Mendelian disorders, defined, **5:**259
    mutations and, **3:**873
    newborn screening for, **5:**433-38
    recombinant DNA and diagnosis of, **6:**523-28
    restriction fragments-length polymorphism analysis and, **6:**524-25
    simply inherited, **3:**819-21
Heremans, **5:**137
Heritability
    defined, **3:**847
    issues in, **3:**857
Hermissendra crassicornis (marine mollusc), **2:**306-16
    and associative learning, **2:**306-10
    membrane processes, **2:**310-16
    trained to respond to light, **2:**306
Heroin addiction, **1:**43-51, **4:**86
Herpangina, **5:**613
Herpes genitalis, **4:**189, 191
    transmission, **4:**450
    trends in, **6:**893-94
Herpes labialis, **4:**189
Herpes simplex, **4:**189-90
    mouth infections, **5:**613
    neonatal infections, **4:**458
Herpes zoster infection, **5:**613
Herpesviridae, **2:**395-98, **4:**187-94
    chemotherapy, **2:**395-98, **4:**193
    cytomegalovirus, **4:**187, 189-90
    definition, **4:**187
    DNA polymerase, **2:**395-98
    Epstein-Barr virus, **4:**187-88, 190
    fate of infected cells, **4:**192-93
    interferon effects, **4:**520
    latent infections, **4:**189, 193
    lytic infections, **4:**188-89
    polymorphism of genomes, **4:**191
    prevention of infection, **4:**193-94
    reactivation, **4:**189, 193
    reproductive cycle, **4:**191-92
    structure and sequence of viral genomes, **4:**190-91
    taxonomy, **4:**187-88
    trans-activation, **4:**192
    transmission, **4:**450
    viral genomes, **4:**190-91

Herpesviridae *(Continued)*
  virion, **4:**187–88, 191
Herpesvirus-6, human, **4:**462
Herpesvirus hominis, **4:**188
Herpesvirus varicellae. *see* Varicella-zoster virus
Herring bodies, **4:**608
Herring meal, toxic, **3:**679
Hertwig's sheath of tooth, **2:**821, 823
Heston, **3:**842
Heterochromatin, **2:**291, 473
  defined, **2:**465, **3:**797
  primate genetic relations and, **3:**884
Heterochrony, **2:**869
Heterocyclic aromatic amines. *see* Aromatic amines, heterocyclic
Heterodimeric peptides, defined, **5:**725
Heterodimers, *fos-jun*, **3:**706
Heterogeneous nuclear RNA. *see* RNA, heterogeneous nuclear
Heterokaryon, **2:**481
Heterokaryotes, embryogenesis of, **3:**306
Heterophyes heterophyes, **2:**380
Heterosexuality
  and human sexual behavior, **6:**869
  and psychological sex differences, **6:**860–61
Heterotherms, **1:**37
Heterotrophic ligands, **4:**137
Heterozygote, **4:**432
  compound, **3:**854
  defined, **3:**135, 709, 813, 823
  heterozygote selection hypothesis, **5:**868
  loss of heterozygosity, **2:**53
Heubner's recurrent artery, **2:**359
1-Hexadecene, **2:**724
Hexamethonium, **2:**522
Hexanedione, **3:**386
Hexokinase
  discovery of, **3:**933
  glucose supply and, **3:**939
  in glycogen formation, **3:**921
  inhibition of, **3:**936
Hexosaminidase, **4:**575
Heymann nephritis. *see* Glomerulonephritis
Heyningen, W. E. van, **3:**734
High-ceiling diuretics. *see* Diuretics, loop
High-density lipoproteins. *see* Lipoproteins, HDL
High-energy phosphate compounds. *see* Phosphate compounds, high-energy
High-energy phosphates. *see* Phosphates, high-energy
High linear energy transfer, **6:**479
High mobility protein 20. *see* Ubiquitin

High performance liquid chromatography. *see* Chromatography, high pressure liquid
Higher brain function, **4:**195–203
  analytical scheme, **4:**216
  bilateral processing, **4:**131–32
  elementary mental operations, **4:**196–97
  functional areas, **4:**197–98
  hemispheric integration, **4:**130–31
  neural implementation, **4:**198–99
  resolution, **4:**215–16
  reverse engineering approach, **4:**196
  sensitivity, **4:**216
  velocity tuning, **4:**200
  versatility, **4:**216–17
  visual motion processing, **4:**199–201
  visual processing of words, **4:**201–2
Higher-order conditioning, Pavlovian, **2:**631
Higher thought processes, **4:**1006
Hilar, defined, **5:**699
Hill, A. V., **5:**181, 183
Hill equation
  for allosteric interaction, **2:**375–76
  for cooperative enzyme models, **3:**429
Hill's coefficient, blood, **4:**137
Hind III fragments
  defined, **5:**191
  exons of, **5:**195
Hindbrain, **2:**3
  and convulsive seizure expression, **6:**784
  as influence on convulsive seizure susceptibility, **6:**784
Hindgut
  anatomy, **3:**66
  defined, **3:**57
Hindgut, fetal, **3:**59
Hinge joints, **1:**374
Hinge region, antibody, **3:**808
Hip fractures, **1:**779–80
Hip joint, **2:**595
Hippocampus, **2:**5, 7, 9, 10, 313, 335, 676, 680, **4:**227–28.
    see also Models, nerve growth factor, septo-hippocampal
  and amnesia, **2:**9
  behavioral functions, **4:**231–32
  commissural connections, **4:**231
  connections, **4:**230–31
  defined, **2:**1
  dyslexia-dysgraphia and, **3:**224
  formation, **4:**227–33
  location and structure, **4:**227–30
  mammalian, network modeling of, **5:**338
  medial temporal amnesia, **4:**677
  nerve growth factor and, **5:**339

Hippocampus *(Continued)*
  sensitivity to disease states, **4:**232-33
  subregions, **4:**227-28
  synapse modification in, **6:**38-39
Hippocrates, **3:**474
Hippocratic Oath, **1:**657
Hirsutism
  characterization, **4:**41
  endocrinopathic, **4:**41
  essential, **4:**41
Hispanic Americans, diet of, **3:**651
Histamine, **2:**13, 16, 23-25, 34, 369
  capillary pressure and, **5:**41
  function of, in allergic response, **1:**188
  gastric acid secretion and, **3:**76
  in gastric blood flow regulation, **3:**760, 761
  gastrins and, **5:**716
  inflammatory role, **4:**470
  in postprandial intestinal hyperemia, **4:**555-56
Histamine receptors. *see* Receptors, histamine
Histamine-releasing factors (HRFs), **1:**190
Histatins, in saliva, **6:**694
Histidine, **2:**13, 14, 536
Histidinohydroxymerodesmosine, **2:**570
Histiocytes, **7:**47
Histiocytoma, malignant fibrous, **7:**702
Histocompatibility antigens, **4:**382-84. *see also* HLA antigens
  in bone marrow transplantation, **1:**792, **5:**704
  genetics, **4:**358-59
  loci controlling histocompatiblity antigens, **1:**792
  minor, **4:**358-59
  sensitization, **4:**360-61
Histocompatibility antigens class II, **2:**753
  autoimmune disease and, **5:**510
  defined, **5:**503
  genetic aspects of, **5:**514
Histocompatibility complex. *see* Major histocompatibility complex
Histocompatibility testing, **3:**148-49
Histocompatiblity antigens class I, **1:**792
Histocompatiblity antigens class II, **1:**792
Histodifferentiation, defined, **3:**281
Histology
  in anatomic pathology, **5:**681
  defined, **5:**681
  of ridged skin, **3:**
Histone-DNA complexes, **4:**245
Histone genes
  cell cycle-regulated, **4:**238-40
  expression, **4:**237-38
  5ps flanking regions, **4:**239-40

Histone genes *(Continued)*
  organization, **4:**238
  regulatory factors that control expression, **4:**239-40
Histone H1
  chromatosome and, **3:**127
  inactive genes and, **3:**132
  linker DNA and, **3:**128
  variation in, **3:**128
Histone H1 kinase, M phase-specific. *see* Maturation-promoting factor
Histone kinase P34(CDC2). *see* Protein P34CDC2
Histones, **2:**292, 419, 420, 426, 432-34, 470, 471, 476, **4:**235-40
  chromatin folding and, **4:**236
  chromatin structure and, **4:**236-37
  coding sequence, **4:**238-40
  core
    chromatin structure and, **3:**133
    defined, **2:**419
    in eukaryotes, **3:**127
  defined, **2:**465, **3:**125
  N1 factor and, **3:**130
  in nucleosome core, **3:**126
  octamer, **2:**419, 471, **3:**129
  properties of, **3:**127-28, **4:**235-36
  role of, **2:**426, 470
  selective synthesis, **4:**237
  synthesis and DNA replication, **4:**237-38
Histrionicotoxins, **7:**599
HIV, **2:**232, 395, 399-401
  and AIDS, **1:**151-62
  CD4 on cell as receptor for, **2:**232-33, 400
  dideoxynucleosides and, **2:**400
  immune responses against, pathways for, **6:**650
  interferon and, **2:**373
  protease and, **2:**400
  recombinant subunit vaccines against, **7:**750
  reverse transcriptase and, **2:**399
  RNase H and, **2:**400
  vaccine for, **5:**703, **6:**647-48
    conceptual approaches for, **6:**650-51
    development of, **6:**651
    progress in research, **6:**649-51
  Zidovudine and, **2:**373, 399-400
HIV-1
  anti-HIV antibodies, **1:**20
  antibody test for, **1:**737-38
  attachment to target cell, **1:**15-16
  bone marrow transplantation for, **1:**797
  cytopathological effects of, **1:**18-19
  heterogeneity, **1:**20-21
  immune disorders caused by, **1:**20

HIV-1 *(Continued)*
    life cycle, **1:**14–18
    morphogenesis, **1:**18
    penetration into cell, **1:**16
    proteins, **1:**11–14, 21
    regulatory regions in 5' long terminal repeat, **1:**17
    replication, **1:**16–17
    stem cell arrest by, **1:**20
    T cell depletion by, **1:**18–20
    tissue infection, **1:**21–22
    translation, **1:**18
    virion structure, **1:**11
HIV-2, testing for, **1:**738, 740
HIV envelope, selected sites on, **6:**649
HIV envelope protein gp41, **1:**12
    syncytia formation, **1:**18–19
HIV envelope protein gp120, **1:**12
    conserved and variable regions in, **1:**21
    effects of, **1:**21
HIV infections
    AIDS and, **1:**151–62
    animal model of, **5:**677
    behavior associated with transmission of, **1:**153
    cofactors, **1:**21–22
    costs for treatment of, **1:**152
    counseling and testing, **1:**157–58
        for infected persons, **1:**159–60
        partner notification in, **1:**158–59
        prevention-oriented follow up programs for infected
            persons in, **1:**159–60
        street outreach programs in, **1:**159
    cytopathological effects, **1:**18–20
    drug use and, **1:**152
    features of, **6:**648–49
    fetal infections, **4:**456–57
    impact of, in United States, **1:**151–53
    Kaposi's sarcoma and, **1:**152
    morbidity and mortality, programs for reducing,
        1:155–56
    neonate infections, **4:**457
    oral manifestations of, **5:**613
    prevention of
        community-based organizations in, **1:**160
        comprehensive programs for, **1:**153–54
        evaluation of, **1:**160
        impediments to, **1:**156
        public health strategies for, **1:**154–56
    trends in, **6:**893
Hives, **4:**410–11
HL-60 cells
    ganglioside GM3 and, **3:**741
    sialylparagloboside and, **3:**734

HLA antigens
    autoimmune diseases and, **3:**148
    melanoma and, **4:**965
    in tissue typing, **3:**148–49
HLA genes, **4:**382–83
HMG-CoA reductase. *see* Hydroxymethylglutaryl CoA
    reductase
Hodgkin's disease, **4:**840, 846–47
    Ann Arbor classification, **4:**847
    bone marrow transplantation for, **1:**795
    classification, **4:**840
    clinical description, **4:**846–47
    natural history of, **4:**846–47
Hoechst 33258 (drug), **2:**458
Hogness box. *see* TATA box
Holandric inheritance, defined, **3:**819
Hold-up time (gas chromatography). *see* Gas hold-up time
Holoclones, **4:**583
Holoenzyme, **2:**527
Holotoxins, **4:**369
Holstege, **5:**45
Holt, L. Emmett, **5:**700
Home bases
    defined, **3:**495
    of early hominids, **3:**496
Homeobox-containing proteins. *see* Genes, homeo box
Homeodomain, identification of, **3:**112–13
Homeokinesis, in adaptational physiology, **1:**37
Homeostasis
    acid-base, and metabolic end products, **5:**877
    of bone, **1:**774
    cardiovascular, **2:**145
    cytokine role, **4:**473
    defined, **3:**931
    hormone regulation of, **3:**326
    hypothalamus role, **4:**308
    maintenance of, **5:**50–51
    peripheral hormones and, **5:**396
    pH. *see* Hydrogen ion concentration
    skeletal, regulation of, **1:**762
    in stomach, **3:**761
    trace element, regulation of, **7:**604–5
Homeothermy, **4:**792
Homicides, media effects on, **1:**1058:**1:**8
Homing behavior, **5:**528
Hominid women, **3:**500–501
Hominids, **2:**189, 593
    African, **3:**490
    anthropological studies of, **6:**878
    Eurasian, **3:**489
    evolution of, **3:**487–93, 495–504
    fossil record of, **6:**142–43

Hominids *(Continued)*
   Miocene, **3:**489
   prehistoric, biocultural sex differences in, **6:**848-49
Homo erectus, **2:**189, 198, **4:**791
   big-game hunters, **2:**198
   discovery of, **3:**492
   H. sapiens and, **3:**492
   sex differences in, biocultural, **6:**848
Homo sapiens, **2:**190
   anatomically modern, **3:**492-93
   appearance of, **3:**491-92
   archaic, **3:**492
      bony pelvis of, **1:**825-31
      evolution of the modern bony pelvis and, **1:**831
      fossil pelvic material of, **1:**827-30
   biological criteria for beginning of human life, **1:**656
   fossil record of, **1:**827, **6:**142-43
   as meat-eaters, **2:**190
   as predators of prey, **2:**190-91
   taxonomy of, **3:**487-88
   what makes us human?, **2:**594, 600
Homo sapiens neanderthalensis. see Neanderthals
Homo sapiens sapiens, **2:**189, 593
   behavior and environment, **2:**798
   and dental caries, **2:**798-800
   Neanderthals and, **3:**502
   origin, unique, **2:**437
   tools of, **3:**499
Homocystinuria
   Down's syndrome and, **3:**182
   newborn screening test for, **5:**437
Homogeneous immunoassay, defined, **3:**208
Homogeneously staining region, defined, **3:**785
Homologous coupling *see* Coupling, homologous
Homology, defined, **3:**281, 881
Homology, serial, defined, **3:**882
Homoplasty
   defined, **3:**881
   in primate evolution, **3:**887
Homosexuality
   and human sexual behavior, **6:**869
   incidence of Kaposi's sarcoma and lymphoma, **4:**345
   and psychological sex differences, **6:**860-61
Homotrophic ligands, **4:**137
Homovanillic acid, **2:**222, 850
   antipsychotic drug, **2:**222
   and mood, **2:**850
Homozygosity, **4:**438
Homozygote, **4:**432
Honey, **2:**104
Hookworm infections, **1:**263-64, **4:**115-17
   anemia and, **4:**116-17

Hookworm infections *(Continued)*
   intestinal blood and iron loss, **4:**115, 120
   reduced food intake and anorexia and, **4:**116
Hopefield formalism, **5:**390
Horizontal cells, **6:**626
Hormone receptors. *see* Receptors, endogenous substances
Hormones. *see also* specific hormones
   abnormal secretion of, **3:**334
   birth control and, **1:**691-94
   in bone cell differentiation, **1:**120-21
   bone cell function-related, **1:**765
   bone remodeling and, **1:**822-23
   cellular effect of, **3:**330
   cervical carcinoma and, **5:**607
   characteristics of, **3:**325-26
   concentration, measurement of, **3:**328-30
   counter-regulatory, glucose synthesis and, **3:**324
   defined, **2:**53, 323, **3:**71, 243, **4:**13, **5:**373, 713
   dyslexia-dysgraphia and, **3:**216
   ectopic, **3:**334-35
   flow cytometry of, **3:**624
   follicular production of, **3:**632
   functions of, **3:**327-27
   half-life of, **3:**325-26
   hypophysiotropic, **2:**11
   hypothermia and, **4:**801
   influences on behavior, testosterone effects on aggression, **1:**115
   local, **4:**590
   mechanisms of action, **2:**322, **3:**330-33
   and menstrual cycle, **6:**556-57
   metabolic regulation and, **4:**1005-6
   metabolism
      in prostate gland, **6:**179
      zinc and, **7:**886-87
   and morning sickness, **6:**125
   in muscle differentiation, **5:**171-72
   neuroendocrine, regulation of, **6:**833-34
   neuroregulators and, **3:**318
   nonclassic, functions of, **3:**323
   oral contraceptives and, **5:**605
   pH$_e$ homeostasis and, **5:**880-81
   prepubescent growth regulation, **3:**351-52
   protein, listed, **3:**321-22
   pulsatile secretion of, **5:**396, 397
   in regulation of bone and calcium metabolism, **1:**800-805
   regulation of steroid synthesizing enzymes, **7:**254
   resistance to, **3:**335
   sexual behavior and, **4:**935
   sexual orientation and, **6:**860-61
   as signals for bone cell differentiation, **1:**788-89

Hormones *(Continued)*
   in storage and release of lipids, **1:**60
   thermal injury and, **7:**455
   tubular H$^+$ and, **5:**878
Horopter, **1:**636
Horseradish peroxidase, **2:**369, 467
Horses
   as animal model for severe combined immuno-
      deficiency, **5:**677
   seasonal intoxication of, **5:**263
Hospital pathology department. *see* Pathology department,
   hospital
Hospitals, pediatric, **5:**700
Host-parasite relations, **4:**385-86, **5:**649-50
Hostility, tricyclic drugs and, **1:**613-14
Howland, John, **5:**701
Hpa II, methylation of, **3:**153, 157
HPLC
   electrochemical detection, **4:**211
   isoenzymes, **4:**577
   resolution, **4:**206
HPRT gene. *see* Genes, HPRT
HPV-11, keratinocytes infection, **4:**585-86
HPV-16, keratinocytes infection, **4:**586
HPV-18, keratinocytes infection, **4:**586
5-HT receptors. *see* Receptors, serotonin
"HTF islands," gene expression and, **3:**872
HTLV-I
   antibody test for, **1:**738
   gene for, **4:**691-92
   tax-1 transcriptional activator protein, **4:**532
   vaccines against, **6:**647
HTLV-III. *see* HIV
hTMa genes. *see* Genes, tropomyosin
hTMnm genes. *see* Genes, tropomyosin
Hubel, **5:**309
Hudspeth, A. J., **3:**235
Hue, **2:**587
   circle, **2:**577, 579
   defined, **2:**575
   extraspectral purple, **2:**576-77
   mutually exclusive, **2:**577-78
   spectral, **2:**576-77
   unique, **2:**577-78
Hull's learning theory, **2:**633
Human chorionic gonadotropin receptors. *see* Receptors,
   LH
Human chorionic gonadotropins. *see* Gonadotropins,
   chorionic
Human engineering, **1:**658-59, **3:**473-80
   and applied science, **3:**476
   and basic science, **3:**476

Human experimentation, **1:**657-58
Human factors. *see* Human engineering
Human Factors Society, **3:**478
Human genome project
   goals of, **3:**832-33, 879
   sequence analysis and, **3:**166
Human immunodeficiency virus. *see* HIV
Human immunodeficiency virus type 1. *see* HIV-1
Human immunodeficiency virus type 2. *see* HIV-2
Human leukocyte antigens. *see* HLA antigens
Human milk. *see* Milk, human
Human Research Society, **3:**475
Human T-cell leukemia virus I. *see* HTLV-I
Humanization of work, **3:**473
Humans. *see* Homo sapiens
Humerus, **4:**264, **7:**38
Humoral hypercalcemia of malignancy, **4:**276-78
Humoral immunity. *see* Antibody formation
Hunger
   appetite and regulation of body weight, **1:**349
   conditioned, **1:**349
   defined, **3:**589
   "Hunger pains" and stomach contractions, **3:**75
   learned controls of, **1:**349
   obesity and, **3:**591
   physiological controls of, **1:**348
   physiology of, **1:**348-49
   and satiating power, **6:**724-25
   termination of, processes identified with, **6:**725-26
   theory, **1:**348
Hunter-gatherers, **2:**190-92
   customary diet, **3:**19-20
   diet of, **3:**500, 501
   early hominids and, **3:**499-500
   home bases of, **3:**496
   hunting
      for calories, **2:**192
      hypothesis, **2:**189-91
   Middle Paleolithic, **3:**501-2
   the new view, **2:**191
   number of groups listed, **2:**191
   social intellect and, **3:**504
   social organization and, **3:**504
Hunter's disease, **1:**796
Hunting reaction, **4:**796
Huntington, George, **3:**830
Huntington's disease, **2:**463, 763, **4:**269-74
   and cell death, **2:**251
   chromosomal mapping of, **3:**138
   DNA marker diagnosis of, **3:**141-42
   familial, **2:**764
   genetic aspects, **3:**830-31

Huntington's disease *(Continued)*
striatum neurotranmsitters in, **4**:271
Hurler's syndrome, **1**:796
HVMLC1, **5**:215
Hyaline membrane disease
advances in prevention of, **5**:705
animal models, **1**:258
incidence of, **5**:706
oxygen toxicity and, **6**:352
Hyaluronan, **2**:642
Hyaluronic acid, **2**:206-8, 577
in gingiva, **4**:35-36
structure of, **6**:258
Hyaluronidase, **3**:603-4
Hybrid cells, **2**:457
defined, **3**:175
human-hamster, **2**:457
human-mouse, **2**:457
Hybrid-gene synthesis, **3**:172-73
Hybridization
defined, **3**:859
in situ, defined, **2**:465, 467
in *in vitro* DNA manipulation, **3**:876
in papillomavirus infection diagnosis, **3**:862
primate genetic relations and, **3**:888
probes, **2**:484
Hybridomas
antibody synthesis, **3**:802-3
defined, **3**:801
monoclonal, cancer immunodiagnosis, **4**:346
and monoclonal antibody technology, **5**:81
Hydatid cyst, **2**:390
Hydralazine, **2**:177
Hydrazone, cytotoxic, **2**:374
Hydride ion transfer, **2**:532
Hydrocarbons
alkane metabolism, **4**:736
chain packing, nomenclature for, **4**:729
characterization, **4**:736
in marijuana smoke, **5**:446
solvents, anesthetic-like properties of, **3**:386
Hydrocarbons, aromatic, polynuclear, **2**:110, 117
Hydrocarbons, substituted
behavior and distribution, **4**:740-41
characterization, **4**:736-37
metabolism, **4**:737-40
Hydrocephalus, normal pressure, **2**:764
Hydrocephalus, X-linked, **1**:699
Hydrochloric acid, **3**:257-58, **5**:827
Hydrochloric acid, gastric. *see* Gastric acid
Hydrochlorothiazide, **2**:276
characteristics of, **3**:97

Hydrocodone, **2**:786
Hydrocortisone, **2**:20, 247, 26, 144, 145, 492, 501, 744, 756, 847
ACTH secretion and, **3**:325
aging and, **3**:903
anti-inflammatory activity of, **7**:112
blood, **2**:144, 145
corticotropin-releasing hormone and, **5**:396
for cortisol deficiency, **3**:353
deficiency of, **3**:353
fetal, **3**:344-45
function, in adrenal glands, **1**:72-74
metabolites, in drug designs, **7**:112-13
muscle protein breakdown and, **3**:324
perinatal, **3**:349
and sensation seeking, **6**:814
synthesis of, **7**:245
Hydrofluoric acid, **7**:451-52
Hydrogen bonding, in DNA binding sites, **3**:118-19
Hydrogen chloride. *see* Hydrochloric acid
Hydrogen-ion concentration
bile salt binding and, **3**:42
carbohydrate metabolism and, **5**:821-28
of carbonic anhydrase medium, **3**:96
in clearance techniques, **3**:758
of feces, **3**:774
fluorescent probes as indicators, **5**:577
of foods, **3**:662
of gastric juice, **3**:76
homeostasis of, **5**:871-81
intracellular, **5**:872-75, 875-81
for nitrosation, **3**:679
regulation of cell, **4**:569
urinary, acetazolamide and, **3**:97
Hydrogen ions. *see* Protons
Hydrogen peroxide
copper-zinc superoxide dismutase and, **3**:184
disposition of, by myeloperoxidase, **5**:429-30
formation of, **3**:914
free radicals and, **3**:894
production of, during respiratory burst, **6**:583
Hydrogenation, of vegetable oils, **3**:570-71
Hydrolases
in bacterial call wall synthesis, **1**:277-78
mannose-6-phosphate receptors and, **5**:875
Hydrolysis
of adenosine triphosphate, **1**:53
bacterial, in colon, **3**:82
in carbohydrate digestion, **3**:79
definition of, **1**:53
in digestion, **3**:79
of ester bonds, **3**:53

Hydrolysis *(Continued)*
 free energy of, **1:**53
 of heat-treated milk fat, **3:**53
 of monoacylglycerols, **3:**53
 peptide bond, **4:**183
 in protein digestion, **3:**80
 saliva and, **3:**74
 of *sn*-2-monoacylglycerol, **3:**53
 of triacylglycerol, **3:**52
Hydrophobia, chemical, defined, **3:**403
Hydrophobic effect, **4:**970
Hydrophobicity
 in aqueous two-phase systems, **1:**356
 defined, **3:**725
 of denatured enzymes, **3:**413–14
 of thermophilic enzymes, **3:**420
Hydrops. *see* Edema
Hydroxocobalamine, **2:**537
N-hydroxy-acetylaminofluorene, **2:**121
N-hydroxy heterocyclic aromatic amine, **3:**677
4-Hydroxyandrostein-3,17-dione, **2:**744
Hydroxyapatites, **2:**90, 98, 773, 822, 823, **3:**481
Hydroxybenzoic acid, **4:**224
25-Hydroxycholesterol, **2:**415
Hydroxydopamines, **2:**13
 in *in vivo* noradrenergic studies, **5:**382–83
 Parkinson's disease and, **5:**670-71
 visual plasticity and, **5:**900
16-alpha-Hydroxyestrone, **2:**57
 and breast cancer, **2:**57
9-(2-Hydroxyethoxymethyl)guanine. *see* Acyclovir
5-Hydroxyindolacetic acid, **2:**716, 836, 845
Hydroxyl radicals
 formation of, superoxide and, **7:**353–54
 production of, during respiratory burst, **6:**583
Hydroxylases
 deficiency in children, **3:**352–53
 estrogen replacement therapy and, **3:**483
Hydroxylysine, **2:**562, 563, 566, 569, 570
3-Hydroxy-3-methylglutaric acid, **2:**411
3-Hydroxy-3-methylglutaric acid reductase, **2:**411, 415, 417
Hydroxymethylglutaryl CoA reductase, **2:**168
 gene amplification in, **3:**788
 GSH/GSSG ratio and, **3:**914
 as inhibitor, **2:**168
 inhibitors, of cholesterol synthesis, **1:**454
Hydroxyproline, **2:**562, 569
 in bone resorption, **3:**481
4-Hydroxy-2-quinolinecarboxylic acid (kynureic acid), **2:**524

Hydroxysteroid dehydrogenase, fetal adrenal cortex and, **3:**344
3-beta-Hydroxysteroid dehydrogenase, in steroid hormone synthesis, **7:**248
20-alpha-Hydroxysteroid dehydrogenase, **2:**756
4-Hydroxytamoxifen, **2:**60
Hydroxytryptamine. *see* Serotonin
5-Hydroxytryptamine. *see* Serotonin
5-Hydroxytryptophan, **2:**13, 847
Hydroxyurea, **3:**102
Hydroxyzine, **2:**787
Hymen, **6:**564
Hyoid bone, laryngeal, **4:**655
Hyoscyamine. *see* Atropine
Hyoscyamus miger, **1:**178
Hyperactivation, sperm. *see* Sperm
Hyperactivity. *see* Hyperkinesis
Hyperacuity, in space perception, **1:**629
Hyperacute rejection, **4:**362
Hyperbilirubinemia, **5:**705
Hypercalcemia, **2:**98
 ectopic hormone production and, **3:**335
Hypercalcemia of malignancy, **4:**275–82
 etiology, **4:**276
 frequency, **4:**275–76
 humoral, **4:**276–78
 hydration, **4:**281
 IL-1 and, **4:**278
 myeloma and hematologic malignancies, **4:**279
 osteolytic metastases and, **4:**278–79
 parathyroid hormone PTH-related protein and, **4:**277
 pathophysiology, **4:**276–79
 serum calcium, **4:**279, 281
 transforming growth factor-alpha and, **4:**277
 treatment, **4:**279–81
 tumor necrosis factor and, **4:**277–78
 vitamin D metabolite and, **4:**278
Hypercapnia, **2:**365
Hypercholesterolemia, **1:**451, **2:**168
 defined, **2:**167
 dietary fiber for, **3:**42
 drug therapy, **1:**454
 endogenous cycle, **1:**451–52
 exogenous cycle, **1:**451–52
Hypercholesterolemia, familial, **1:**452, **2:**665–66
 description, **2:**665
 distribution, geographic, **2:**665
 founder effect by immigrants, **2:**665
 genetic aspects of, **3:**820
 genetic defect in, **1:**453, **2:**665
 levels, **2:**665
 lipoprotein, low-density, in excess, **2:**665

Hypercholesterolemia, familial *(Continued)*
  model in animal. *see* Watanabe
Hypercorticism, **2:**20, 851, **3:**897
Hyperemia, intestinal
  absorption, **4:**555
  gastrointestinal hormones and peptides, **4:**555
  intestinal blood flow regulation, **4:**552
  local tissue chemicals, **4:**555-56
  localization, **4:**554
  luminal stimuli responsible for, **4:**554
  mucosal neural receptors and enteric nervous system,
    4:555
  postprandial mechanisms, **4:**554-57
  tissue metabolism, **4:**556-57
Hyperemia, reactive, in stomach, **3:**761
Hyperextension, **5:**133
Hyperglycemia
  in diabetic ketoacidosis, **5:**827
  drugs inducing temporary, **7:**572
  viscous dietary fiber for, **3:**42
Hyperglycemic action
  defined, **5:**725
  of glucagon, **5:**733
Hypergonadotropic hypogonadism, **3:**358
Hyperinsulinism, in children, **3:**353
Hyperkalemia, **3:**99
Hyperkinesis, **2:**718
  in fragile X syndrome, **3:**713-14
Hyperkinetic syndrome. *see* Attention deficit disorder with
    hyperactivity
Hyperlipidemia, **1:**451
Hyperlipidemia, familial, combined, **2:**661
Hyperlipoproteinemia type III, **2:**666
Hypermutability hypothesis, **5:**868
Hypermutation, **2:**235
Hypernasality, **7:**159
Hyperoxia, **4:**125
Hyperparathyroidism, **1:**808
  conditions causing, **5:**665
  hypercalcemic etiology, **4:**276
  treatment of, **5:**665-66
Hyperphagia
  defined, **3:**243
  immunity impairment and, **5:**504, 505
Hyperplasia
  defined, **5:**173
  of endocrine organs, in the aged, **3:**902
Hyperplasia, adrenal, congenital. *see* Adrenal hyperplasia,
    congenital
Hyperplasia, gastric mucosal. *see under* Gastric mucosa
Hyperplasia, psoriatic epidermal. *see* Psoriasis

Hyperpolarization, **2:**521
  defined, **2:**305
Hyperpotassemia. *see* Hyperkalemia
Hyperprolactinemia, **2:**28
Hypersecretion, hormonal. *see under* Hormones
Hypersensitivity. *see* Allergies
Hypersomnia, **2:**511, 517, **7:**79-80
Hypertension, **2:**167, **4:**283-94
  adrenal gland disorders and, **1:**78
  aging and, **3:**896
  atrial natriuretic factor and, **1:**476-77
  and calcium antagonist, **2:**87-88
  cigarette smoking and, **7:**518
  as complication of kidney dialysis, **1:**409
  consequences of, **2:**167
  diagnosis of, **4:**288-89
  enzyme inhibitors and, **3:**396
  essential, **4:**289-92
  genetic predisposition to, **4:**290
  hemorrhagic stroke and, **5:**602
  isolated systolic, **4:**290
  kinins and, **4:**599
  malignant, **4:**290
  pathophysiology of, **4:**290-92
  pregnancy-induced, **4:**919
  primary, **4:**289
  as psychophysiological disorder, **6:**336-37
  renin-aniogtensin-aldosterone system and, **5:**708
  renovascular, **4:**289
  research dues to kidney dialysis, **1:**410
  secondary, **4:**289
  therapy, **4:**292-94, 293
  thiazides and, **3:**97
  types of, **4:**289-90
Hypertension, pulmonary
  in infants, **5:**705
  lung injury and, **6:**352-53
Hyperthermia. *see also* Fever
  and cancer, **4:**295-301
  with chemotherapy, **4:**298, 299
  as coinducer, **4:**300-301
  as disturbance in thermoregulation, **1:**760
  effect of, on radiation of skin, **6:**465
  localized, **4:**297
  as mutagen, **4:**300
  regional, **4:**298-99
  as teratogen, **4:**300
  treatments, history of, **4:**296-97
  from ultrasonic waves, **7:**729
  whole-body, **4:**298-99
  with X-irradiation, **4:**297-98

Hyperthyroidism, **2:**20
  antibodies and, **4:**408
  in children, **3:**352
  neonatal, **3:**340, 348
Hypertrichosis, **4:**41
Hypertrophy, defined, **1:**57, **2:**167
Hypertropia. *see* Strabismus
Hyperuricemia, **6:**392–93
Hyperventilation, **5:**825
Hypnosis
  coping with stress, **7:**619
  dissociation in, **7:**618–19
  of phobic patients, **7:**620
  rating of, **7:**616
  relevance of measurement, **7:**617
  self, as coping mechanism for stress, **7:**621–23
  social-psychological forces affecting, **7:**617
  susceptibility to, **7:**616, 618, 620
Hypocalcemia
  neonatal, **3:**349
  parathyroidectomy and, **5:**665
Hypochloremia, **3:**93
Hypochondriasis, **4:**998
Hypochromic anemia. *see* Anemia, hypochromic
Hypodermoclysis, **5:**700
Hypogeusia, **2:**649
Hypoglycemia, **2:**34, 106
  in children, **3:**353
  and criminality, **2:**717
  headaches associated with, **4:**58
  and insulin, **2:**717
  pancreatic glucagon and, **5:**719
Hypogonadism
  in delayed puberty, **3:**358
  hyperprolactinemic, **2:**29
  hypogonadotropic, **2:**28, **3:**358
  hypogonadotropic, idiopathic, **2:**498
  secondary, **5:**398–99
Hypokalemia
  defined, **3:**93
  diuretics and, **3:**99
Hypomania, **2:**831, 832
  pharmacological, **2:**833
  and tricyclic antidepressants, **2:**833
Hypomenorrhea. *see* Menstruation disorders
Hyponasality, **7:**159
Hyponatremia, defined, **3:**93
Hypoparathyroidism, **1:**807–8
  character of, **3:**353
  conditions causing, **5:**664–65
  congenital, **3:**346, 349
  treatment for, **5:**665

Hypophonia, **2:**763
Hypophysectomy, partial, transsphenoidal, **5:**400
Hypophysis. *see* Pituitary gland
Hypopotassemia. *see* Hypokalemia
Hyposecretion, hormonal. *see under* Hormones
Hyposecretion, primary adrenal. *see* Addison's disease
Hyposmia, **5:**529
Hypotension, **2:**144
Hypotension, orthostatic, **2:**151
  in astronauts, **2:**148
  defined, **2:**147
Hypothalamic area, lateral, **4:**720–21
Hypothalamic area, posterior, **4:**719
Hypothalamic gonadoliberin. *see* Gonadorelin
Hypothalamic hormones, **2:**17, **5:**395, 19
Hypothalamic-hypophyseal-portal system, **4:**931
Hypothalamic neoplasms, **5:**398
Hypothalamic nucleus, **4:**607–8
Hypothalamic-pituitary axis, **3:**632
Hypothalamic-pituitary-gonadal axis, **4:**931
Hypothalamic releasing hormones, **5:**395–96
Hypothalamo-hypophyseal system
  anatomy, **5:**394
  at birth, **3:**349
  fetal, development of, **3:**337–38, 342–45
Hypothalamo-pituitary-gonadal system
  at birth, **3:**348–49
  fetal, development of, **3:**342
  in puberty, **3:**355
Hypothalamo-pituitary-thyroid system
  at birth, **3:**347–48
  fetal, development of, **3:**339–40
Hypothalamus, **2:**5, 9, 11, 12, 18, 20, 31, 32, 35–39, 335,
    336, 493, 494, 831, 834, **4:**303–12, **5:**349
  adrenal medulla and, **1:**76–77
  anatomy, **4:**303–5, **5:**394
  appetite behaviors, **4:**309–10
  autonomic control, **4:**310
  autonomic outputs, **4:**307–8
  at birth, **3:**347
  cardiovascular system, **4:**310
  caudal, projections to spinal cord, **4:**714
  circadian rhythms, **4:**311
  cortical-behavioral outputs, **4:**307
  cortical-limbic inputs, **4:**306
  defined, **2:**31, **5:**393
  dysfunction, **2:**28
  emotional expression, **4:**311
  endocrine control, **4:**311–12
  fetal, development of, **3:**337–38
  food intake and, **5:**519
  functions, **4:**303, 308–12

Hypothalamus (*Continued*)
  homeostasis and control theory, **4**:308
  humoral inputs, **4**:306–7
  immunization and, **5**:385
  inputs, **4**:305–7
  integrative aspects, **4**:312
  intrinsic connections, **4**:308
  median eminence of, anatomy of, **6**:556
  mediobasal, in puberty onset, **3**:355
  as neuroendocrine organ, **3**:318
  neurotransmitters, **4**:308
  outputs, **4**:307–8
  peptides secreted by, **6**:14
  and pituitary, anatomical relationship of, **6**:10
  preoptic area, **4**:242–43, 304
  primary sensory inputs, **4**:306
  in puberty, **3**:355
  somatic-sensory inputs, **4**:306
  somatostatin and, **5**:733
  thermoregulation, **4**:310–11
  visceral function, **4**:310–11
  visceral inputs, **4**:305–6
  weight loss and gonadotropin secretion, **1**:743
Hypothermia, **4**:791–809, 792–93, 799–806
  acidosis, **4**:801
  afterdrop, **4**:804
  alcohol and, **4**:803
  chronic effects of peripheral nonfreezing temperature,
      4:798–99
  core effects of freezing temperature, **4**:808
  core temperature measurement, **4**:793–95
  definitions, **4**:799
  dehydration, **4**:801
  dysarthria, **4**:800
  heat loss, **4**:792
  hormones and, **4**:801
  hypothermic-cardioplegic arrest, **4**:92
  hypoxia, **4**:792
  immersion, **4**:802
  metabolic icebox, **4**:800, 802
  nervous system, **4**:800–801
  oxygen consumption, **4**:800
  pediatric, **4**:803
  peripheral effects of freezing temperature, **4**:806–8
  peripheral response to nonfreezing temperature,
      4:795–98
  physiological basis of thermoregulation, **4**:793
  prevention, **4**:803
  primary, **4**:800
  psychological changes, **4**:797
  rewarming, **4**:803–6
  secondary, **4**:800, 803

Hypothermia (*Continued*)
  types, **4**:802–3
Hypothesis generation, **1**:619–20
Hypothesis testing, **1**:619, 621
  biases in, **1**:615 or 622
  epistemic needs of individual in, **1**:617
  evaluation of, **1**:619
  humans as testers, **1**:615–16
  hypothesis generation in, **1**:618
  inference rules in, **1**:617, 618–19
  information evaluation in, **1**:619
  information-processing sources of bias in, **1**:617–18
  model for, **1**:616–17
  motivational sources of bias in, **1**:619
  rationality of people in, **1**:621
Hypothyroidism, **2**:20, 832, 834
  antibodies and, **4**:408
  in children, **3**:351
  congenital, **3**:348, 352, **5**:436
  and dental caries, **2**:813
Hypotonia, in fragile X syndrome, **3**:714
Hypovolemic shock. *see* Shock
Hypoxanthine
  as oocyte maturation inhibitor, **5**:556
  purine nucleotide biosynthesis and, **3**:806
  in sudden infant death syndrome, **7**:318–19
Hypoxanthine phosphoribosyltransferase, deficiencies,
      3:855
Hypoxia. *see* Anoxia
Hypsarrhythmia, in infantile spasms, **3**:458

I

IA antigens. *see* Histocompatibility antigens class II
Ibuprofen, **2**:785
ICAM-1 molecule, **3**:450
Ice Age, hominid evolution in, **3**:501–2
Ice crystal formation, **2**:723, 724
Iconic memory, **4**:673, 676
Icosahedron, **7**:772
Icthyosis, X-linked, **7**:254
Id, **2**:853, 854
  and libido, **2**:854
  "where id was, ego shall be," **2**:854
ID, defined, **6**:307
Identical twins. *see* Twins, monozygotic
Identification, **1**:493
Identity crises. *see also* Self-concept
Identity crises, during adolescence, **1**:67–68
Ideomotor theory of biofeedback, **1**:663–64
Idiocy. *see* Mental retardation
Idiopathic hypertrichosis, **4**:41

Idiopathic venous thromboembolism. *see* Thromboembolism, venous

Idiotypes, **4:**313–18, 314
  applications, **4:**317
  cross-reactivity, **4:**315
  defined, **7:**751
  future prospects, **4:**317–18
  historical overview, **4:**314–15
  immunotherapy with, **4:**348–49
  internal image, **4:**314, 316
  network, **4:**316–17
  self-recognizing V domains, **4:**316
  structural basis, **4:**315–16
  V domains, **4:**313–14

Idoxuridine. *see* 5-Iododeoxyuridine

IFNs. *see* Interferons

IgA, **1:**677
  induction of, **5:**139–40
  structure of, **1:**564, **5:**137–38

IgA, polymeric
  synthesis of, **5:**141
  transport of, **5:**138–39

IgA, secretory, **5:**137, **6:**694–95

S-IgA antibodies, **5:**141

IgA-committed cells, **5:**137

IgD, **1:**567, 677

IgE, **1:**677
  antibody, **1:**299
  production in helminth parasitic diseases, **4:**387
  receptors, **1:**184
  regulation of synthesis, **1:**185
  structure of, **1:**183–84
  sulfite-induced asthma and, **7:**339
  synthesis of, **1:**184
  testing for, in allergic patient, **1:**193

IgG, **1:**677
  autoantibody overproduction in autoimmune disease, **1:**302
  structure of, **1:**564

IgM, **1:**677
  in beta cell triggering, **1:**576
  fetal, as indication of fetal blood sampling, **1:**731–32
  structure of, **1:**564

Ileocecal sphincter, **3:**73

Ileostomy patients, **3:**43

Ileum
  active transport in, **3:**81
  anatomy, **3:**65, 76

Ileus. *see* Intestinal obstruction

Illicit drugs. *see* Street drugs

Illiteracy. *see* Dysgraphia; Dyslexia; Literacy

Illusions
  examples of, **7:**830
  in LSD use, **5:**448

Illusions, visual, **7:**829–33
  cognitive-judgmental contributions to, **7:**832–33
  lateral neural interactions as factor in, **7:**831–32
  mechanisms of, **7:**830
  neural contributions to, **7:**830–32
  optical blur as factor in, **7:**831
  optical contributions to, **7:**830–32
  quantitation of cognitive contributions to, **7:**833
  quantitation of physiological processes to, **7:**833
  spherical aberration and, **7:**831
  structural mechanisms and, **7:**832
  varieties of, **7:**829–30

Image, hallucinatory, of mother, **2:**854

Iminodibenzyl derivatives, **1:**308–9

Imipramine, **1:**613–14, **2:**836, 843

Imitation-modeling hypothesis, **1:**1058:1:2

Immediate cause of death, **5:**99

Immersion injury, **4:**798–99

Immnocytochemistry. *see* Immunohistochemistry

Immobilization, **1:**765

Immortalization, in Epstein-Barr virus, **3:**463, 467–70

Immune-associated antigens. *see* Histocompatibility antigens class II

Immune clearance, **1:**299

Immune complex disease. *see also* Antigen-antibody complexes
  Arthus reactions, **4:**409–10
  mechanism of induction, **1:**300

Immune network, **4:**313, 316–17

Immune response, **2:**392, 399, 751–62, **4:**324–25
  antibody-antigen complexes in, **1:**301–3
  clinical implications of conditioning studies on, **6:**324–25
  and cytokines, **2:**751–62
  double-edged sword of, **4:**407
  effect of conditioning on, **6:**322–23
  effect of neural influences on. *see* Psychoneuroimmunology
  to *Entamoeba histolytica* infection, **1:**226–28
  genes controlling, **4:**379
  and heat-shock proteins, **4:**105–6
  lymphocytes, **4:**830
  parasite infections and, **5:**647–49
  platelet-activating factor and, **6:**48–49
  reexposure to microorganisms, **4:**325
  regulation of, **5:**144–47
  retinoids and, **4:**824–28
  to vaccines, interleukin-2 effects, **4:**533

Immune response genes. *see* Genes, immune response

Immune surveillance, **4:**319-22
 concept of, **4:**319-20
 definition, **4:**319
 possible existence, **4:**321
 validation, **4:**320-21
Immune system, **4:**323-36
 adaptive, functional abilities, **4:**406
 aging, **3:**903
  cell-mediated immunity and, **6:**76
  compartmentalization of, **6:**77-78
  fibroblasts and, **6:**76
  humoral immunity and, **6:**77
  polymorphism, **6:**75-78
  progression of, **6:**75-76
  thymus and, **6:**76
 altered, in inducing autoimmunity, **1:**525-26
 amplification and renewal, **4:**331-35
 antigens, **4:**393
 basic functions of, **7:**395
 basis of self-recognition in, **1:**520-22
 behavioral manipulation and, **6:**321-22
 biology and paradigms, **4:**323-26
 chronic disease of, **1:**300-301
 clonal selection theory, **4:**332-33
 completeness, **4:**328-29, 335
 defined, **5:**703
 degeneracy, **4:**328-29
 diet and, **5:**503-5
 differentiation of effector functions, **4:**333
 discrimination between self and nonself, **1:**520-26, **4:**324
 diversity, **4:**326-28, 392
 effector functions, **4:**325-26
 endogenous selection of repertoires, **4:**332
 exogenous antigen detection, **4:**393-94
 general characteristics of, **1:**513
 genetic defects, **4:**837
 graft rejection, **4:**362
 immunoregulatory interference, **1:**300-301
 inactivation reactions, **4:**406-8
 individuality, **4:**329-31
 levels of organization, **4:**323
 lithium effects, **4:**752
 loss of shape-space homeostatic functions, **4:**325
 marijuana and cannabinoid effects, **4:**909
 metadynamic selection of repertoires, **4:**332
 molecular recognition, **4:**323
 in neural-immune interactions, **5:**373-86
 neural influences on. *see* Psychoneuroimmunology
 nonresponder, **4:**330
 organization of, **1:**676-77
 preimmune repertoires, **4:**332

Immune system *(Continued)*
 secondary responses, **4:**325
 shape-space homeostasis, **4:**325-26
 specificity, **4:**328
 starting repertoires, **4:**330-31
 stress and, **6:**320-22
 surveillance, **4:**319-22
 system at work, **4:**335-36
 T-cell receptors, **4:**329
 trichothecenes and, **7:**663, 664-65
 tumor cell interactions, **4:**1012
 ultraviolet radiation and, **7:**68
 V-region amplification, **4:**333
 V-regions, **4:**326-27, 330-31
 vaccines, **4:**326
 vitamin D and, **7:**868
 zinc and, **7:**885-86
Immunity
 acquired, macrophage role, **4:**862
 aging effects, **4:**786
 animal studies on stress and, **6:**320-22
 cell mediated, and aging immune system, **6:**76
 characteristics, **4:**406
 classification, **4:**407
 concomitant, **4:**386, 389-90
 conditioned changes in, **6:**322-23
 effects of radiation on, **6:**420-23
 endocrines effects, **6:**321-22
 herd, **4:**444
 humoral, and aging immune system, **6:**77
 immune tolerance, **4:**325
 innate vs. adaptive, **4:**402
 and learned helplessness, **4:**669
 longevity and, **4:**789
 lymphoid cell-mediated, **4:**405
 mucosal, **5:**137-42
 nonsterilizing, parasites, **4:**386
 nutrition and, **5:**503-16
 in protein-energy malnutrition, **4:**895-96
 stress and, **6:**320-22
 in upper respiratory tract, **5:**692
 to viruses, **7:**780-81
Immunity, cellular. *see* Cell-mediated immunity
Immunity, humoral. *see* Antibody formation
Immunity, natural
 aging and, **3:**896
 in famines, **3:**563
 macrophage role, **4:**862
 oral contraceptives and, **5:**608
Immunization
 active, **4:**464
 for child care in developed countries, **4:**465-66

Immunization (*Continued*)
  passive, **4:**464
  process of, **4:**402–3
  recommended schedule for, **5:**702
  strategies, **1:**299
Immunoadsorbent techniques. *see* Immunosorbent
    techniques
Immunoassay. *see also* Radioimmunoassay
  in drug analysis, **3:**207–9
  in food monitoring, **3:**669
  heterogeneous, **3:**208
  nonradionucleotide, **4:**351–56
  parasite detection, **4:**388–89
  superoxide dismutase measurement using, **7:**358
Immunobiology of transplantation. *see* Transplantation
    immunology
Immunocompetence, and sudden infant death syndrome,
    7:318
Immunocompromised host, **6:**410
Immunoconjugates, **4:**367–78, 419–20
  accumulation in tumor tissue, **4:**377
  chemical synthesis, **4:**375–76
  chemoimmunoconjugates, **4:**370
  effector molecules, **4:**367–73
  immunogenicity, **4:**377
  monoclonal antibodies as, **1:**688–89
  synthesis of, **4:**375–76
  targeting moieties, **4:**374–75
  targets for, **4:**373–74
Immunodeficiency
  bone marrow transplantation for, **1:**793–94
  in children, **5:**706
  diseases, **4:**837
  Epstein-Barr virus and, **3:**464, 465–66
  experimental metastasis, **4:**1008
Immunodeficiency disease, acquired. *see* Acquired immu-
    nodeficiency syndrome (AIDS)
Immunodeficienty, severe combined. *see* Severe combined
    immunodeficiency
Immunodiagnosis. *see* Immunologic tests
Immunofluorescence microscopy. *see* Microscopy, fluor-
    escence
Immunofluorescence technique. *see* Fluorescent antibody
    technique
Immunogenetics, **4:**379–84
  allotypes, **4:**381–82
  blood group antigens, **4:**379–81
  complement components, **4:**384
  histocompatibility antigens, **4:**382–84
  immunoconjugates and, **4:**377
Immunogens. *see* Vaccines
Immunoglobulin, alpha chain, **4:**404

Immunoglobulin, delta chain, **4:**404
Immunoglobulin, epsilon chain, **4:**404
Immunoglobulin, gamma chain, **4:**404
Immunoglobulin, mu chain, **4:**404
Immunoglobulin receptors
  in beta cell triggering, **1:**567
  on beta cells, **1:**567
Immunoglobulins, **2:**754. *see also* Antibodies
  amino acid sequence, **4:**833–34
  on B cells, **4:**831–32
  B-lymphocyte bound, and study of tumor clonality,
    7:691
  cDNA clones of, **3:**807
  defined, **4:**404
  DNA methylation and genes, **3:**160
  enzyme binding, **4:**580
  as gene family, **3:**798
  gene rearrangement, **2:**754
  heavy-chain structural genes, **4:**833–34
  length of V, D, J and C segments of genes, **7:**396
  properties of, **3:**809
  protection, in leukemia treatment, **4:**695
  purified, **4:**447
  receptors, antigen binding on, **6:**520–22
  serum, **4:**335
Immunohistochemistry, **3:**277
Immunologic diseases
  bone marrow transplantation for, **1:**793–94
  diagnosis of, **5:**684
  immune effector selection and, **4:**397–98
Immunologic privilege, **4:**334, 392
Immunologic tests, **4:**324
  bispecific monoclonal antibodies in, **5:**87
  human monoclonal antibodies in, **5:**85–86
  markers, in autoimmune disease, **1:**338–39
  mouse monoclonal antibodies in, **5:**83–84
Immunologic tolerance, **3:**312
Immunological memory, **7:**747
Immunological techniques, **4:**577
Immunology, regional, **4:**391–400
  autoimmunity avoidance, **4:**394
  collision with systemic immune concerns, **4:**398
  components, **4:**398
  differentiated lymphoid organs and tissues, **4:**393
  differentiated tissue function maintenance, **4:**394–95
  draining lymphoid organs and tissues, **4:**397
  elements, **4:**397
  existence of uniquely differentiated nonlymphoid tis-
    sues and organs, **4:**392–93
  immune effector response amplification, **4:**394
  regional spheres, **4:**398–400
  regionally located antigens, **4:**394

Immunology, regional *(Continued)*
    resident antigen-presenting cells, **4**:396
    selection of immune effectors, **4**:397-98
    specialized endothelium of postcapillary venules,
        4:396-97
    thesis, **4**:395
    tissue-seeking lymphocytes, **4**:395-96, 399
    unique spectra of immune effectors, **4**:397
    uniquely differentiated parenchymal cells, **4**:396
Immunology, transplantation. *see* Transplantation immunology
Immunology laboratories. *see* Laboratories, immunology
Immunomodulators. *see* Adjuvants, immunologic
Immunopathology, **4**:401-13
Immunoperoxidase staining, **6**:197
Immunopoiesis, **4**:841-44
Immunopotentiating agents. *see* Adjuvants, immunologic
Immunoproliferative small intestinal disease, **3**:769-70
Immunoregulation, **4**:474
Immunosorbent techniques, **3**:408
Immunosorbents, **1**:379-80
Immunostaining, **5**:682
Immunosuppression, **2**:88. *see also* Transplantation immunology
    behaviorally conditioned, **6**:324
    and calcium antagonist, **2**:88
    glucocorticoids and, **7**:269-70
    interleukin-2 effects, **4**:533
    nonspecific, preventing graft rejection, **4**:363-64
    specific
        preventing graft rejection, **4**:364-65
        in treatment of autoimmune disease, **1**:532-33
    in treatment for multiple sclerosis, **5**:148
    tumor incidence, **4**:320
Immunosuppressive agents, **1**:684-85
    adriamycin, **1**:686
    azathioprine, **1**:685
    cyclophosphamide, **1**:685-86
    cyclosporine A, **1**:685
    steroids, **1**:684-85
Immunotherapy
    active, nonspecific, **1**:678
    active, specific, **1**:678
    adoptive, with IL-2-activated LAK cells, **1**:679
    bispecific monoclonal antibodies in, **5**:86-87
    cancer, with interleukin-2, **4**:532-33
    in carcinoma, **4**:337-38, 346-49
        active, **4**:346-48
        passive, **4**:348-49
    chimeric monoclonal antibodies in, **5**:88
    human monoclonal antibodies in, **5**:86
    mouse monoclonal antibodies in, **5**:84

Immunotherapy *(Continued)*
    for multiple sclerosis, **5**:148
    single-chain monoclonal antibodies in, **5**:89
    in treatment of allergy, **1**:195
Immunotoxins, **4**:367-69, 415-22
    antibodies, **4**:416
    cell adhesion mechanisms, **4**:427-28
    clinical applications, **4**:421-22
    immunoconjugates, **4**:419-20
    in vitro characterization, **4**:420
    in vivo characterization, **4**:420-21
    lethal to human tumors, **4**:420
    linkage, **4**:418-19
    metal, **7**:553
    ricin, **4**:417-18
    theory of cell-selective agents, **4**:415-16
    toxicity, **4**:421
    toxin proteins, **4**:416-18
Impact injury, **2**:648
    in sports, **2**:648
Impedance, electrical. *see* Electrical impedance
Impedance (acoustics)
    defined, **3**:229
    in middle ear, **3**:230-31
Impedance analysis. *see* Tetrapolar bioelectrical impedance analysis
Implantation, blastocyst. *see* Ovum implantation
Impotence, **2**:29, **6**:873-74
Impressions
    formation
        averaging model of, **5**:794
        on-line vs. memory-based evaluations in, **5**:796-97
        person perception and, **5**:793-94
    management, person perception and, **5**:785-86
    structure of, person perception and, **5**:794
    trait-behavior associations in, **5**:795
Imprinting (genetics), **2**:475
    defined, **2**:475
    methylation and, **3**:795-96, 872
Improved Standards for Laboratory Animals Act, **1**:253
Impulse control disorders, **4**:998
Impulse-timing, as method of motor control, **5**:111-12
Impulsive behavior, **5**:770
In vitro fertilization. *see* Fertilization *in vitro*
Inborn errors of metabolism. *see* Metabolism, inborn errors
Inbreeding
    genetic load, **4**:438-39
    isogeneic animals, **4**:338
Inbreeding, human, **4**:431-41
    close versus remote, **4**:433-34
    cultural factors, **4**:438
    finite population size, **4**:438

Inbreeding, human *(Continued)*
  gene frequencies, **4:**435
  genealogy, **4:**434
  genetic load, **4:**438–39
  geographic isolation, **4:**438
  history, **4:**431
  incest consequences, **4:**440
  interpopulation variation, **4:**436–37
  isonymy, **4:**434–35
  measurement, **4:**434–36
  migration matrices, **4:**435–36
  morbidity, **4:**439–40
  mortality, **4:**439
  natural selection and, **4:**440
  random and nonrandom, **4:**433
  in subdivided populations, **4:**434
  temporal trends, **4:**436–38
Inbreeding coefficient, **4:**432–33, 436–37
Incest
  genetic consequences, **4:**440
  and sexuality, **6:**886–87
Incisor, **2:**820, 821
Inclusion bodies. *see* Cellular inclusions
Inclusion cysts, **5:**614
Incoherence problem, **2:**647, 556
Incubators, infant, **5:**705
Indalpine, **2:**846
Independent assortment (genetics), **3:**849
India, famines in, **3:**564
Indians, American, **2:**796–97
  diet of, **3:**651
  origin based on teeth, **2:**796–97
  and shoveling, dental, **2:**796
Indium-111 isotope, **4:**371
Individual vs. society, **4:**704–5
Individualism, in behavior, **1:**574
Indolamine, **2:**337, 843
Indole-3-carbinol, **3:**681
Indole alkaloids, **1:**180
Indolizidines, **7:**599
Indomethacin, **4:**281
Indoramine, **2:**177
Inducer cells. *see* Helper cells
Inductions, embryonic. *see* Embryonic induction
Inductive interactions, between bone and cartilage cells,
  1:788
Inductive reasoning
  characteristics of, **6:**498–99
  role of statistics in, **6:**495
Industrial engineering, **3:**473
Industrial hygiene, **3:**473
Industrial psychology. *see* Psychology, industrial

Infancy
  growth, **4:**26
  infectious diseases of, **4:**459–62
  language acquisition, **4:**635–36
  sensory-motor changes in, **6:**819–27. *see also* Newborn, sensory-motor activity and changes in
Infant
  carrying of, by early hominids, **3:**496–97
  development
    of hearing, **5:**748
    perceptual functions in, **5:**747–48
    visual acuity in, **5:**748
  glutathione levels of, **3:**911
  and mother, **2:**854
  Neolithic face pattern and, **3:**227
  sudden infant death syndrome and, **7:**320–21
  zinc metabolism in, **7:**888–89
Infant, newborn
  changes in drug kinetics in, **5:**853–54
  effects of antimicrobial agents. *see under* Antimicrobial agents
  eye-hand coordination in, experimental study of, **6:**822
  herpes simplex, **4:**189–90
  infectious diseases of, **4:**457–59
  intestinal microflora. *see under* Intestinal microflora
  lactase activity, **4:**615–16
  laryngeal reflexes, **4:**651–52
  lipolysis in, **3:**81
  locomotion in, **4:**779–81
  mammalian, respiration in, **6:**576–77
  motor activity in, **6:**820–21
  nervous system function of, changes in, **6:**823
  pain in, assessment of, **5:**630–31
  pain transmission in, **5:**630
  parity effect, **4:**28
  postural maturation in, **6:**822–23
  postures of, most frequent, **6:**821
  salt preference and, **6:**719–20
  sensory activity in, **6:**820
  sensory-motor activity and changes in, **6:**819–27
  sensory-motor coordinated behaviors of, **6:**821–22
  sensory-motor development
    bases and determinants of, **6:**822–24
    course and processes of, **6:**824–27
    theories of, **6:**825–27
Infant, newborn, screening, **5:**433–38
  diseases available for, **5:**436–38
  false-negative results in, **5:**434
  false-positive results in, **5:**434
  future directions of, **5:**438
  medical legal issues and, **5:**435
  new technology for, **5:**438

Infant, newborn, screening *(Continued)*
    outcomes of, **5**:435
    phenylketonuria as model for, **5**:435–36
    principles of, **5**:434
    purpose of, **5**:433–34
    target population for, **5**:434
Infant, premature
    intestinal microflora, **1**:320
    triacylglycerol digestion of, **3**:52
    twin birth and, **3**:314
Infant botulism, **7**:318
Infant care, as primate behavior, **6**:130–31
Infant food, bioavailability of, **3**:47
Infant mortality
    Black, **5**:103
    Boston, nineteenth century, **5**:700
    in developing countries, **5**:701
    United States
        eighteenth century, **5**:699–700
        twentieth century, **5**:701, 702
Infant nutrition
    in developing countries, **3**:26
    history of, **5**:700–701
    of neonates, **5**:704–5
Infantile spasms. *see* Spasms, infantile
Infantilism, genital. *see* Hypogonadism
Infection
    amoebic. *see* Amebiasis
    anerobic. *see* Anaerobic infections
    bacterial. *see* Bacterial infections
    congenital, **1**:731–32
    dental, **2**:784
        antibiotics for, **2**:784–85
        bacteria, prominent in, listed, **2**:784
    headaches associated with, **4**:57–58
    lymphocyte response, **4**:830
    persistent, **7**:642
    reproductive rate of, **7**:893
    role in protein-energy malnutrition, **4**:896
    subclinical, **7**:641–42
    xenotic, **7**:898
    zoonotic, **7**:898–99. *see also* Zoonoses
Infectious diseases, **4**:443–51
    animal, antibiotics for, **1**:294
    antigenic variation as factor in transmission of, **7**:643
    blood supply testing for, **1**:737–40
    colonization as factor affecting transmission of, **7**:641–42
    control, among children, **5**:701–3
    of the developed world, **4**:450–51
    of developing countries, **4**:447–50

Infectious diseases *(Continued)*
    diagnosis of, recombinant DNA technology in, **6**:530–31
    disease models, animal, **1**:255–56
    environment as factor in transmission of, **7**:643
    in famines, **3**:563
    fetal infections, **4**:456–57
    helminth, **7**:898
    incidence rate of, **7**:639
    of infants and toddlers, **4**:459–62
    longevity and, **4**:789
    macrophages as reservoirs, **4**:862
    measures of occurrence and outcome in, **7**:639–40
    mortality rate and, **7**:639
    of neonates, **4**:457–59
    pediatric, **4**:453–57
    of preschool children, **4**:462–64
    prevalence rate of, **7**:639
    prevention, **7**:643–44
    protozoal, **6**:267–68, **7**:897–98
    recombinant DNA technology in diagnosis of, **6**:530–31
    of school-age children, **4**:464–65
    transmission, **7**:639–44
        amoebiasis, **1**:218–19, 229–30
        factors affecting, **7**:640–43
        horizontal, **7**:640
        modes of, **7**:640
        persistent infection as factor in, **7**:642
        population size as factor in, **7**:642–43
        subclinical infection as factor affecting, **7**:641–42
        vertical, **7**:640
        virulence as factor affecting, **7**:641
    vaccine preventable, **4**:444–47
    zoonoses and, **7**:643
Infectious human wart virus. *see* Papillomaviruses
Infectious mononucleosis
    and depression, **2**:833
    EBNA-2 antibodies in, **3**:468
    Epstein-Barr virus and, **3**:465–66
    fatal, **3**:465–66
Infectious mononucleosis-like syndrome, chronic. *see* Fatigue syndrome, chronic
Inference rule generation and information search, **1**:620
Inference schema. *see* Reasoning, conditional
Inferior colliculus. *see* Colliculus, inferior
Infertility, **2**:29. *see also* Amenorrhea
Infertility, female, **3**:365
Infiltrating gastric carcinomas. *see* Diffuse-type gastric carcinomas
Inflammation, **2**:785, **4**:469–77
    acute, **4**:405
    amplification of response, **4**:394, 472

Inflammation (*Continued*)
  antibody-mediated, **4**:403–5
  bradykinin role, **4**:470
  cell-mediated, **4**:405
  characteristics of, **1**:311
  chronic, **4**:405–6, 469–70, 472, 862
  clinical signs of, **1**:25
  complement activation of response, **1**:298
  complement cascade activation, **4**:471
  contribution of polymorphonuclear neutrophils to, **5**:431
  cytokine role, **4**:473–75
  drug therapy, **4**:476–77
  Fc receptors, **1**:298
  glucocorticoids and, **7**:269–70
  headaches caused by, **4**:54–55
  histamine role, **4**:470
  initiation and progression, **4**:469–72
  kinin system role, **4**:470–71
  kinins and, **4**:598–99
  lipid mediators, **4**:472–73
  macrophage margination during, **5**:819
  mediators, **1**:314, **4**:470
  neutrophils and, **5**:813
  phagocytes in, **5**:818–189
  phospholipase A_2, **4**:472
  phospholipase C, **4**:472
  platelet-activating factor and, **4**:470–71, 473, **6**:44
  process of, **4**:403
  prostaglandins and, **2**:785
  proteoglycans and, **6**:263
  in soft tissue repair, **7**:501
  steroid effects on, **1**:312–13
  suppression of, **1**:312
  thermal injuries and, **7**:450
  tumor necrosis factor role, **4**:474–75
  vascular phase, **4**:473
  wheal-and-flare reaction, **4**:470
Inflammatory bowel disease, **4**:618
Inflammatory headaches, **4**:54–55
Influence, defined, **5**:799
Influenza, **4**:486–91
  clinical manifestations, **4**:488–89
  complications of, **4**:490
  diagnosis, **4**:490
  host response to, **4**:489–90
  pathogenesis, **4**:489
  treatment of, **4**:490
  vaccines, **2**:491, **4**:446
Influenza A, **7**:783
Influenza viruses, **2**:399
  and amantadine, **2**:399

Influenza viruses (*Continued*)
  classification, **4**:480–81
  epidemics, **4**:487–88
  hemagglutinin, **4**:482–83
  M1 protein, **4**:484
  M2 protein, **4**:484
  neuraminidase, **4**:483–84
  nomenclature, **4**:480–81
  nonstructural proteins, **4**:485
  nucleocapsid, **4**:484–85
  nucleoproteins, **4**:485
  pandemics, **4**:487–88
  polymerase proteins, **4**:485
  replication cycle, **4**:486–87
  and rimantadine, **2**:399
  RNA, **4**:485
  staining of, **3**:278
  vaccines, **4**:491
  viral envelope, **4**:481–82
  virulence of, **4**:489
Information
  attitudes and, **1**:487, 495
  complexity, affective consequences of, **1**:106
Information dissemination, in health care technology, **4**:66
Information feedback. *see* Feedback, information
Information processing, **2**:547–52, 853. *see also* Encodingism
  cellular, **4**:558
  by computer, **2**:547
  empty symbol problem, **2**:552
  flow from environment to perception to cognition to language, **2**:548
  higher brain function, **4**:195
  interpretation problems, **2**:552
  olfactory. *see under* Smell
  parallel distributed. *see* Connectionism
  schematic, **2**:853, 856
Information processing, human. *see* Mental processes
Information systems, for molecular sequence data, **3**:163–67
Informed consent, **4**:76–77
Informed Consent Doctrine, **1**:136
Infrared rays, **5**:589
Infusions, intravenous
  for dehydration, **5**:707
  fructose in, **5**:827
Infusions, parenteral, **3**:577
Ingestive behavior. *see* Feeding behavior
Inhalation injury, smoke. *see* Smoke inhalation injury
Inheritance, complex. *see* Complex inheritance
Inheritance, maternal. *see* Maternal inheritance
Inheritance, mitochondrial. *see under* Mitochondria

Inheritance, polygenic. *see* Polygenic inheritance
Inheritance, simple. *see* Simple inheritance
Inhibin, **2**:21
    corpus luteum and, **3**:634
    FSH secretion and, **3**:632
    function of, **6**:91
    gonadoliberin and, **5**:732
    in puberty, **3**:355
    Sertoli cell secretion of, **7**:424-25
Inhibition (psychology), dietary restraint and, **3**:689
Inhibition system, behavioral (BIS), **2**:711, 715, 716
Inhibitors, protease. *see* Protease inhibitors
Initial velocity assumption, defined, **3**:427
Initiation, carcinogenic, **3**:671
Initiation signal, defined, **5**:205
Injection devices, HPLC, **4**:209-10
Injuries. *see* Wounds and injuries
Innatism, **2**:549
Innervation, **2**:31-33. *see also* specific organs
    afferent, **2**:32, 33
    of bladder, **5**:43-44
    of central auditory system, **3**:240
    cranial meninges, **4**:987
    definition, **2**:357
    efferent, **2**:32, 33
    extrinsic, of gut, **2**:31-33
    of eye, **3**:549
    of foregut, **3**:64
    fusimotor, **3**:721-22
    of hindgut, **3**:66
    of intrinsic external urethral sphincter, **5**:43-44
    of midgut, **3**:65
    myogenesis and, **5**:171
    "pulse-step," **3**:558
Innervation ratio, defined, **5**:176
Inositol 1,4,5-triphosphate, in Ca^{2+} messenger system, **3**:333
Inositol 1,4,5-trisphosphate, as cellular messenger, **7**:20
Inositol phosphates, **2**:304
    cytosolic Ca^{2+} and, **4**:17
Inositol phospholipid, **2**:304
Insecticides
    organochlorine, **7**:561
    organophosphorate, **3**:252-53
Insects
    development, segmentation in, **3**:287
    gas convection in, mechanisms of, **6**:568-69
Insemination
    artificial, **3**:602-3
    intrauterine, **3**:602-3
Insertion elements, DNA. *see* DNA insertion elements
Insertional mutagenesis. *see* Mutagenesis, insertional

Insertions, of muscles, **4**:260-61
Insomnia, **2**:835. *see also* Sleep disorders
    brain sites associated with, **7**:72
    characteristics and treatment of, **7**:75
    characterization, **4**:998
    drug and alcohol dependency as causes of, **7**:78-79
    psychiatric disorders and, **7**:78
    psychophysiological, **7**:78
    as sleep disorder, **7**:77-79
Insulin, **2**:23, 54, 62, 105, 106, 259, 301, 323, 415, 658, 717, 834, **4**:493-501,
    actions *in vivo*, **4**:494-96
    in aerobic exercise, **3**:511
    and age, **2**:106
    cell membrane receptors for, **6**:517-18
    and criminality, **2**:717
    detection by radioimmunoassay, **6**:459
    diabetic state, **4**:498
    effect on feeding behavior, **4**:248
    in energy metabolism, **3**:327
    in fatty acid uptake, **3**:588
    fetal, **3**:339, 346
    food intake and, **5**:520
    function of, **5**:732-33, **6**:89
    galanin and, **5**:721
    in glucose regulation, **3**:324
    glycogen metabolism and, **3**:926-27
    in hunger and meal initiation, **1**:349
    and hypoglycemia, **2**:23
    in keratinocyte cultures, **3**:447
    levels in protein-energy malnutrition, **4**:894
    and lipid regulation, **1**:60
    in lipogenesis, **3**:575
    lipoprotein lipase activity and, **3**:587
    maintenance of lactation, **4**:607
    mechanism of action, **4**:498-500
    metabolic regulation role, **4**:1005
    in nutritional utilization, **5**:756
    pancreatic glucagon and, **5**:719
    pancreatic polypeptide and, **5**:719
    perinatal, **3**:350
    pH and, **5**:824
    prepubescent, **3**:351
    in protein kinase cascade, **3**:333
    in protein phosphorylation, **6**:204-5
    receptors, **2**:301, **5**:732, **6**:94
    sources of, **5**:734
    structure, **4**:494
    synthesis, **3**:327
    thiazide therapy and, **3**:98
    vitamin D and production, **7**:868

Insulin-like growth factor I
    antibody fusion proteins and, **3:**809
    estrogen and, **3:**484, 485
    fetal, **3:**339
Insulin-like growth factor II, fetal, **3:**339
Insulin-like growth factors. *see* Somatomedins
Insurance claims analysis, **4:**63–64
Integrain, **2:**270
Integration of functions, behavioral development and, **1:**582
Integrin receptors. *see* Receptors, integrin
Integrins
    characterization, **4:**630
    effect of transforming growth factor on, **7:**637
    role in neural cell response to laminin, **4:**630–31
Integument. *see* Skin
Intellect, social. *see* Social intellect
Intellectual behavior, development
    birth to 3 months, **1:**574
    3 to 10 months, **1:**575–76
    10 to 20 months, **1:**576–77
    20 to 36 months, **1:**578
    3 to 5 years, **1:**578–79
    5 to 8 years, **1:**580
    8 years to preadolescence, **1:**581
    differentiation vs. integration of processes, **1:**582
    general stages vs. specific cases, **1:**582
    role of shared heredity, **1:**582
Intelligence. *see also* Wechsler scale
    of animal, **2:**633
    in children, **5:**709
    in fragile X syndrome females, **3:**715
    in fragile X syndrome males, **3:**714–15
    quotient, **2:**719, 871
    test, **2:**719
    X chromosomes and, **3:**840
Intelligence tests, development of, **3:**474, 838
Intensity, exercise, defined, **3:**515
Intensive care units, neonatal, advances in, **5:**705
Intention, in human information processing, **5:**803
Intentions, aggressive, **1:**114
Interaction, neuronal, heterosynaptic, **2:**305
    defined, **2:**305
Interactivism, **2:**547, 555–57
    between system and environment, **2:**555
Interarytenoid muscle, **4:**657
Intercellular adhesion molecule 1
    inflammation and, **4:**474
    inflammation role, **4:**476
    LFA-1 and CR3 binding, **4:**475–76
Interdental papilla, **4:**33–34
Interference (genetics), **3:**823, 825

Interferon, immune. *see* Interferon type II
Interferon alpha. *see* Interferon type I
Interferon-beta. *see* Interferon type I
Interferon-gamma. *see* Interferon type II
Interferon receptors, **4:**515
    gene dosage effect of, **3:**177
    genes, **4:**515
    and mechanism of interferon action, **4:**516–17
    properties, **4:**515–16
Interferon type I, **1:**680–81, **2:**759–60
    antibody conjugates with, **4:**371
    gene for, **4:**514
    for papillomavirus infection, **3:**866
Interferon type II, **1:**680–81, **2:**753, 759–60, **4:**836
    in bone resorption, **1:**807
    effects on bone cell function, **1:**765
    gene for, **4:**514
    genetic locus of, **3:**183
    interleukin-2 and, **5:**731
    keratinocyte growth and, **3:**448–49, 450
    mediating graft injury, **4:**362
Interferons, **2:**249, 373, 399, 752, **4:**513–26
    2-5 A system and, **4:**524
    anticellular actions of, **4:**523–24
    antiviral, **2:**759
    antiviral actions, **4:**518–21, 524–25
    and AZT synergism, **2:**373
    biological effects, **4:**517–18
    cellular mRNAs and proteins regulated by, **4:**517
    cellular processing, **4:**516–17
    chemical properites of, **4:**514
    control of virus infections, **7:**789–90
    functions, **4:**823
    genes, **4:**514–15
    genes activated by, **4:**515
    genetic locus of, **3:**183
    immunotherapy with, **4:**348
    increase of natural killer cell activity by, **5:**298
    in inflammation, **1:**313, 314
    internalization, **4:**516–17
    lymphoma treatment with, **4:**850
    nutrition and, **5:**511
    P1 protein kinase system and, **4:**524–25
    properties of, **2:**757, **4:**514
    role of enzymes in, **4:**517
    synthesis inhibition, **4:**525
    treatment of neurologic disorders with, **4:**523
    treatment of viral infections with, **4:**521–24
Interleukin-1, **1:**682–83, **2:**35, 256, 649, 650, 753–61, **4:**835–36
    in acute phase response, **1:**28, 29, 32
    antiradiation effects of, **6:**423

Interleukin-1 *(Continued)*
  biological products, **4:**860
  bone cell function and, **1:**765
  in bone remodeling, **1:**823
  in bone resorption, **1:**807
  bone resorption and, **4:**474
  estrogen and, **3:**485
  in fever, **1:**759
  functions of, **5:**731
  hypercalcemia of malignancy etiology, **4:**278
  immunization and, **5:**385
  in inflammation, **1:**313, **4:**473-74, 314
  lymphocyte-activating, **2:**755
  and macrophage, **2:**755
  PGE$_2$ and, **5:**511
  production, **4:**860
  properties, **2:**756, **4:**473
  pyrogenic, **2:**755
  and T-cell, **2:**755
  in tobacco-associated cardiovascular disease, **7:**517
Interleukin-2, **1:**683, **2:**753-61, 835-36, **4:**527-33
  activated killer cells, **4:**340
  activation of LAK cells, **1:**679-80
  and B-cell stimulation, **2:**756
  food intake and, **5:**508, 510
  functions of, **5:**731
  genes, structure and expression, **4:**531-32
  as growth factor, **1:**686
  helper T cell secretion, **4:**361
  immunotherapy with, **4:**347
  in inflammation, **1:**313, 314
  interaction with interleukin-2 receptor, **4:**530-31
  lymphokine-activated killer and, **5:**384-85
  properties, **2:**756
  receptors. *see* Receptors, interleukin-2
  signal transduction, **4:**530-31
  structure, **4:**528-29
  and T-cell growth factor, **2:**753
  T-cell receptors for, **6:**518-19
  therapeutic implications, **4:**532-33
Interleukin-3, **1:**683, **2:**753, 756, 761
  clinical applications, **1:**687
  and hematopoiesis, **2:**756
  monocyte production and differentiation promotion, **4:**854
  properties, **2:**756
  receptors. *see* Receptors, interleukin-3
  and thymocyte, human, **2:**756
Interleukin-4, **1:**683, **2:**753-61
  and B-cell stimulation, **2:**756
  lymphoblastoid cell lines and, **3:**464
  in maturation of mast cells, **1:**186

Interleukin-4 *(Continued)*
  properties, **2:**756
  in regulation of beta cell response, **1:**569
  in regulation of IgE synthesis, **1:**185
  and T-cell, **2:**757
  and thymocyte, **2:**757
Interleukin-5, **1:**683, **2:**753-55, 758, 761
  B-cell growth factor II, **2:**758
  and immunoglobulin M secretion, **2:**758
  properties, **2:**757
Interleukin-6, **1:**684, **2:**753-55, 758, 761
  in acute phase response, **1:**28, 32
  B-cell activation, **2:**758
  bone resorption and, **4:**279
  hepatocyte activation, **2:**758
  in inflammation, **1:**313
  and plasmacytoma growth, **2:**758
  properties, **2:**757
  in regulation of beta cell response, **1:**569
  and thymocyte proliferation, **2:**758
Interleukin-8, **2:**753, 759, 762
  in acute phase response, **1:**28
  and lymphocyte accumulation, **2:**759
  as neutrophil-activating peptide, **2:**759
  properties, **2:**757
Interleukins
  in epidermis, **3:**449-50
  function of, **7:**482
  functions of, **4:**823
  hemopoietic cells and, **4:**173-75
  IgA and, **5:**140
  targeting moiety of immunoconjugates, **4:**374-75
Intermediate-density lipoproteins. *see* Lipoproteins, IDL
Intermediate-term memory, **4:**673, 676
Intermedins. *see* MSH
Intermodulation products, **3:**239
Internal images, **4:**317
Internalization, **1:**493
International Classification of Diseases, **5:**99-100
International Classification of Epilepsies and Epileptic Syndromes, **3:**457
International Classification of Epileptic Seizures, **3:**455-56
International Ergonomics Association, **3:**479
International Society for Developmental Psychobiology, **2:**870
International Union of Biochemistry Nomenclature Committee, **3:**726
Interneurons, **2:**677, **5:**524
Interosseous muscle, first dorsal. *see* First dorsal interosseous muscle

Interpersonal attraction, 4:535-45
    adult relationship development theories, 4:543-44
    affiliation, 4:537-38
    altruistic love, 4:542-43
    as an attitude, 4:536-37
    attachment, 4:542
    balance theory, 4:538-39
    biological roots, 4:535-36
    childhood relationships, 4:545
    close relationships, 4:544
    cognitive consistency theories, 4:538-39
    as emotion and feeling, 4:537
    emotional physiological linkage, 4:544
    familiarity, 4:542
    liking, 4:540-42
    physical attractiveness, 4:541-42
    physical proximity, 4:541
    romantic love, 4:543
    self-evaluation maintenance theory, 4:544-45
    similarity, 4:541
    social exchange theories, 4:539-40
Interphase
    defined, 5:65
    DNA synthesis in, 5:66
    of mitosis, 2:243-44, 291
Interplexiform cells, 6:626-27
Interstitial cell-stimulating hormone. *see* LH
Interstitial edema. *see* Edema, interstitial
Interstitium
    defined, 5:31
    structure of, 5:35-36
Intertwin transfusion syndrome. *see* Fetofetal transfusion
Intervening sequences. *see* Introns
Interventional radiology. *see* Radiology, interventional
Intestinal absorption, 2:37-38
    aging and, 3:902
    ascorbic acid, 1:430
    in colon, 3:82
    defined, 3:79
    of electrolytes, 3:81
    of fats, 3:81, 572-73
    of fluids, 3:81
    of human milk lipids, 3:47-56, 151-62
    and hypothalmus, 2:37
    of lipids, 3:572
    of protein, 3:80
    of sweetening agents, 3:80
Intestinal diseases
    ascariasis, 1:269-70
    Entamoeba histolytica-associated, 1:222(t), 222-23
    hookworm anemia, 1:263-64
    pinworms, 1:269

Intestinal diseases *(Continued)*
    roundworm infections, 1:269-70
    snail fever, 1:263
    trichuriasis, 1:269-70
Intestinal microflora, newborn infants
    anaerobic bacteria, 1:318
    attachments, 1:318
    bacteriocins, 1:318
    breast feeding vs. cow milk effects, 1:319
    development of, 1:319
    ecology of, 1:318-19
    functions of, 1:319
    mode of delivery effects, 1:319-20
Intestinal motility, 2:38, 4:553
Intestinal mucosa
    cancer in, 3:768
    microcirculation, 4:548-49
Intestinal neoplasms, 2:116
Intestinal obstruction, 3:3
Intestinal secretions, 2:37-38
Intestinal-type gastric malignant tumors
    defined, 3:769
    precancerous state of, 3:770
Intestine, large
    adenocarcinomas of, 3:769
    anatomy, 3:82
    external radiation and, 6:478
    localized cancer in, 3:775
    microbial flora, in formation of secondary bile acids, 1:625
    precancerous state of, 3:770-71
    radiation damage, characteristics of, 6:478
    radiosensitivity of, 6:469-80
    secondary bile acid formation in acute phase response, 1:625
    seed mucilage and, 3:41
    short-chain fatty acids in, 3:43
    stress and motility in, 7:290
    structure and function of, 6:477-78
Intestine, small
    adenocarcinomas of, 3:769
    alcohol consumption and, 1:175
    anatomy, 3:65, 76-77
    autonomic control of fluid and electrolyte secretion, 1:547-48
    biological parameters of, effect of sublethal radiation on, 6:476
    endogenous secretions, 3:77
    galanin in, 5:721
    hormonal activity in, 5:716
    motility of, 3:77-78
    mucosal villi, 3:76

Intestine, small *(Continued)*
    peristaltic reflex, **1:**550
    radiation effects, experimental studies and, **6:**472–73
    radiation of, clinical studies and, **6:**475–77
    radiosensitivity of, **6:**469–80
    stress and motility in, **7:**289–90
    villi, anatomy, **3:**76
    viscous foods and, **3:**41
Intestines, **2:**39
    anaerobic bacteria found in, **1:**241
    anaerobic bacterial infections, **1:**244, 246
    blood flow regulation, **4:**548–57
        absorption influence, **4:**552
        autoregulation, **4:**552
        autoregulatory escape, **4:**551
        cardiovascular response to meals, **4:**553–54
        circulating chemicals, **4:**551–52
        intestinal hyperemia, **4:**554–57
        intrinsic mechanisms, **4:**552
        motility and, **4:**553
        oxygen concentration, **4:**556–57
        parasympathetic neural regulation, **4:**551
        resistance changes, **4:**548
        rhythmic contractions and, **4:**553
        secretion and, **4:**552–53
        splanchnic organs and, **4:**551
        sympathetic neural regulation, **4:**551
        techniques and importance, **4:**550–51
        tissue metabolism, **4:**552
        tonic contractions, **4:**553
        venous-arteriolar response, **4:**552
    bowel bacterial overgrowth syndrome, **1:**244, 246
    brain-gut interaction, **2:**32–38
    calcium metabolism in, **5:**661–62
    circulation and microcirculation, **4:**549–50
    colonization by *Entamoeba histolytica*, **1:**223–24
    in electron microscopic imagery, **3:**274
    gut dimensions, in mammals, **3:**60
    gut smooth muscle, autonomic control of, **1:**549–50
    innervation, **2:**32–33
    irritants, **2:**38
    neuronal circuits in, **1:**549
    phosphate metabolism in, **5:**662
    spinal cord dysfunction and, **7:**231–32
    tissue layers, **4:**550
    tissue metabolism, **4:**556–57
    vitamin D effect on, **7:**867
    zinc metabolism in, **7:**881
Intimacy, **4:**544
Intimidation. *see* Coercion
Intoxication. *see* Alcoholic intoxication
Intracellular messengers. *see* Messengers, intracellular

Intrafusal muscle fibers
    defined, **3:**719
    sensory endings of, **3:**721
    spindles and, **3:**719, 720
Intrahemispheric fiber tracts. *see* Fiber tracts (white matter), intrahemispheric
Intramuscular electromyographic activity, defined, **3:**691
Intrauterine cross-transfusion. *see* Fetofetal transfusion
Intrauterine diagnosis. *see* Prenatal diagnosis
Intrauterine fetal growth. *see* Fetal development
Intrauterine growth retardation. *see* Fetal growth retardation
Intrauterine insemination. *see* Insemination, intrauterine
Intrauterine retention period, defined, **3:**293
Intravascular contrast agents, **3:**4
Intravenous infusions. *see* Infusions, intravenous
Intrinsic factor, in gastric juice, **3:**76
Intrinsic feedback. *see* Feedback, intrinsic
Introns, **2:**294, 295, 479. *see also* Exons
    defined, **3:**791, **4:**13, **5:**205, **6:**685
    of DMD gene, **5:**196
    evolution, **2:**479–80
    of *fos* gene, **3:**702
    gene expression and, **3:**871
    intron-exon borders of DMD gene, **5:**200
    intron-exon ratio of elastin, **3:**265
Intropism, cardiac. *see* Myocardial contraction
Introversion, biological basis of, **5:**774
Inverse PCR. *see* Polymerase chain reaction
Inversion, as joint movement, **5:**133
Inversion, chromosomal, **2:**445
    defined, **3:**822
Invert sugar, **2:**101
Inverted repeat sequences. *see* Repetitive sequences, nucleic acid
Iodine
    biological function of, **7:**606
    deficiency
        consequences of, **7:**609
        diseases associated with, **7:**486
    maternal, fetal thyroid gland and, **3:**340
    metabolism, thyroid follicular cells and, **7:**485–87
    requirements in the United States, **7:**486
    significance of, in body, **5:**49
    thyroid hormone removal of, **7:**490
    toxicity, **7:**609
Iodine radioisotope, **4:**371–72
Iodine radioisotopes, **6:**457–58
5-Iododeoxyuridine, **2:**396
    and adenosine deaminase, **2:**397
Iodoquinol, **2:**382
Iodothyronines, **3:**339

Ion channel receptors, ligand-gated. *see* Receptors, ligand-gated ion channel
Ion channels, **2:**236, 287–89, **4:**559. *see also* Biological transport, active
    astrocyte, **1:**438–39
    calcium channel, **2:**310–12
    and communication, **2:**287–89
    compared to pumps, **4:**559
    defined, **2:**279, **5:**415, 305
    G proteins and, **4:**17, **5:**417, 420
    5-HT$_3$ receptors and, **5:**422
    intercellular, **2:**287
    intracellular, **2:**287–89
    ion transport, **4:**558
    Nicotinic acetylcholine receptor and, **5:**416
    for sodium, **2:**288–89
    transmembrane channels, **4:**973
    voltage-sensitive, **5:**420–21
Ion exchange, **2:**89, **4:**218
Ion-exchange chromatography, **4:**208
Ion pump. *see* Biological transport, active
Ion-selective electrodes, **3:**202
Ionization potential, defined, **5:**903
Ions
    binding to transport proteins, **4:**560–61
    and calcium, **2:**284–86
    defined, **2:**279, **3:**317
    intracellular, in sperm capacitation, **3:**602
    and ion concentration gradient, **2:**283
    for sodium/potassium. *see* Calcium
    transport in cystic fibrosis, **2:**732
Iontophoresis, **2:**306. *see also* Microiontophoresis
4-Ipomeanol, **7:**549
Iprindole, **2:**843
Iproniazid, **2:**843
Iris
    anatomy, **3:**540
    autonomic innervation, **1:**544–45
Iron
    absorption, **2:**656
    and anemia, **2:**654
    in antiquity, **2:**653
    biochemistry, **2:**649–60
    biological function of, **7:**606
    in bone marrow cell, **2:**656
    in catalase, **2:**653
    in chlorosis, **2:**653
    in cytochrome, **2:**653
    and desferrioxamine for chelation of, **2:**655
    excess, **2:**654–55
        in African beers, **2:**655
        and cirrhosis, alcoholic, **2:**655

Iron
    excess *(Continued)*
        and hemochromatosis, **2:**654–55
        and idiopathic pulmonary hemosiderosis, **2:**655
        and nocturnal, hemoglobinuria, **2:**655
        and rheumatoid arthritis, **2:**655
        in thalassemia, **2:**655
    and hemoglobin, **2:**653
    and hemorrhage, **2:**653
    hypoferremia and, **2:**650
    imbalance, immunity and, **5:**513
    and Maillard browning, **2:**656
    metabolism, **2:**649–60
    in oxidation reduction reactions, **5:**52
    significance of, in body, **5:**49
    therapeutic use, **2:**653
    toxicity, **1:**378, **7:**610
    use and excretion of, **5:**52
Iron, dietary. *see* Dietary iron
Iron chelation, **2:**655
Iron deficiency, **2:**653–54
    and anemia, **2:**654
    consequences of, **7:**610
    as diet-related disease, **5:**487
Iron deficiency anemia. *see* Anemia, hypochromic
Iron-protoporhyrin IX, **4:**135–36, 141
Ischemia, **2:**167
    defined, **2:**167
    free radical damage during, **7:**358–59
    intestinal blood flow and, **4:**553
    tissue, **7:**505
Ischemic heart disease. *see* Coronary disease
Ischium. *see* Pelvis
Islet-activating protein. *see* Pertussis toxins
Islet cell tumor, ulcerogenic, symptoms of, **5:**721–22
Islets of Langerhans
    beta-cell destruction in, **7:**571–72
    carcinogenesis and, **7:**573
    disorders, **3:**353
    drug-induced diabetes in, **7:**571–72
    induction of glucose intolerance and, **7:**572–73
    mechanisms of toxicity in, **7:**571
    response to injury of, **7:**571–73
    structure and function of, **7:**570–71
    toxicity in, **7:**570–73
Isoascorbic acid. *see* Erythorbic acid
Isobologram of drug, **2:**371
Isocarboxazid, **1:**309–10
Isochores, **4:**253–55
    characteristics of, **3:**874
    chromosome bands and, **4:**254–55
    functional aspects, **4:**256

Isochores *(Continued)*
  genes and, **4:**254
  genome evolution and, **4:**256-57
  genome functions and, **4:**255-56
  integration of mobile and viral DNA sequences, **4:**255
  translocation breakpoints and fragile sites, **4:**255-56
Isochromosome, **2:**1, 4, 8, 437, 440, 441, 444
Isocitrate, **2:**508-9
Isocitrate dehydrogenase, **2:**509
Isocitrate lyase, **2:**508
Isocortex, **2:**673-77, 682
Isocratic liquid chromatography, **4:**212-13
Isodesmosine, **2:**206
Isodisomy, **2:**439
Isoelectric focusing, **3:**408
Isoenzymes, **4:**573-80
  amino acid sequencer, **4:**576
  analytical methods, **4:**577
  applications of analysis, **4:**577-80
  in cancer tissue, **4:**578-79
  chromosomal locations, **4:**574
  deficiency, **4:**578-79
  derivation, **4:**573-75
  developmental aspects, **4:**576-77
  function, **4:**576
  as genetic markers, **3:**828
  glucose-6-phosphatase dehydrogenase deficiency, **4:**578-79
  inborn errors of metabolism, **4:**578-80
  oligomeric, **4:**574-75
  organ and tissue specificity, **4:**578
Isoforms
  defined, **5:**55, 161, 205
  neonatal, secondary myotubes and, **5:**164
Isoforms, muscle-specific
  functional differences between, **5:**164-65
  functions of, **5:**163-65
  temporal/spatial specificity of, **5:**164
Isoleucine, **2:**507
3-Isoleucine,8-leucine vasopressin. *see* Oxytocin
Isoleucyl, leucyl vasopressin. *see* Oxytocin
Isoluminance, **2:**580, 581, 589
Isomere, dental, **2:**789
Isomerism, drug kinetics and, **5:**848-49
Isomerism, geometric. *see* Geometric isomerism
Isomers, optical. *see* Optical isomers
Isometric contraction
  aging and, **3:**904
  cardiovascular adjustments to, **3:**520
  defined, **3:**515, 691
  muscle activation in, **3:**692-96

Isometric contraction *(Continued)*
  skeletal muscle endurance and, **7:**31
  torque and, **3:**697
  voluntary movement and, **3:**697-98
Isometric tension, maximum. *see* Maximum isometric tension
Isoniazid
  mechanism of action, **1:**327
  medical uses of, **1:**329
  for tuberculosis, **5:**706
Isoniazid acetylation polymorphism, **5:**832
Isonicotinic acid hydrazide, **2:**569. *see* Isoniazid
Isonymy, inbreeding measurement, **4:**434-35
Isopentenyl, **2:**415
  pyrophosphate, **2:**415
  tRNA, **2:**415
Isoprenaline, **2:**848
  and airway resistance, **6:**362
Isoproterenol, **5:**382
Isoquinoline alkaloids, **1:**179-80
Isosceles tolerance triangle. *see* Acceptance triangle
Isosexual sexual precocity, defined, **3:**357
Isosorbide, **3:**100
Isosorbide dinitrate, **2:**172
Isospora infection, **6:**279
Isotonic contraction
  defined, **3:**691
  skeletal muscle endurance and, **7:**31
Isotopes, **3:**208
Isotretinoin, birth defects caused by, **1:**701
Isotretinoin teratogenicity syndrome, **1:**698
Isotypes, defined, **3:**801
Ispaghula husk. *see Plantago ovata*
Isradipine, **2:**172
Issue-relevant thinking. *see under* Thinking
Isteria monocytogenes, in food, **3:**665
Ivermectin, **1:**265, **2:**387, 388, 392-93
Ixodes ticks, and malaria, **6:**277

J

Jacobi, Abraham, **5:**700
James' spots, **4:**875
Japan, concept of individualism in, **4:**703
Jaundice, **6:**121
Jaw
  embryonic, diagram of, **2:**817
  movement, **2:**780
Jejunum, **2:**418
  absorption in, **3:**79
  anatomy, **3:**65, 76
  fat absorption in, **3:**573

Jejunum *(Continued)*
    passive transfer in, **3:**81
    proximal, ascorbic acid absorption, **1:**430
Jellyfish toxins, **7:**595
Jendrassik maneuver, defined, **3:**723
Jet lag, **2:**494
    ACTH secretion and, **5:**398
    and sleep disorders, **7:**81
Jews, Gaucher disease in, **3:**779–80, 781
Joint angle
    muscle mechanical advantage and, **3:**695–96
    voluntary movement and, **3:**697
Joint moment
    at ankle, **5:**189
    defined, **5:**187
Joints
    afferents, reflexes involving, **6:**159–60
    between bones. *see* Articulations
    movement. *see* Proprioceptors
    shapes of, **1:**789
    skeletal muscle and, **5:**131–33
    synovial, **2:**P213
Judgment, mood effects, **1:**110
Jump, vertical. *see* Vertical jump
Junction, **2:**261–78
    adherens, **2:**268
    anchoring, **2:**268–70
    cellular, **2:**267–78
    dentogingival, **2:**777–78
    desmosome, **2:**267, 268
    gap-, **2:**274–78
    intermediate, **2:**132
    occluding, **2:**271–73
    septate, **2:**268
        structure, **2:**268
    tight, **2:**271–73
Junctional region of antibody molecules, **1:**564
Jung, Carl Gustav, **3:**193
Junk DNA, **2:**478
Juvenile delinquency, **3:**844–45
Juxtaglomerular cells, defined, **5:**725

K

K cells, cytotoxic destruction of antigenic target cell by, **1:**299
K^+ channels, G proteins and, **4:**17
K^+ current, in olfactory receptor cells, **5:**526
K^+ voltage-sensitive channels, **5:**421
85K protein *see* Proteins, 85K
Kainate, **2:**313, 314, 339, 523, 525

Kainate glutamate receptors. *see* Receptors, glutamate, kainate subtype of
Kainic acid model, of Huntington's disease, **4:**272
Kallidin, **5:**730
Kallikrein-kininogen-kinin system, **4:**589
Kallikreins
    characterization, **4:**590–91
    salivary, function of, **6:**695
Kanamycin nucleotidyl transferase, **3:**420
Kanjii, dyslexia-dysgraphia and, **3:**221
Kanner's syndrome. *see* Autism, infantile
Kaposi's sarcoma, in homosexuals, **4:**345
Karatinocyte, **2:**753
Karyokinesis. *see* Mitosis
Karyotyping, **2:**437, 445–48, 475
    abnormalities in leukemia, **4:**692
    of atomic bomb victims, **5:**248
    defined, **2:**465, **3:**135, 175
    in Down's syndrome diagnosis, **3:**176
    male, human, stained, **2:**466–67
    restriction fragment-length polymorphisms and, **3:**149
Kaschin-Beck disease, **6:**796
Katz Adjustment Scale, **1:**611
Kawasaki syndrome, **4:**462
Kayser-Fleischer ring in cornea, **2:**651
Kearns syndrome, **5:**62
Keloid, **2:**574
Keratan sulfate, **6:**258
Keratin, **2:**780
    formation, role of vitamin A in, **7:**850
    orientation of microfibrils in, **7:**43
    properties of, **5:**224
    structure of, **6:**221
Keratin sulfate, **2:**128, 206, 207, 573
Keratinization, hair, **4:**40
Keratinocytes
    Ad12-SV40 hybrid virus, **4:**585
    adenovirus E1A, **4:**585
    chemical carcinogens and physical agents, **4:**586
    cultivation, **4:**582
    cytokine expression of, **3:**449–50
    defined, **3:**445
    differentiation, **4:**582
    disaggregated cultures, **4:**582–83
    epidermal cell grafts, **4:**583
    explant cultures, **4:**582
    holoclones, **4:**583
    local regulation of, **3:**446
    malignant cell transformation, **4:**583–84
    malignant transformation, **4:**584–86
    meroclones, **4:**583
    murine, **3:**447

Keratinocytes *(Continued)*
    papillomaviruses and, **3:**861, 864, **4:**585–86
    paraclones, **4:**583
    signal transduction mechanisms of, **3:**452
    simian virus 40, **4:**585
    TGF-alpha and, **3:**447
    transfection with SV40, **4:**585
    transformation, human, **4:**581–86
Keratoses, **7:**48
Kernel of truth hypothesis, **7:**234
Keshan disease
    in China, **3:**21
    and selenium deficiency, **6:**795–96
Ketalar. *see* Ketamine
Ketamine, **2:**534
    phencyclidine and, **5:**446
Keto acids, **3:**563
Ketoacidosis, diabetic. *see* Ketosis, diabetic
Ketoconazole, **1:**327, **2:**745
Ketogenesis, **3:**563
Ketone bodies
    defined, **3:**561
    extracellular acidosis and, **5:**827
    fatty acids and, **3:**324
    in fetal nutrition, **3:**48
Ketones
    defined, **3:**317
    neurotoxicity of, **7:**548–49
Ketoprofen, **2:**785
Ketose phosphate. *see* alpha-Ketol
Ketosis, causes of, **3:**575
Ketosis, diabetic, pH and, **5:**824
17-Ketosteroid reductase, **7:**252
Ketosteroids, **2:**56
Ketosuccinates. *see* Oxaloacetates
Kidney, **2:**141–44
    acid-base homeostasis and, **5:**828
    aging of, **3:**902
    anatomy of, **7:**733–34
    artificial. *see* Artificial kidney
    atrial natriuretic factor and, **1:**473
    autoimmunity and, **1:**529
    autonomic innervation of, **1:**543
    calcium, parathyroid hormone regulation of, **5:**661
    capillaries of, **5:**32, 38
    dialysis. *see* Dialysis; Hemodialysis; Peritoneal
        dialysis
    dysfunction in essential hypertension, **4:**291–92
    filtration of blood, **4:**570–71
    function, mineralcorticoids and, **7:**270–72
    heart-lung machine complications, **4:**95–96
    in lactate disposal, **5:**827

Kidney *(Continued)*
    organic chemicals toxic for, **7:**550
    pH homeostasis and, **5:**878–79
    phosphate, parathyroid hormone regulation of, **5:**662
    platelet-activating factor and, **6:**50
    radiation effects, **6:**413
    secondary hypertension, **4:**289
    vitamin D metabolism in, **7:**862
    zinc metabolism in, **7:**882–83
Kidney, polycystic, **3:**142
Kidney diseases. *see also* Balkan nephropathy
    in children, **5:**708–9
    mercury-induced, **3:**387
    vitamin D and, **7:**870
Kidney failure
    artificial cells in treatment of, **1:**379
    changes in drug kinetics and, **5:**854
    hyperkalemia and, **3:**99
Kidney neoplasms, **2:**451
    renal-cell carcinoma, **4:**347
    renal cell tumor, **2:**118
Kidney transplantation, in children, **5:**709
Kidney tubules. *see also* Tubular reabsorption; Tubular
        secretion
    at birth, parathormone and, **3:**
    pH_i and, **5:**875
Kidney tubules, proximal, **3:**95
Killed vaccines. *see* Vaccines, inactivated
Killer cell helper factor. *see* Interleukin 2
Killer cells, **2:**756
    activated, **5:**293, 302
    assays for, **5:**291–92
    cell-mediated cytotoxicity of, **5:**291–92
    major histocompatibility complex restricted, **5:**292
    mechanisms of lysis by, **5:**292
    non-major histocompatibility complex restricted, **5:**292
    targets, **5:**291
Killer cells, lymphokine-activated, **1:**677 679–80, **2:**756,
        4:340
    adherent, **1:**680
    adoptive immunotherapy with, **1:**679
    characteristics of, **5:**302
    derivation of, **5:**293
    IL-2-activated, **1:**679–80
    neuropeptides and, **5:**384–85
    tumor rejection, **4:**340
Killer cells, natural, **1:**677, **2:**756
    characteristics of, **4:**166, 812, **5:**299
    conjugates formed by, **5:**298
    as early defense mechanisms, **5:**294
    functions, **4:**166, **5:**302–3
    general properties of, **5:**295–99

Killer cells, natural (*Continued*)
    genetics of, **5**:300
    graft rejection, **4**:362
    history of, **5**:294
    interleukin-1 and, **5**:385
    kinetics for lysis by, **5**:299–300
    lineage of, **5**:296–98
        viewpoints concerning, **5**:294–95
    as lymphokine producers, **5**:293–94
    lytic mechanisms of, **5**:301
    lytic molecules produced by, **5**:299
    mechanisms of activity, **5**:298–99
    and natural cytotoxic cells, surface phenotype differences between, **5**:301
    ontogeny and maintenance of, **5**:300
    and other effector cells, **5**:291–304. *see also* Killer cells
    phenotypic properties of, **5**:295–96
    receptors utilized by, **5**:298–99
    regulation of, **5**:301
    regulation of activity, **5**:298
    surface antigens of, **5**:296
    surface-marker phenotype of, **5**:297
    surface phenotype of, **5**:301
    target antigens for, **5**:299
    target type for, **5**:300
    terminology, **5**:292–93
    THC effects, **4**:909
    tumor cell destruction, **4**:1012
    tumor rejection, **4**:339–40
    virus-infected targets of, **5**:301–2
Killer phenotype. *see* Phenotype
Kilobases, defined, **3**:135
Kimura figures-instruction
    defined, **3**:215
    theta activity and, **3**:216
Kin selection, as sociobiological mechanism, **1**:575
Kinase. *see* Protein kinases
Kindling. *see* Affect
Kinesin, **2**:242
    in anterograde axonal transport, **5**:19–20
Kinesthesia. *see* Proprioception
Kinetechores, defined, **5**:56
Kinetic dysgraphia. *see* Dysgraphia, efferent
Kinetics
    of DNA binding, **3**:120
    of enzyme function, **3**:425–28
    of nuclear protein transport, **5**:456
    thermodynamic aspects of, **3**:424–25
Kinetochores, **2**:244, 473
    defined, **5**:70
    fibers, chromosomes and, **5**:70

Kinetochores (*Continued*)
    in metaphase, **5**:70
    traction fiber theory on, **5**:71
Kininases, **4**:592–93
Kininogens, **4**:590, 591–92
    characteristics of, **5**:730
Kinins, **4**:589–601
    in allergic response, **1**:189–90
    assays, **4**:593–94
    blood flow regulation by, **4**:594
    and blood pressure regulation, **4**:595
    central nervous system effects, **4**:595–96
    disease process and, **4**:591
    formation, **4**:590, 592
    functions, **5**:730
    gastrointestinal functions, **4**:594
    inflammation role, **4**:470–71
    ion/water transport and, **4**:594
    in pathology, **4**:597–601
    PNS effects, **4**:595–96
    smooth muscle effects, **4**:595
    structure, **4**:590
    vascular tissue and myocardium and, **4**:594–95
Kinky hair syndrome, **2**:850–51, **3**:269
Kinurenic acid, **2**:313
Kirsten murine sarcoma virus, **4**:585
Kjeldahl method, **6**:192
Klebsiella, **4**:450
Kleitman, Nathaniel, **3**:193
Klenk, Ernst, **3**:725, 737, 732
Kleptomania, **4**:998
Klinefelter's syndrome, **2**:713, **3**:821
Knee joint, **1**:376
Knee ligaments, **3**:12
Kneecap, **7**:38
Knight, **5**:139
Knowledge
    abstract, **2**:556
    and language, **2**:556
    and mathematics, **2**:556
    source of, **2**:549
    studying and learning strategies, **7**:310–11
Knowledge bases (computer). *see* Artificial intelligence
Koilocytes, defined, **3**:861
Koilocytosis, **3**:862
Kokuten-byo. *see* Black spot disease, cereal
Kolmogorov forward equations, **4**:924
Korotokoff sounds, **4**:289
Korsakoff syndrome, **2**:764, **4**:677
Koshland, **5**:138
Koshland, Daniel E., Jr., **5**:72–73
85Kr, as tracer, **3**:758

Krebs, Hans, **3**:932
Krebs cycle. *see* Citrate cycle
Kringle domains, **4**:184
Kuhl, W., **3**:778
Kuhn, R., **3**:726
Kupffer cells, **6**:608
Kuru, **1**:214
Kwashiorkor, **4**:894
    defined, **3**:17
    protein undernutrition and, **6**:235
    in young children, **3**:26
Kynurenic acid, **2**:524, 525

L

L-aminophosphonobutyric acid receptors. *see* Receptors,
        L-aminophosphonobutyric acid
L-DOPA. *see* Levodopa
L-gamma-glutamyl-L-alpha-aminobutyrylglycine, **3**:910
L-region, in micturition control, **5**:45
Labeling, food. *see* Food labeling
Labia, anatomy of, **6**:564
Labor, **2**:403–9
    beta-Lactamase, **2**:372
    obstructed, **4**:916, 918–19
    and penicillin, **2**:372
    perinatal endocrinology and, **3**:347–50
    stages of, **2**:403–8
    timing of, in rodents, **2**:498
Labor, premature, **2**:87
Labor onset, **3**:338
Labor-saving devices, **3**:22
Laboratories, immunology, **5**:686
Laboratories, molecular diagnostic, **5**:686
Laboratory medicine. *see* Pathology, clinical
Lacrimal apparatus
    anatomy, **3**:547
    autonomic innervation of, **1**:547–48
Lactalbumin, **4**:605
Lactam antibiotics. *see* Antibiotics, lactam
Lactams, beta. *see* Beta-lactams
Lactase deficiency
    congenital, **4**:617–18
    secondary, **4**:618
Lactases, **2**:104
    biosynthesis, **4**:616–17
    digestibility, **2**:105–6
    genetics of persistence, **4**:616
    intolerance, **2**:104–5
    processing, **4**:616–17
Lactate dehydrogenase, **4**:574–76
    pattern in tumors, **4**:578

Lactates, **2**:106
    bicarbonate and, **5**:825
    in fifth-stage glycolysis, **3**:938–40
    glucose entry and, **5**:824
    hepatic intracellular pH and, **5**:824
    kidney and, **5**:827
    in membrane transport, **5**:823
    pyruvates and, **5**:822
Lactation, **4**:603–9. *see also* Milk, human
    adrenal steroids, **4**:606
    bile salt-stimulated lipase synthesis and, **3**:50
    diet and, **3**:659, **4**:605
    drug excretion and, **5**:847
    fertility and, **4**:607–8
    growth hormone, **4**:606
    hormones and initiation, **4**:605–6
    insulin and, **4**:606
    maintenance, **4**:606
    milk-ejection reflex, **4**:607–8
    milk synthesis, **4**:604–7
    prolactin, **4**:606
    thyroid hormones, **4**:606
    zinc and, **7**:888
Lactic acid
    defined, **3**:317
    in food spoilage, **3**:663
    in glycolysis, **3**:932
    in yogurt, **3**:667
Lactic acid bacteria, **3**:668
Lactobacillus bulgaricus, **3**:667
Lactoferrin, **2**:655
    salivary, function of, **6**:695
Lactogen, placental, **2**:259
Lactogenic hormone, pituitary. *see* Prolactin
Lactoperoxidase, **2**:655
    antibody conjugates with, **4**:372–73
    in milk preservation, **3**:663
Lactose, **4**:611–21
    in breast milk, **4**:605
    consumption
        adaptation to, **4**:619
        lactase activity and, **4**:617
    digestion and absorption, **4**:612–13
Lactose intolerance, **4**:613–15, 618–19
    Ascaris infection and, **4**:117
    diagnosis, **4**:619–20
    management, **4**:620–21
    relation to intolerance, **4**:618–19
Lactose synthetase, **4**:605
Lactotrophs, **4**:606, **6**:14–16
S-lactoyl glutathione, **3**:915
Laetoli site, fossils of, **3**:490

Lafora disease. *see* Epilepsy, myoclonus
LAK cells. *see* Killer cells, lymphokine-activated
Lambda repressor, recognition sequences for, **3:**122, 313
Lambert-Eaton syndrome, **5:**258
Lamin, **2:**641, 642
Lamina cribrosa, **3:**550
Lamina fusca, **3:**536
Laminae, **2:**295-96
   chemoarchitecture
      of dorsal horn, **7:**202-3
      of intermediate gray matter, **7:**203-4
      of motor nuclei, **7:**205
   in prometaphase, **5:**70
   spinal cord
      classification of, **7:**197-98
      segmental variations of, **7:**198-99
Laminin, **2:**566, **4:**623-31
   in basement membranes, **4:**624
   biological activities, **4:**626-28
   biologically active sites on, **4:**627-28
   in cell migration, **3:**524
   cellular receptors, **4:**630-31
   effects on neuronal cells, **4:**628-30
   elastin and, **3:**267
   functions of, **3:**526, 528
   in gingiva, **4:**36
   integrins and, **4:**630-31
   localization, **4:**626
   nerve regeneration and, **5:**345
   as platelet receptor, **6:**56-57
   structure, **4:**625-26
   tumor cell receptors, **4:**1013
Landmarks, **2:**689
   chromosomal, **2:**445
   defined, **2:**689
Landsteiner, K., **3:**725
Langer-Giedion syndrome, **2:**440
Langerhans cells
   radiation effects, **6:**467
   regional immunology, **4:**396
Language, **2:**9, 557, 598, 623, 628, 874, **4:**633-39
   and aging, **1:**119-27, **4:**637
   anatomy for, **2:**598
   animal models, **1:**258
   brain systems and, **3:**224-27
   Broca's area, **2:**862
   competence, in older adults, **1:**124-27
   evolution, **4:**633-39
   formulation, disorders of, **7:**161
   frontal lobe activation, **1:**483
   hominid, **3:**503
   lateralization, **4:**634

Language *(Continued)*
   as measure of population differentiation, **6:**103-4
   neocortical lateralization and, **5:**313, 314
   in older adults, **1:**119-21
   origins of, **4:**635
   pathology, **4:**202, **7:**153-63
   performance, variability in older adults, **1:**124-27
   and plasticity, hemispheric, **2:**861
   Wernicke's area, **2:**862
Language development, **7:**154
   birth to 3 months, **1:**575
   3 to 10 months, **1:**575
   10 to 20 months, **1:**576
   20 to 36 months, **1:**577-78
   3 to 5 years, **1:**578
   5 to 8 years, **1:**580
   8 years to preadolescence, **1:**581
   biological bases of, **4:**634-35
   differentiation vs. integration of processes, **1:**582
   general stages vs. specific cases, **1:**582
   in infants, **4:**635-36
   meaning, **4:**636
   oral skills, **4:**637
   role of shared heredity, **1:**582
   syntax, **4:**636-37
   theories of, **4:**633-34
   through life span, **4:**637
   written skills, **4:**637
Language development disorders
   acquired, **4:**638
   congenital, **4:**638
   in fragile X children, **3:**716
Language disorders, **7:**160-62
   acquired, **4:**638
   childhood, **7:**160-62
   childhood, acquired, **7:**162
   congenital, **4:**638
   expressive, **7:**161-62
   in older adults, **1:**121-24
   receptive, **7:**160
   relationship of autism to, **1:**507
Language therapy, **3:**716
Lanosterol, **2:**745
Lanugo hairs, **4:**40
Laparoscopy
   defined, **3:**361
   in endometriosis diagnosis, **3:**366
   in endometriosis surgery, **3:**367
Laparotomy, **3:**367
Large intestine. *see* Intestine, large
Larva migrans, **2:**388

Laryngeal muscles
    extrinsic, **4:**657
    intrinsic, **4:**656–57
Laryngeal neoplasms, **1:**123
Laryngeal nerves, **4:**657–58
Laryngeal paralysis. *see* Vocal cord paralysis
Laryngeal reflexes, **4:**647–52
    cardiovascular, **4:**650
    deglutition, **4:**649–50
    developmental changes in, **4:**651–52
    phonatory, **4:**650–51
    respiratory, **4:**647–49
Laryngeal ventricle, **4:**654
Laryngospasm, **4:**649
Laryngotracheitis, **4:**460
Larynx, **4:**653–63
    airflow regulation by, **4:**647–48, 660
    airway obstruction of, **6:**358–59
    anatomy, **4:**653–58, **6:**589–90
    closure of, **4:**650
    development, **4:**657–58
    elevation of, **4:**650
    embryology, **4:**657–58
    functions, **4:**660–63
    mucosal cover, **4:**657–58
    phylogeny, **4:**659–60
    quadrangular membrane, **4:**656
    resonance, **4:**662
    segmentation, embryonic, **4:**658
    speech function in, **6:**359
Laser Doppler technique, **3:**759
Laser flow cytometry. *see* Flow cytometry, laser
Lassitude. *see* Fatigue
Late promoters. *see* Promoter regions (genetics)
Latent infection membrane protein. *see* Proteins, latent
    infection membrane
Latently obese. *see under* Obesity
Lateralization, cerebral, **2:**351, 871–74
    confused in schizophrenia, **2:**865
    dysfunction, **2:**861
    evidence, anatomical, **2:**862–63
    handedness, **2:**862
    in schizophrenia, **2:**861
    and tachistoscopy, **2:**862
Lateralization, neocortex
    in animals, **5:**313–14
    defined, **5:**305
    functional consequences of, **5:**313–14
Latex bead test, **1:**555
Lathyrism, **2:**346, 347, 571
    as cause of amyotrophic lateral sclerosis, **1:**233
    spastic paralysis of leg, **2:**347

Lathyrus odoratus (sweet pea), **2:**564, 571
Lathyrus sativus, **2:**346, 347
    aggression, **2:**839
    lead, **2:**346
    manganese, **2:**346
    methyl alcohol, **2:**346
    methyl bromide, **2:**345
    methyl mercury, **2:**345, 346
    nitrate, **2:**346
    organophosphate, **2:**346
    physical, **2:**341
    tetrodotoxin, **2:**345
    triethyltin, **2:**345
    triorthocresyl phosphate, **2:**346
    vincristine, **2:**346
    in work place, **2:**342
Lats receptors. *see* Receptors, thyroid hormone
Laundry starch. *see* Starch, laundry
LDL apheresis, **1:**454
Lead, **2:**345
Lead poisoning, **2:**345
    of children, **5:**710–11
    effects of, **3:**386–87
    entry routes of, **3:**386
    pyrimidine metabolism and, **6:**398
Leader sequences, defined, **3:**777
Leader signal peptides. *see* Signal peptides
Leakey, M. D., **3:**496
Learned helplessness. *see* Helplessness, learned
Learning, **2:**550
    and advertising, **2:**636
    associative, of *Hermissendra*, **2:**306–10
        defined, **2:**305
        disorder, genetic, **2:**665–71
        metabolism, **2:**664–71
    biofeedback parameters affecting, **1:**664
    classical conditioning, **1:**661–62
    defined, **4:**674
    effects, defined, **5:**122
    generalized motor programs and, **5:**126
    information feedback for, **5:**126–28
    and memory, **4:**673–78
    mouse model, **2:**672
    network modeling of, **5:**390–91
    operant conditioning, **1:**661–62
    particles, **2:**664–71
    perception and development of, **5:**748
    prenatal, **5:**806
    preserved, and amnesia, **5:**402–3
    principles of, **5:**123–28
    receptor, **2:**416–17, 665–66
    scripts for social behavior, **1:**8–9, 1058

Learning *(Continued)*
"supervised," in back-propagation, **5**:391
and Tangier disease, **2**:667
tests of, **5**:122–23
Lecithin-cholesterol acyl transferase, **3**:573
Lectins
cell membrane carbohydrates as receptors for, **6**:519
Entamoeba histolytica, **1**:218, 223
as plant toxins, **7**:594
Lectins, triticum vulgare. *see* Wheat-germ agglutinins
Ledeen, **3**:726
Leeks, NSP content of, **3**:40
Leg, muscles of, **4**:266
Legal medicine. *see* Forensic medicine
Legislation. *see also* Regulatory standards, toxicological
Legislation, food, advances in, **3**:29
Legumes, **3**:654–55
Leguminous seeds, nonstarch polysaccharides in, **3**:37
Leiomyoma
defined, **3**:765
of gastrointestinal tract stromal tissues, **3**:770
Leiomyosarcoma, **7**:702
defined, **3**:765
of gastrointestinal tract stromal tissues, **3**:770
Leishmania, **4**:448–49
biochemistry, **4**:682–83
cell biology, **4**:681–82
cultivation in vitro and in vivo, **4**:681
host cell entry by species of, **5**:646–47
molecular mechanism of parasitism, **4**:684–85
taxonomy, **4**:681
Leishmania tropica, **5**:651
Leishmaniasis, **4**:448–49, 679–85, **6**:271–72
causative agents, **4**:681–84
control of, **4**:680–81
cutaneous, **5**:655, **6**:272
epidemiology, **4**:679–80
immune deviation, **4**:388
immunology, **4**:683–84
mucocutaneous, **6**:272
pathology, **4**:680
vaccination against, **4**:390
visceral, **6**:272–73
Lejeune, Jerome, **3**:176, 182
Lemniscus, lateral, **3**:240
Length-tension curve, muscle. *see* Muscle, skeletal, length-
tension curve of
Lennox-Gastaut syndrome, epilepsy and, **3**:458
Lens, crystalline, **3**:544–45
Lens proteins. *see* Alpha-1 crystallins
Lentigo maligna melanoma, **4**:890
Lentiviruses, **6**:646–47

Leprosy, **4**:449
antimicrobial drugs, **1**:326, 329
disease models, animal, **1**:255–56
Leptomeninges
characterization, **4**:985
defined, **4**:979
functions, **4**:990
Lesch-Nyhan syndrome
genetic aspects of, **3**:855
mutation rate of, **5**:245
Lethal dose 50 (LD$_{50}$), for radiation sickness, **6**:411
Lethal equivalents, inbreeding, **4**:439
Leucine zippers
defined, **3**:701
in heterodimer formation, **3**:706
in transacting factors, **3**:113
Leukemia, **4**:687–96
in children, **5**:704
classification and types, **4**:688–89
clinical description, **4**:694–95
diagnosis, **4**:693–94, **6**:529
Down's syndrome and, **3**:183
etiology, **4**:691–92
growth kinetics, **4**:922–23
hypoglobulinemia, **4**:693
immunotoxin therapy, **4**:421–22
incidence, **4**:689
and atomic bombs, **6**:417
in utero diagnostic radiation and, **6**:416
kinetics, **4**:692–93
morphology, **4**:693
radiation-induced, **4**:344
smoldering, **4**:694
therapy, **4**:695–96
lithium effects, **4**:751–52
Leukemia, hairy cell, **4**:751
Leukemia, lymphoblastic, acute. *see* Leukemia, lymphocytic, acute
Leukemia, lymphocytic, **4**:677, 681, 695
Leukemia, lymphocytic, acute, **1**:794, **4**:688, 689
Leukemia, lymphocytic, chronic, **4**:688, 693, 695
Leukemia, myelocytic, acute
bone marrow transplantation for, **1**:794–95
genetic locus of, **3**:183
Leukemia, myelocytic, chronic, **4**:676, 679–80, 695
Leukemia, myelogenous, acute. *see* Leukemia, mye-
locytic, acute
Leukemia, myeloid, acute, **4**:688–89
Leukemia, myeloid, chronic, **1**:794–95, **4**:692
Leukemia, T-cell, **4**:342, 532, **6**:647
Leukemia inhibitory factor (LIF), **1**:28, 32
Leukemia retroviruses, **6**:644–46

Leukocyte adhesion deficiency, **4**:475
Leukocyte antigens. *see* HLA antigens
Leukocyte-endothelial cell interaction, **4**:475–76
Leukocytes
    accumulation in inflammation, **4**:476
    flow cytometry of, **3**:624
    heart-lung machine complications, **4**:96
    lithium effects, **4**:749–50
    nitric oxide and, **3**:375
    passenger, **4**:361
Leukocytes, polymorphonuclear. *see* Neutrophils
Leukocytosis, **1**:30–31
Leukodystrophy, as myelin disorder, **5**:282
Leukopenia, defined, **5**:259
Leukoplakia, hairy, **5**:613
Leukotriene C4 (LTC4), **1**:186–89
Leukotrienes
    in acute phase response, **1**:26, 27
    functions of, **5**:507–8
    in gastric mucosal blood flow regulation, **3**:760
Levallois technique, **3**:497–98
Levamisole hydrochloride
    ascariasis treatment with, **1**:270
    immunopotentiating effects, **1**:684
Levene, P. A., **3**:725
Levers
    in body, operation of, **5**:132
    skeleton movements as examples of, **5**:131–33
Levodopa, **2**:12, 13, 23, 28, 218, 833
    for Parkinson's disease, **5**:672
    synthesis, in adrenal medulla, **1**:643
Levulose. *see* Fructose
Lewis antigen, **4**:381
Lewy bodies, in Parkinson's disease, **5**:668
Leydig cells, **6**:839
    characteristics of, **7**:432–34
    electron micrograph of, **7**:432
    function, luteinizing hormone and, **7**:432–34
    testosterone synthesis by, **7**:253
LFA-1, **4**:475–76
LH, **2**:20, 497, 21, 26, 27, 247, 498
    effect on fertilization, **1**:1–4
    fetal, **3**:342
    and folliculogenesis, **7**:346–47
    function of, **6**:88
    lactational amenorrhea, **4**:608–9
    late-stage follicles and, **3**:634
    and Leydig cell function, **7**:432–33
    and menstrual cycle, **2**:498, **6**:556–57
    in menstruation, **7**:739
    oocyte maturation induction and, **5**:557
    perinatal, **3**:348

LH (Continued)
    in polycystic ovarian disease, **3**:630
    secretion
        effect of hypothalmic dysfunction on, **1**:743
        serotonin and, **6**:833–34
        steroid secretion and, **7**:263–64
        surge, **3**:633–34
LH/Choriogonadotropin hormone, **1**:86
LH-FSH releasing hormone. *see* Gonadorelin
LH-releasing hormone. *see* Gonadorelin
Lichen planus, **5**:614
Lieb, **3**:777
Life cycle, medicalization of, **4**:706–8
Life-cycle transitions, **4**:697–710. *see also* Menopause
    cultural construction
        of the person, **4**:702–4
        of time, **4**:698–700
    culture of oppositions, **4**:704–5
    dual aspects, **4**:701–2
    ecological and structural times, **4**:700
    fragmented life cycle, **4**:705
    identity and aging, **4**:705–6
    Japanese concepts, **4**:703–4
    medicalization, **4**:706–8
    rituals, **4**:702
Life cycles
    fragmented, **4**:705
    Plasmodium, **4**:873–74
Life expectancy
    as aging process criterion, **5**:477
    average, defined, **3**:891, **5**:99
    defined, **3**:891, **5**:99
    language development through, **4**:637
    maximum life span, **4**:783, **5**:477–78
    for selected countries, **5**:101
    in twentieth century, **3**:896
    of U.S. females, **5**:102
Life tables
    defined, **5**:99
    types of, **5**:102
Ligaments
    arcuate popliteal, **1**:376
    characterization, **1**:373
    cruciate, **1**:376
    denticulate, **4**:990
    fibular collateral, **1**:376
    function, **1**:373
    knee. *see* Knee ligaments
    oblique popliteal, **1**:376
    patellar, **1**:376
    periodontal, **1**:372
    repair, growth factors and, **7**:507–8

Ligaments *(Continued)*
  synovial joints, **1**:373
  tibial collateral, **1**:376
Ligand receptors. *see under* Receptors, endogenous sub-
    stances
Ligands
  defined, **3**:403
  heterotrophic, **4**:137
  homotrophic, **4**:137
  interleukin-1 forms as, **5**:731
  opioid peptides as, **5**:731-32
Ligation, one-pot. *see* One-pot ligation
Light
  absorption, **5**:587-92
  attenuation coefficient of, **5**:592
  in melatonin synthesis, **6**:5-6
  therapeutic use, **5**:594
  visible, chemical reactions and, **6**:443
  wavelength, **2**:578-79
  wavelength, complementary, **2**:578
Light scattering
  coefficient of, **5**:591
  consequence of high tissue, **5**:592
  defined, **3**:621
  fundamentals of, **5**:587
  in tissues, **5**:590-91
Lignified plant tissues
  bile salts and, **3**:42
  composition of, **3**:39
  dietary fiber from, **3**:36
  fermentability of, **3**:44
  of wheat grain, **3**:40
Lignin
  in aleurone layer, **3**:40
  defined, **3**:35
  dietary, **3**:36
Likert scale, **1**:499
Limb-withdrawal reflex, **5**:808
Limbic system, **2**:217, 673, **4**:231-32, 711-24
  and convulsive seizure involvement, schematic of,
    6:785
  in convulsive seizures, **6**:781-82
  dyslexia-dysgraphia and, **3**:224
  histamine receptors in, **5**:423
  5-HT receptors in, **5**:422, 423
  hypothalamal outputs, **4**:307
  inputs to hypothalamus, **4**:306
Limbs. *see* Extremities
Limit dextrin. *see* Dextrin, limit
Lincomycin, **1**:290(t)
Line Bisection Test, **1**:611
Linear programming. *see* Programming, linear

LINES, **4**:252, 255
Lineweaver-Burk plots
  for cooperative enzymes, **3**:428
  for enzyme inhibitor potency determination, **3**:395
  for enzyme inhibitors, **3**:430
  equation for, **3**:428
Linkage (genetics)
  alpha-linkages, **3**:35
  analysis of, **3**:828-30
    in cystic fibrosis, **3**:831
    in disease, **3**:875
  behavior disorders and, **3**:843
  beta-linkages, **3**:35
  defined, **3**:823, 835, 847
  in muscular dystrophy, **5**:202
  principles of, **3**:849-50
Linkage mapping. *see* Chromosome mapping
Linker DNA. *see* DNA, linker
Linkers, oligonucleotide, **3**:169
Linoleic acid
  in breast milk, **4**:605
  deficiency of, **3**:576-77
    in steatorrhea, **3**:573
  functions of, **3**:577
  sources of, **5**:533-34
Linolenic acid
  in animals
    dietary role of, **5**:537-38
  desaturation of, **5**:534-35
  elongation of, **5**:534-35
  omega-3, functions of, **3**:577-78
  utilization of, **3**:50
Liothyronine. *see* Triiodothyronine
Lipase
  defined, **3**:47
  function of, **3**:52-54
  of gastrointestinal fat digestion, **3**:48-51
  hepatic, cholesterol and, **3**:586-87
  hormone-sensitive, in lipolysis, **3**:575
  in lipid digestion, **3**:80-81
  in milk triacylglycerol utilization, **3**:52
  pancreatic, in fat digestion, **3**:81
  physiological importance of, **3**:50-51
Lipid A, **1**:274
Lipid bilayers, **3**:726, 728
Lipid-membrane, artificial cells, **1**:381
Lipid peroxidation
  CuZnSOD and, **3**:186
  zinc and, **7**:884
Lipids, **4**:725-48
  absorption of, **3**:572
  of astrocytes, **1**:440

Lipids (*Continued*)
    asymmetry, in red cell membrane, **6**:541–42
    in breast milk, **4**:605
    classification, **4**:732–34
    composition, in fat cells, **1**:60–61
    digestion of, **3**:51
    effect of birth control pill on, **1**:693
    enzymatic modifications in Golgi apparatus, **4**:4
    extraction, **4**:733–36
    extraction and analysis, **4**:733–36
    fatty acid composition of, **1**:60
    food analysis, **4**:222–23
    in gingiva, **4**:36
    in Golgi apparatus, **4**:6–7, 9
    hair, **4**:46
    human milk, digestion of, **3**:47–56
    in infant nutrition, **3**:48
    liquid crystalline state of, **4**:727–28
    liquid state of, **4**:728–31
    mediators of inflammation, **4**:472–73
    membrane, **4**:970–71
        amphipaths, **4**:970
        biogenesis, **4**:976–77
        classes, **4**:970
    metabolism, and dietary fiber, **3**:42
    micellar solutions, **4**:729
    nutritional quality of, **5**:499–500
    ordered fluid structure, **4**:730
    packing in solid state, **4**:727
    phase changes, **4**:731–32
    in postprandrial intestinal contents, **3**:54–55
    red cell membrane, **6**:538–42, 541
    release from fat cells, **1**:60
    solubility, and drug transport in brain, **1**:723
    storage, in fat cells, **1**:60
    surface behavior, **4**:731–32
    suspensions, **4**:728–31
    transitions, **4**:731–32
    transport, **3**:573–74
    transport, in brain, **1**:716
Lipfuscin, **3**:900
    as index of aging, **3**:892, 894
Lipolysis
    gastric, fat absorption and, **3**:51
    gastric lipase and, **3**:53
    of milk fat, bile salts and, **3**:52
    mixed micelles and, **3**:54
    model of, **3**:54
    in the newborn, **3**:81
    rate of, **3**:54
    in white adipocyte, **3**:574–75
Lipophosphoglycans, **4**:682

Lipopolysaccharides
    as adjuvant for subunit vaccines, **7**:753
    of bacterial cell wall, **1**:274–75
    characterization and functions, **4**:748
    salmonella, structure of, **6**:703–4
Lipoprotein-associated coagulation inhibitor, **7**:497
Lipoprotein kinase
    fatty acid uptake and, **3**:586–87
    lipoprotein triacylglycerol and, **3**:585
Lipoprotein lipase
    functions of, **3**:587
    in lipoprotein metabolism, **6**:34
    in storage and release of lipid, **1**:60
    in white adipocyte lipolysis, **3**:574
Lipoprotein triacylglycerol, **3**:585
Lipoproteins
    atherosclerosis-linked variations in, **1**:453
    characteristics of, **6**:24–27
    classified, **3**:573–74
    defined, **3**:567
    esterified fatty acids and, **3**:585
    metabolism, enzymes and transfer factors involved in,
        6:34
    metabolism disorders, immunity and, **5**:506
    oral contraceptives and, **5**:604
    plasma, **6**:23–25
        characteristics of, **6**:25
        chylomicron as, **6**:27–28
        chylomicron remnants as, **6**:28
        classification of, **6**:24
    receptors, **6**:32–34
    specialized, **6**:31–32
    synthesis and catabolism
        hepatically-derived, **6**:30
        intestinally-derived, **6**:29
Lipoproteins, HDL, **1**:451, **6**:31
    coronary heart disease and, **3**:573–74
    protective role of, **1**:452
Lipoproteins, IDL, **1**:451, **6**:29–30
Lipoproteins, LDL, **6**:30–31
    characterization, **1**:451
    clearance from the blood, **1**:451–52
    LDL apheresis, **1**:454
    receptor defects, **1**:453
    receptor-mediated endocytosis, **1**:452
    receptors for, **1**:452
    uptake by macrophages, **1**:453
Lipoproteins, VLDL, **6**:28–29
    defined, **3**:318
    physiology of, **3**:573
    secretion, **1**:451
Liposarcoma, **7**:702

Liposarcoma, myxoid, **2:**451
Liposomes
    antibody conjugates with, **4:**373
    as artificial cells, **1:**381–82
    protein degradation and, **3:**419
    as vehicle for synthetic vaccine, **7:**755
Lipoxygenase pathway, **6:**351
Lipoyl enzymes, **2:**540
Lips, **2:**599
Liquid chromatography. *see* Chromatography, liquid
Liquid crystals, **4:**730
Liquid food reservoirs, **3:**594–95
Liquid-liquid extraction, **3:**200
Liquid-liquid two-phase systems. *see* Aqueous two-phase
    systems
Liquid-solid extraction, **3:**200
Lisiniprol, **2:**175, 179
Lissauer's tract, **7:**206
Listening, dichotic technique, **2:**353
Listeria monocytogenes, **4:**458
Lisuride, **2:**28
Literacy, **3:**226
Lithium, **2:**493, **4:**749–53
    in affective disorder, **2:**493
    antidepressant effects, **1:**309
    autism and, **1:**510
    bipolar depression and, **5:**93
    and caffeine, **2:**79
    drug addiction and, **1:**49
    exposure, birth defects caused by, **1:**701
    granulopoietic effects, **4:**750–51
    immunologic effects, **4:**752
    mania and depression and, **1:**95
Lithium carbonate, **2:**843
Liver, **2:**33, 103, 105
    alcohol and, **1:**173–74
    anatomy, **3:**63, 78
    autonomic innervation of, **1:**543
    B cell development in, **1:**565
    bicarbonate and ammonium homeostasis and, **5:**879–80
    and bile salt, **2:**414
    capillaries of, **5:**32, 39
    edema safety factors in, **5:**40
    fatty acid esterification and, **3:**583–84
    glutathione and, **3:**908
    glycogen metabolism and, **3:**920, 929
    glycolysis in, **3:**939–40
    in hunger and meal initiation, **1:**348
    in immune system aging, **6:**76
    insulin in, **3:**324
    lactate disposal by, **5:**825–26
    light scattering in, **5:**590

Liver *(Continued)*
    organic chemicals toxic for, **7:**550
    pH of, **5:**824
    physiology, **3:**79
    in protein digestion, **6:**232
    radiation effects, **6:**414
    rat, nuclear pores of, **5:**452
    TEM micrograph of, **3:**273
    very low density lipoproteins and, **3:**573
    vitamin metabolism in,
        vitamin A, **7:**848
        vitamin D, **7:**862
    zinc metabolism in, **7:**882
Liver, fatty. *see* Fatty liver
Liver abscess, *Entamoeba histolytica*-associated,
    1:222–23, 229
Liver cirrhosis, **2:**574, 651
Liver diseases
    animal, *N*-nitrosodimethylamine and, **3:**679
    changes in drug kinetics in, **5:**854
    chronic, longevity and, **4:**789
    vitamin D and, **7:**869–70
Liver enzymes, **3:**403–4
    and drug metabolism, **5:**832
    glucocorticoid induced, **7:**269
Liver failure, artificial cells in treatment of, **1:**379
Liver neoplasms, **4:**755–68
    aflatoxin B1 and, **3:**676
    aflatoxin M1 and, **5:**265
    cell tumor in mouse, **2:**117, 118, 241, 253, 254, 663
    control of, **4:**767
    geographic correspondence of HBV infections and
        4:761
    hepatitis B virus and, **4:**342
    incidence in hepadnavirus-infected populations,
        4:761–62
    oral contraceptives and, **5:**607
    presence of hepadnaviruses in, **4:**762–63
Livestock. *see* Animals, domestic
Loa loa, **2:**388
Local inductions. *see* Inductions, secondary
Locomotion, **2:**593, **4:**769–81
    components, **4:**771
    descending mesencephalic projections and, **4:**716–18
    elastic energy storage in, **5:**185
    equilibrium maintenance, **4:**777–78
    initiation of, **4:**772
    spinal-cerebellum interactions, **4:**776–77
    spinal locomotor network, **4:**773–76
    step cycle
        gating of sensory information, **4:**776
        sensory control, **4:**774–76

Locomotion *(Continued)*
  visuomotor coordination in, **4:**771, 778-79
Locus, genetic. *see* Genetic locus
Locus coeruleus, **2:**217, 218, 686
Locust, **2:**311-12
Log kill function, **4:**922
Logic
  of conditionals, **2:**620
    truth, **2:**620-22
  and "if," **2:**622
  and inference, **2:**620
  and proposition, **2:**620
  and reasoning, **2:**622-23
Lomboroso's concept of the "born criminal" (1876), **2:**710
Loneliness, **4:**538
Long-latency reflex, defined, **3:**697
Long-term memory, **4:**676-77
Long terminal repeats. *see* Repetitive sequences, nucleic acid
Longevity, **4:**783-90
  aging effects, **4:**786-88
  caloric restriction and, **3:**894-95
  as criterion for aging process, **5:**478
  defined, **3:**891
  enhancement of, **4:**788-89
  genetic aspects of, **3:**892
  limits on, **5:**102-3
  predictors of, **4:**783-84
  theoretical causes of, **4:**785-86
  theories of, **3:**895
Looming, as factor in space perception, **1:**632-33
Loop diuretics. *see* Diuretics, loop
Loquaciousness, in older adults, **1:**121
Lordosis reflex, **4:**715-16
Lousiness, **2:**390
Lovastatin, **2:**168
Love
  altruistic, **4:**542-43
  romantic, **4:**543
Lovejoy, **3:**500
Low-density lipoproteins. *see* Lipoproteins, LDL
Low-sodium diet. *see* Diet, sodium-restricted
Low temperature effects. *see* Hypothermia
Lowicryl, **2:**727
Lowry method, **6:**192
LTB$_4$ inflammation, **4:**473
Lucretius, *De Rerum Naturae*, **3:**195
Luliberin. *see* Gonadorelin
Lumbar punctures, **4:**989
Lumbar vertebrae, **2:**596, **7:**768
Lumbosacral plexus, **5:**354-55
Lumen, defined, **3:**71, 93

Lumichrome, **2:**529
Lumiflavin, **2:**529
Luminance
  axis, **2:**581
  and brightness, **2:**579-80
  channel, **2:**588
  defined, **2:**575
  property, additive, **2:**581
Lundborg-Unverricht syndrome. *see* Epilepsy, myoclonus
Lung, **2:**180. *see also* terms beginning with Pulmonary
  aging of, **3:**902
  asbestos fibers and
    EDAX analysis, **1:**423
    SAED analysis, **1:**423
  asbestos toxicity to, **7:**556
  blood and air distribution to, **5:**619
  capillaries of, **5:**32, 39
  carbon dioxide and oxygen exchange in, disorders of, **6:**370-74
  carbonic anhydrase and, **5:**
  changes, space travel and, **7:**149
  cigarette smoke and, **7:**515-16
  circulation. *see* Pulmonary circulation
  compliance
    effect of disease on, **6:**601
    measurement of, **6:**369-70
  complications of cardiopulmonary bypass, **4:**95
  congestion, in heart failure, **2:**174
  cyclo-oxygenase pathway in, **6:**351
  cytochrome P450 pathway in, **6:**351
  eicosanoids and, **6:**350-51
  endometriosis in, **3:**369
  failure, **4:**97
  fetal, development of, **3:**617-18
  fibrosis, **6:**373
  gross morphology of, **6:**591
  injury
    endothelial cells in, **6:**351-53
    smoke inhalation and, **7:**453
  innervation of, **6:**346-47
  lipoxygenase pathway in, **6:**351
  lobe, measurement of pressure-volume curve of, **6:**601
  marijuana and, **4:**908, **5:**446
  maturation of, **6:**345-46
  mechanical properties of, respiratory adaptation and, **6:**573
  and mechanics of breathing, **6:**600-601
  neurohumoral development in, **6:**346-47
  noxious agents and, **6:**348
  particle retention by, **7:**554
  pathophysiology of, organic chemicals and, **7:**549
  in pH homeostasis, **5:**877-78

Lung *(Continued)*
plain film radiography of, **3:**2
platelet-activating factor and, **6:**50
primary lobule of, **6:**592
as pulmonary tree, **6:**591-92
pulmonary ventilation of, **6:**595-97
radiation effects, **6:**414-15
space travel effects, **7:**149
structure, respiratory adaptation and, **6:**572-73
and sudden infant death syndrome, **7:**318-19
surgery, **4:**96
toxicology, **7:**575-85
transport of gases in, respiratory adaptation and,
6:574-75
vasomotor tone of, factors affecting, **6:**347-48
ventilation, **5:**618, **6:**595-97
respiratory adaptation and, **6:**374, 573
Lung diseases, **6:**357-77
air pollution and, **3:**387-88
airway obstruction, **6:**358-64
alveolar volume and, abnormal changes in, **6:**364-70
in children, **5:**705-6
disorders in gas transfer, **6:**370-74
gas exchange and, **6:**373-74
as psychophysiological disorder, **6:**337
Lung diseases, obstructive
longevity and, **4:**789
tobacco and, **7:**516-17
Lung neoplasms, **2:**116, 226, 451
asbestos-related, **1:**420-21, **2:**116
and cigarette smoking, **2:**748, **7:**515
female mortality and, **5:**103
radiation induced, **6:**424
smoking and occupational exposure to asbestos,
1:420-21
Lung volume
changes, pathophysiological aspects of, **6:**369
factors affecting, **5:**618-19
muscular dystrophy and, **5:**192
obstruction, pathophysiologic aspects of, **6:**369
and pulmonary pathology, **6:**368
regulation of, **6:**575-76
restriction of, **6:**369
Lung volume measurements, **6:**597
Lupus erythematosus, discoid, **5:**615
Lupus erythematosus, systemic, **2:**616
autoantibodies and antigens in, **1:**334
as autoimmune disease, **1:**529
beta-adrenoceptor antagonists and, **5:**383
characteristics of, **1:**332-35, **4:**410
defined, **2:**601

Lupus erythematosus, systemic *(Continued)*
as factor in spontaneous abortion, **1:**5
nutrition and, **5:**509-10
Luria, **3:**222, 838
Luria-Delbruck model, **4:**922-24
Luria-Nebraska Neuropsychological Battery, **1:**611,
2:865-66
Luteinization
corpus luteum formation and, **3:**634-35
defined, **3:**627
Luteinizing hormone receptors. *see* Receptors, LH
Luteinizing hormone-releasing hormone. *see* Gonadorelin
Luteinzing hormone. *see* LH
Luteotropin. *see* Prolactin
Lutropin. *see* LH
Lutropin receptors. *see* Receptors, LH
LYDMA, defined, **3:**470
Lyme disease
tests for, **1:**740
as zoonosis, **7:**894-95
Lymph, **4:**830
Lymph nodes
draining tissues, **4:**399
metastasis, **4:**891
noradrenergic innervation of, **5:**377, 381
of stomach, **3:**757
T lymphocytes in, **4:**831
Lymphadenopathy-associated virus. *see* HIV
"Lymphatic insufficiency," **5:**41
Lymphatic sheath, periarteriolar
model of, **5:**378
noradrenergic innervation of, **5:**381
tyrosine hdyroxylase-positive nerve terminals in, **5:**379
Lymphatic system, **2:**186-87
function of, **5:**36
interstitium and, **5:**35-36
in solvent exchange, **5:**37
Lymphoblast cell, human, **2:**456
Lymphoblastic leukemia. *see* Leukemia, lymphocytic
Lymphoblastoid cell lines
EBNA-2 and, **3:**468-69
EBNAs and, **3:**470
Epstein-Barr virus and, **3:**464
Lymphocyte activating factor. *see* Interleukin 1
Lymphocyte-detected membrane antigens. *see* LYDMA
Lymphocyte function-associated antigen-1, **3:**183
Lymphocyte-mediated cytotoxicity, **4:**811-17
Lymphocyte mitogenic factor. *see* Interleukin 2
beta-Lymphocyte-stimulating factor, **1:**28
Lymphocytes, **2:**4-5, 7, 256-57, 259, **4:**829-37. *see also*
B-lymphocytes; Immunoglobulins; T-lymphocytes
activity, **4:**334

Lymphocytes (*Continued*)
antigen recognition, **4**:832–34
in autoimmune disease, **1**:520–22
cell receptors on, **6**:515
cellular and humoral, **4**:830–31
characterization, **4**:165–66
chromosomal variations in, **5**:248
cytolytic, **4**:811–13
decay, **4**:332
development and migration, **4**:831
diseases of immunological aberrations, **4**:837
DNA rearrangement in, **3**:798
doubling times, **4**:333
functions, **4**:165–66
retinoids and, **4**:824–26
hormonal signaling of, **5**:376
HPRT in, **5**:248–49
in human responses, **4**:830–32
idiotypes, **4**:313–14
in immune system aging, **6**:76
immunocompetent, **4**:331
immunological self-tolerance, **4**:837
in inflammation, **1**:312
large granular, and natural killer cells, **5**:295–96
life span, **4**:331–32
lithium effects, **4**:752
lymphoid cell-mediated immunity, **4**:405
lymphoid nodules, solitary, **5**:139
major histocompatibility complex, **4**:834–35
membrane receptors, for antigen, **6**:519–22
neuropeptides and, **5**:383
norepinephrine and, **5**:382
nuclear pores of, **5**:452
organization of, **6**:606–7
in Peyer's patches, **5**:140
phospholipids in, **5**:508
progenitors, **4**:169–71
radiation effects, **6**:421–22
responses to retinoids, **4**:819–28
somatostatin and, **5**:384
specificity, **4**:332
stem cells, in bone marrow transplantation, **1**:791–92
from thymus, **4**:830–31
tissue-seeking, **4**:395–96, 399
tumor-infiltrating, **1**:677, 680
vasoactive intestinal peptide and, **5**:384
beta-Lymphocytes
activation, nonspecific, **1**:568
in autoimmune disease, **1**:520–22
development of, **1**:565
differentiation to antibody-forming cell clones, **1**:568–69

beta-Lymphocytes (*Continued*)
factors affecting responsiveness, **1**:571–72
general characteristics of, **1**:563
memory responses, origins of, **1**:569–70
proliferation and differentiation, **1**:568–69
and recognition of antigenic determinants, **1**:565
regulation of responses, **1**:569
self-nonself discrimination by, **1**:571–72
subpopulations of
CD.5 beta cells as, **1**:566
conventional primary beta cells as, **1**:566
memory beta cells as, **1**:566–67
triggering of, **1**:567–69
Lymphocytes, T. *see* T lymphocytes
Lymphocytes, tumor-infiltrating, **4**:1012
Lymphoepithelium, in mucosal immunology, **5**:139
Lymphogranuloma venereum, **4**:450
Lymphoid cells. *see* Lymphocytes
Lymphoid organs
differentiated, **4**:393
draining, **4**:397
innervation of, **5**:376–81
primary, noradrenergic innervation of, **5**:380–81
secondary, chemical sympathectomy of, **5**:383
vascular innervation, **5**:378
Lymphokine-activated killer cells. *see* Killer cells, lymphokine-activated
Lymphokines, **4**:835–36
characterization, **4**:811, 823
cloning and production of, **5**:293–94
functions, **4**:823
helper T cell secretion, **4**:361
immune system, **4**:333
immunity and, **5**:510
in *Plasmodium* infections, **4**:879
production, **2**:751
Lymphoma, **2**:450, **4**:839–51
bone marrow transplantation for, **1**:795
chemical agents effects, **4**:844–45
childhood protection by breast feeding, **4**:342
chromosomal abnormalities causing, **4**:845
classification, **4**:840–41
diagnosis, recombinant DNA technology in, **6**:529
diseases associated with, **4**:846
dysmorphology, **4**:845–46
entry into circulation, **4**:1011
etiology, **4**:844
gastrointestinal, **3**:769–70
heterogeneity in, **4**:848–49
hypercalcemia of malignancy and, **4**:279
immunology, **4**:846
immunotoxin therapy, **4**:421
incidence in homosexuals, **4**:345

Lymphoma (Continued)
   lymphoma-associated diseases, **4:**846
   malignant, **2:**450
   staging of, **4:**849
   treatment of, **4:**849–50
Lymphoma, B-cell
   Epstein-Barr virus, **4:**190
   in immunosuppressed patients, **4:**320
Lymphoma, Burkitt's. see Burkitt's lymphoma
Lymphoma, non-Hodgkin's, **1:**795, **4:**840–41
   Ann Arbor classification, **4:**847
   classification, **4:**840–41
   clinical description, **4:**847–48
   natural history of, **4:**847–48
   working formulation, **4:**843
Lymphopoietic cells, **3:**624
Lymphoproliferative disorders, **3:**465–66
Lymphoreticular tissues, bronchus-associated. see
      Bronchus-associated lymphoreticular tissues
Lymphoreticular tissues, mucosal-associated. see
      Mucosal-associated lymphoreticular tissues
Lymphotoxin, **1:**807, **4:**836
Lyon, Mary, **3:**855, **5:**192
Lyonization
   of chromosome X, **2:**443
   defined, **3:**835
   in fragile X syndrome, **3:**715, 841
Lysergic acid diethylamide
   characteristics of, **1:**180–81
   marijuana and, **5:**444
   sources of, **5:**447
Lysine, **2:**507, 564, 569
   on acetylated alpha-tubulin, **5:**18
   in elastin, **3:**261
   radioactive, **3:**268
Lysosomes, **2:**238, 239, 416
   enzymes, **2:**240
      in gingiva, **4:**36
   fusion with phagosome, **4:**857
   ganglioside degradation in, **3:**731–32
   Golgi apparatus, in formation, **4:**4, 11
   pH of, **5:**875
   primary, macrophage, **5:**813
   secondary, macrophage, **5:**813
   in sorting and transport of proteins, **6:**254
   storage disorders, **4:**578–79, **5:**677–78
Lysosphingolipids
   biochemical targets of, **7:**187
   cellular location of, **7:**185
   mechanism of action of, **7:**186–88
   metabolism of, **7:**185
   structure of, **7:**185

Lysozyme. see Muramidase
Lysyl hydroxylase, **2:**562, 563
   ascorbic acid as cofactor for, **1:**431
Lysyl oxidase, **2:**541, 564, 569, 571, 650, 652, **3:**262
Lysyl residues, **3:**261
Lytic cycle, viral. see Viral lytic cycle

M

M-1 reflex. see Reflex, stretch
M-2 reflex. see Long-latency reflex
M-BACOD antineoplastics, **4:**850
M-band protein, **5:**169
M cells
   in mucosal immunology, **5:**139
   of retina, **7:**804–5
M fee
   dietary restriction and, **5:**515
   excess dietary lipids and, **5:**508
   retinoids and, **5:**512
M-line proteins. see Proteins, M-line
M phase. see Mitosis
M phase-promoting factor. see Maturation-promoting
      factor
M-region (pontine micturition center), **5:**45
Mab AbHJM1, **4:**963
Mab CL203.4, **4:**960
Mab LM609, **4:**960
Macaque monkey, **2:**593–600
   comparative anatomy, **2:**593–600
MacLeod, Colin, **3:**933
Macroglobulins, **2:**659
   targeting moiety of immunoconjugates, **4:**374–75
Macromolecular synthesis, viral, **4:**518–19
Macromolecular systems, **3:**179
Macromolecules
   microvessel exchange of, **5:**34–35
   partial specific volume of, ultra-centrifugation and,
      7:719
   size, and ultracentrifugation, **7:**719
Macronutrients
   appetites for, **1:**347
   defined, **5:**471
   heats of combustion of, **5:**475
Macroorchidism
   defined, **3:**709
   in fragile X syndrome, **3:**711
Macrophage colony-stimulating factor, **1:**686–87. see also
      Colony-stimulating factor
Macrophage inflammatory peptides, **1:**28, 29
Macrophages, **2:**417, 751, **4:**853–62. see also Interleukins
   accessory cell activity, **4:**859

Macrophages (*Continued*)
  activation
      and cytotoxicity, **5:**817-18
      and stimulation, **4:**857-58
      by TNF, **4:**474-75
  adrenergic agents and, **5:**382
  alveolar, function of, **6:**366
  antigen presenting, **5:**818
  attachment, **4:**856
  autocrine effects, **4:**860
  bone marrow, **6:**609
  cancer prevention or cure, **4:**340
  characteristics of, **4:**166, **5:**813
  chemotaxis, **4:**856, **6:**610
  in chronic infections, **6:**611
  development of, **6:**606-07
  extracellular killing, **4:**858
  function, **4:**166, 853, 402-3
  in inflammation, **1:**312, **6:**606
  ingestion, **4:**856-57
  interrelationship with host defense, **6:**611-12
  intracellular fate, **4:**857
  life history and distribution, **4:**854
  lithium effects, **4:**752
  liver, **6:**608
  lymph node, **6:**609
  monocyte, myeloid maturation stages of, **5:**428
  morphology, **4:**854-55
  in pathology, **4:**861-62
  peritoneal, **6:**608
  phagocytosis, **4:**855-57
  physiological roles, **4:**861
  platelet-activating factor and, **6:**48-49
  pulmonary, **6:**608
  radiation effects, **6:**421
  receptors, in phagocytosis, **6:**610
  in regulation of immune response, **6:**612
  secretion, **4:**858-59, 860-61, **6:**610
  as source of platelet-activating factor, **6:**45-46
  spleen, **6:**608-09
  suppressor, **4:**859
  surface markers, **4:**855-56
  surface receptors, **4:**855-56, **5:**815-16
  in testicular cells, **7:**434
  tissue. *see* Phagocytes
  tumor cell destruction, **4:**1012
  in tumor cell destruction, **6:**611
  tumor rejection and, **4:**340
  wound and tissue repair, **7:**505
Macula communicans, **2:**274
Macular lutea, **3:**542
Madder, **2:**699

Magnesium
  as element essential for life, **5:**48-49
  imbalance, immunity and, **5:**514
  in modulating vascular tone, **6:**353
  and parathyroid hormone secretion, **5:**663
Magnetic field, **4:**864-65
Magnetic moment, **4:**864-65
Magnetic polarity, **3:**492
Magnetic resonance imaging, **4:**863-71
  for atherosclerosis imaging, **1:**449
  axial spin-echo image, **4:**868, 870
  clinical applications, **4:**867-70
  contrast enhancement, **4:**870
  defined, **3:**1, 671
  in diagnosis, **3:**6
      of senile dementia, **1:**131
      of strokes, **7:**304-5
  in drug analysis, **3:**202, **5:**839
  in ganglioside study, **3:**731
  histone study and, **3:**127
  image display, **4:**866-67
  image formation, **4:**866-67
  imaging parameters, **4:**867-69
  instrumentation, **4:**867
  localization of cognitive functions, **1:**480
  in metabolism study, **3:**378
  normal anatomy, **4:**869-70
  pathological conditions, **4:**869-70
  in pH assessment, **5:**873
  radio frequency pulse effect, **4:**864-65
  signal intensity and tissue type, **4:**867-68
  spin-echo sequence, **4:**866, 868-69
  structure of matter and magnetic field effects,
      4:864-65
  T1- and T2-weighted images, **4:**869-70
  T1 and T2 relaxation, **4:**865-66
  T1 and T2 relaxation time measurement, **4:**866
Magnetization, **4:**864-66
Magnetosphere, **6:**428
Magnocellular neurons, **4:**608
  defined, **5:**394
Major basic protein (MBP), **1:**189
Major histocompatibility complex, **2:**225-26, 230,
      4:382-84
  antigen presentation, **4:**834-35
  in autoimmune disease, **1:**520, 531-32
  classes, **4:**359
  defined, **2:**225
  genetic aspects of, **5:**514
  genetic map of gene frequencies, **6:**82
  as genetic marker for affective disorders, **1:**99
  immune system and, **4:**329

Major histocompatibility complex (*Continued*)
    importance in transplantation, **4**:358
    induction of allogeneic genes, **4**:365
    matching and graft survival, **4**:359-60
    molecular basis, **4**:359
    olfaction and, **5**:529
    in synthetic vaccine design, **7**:748
    in transplantation, **4**:359-60
    tumor cell destruction, **4**:1012
Major histocompatibility complex, human. *see* HLA antigens
Malabsorption syndromes
    bile salt-stimulated lipase and, **3**:51
    in cystic fibrosis, **5**:706
    fat-related, **3**:573
    hookworm infection and, **4**:116
    low lipase/bile salt concentrations and, **3**:51
    strongyloidiasis, **4**:118-19
Malaria, **1**:267, **2**:382-84, **4**:873-85, **6**:275-77
    altered hemoglobin, **4**:143-44
    antigenic variation, **4**:388
    antimalarial drugs, **4**:883-84
    cell invasion by parasite, **5**:646
    control of, **4**:449
    in developing countries, **4**:449
    disease models, animal, **1**:255
    distribution, **4**:876
    drug resistance mechanisms in, **5**:651
    epidemiology, **4**:876
    erythrocyte modification in, **5**:648
    genetic polymorphism and, **4**:881-83
    host defense mechanisms, **4**:878-79
    and human population movements, **4**:876-78
    hyperreactive malarious splenomegaly, **4**:389
    immune deviation, **4**:388
    impact on biology, **4**:885
    parasite defense mechanisms, **4**:879-80
    pathological effects, **4**:878
    pigment, **2**:391
    prevention with drugs, **2**:383-84
    and quinine, **2**:382-83
    TNF-a role, **4**:862
    vaccines for, **4**:449, 884-85, **5**:655-56
    vectors, attacks on, **4**:883-84
Malate, **2**:504, 506, 508
Malathion, **2**:390, 393, **7**:546-47
Males. *see also* Men
    delayed puberty in, **6**:343
    genes, on Y chromosome, **6**:838-39
    gonadal development in, **6**:343
    identity, in childhood, **6**:856-57
    maturation, **4**:27

Malformations. *see* Abnormalities
Malignant hyperthermia, **5**:829
Malnutrition. *see* Nutrition disorders
Malnutrition, protein-calorie. *see* Protein-calorie malnutrition
Malolactic fermentation, in wine production, **3**:667
Maltase, **2**:105
Malting, defined, **3**:17
Maltose, **2**:103, 104
Maltotriose, **2**:104
Mambas, toxins produced by, **7**:600
Mammalian cell lines
    drug resistance in, **3**:788
    gene expression in, **3**:806
    gene transfection of, **3**:805
Mammals
    adapted to special environments, respiration in, **6**:577-78
    body pattern formation in, **3**:284, 290
    dreaming by, **3**:195
    genetic transmission patterns in, **3**:850
    oxygen consumption of, size and metabolism as factors in, **6**:571-72
    respiratory system design in, **6**:571
Mammary gland. *see* Breast
Mammary tumor virus, murine, **2**:57
Mammography
    advances in, **3**:5
    defined, **3**:1
    as method of X-ray imaging, **6**:439-40
    and risk of breast cancer from, **6**:415
Mammotropic hormone, pituitary. *see* Prolactin
Mammotropin. *see* Prolactin
Mandible, **2**:705, 706, 778, **5**:363
Manganese
    biological function of, **7**:607
    deficiency, consequences of, **7**:610
    imbalance, immunity and, **5**:514
    mining, Parkinson's disease and, **5**:669
    and Parkinsonism, **2**:346, 347
    significance of, in body, **5**:49
    toxicity, **7**:610
Manic-depressive psychosis. *see* Bipolar disorder
Manic disorder, **2**:832, 833, 842
    and behavior, **2**:842
    and carbamezapine, **2**:843
    characterization, **4**:997
    from cocaine abuse, **5**:441
    diagnostic criteria for, **1**:94, 95
    and haloperidol, **2**:843
    and lithium carbonate, **2**:843

Manic disorder *(Continued)*
as mood disorder, **5**:94–95
Mannitol, **3**:100
Mannose, **2**:561
Mannose-6-phosphate, **2**:240
binding proteins, **4**:976
Mannose-6-phosphate receptors. *see* Receptors, mannose-6-phosphate receptors
Manserine, **2**:843
Mansonella ozzardi, **2**:388
Manual dominance. *see* Dominance, manual
MAP kinases, **5**:896
MAP-1 (neuron protein), phosphorylation of, **5**:897, 898
MAP1C. *see* Dynein
MAP-2 (neuron protein)
calcium-phospholipid-dependent protein kinase and, **5**:896
CaM-kinase II and, **5**:895
cAMP-dependent protein kinases and, **5**:895
phosphorylation of, **5**:897, 898
Maple syrup urine disease, **5**:437
Mapping, chromosomal. *see* Chromosome mapping
MAPs (neuron proteins). *see also* Microtubule-associated proteins
casein kinase II and, **5**:896
function of, **5**:894
tubulin isoforms and, **5**:897
Marasmus. *see* Protein-calorie malnutrition
Marcaine (bupivacaine), **2**:784
in dentistry, **2**:784
Marchand, **3**:777
Marfan syndrome
defined, **3**:261
fragile X syndrome and, **3**:713
inherited, **1**:698
Margarine
advent of, **3**:21
P/S ratio of, **3**:572
Margination, **5**:819
Marijuana and cannabinoids, **4**:905–11
abuse of, **5**:444–46
addiction, **1**:43–51
alcohol and, **3**:212–13
in pregnancy, **3**:212–13
analytical determination, **4**:906
behavioral effects, **4**:907–8
cardiovascular effects, **4**:908
consequences of abuse, **5**:445
endocrine and reproductive system effects, **4**:908–9
hashish, **4**:906
immune system effects, **4**:909
LSD flashbacks and, **5**:448

Marijuana and cannabinoids *(Continued)*
mechanisms of action, **4**:907
metabolism and pharmacokinetics, **4**:906–7
patterns of use, **4**:907
plant and its components, **4**:905–6
pulmonary effects, **4**:908
therapeutic aspects, **4**:909–11
tolerance and dependence, **4**:909
toxicity, **4**:907
Marine toxins, **3**:673
Marker X. *see* Fragile X
Markers, biological. *see* Biological markers
Markers, microscopic. *see* Microscopic markers
Marketing, **5**:800
Marr, David, **5**:388
Marriage
attitudes toward, self-monitors and, **6**:806
behavioral genetics methodology and, **3**:837
extramarital affairs and, **6**:885–86
as factor against suicide, **7**:331
practices of, sexuality and, **6**:885
successful, trait descriptions and, **5**:771
Marriott, McKim, **5**:701
Mass media, persuasion in, **5**:800
Mass spectrometry. *see* Spectrum analysis, mass
Mass-spring, as method of motor control, **5**:112–13
Mast cell growth factor-2. *see* Interleukin 4
Mast cells
in allergic response, **1**:185–86
connective tissue, **1**:186
defined, **3**:755
degranulation, biochemical events of, **1**:187
food hypersensitivity and, **3**:673
functions, **4**:402–3
gut-associated lymphoid tissue and, **5**:381
in inflammation, **1**:313
products of, **1**:188
skin, **1**:186
Mastication
in digestive process, **3**:74
monitoring of, **3**:595
muscles for, **4**:261
neural basis of, **5**:370–71
Masticatory muscles, **7**:678
Mastocytes, defined, **7**:47
Mastoparan, **2**:94
Masturbation, **6**:870
Materials testing, **3**:669
Maternal age, and fetal chromosome abnormalities, **1**:699–700, **3**:814
Maternal-fetal exchange, **3**:612–13

Maternal inheritance
  defined, **3:**819, 847
  principles of, **3:**858
Maternal mortality, **4:**913–20
  case studies, **4:**915–16
  current status in developing world, **4:**916–17
  developing countries, **4:**915–20
  history of, **4:**914–15
  illegal abortion, **4:**916–17
  obstructed labor, **4:**916, 918–19
  possible interventions, **4:**918–20
  postpartum hemorrhage, **4:**916, 919
  postpartum infection, **4:**919–20
  pregnancy-induced hypertension, **4:**919
  in rural areas, **4:**918–19
  septic abortion, **4:**920
Maternity waiting home, **4:**918
Mathematical ability
  in fragile X syndrome children, **3:**715
  of neural networks, **5:**392
Mathematical nutrition education. *see under* Nutrition
    education
Mathematics, **2:**549
Mating behavior. *see* Sex behavior
Matrix, extracellular. *see* Extracellular matrix
Matrix of articular cartilage and intervertebral disc. *see*
    Articular cartilage and intervertebral disc
Matrix space, defined, **5:**55
Maturation, **2:**873, **4:**25–26. *see also* Growth
  and experience, **2:**874
  females, **4:**27
  males, **4:**27
  and motor activity, **2:**873
  psychosocial development during adolescence, **1:**67–68
  secular trends, **4:**30
Maturation-promoting factor
  functions of, **5:**66, 67
  microtubule protein phosphorylation and, **5:**901
Maturational lag. *see* Child development disorders
Maurer's clefts, **4:**875
Maurer's spots, **4:**875
Maxam-Gilbert sequencing, **3:**877
Maxilla, **2:**778–79
Maximal heart rate. *see* Heart rate, maximal
Maximal oxygen consumption. *see* Oxygen consumption,
    maximal
Maximum isometric tension
  defined, **5:**173
  in sliding filament theory, **5:**180
Maximum life expectancy. *see* Life expectancy
Maze experiment. *see* Hull, C. L.
McArdle's disease. *see* Glycogen storage disease type V

McCulloch-Pitts formal neuron, **5:**387
McGuire, William, **5:**800
McKay effect
  in *Caenorhabditis elegans*, **3:**895
  defined, **3:**894
McKusick, V. A., **3:**850
McReynolds constants, **3:**204
MDA. *see* 3,4-Methylenedioxyamphetamine
Meads's theory of the reflexive self, **2:**857
Measles, **4:**939–48
  cause of, **4:**940–41
  defined, **4:**940
  epidemiology, **4:**944–46
  immunocompromised children and, **5:**706
Measles, German. *see* Rubella
Measles vaccines, **4:**445, 466, 943–44
Measles virus
  characterization, **4:**940–41
  encephalitic disease and, **4:**943
  eradication of, **4:**947–48
  florid period, **4:**942
  origin of, **4:**947
  penetration and incubation, **4:**942
  prodomal period, **4:**942
  secondary effects, **4:**942–43
Measurement cell constant, defined, **3:**9
Meat
  amino acid pyrolysis products in, **3:**677
  in "basic four" food groups plan, **3:**654
  cured, nitrite in, **3:**679
  in hunter-gatherer diet, **3:**500
  maternal aversion to, **3:**33
  nutrient composition of, **5:**497
  spoilage of, **3:**663, 668
Meat-eating. *see* Carnivorousness
Mebendazole, **2:**387, 388, 390
  ascariasis treatment with, **1:**270
  pinworm treatment with, **1:**269
  in treatment of trichinosis, **7:**655
  trichuriasis treatment with, **1:**270
Mebubarbitol. *see* Pentobarbital
Mebumal. *see* Pentobarbital
Mechanical ventilation. *see* Respiration, artificial
Mechanochemical enzymes. *see under* Enzymes
Mechanoreceptors, **2:**683
  morphology, visceral afferent systems and, **7:**796
  muscle spindles as, **3:**719
  orofacial, and touch, **5:**362–64
Mechanotransduction
  in cochlear hair cells, **3:**234–35
  outer hair cell motility generation and, **3:**237
Meckel-Gruber syndrome, **1:**699

Meclastine, **2:**23
Meconium, defined, **3:**609
Meconium aspiration, **5:**705
Meconium ileus, **2:**735
Medawar, Peter, **3:**892
Media, influence of, **7:**236
Media effects on aggression. *see also* Violence, observed
    archival analyses, **1:**7–8,1058
    field experiments, **1:**5, 1058
    longitudinal studies, **1:**5–7, 1058
    priming of aggressive associations, **1:**9, 1058
    script learning, **1:**8–9, 1058
Median eminence
    anatomy, **5:**394
    physiology, **5:**395–96
Mediastinum, **3:**2
Medical research. *see* Research
Mediterranean lymphoma. *see* Immunoproliferative small
    intestinal disease
Medroxyprogesterone acetate, **2:**111, **3:**368
Medulla, **2:**3
    hair, **4:**44
    somatosensory pathways from receptors to, **7:**122–23
Medulla oblongata, **2:**3, **5:**350
Medullary arteries of spinal cord, **7:**193–94
Medullary medial reticular formation, **4:**712–14
Medullary medial tegmentum, **4:**714–15
Mefenamic acid, **4:**556–57
Mefloquine, **2:**384, 301–92
Megakaryocytes
    characterization, **4:**167
    functions, **4:**167
Meganeurites, **3:**733
Meiosis, **2:**244, 438, **4:**949–56
    chromosome synapses, **4:**951–52
    crossing-over in, **3:**850
    defects, **4:**956
    defined, **3:**823, 847
    divisions, **4:**954–56
    gametogenesis, **4:**949–51
    gene conversion, **4:**953–54
    oocyte, **5:**552
        after activation, **3:**606
        LH surge and, **3:**633, 634
    oogenesis, **4:**950–51
    recombination, **4:**952–53
    sperm, **7:**427–28
    spermatogenesis, **4:**950
MeIQx, **3:**677
Meissner corpuscles, **7:**121
Meister, A., **3:**913, 917
MEL-14, **4:**476

Melancholia. *see* Depression
Melanesian origin, **2:**797
Melanins, **7:**44
    in hair, **4:**45
    light absorption by skin, **5:**589
    Parkinson's disease and, **5:**668
    skin color and, **5:**594
    synthesis, **4:**887
    ultraviolet absorption by, **5:**593–94
Melanocyte-stimulating hormones. *see* MSH
Melanocytes, **4:**35
    adrenocorticotropic hormone and, **5:**729
    aging and, **3:**903
    defined, **3:**533
    of epidermis, **7:**44
    hair, **4:**41, 45, 45
    in keratinocyte growth regulation, **3:**451
Melanoma, **4:**887–91
    acral-lentiginous, **4:**890
    chondroitin sulfate proteoglycan functional role, **4:**959
    Clark levels, **4:**957
    distant metastasis, **4:**891
    embryology, **4:**887
    epidemiology, **4:**887–88
    etiology, **4:**888
    exposure to ultraviolet radiation, **4:**888
    ganglioside antibodies and, **3:**733
    ganglioside content, **4:**962–63
    genetics, **4:**888
    host-tumor relationship, **4:**888–89
    human cutaneous, **4:**957
    immunogenicity, **4:**344
    incidence, **4:**958
    lentigo maligna, **4:**890
    localized, **4:**891
    malignant, oral contraceptives and, **5:**607–8
    nodular, **4:**890
    prognosis, **4:**891
    radial growth phase, **4:**957
    regional lymph node metastasis, **4:**891
    staging, **4:**890–91
    superficial spreading, **4:**889–90
    vertical growth phase, **4:**889, 957
Melanoma antigens and antibodies, **4:**957–67
    calcium-binding S-100 protein, **4:**962
    cell substrate-interacting glycoproteins, **4:**958–61
    expression during tumor progression, **4:**965–66
    functional properties of cell-adhesion molecules,
        4:960–61
    ganglioside antigens, **4:**962–65
    growth factor receptors, **4:**965
    HLA class II antigens, **4:**965

Melanoma antigens and antibodies (*Continued*)
  melanoma-associated cell-adhesion molecules, **4:**960
  melanoma-associated high-molecular weight antigens,
      4:959
  melanotransferrin, **4:**961-62
  shedding during tumor progression, **4:**966
Melanophores, fish, **5:**900
Melanosomes. *see* Melanocytes
Melanotransferrin, **4:**961-62
Melanotropin. *see* MSH
Melatonin, **2:**495, 499, 849
  blood levels of, **6:**3-5
  and depression, **2:**833
  excretion of, **6:**5
  gonad effects, **4:**931
  metabolism of, **6:**5
  secretion, and sleep, **7:**75
  synthesis
      by pineal body, **6:**3
      role of light in, **6:**5-6
Melittin, **2:**94
Melphalan, **2:**114, **4:**370
Membrane attack complex. *see* Complement membrane
    attack complex
Membrane channels. *see* Ion channels
Membrane fluidity, **5:**507
Membrane glycoproteins
  defined, **5:**191
  dystrophin and, **5:**198, 203
Membrane lipids, **4:**970-71, 1003
Membrane proteins, **2:**237-38, **4:**971-73
  activity, **4:**972
  amino acid sequence, **4:**972
  anchor sequence, **4:**976
  assembly, **6:**240
  association of, **4:**971-72
  biogenesis, **4:**976
  classes of, **6:**240
  composition, **4:**972
  disposition, **4:**972-73
  energy barrier to translocation, **4:**972-73
Membranes, **4:**969-77
  bilayer diffusion, **4:**971, 973
  bilayers, **4:**970, 971
  biogenesis, **4:**976-77
  connections, **4:**975-76
  coupled transport, **4:**974
  cross-partitioning, **1:**362
  depolarization, **2:**134
  drug transport through, **5:**839
  electrochemical gradient, **4:**973
  fluid bilayer, **4:**970

Membranes (*Continued*)
  fusion and fission, **4:**971
  G proteins in, **4:**23
  gangliosides in, **3:**726, 728-29, 734
  gap junctions, **4:**973
  mitochondrial, **4:**1004
  permeability determinants of, **5:**839
  processes in *Hermissendra*, **2:**310-16
  receptor regulation, **4:**975
  red cell. *see* Erythrocyte membrane
  resting electrical potential, **2:**135
  separation and fractionation in aqueous two-phase
      systems, **1:**361-62
  signal transduction mechanisms, **4:**975
  and spectrin, **2:**41
  stimuli and receptors, **4:**974-75
  structure, **2:**41, **4:**1004, 413-14
  transduction, **4:**974-75
  transmembrane complexes, **1:**459-60
  transmembrane pore, **2:**612
  transmembrane signaling
      mechanism of, **2:**326-29
      platelet-activating factor and, **6:**49
  transport. *see* Biological transport
  transverse asymmetry, **4:**971
  viscosity, **4:**971
Memory, **2:**8, 9, 305-10, 549, 763-65
  and Alzheimer's disease, **4:**677-78
  amnesia as dysfunction of, **5:**402-4
  anatomy of, **5:**404-5
  and behavior, **2:**306
  beta cell response, **1:**569-70
  beta cells, **1:**566-67
  brain neuroanatomy and, **4:**674-75
  cellular, **2:**305-10
  declarative, **2:**8, 9
  declarative vs. nondeclarative, **5:**403
  deficits, auditory sequential, **7:**156
  defined, **4:**674
  diencephalic amnesia, **4:**677
  of dreams, **3:**196-97
  dysfunction, **5:**402-4
  episodic, **2:**765
  formation, behavioral changes and, **6:**40
  functional localization, **4:**674
  generalized central capacity and, **5:**152
  in *Hermissendra* model, **2:**306-10
  hippocampal functions, **4:**232
  hippocampus and, **4:**677
  Korsakoff's syndrome, **4:**677
  and learning, **4:**673-78
  medial temporal amnesia, **4:**677

Memory *(Continued)*
    modulatory systems for, **4**:676
    mood and, **1**:109–10
    mood-congruent, **1**:109
    mood-incongruent, **1**:109–10
    network modeling of, **5**:390–91
    neural networks and, **4**:674–76
    neurology of, **5**:401–5
    positron emission tomographic studies of, **3**:225
    procedural, **2**:8, 9, 763, 765
    procedural knowledge and, **4**:676
    semantic, **2**:765
    short-term, **2**:8
    spatial, **2**:8, 9
    storage, **4**:674, **5**:401–2
    taxonomy of, **5**:403
    as topic for neuroscience and psychology, **5**:401–5
    traces, **4**:675–76
    transformation of perceptions into, **5**:404
    types of, **4**:676–77
    verbal, **2**:8, 9
Men. *see also* Males, Sexes
    anatomy of reproductive system, **6**:559–62
    interest patterns in, **6**:857–58
    rates of psychopathology among, **6**:861–62
    reproduction and nutrition, **1**:748–49
    sex differences. *see* Sex characteristics
    sexual development, **6**:837–39
    sexual response, adult, **6**:865–67
    and women
        personality differences between, **6**:857–58
        scholastic differences, **6**:858–59
        specific ability differences, **6**:858–59
Menarche
    body fat and fertility, **1**:741–50
    components of weight at, **1**:744–45
    delayed, physical exercise and, **1**:746–47
    fatness as minimal determinant for, **1**:745–46
Mendelian heredity. *see* Genetics
Mendlewicz, **3**:843
Menhaden fish oil, **5**:509
Meninges, **4**:979–91
    brain and spinal cord, **5**:351
    cranial, **4**:980–87
        blood supply of, **4**:987
        innervation, **4**:987
    functions of, **4**:990
    spinal, **4**:988–90
    spinal cord, **7**:195
    tumors of, **4**:990–91
Meningioma, **2**:451, **4**:979, 990–91

Meningitis
    bacterial, causes of, **1**:555
    in infants and toddlers, **4**:461–62
Meningitis, tuberculous. *see* Tuberculosis, meningeal
Meniscus, **1**:376
Meniscus, lateral, **1**:376
Meniscus, medial, **1**:376
Menkes' disease. *see* Kinky hair syndrome
Menopause, **4**:707–8
    bone loss and, **3**:482
    as deficiency disease, **4**:708–9
    defined, **3**:627
    distress, psychosocial explanations, **4**:709
    non-Western societies, **4**:707
    radiation-induced, **6**:416
    sexuality and, **6**:888–89
Menstrual age. *see* Gestational age
Menstrual cycle, **2**:247–48, 498
    defined, **3**:627
    fatness as minimal determinant for restoration of, **1**:745–46
    human, **2**:247–48, 498
    in mammals, **2**:498
    neurochemical control of, **6**:556–57
Menstruation, **2**:21, **4**:932
    anorexia nervosa and, **3**:246
    breast response to oral contraceptives and, **5**:606
    bulimia nervosa and, **3**:247
    endocrine control of, **7**:739
    obesity and, **3**:248
    retrograde
        defined, **3**:361
        endometriosis and, **3**:363
    role of progesterone during, **7**:276
    taboos, and sexuality, **6**:886
Menstruation disorders
    in endometriosis, **3**:365
    MZ twinning and, **3**:314
    oral contraceptives and, **5**:608
Mental deficiency. *see* Mental retardation
Mental disorders, **2**:859, 877, **4**:993–98. *see also* specific
    disorders
    affective, **2**:865
    classification of, **3**:838
    componential model, **1**:608–9
    diagnosis of, **4**:994–96
    diagnostic categories of, **4**:996–98
    DNA marker diagnosis of, **3**:148
    emotional, **2**:865
    genetic aspects of, **3**:835–45
    headache associated with, **4**:55
    insomnia and, **7**:78

Mental disorders *(Continued)*
    learning disorder, **2:**864
    manifestations, **4:**993–94
    medical model, **1:**608
    mercury-induced, **3:**387
    oral contraceptives and, **5:**608
    as predictors of suicide, **7:**331
    schizophrenia, **2:**865
Mental processes, **5:**809–10. *see also* Thinking
Mental retardation, **2:**863–64
    autism and, **1:**506–7
    characterization, **4:**997
    defined, **3:**709
    in Down's syndrome, **3:**175
    fragile X syndrome diagnosis and, **3:**714
    gamma-glutamyl transpeptidase deficiency and, **3:**916
    genetic aspects of, **3:**821, 838–39
    incest and, **4:**440
    infantile spasms and, **3:**458
    and radiation, **6:**413
    X-linked, defined, **3:**709
Mepacrine, **4:**884
Meperidine, **2:**786
Mepivacaine (carbocaine), **2:**784
Meprobamate, **2:**787
Mercapto compounds. *see* Sulfhydryl compounds
beta-Mercaptoethylamine, **2:**539
7-Mercaptoheptanoylthreonine phosphate, **2:**543
6-Mercaptopurine, **2:**374
Mercapturic acid, **3:**915
Mercurials, and interferons, **4:**525
Mercury poisoning
    effects of, **3:**387
    tissue distribution and, **7:**551
Merkel cells, **7:**676
Meroclones, **4:**583
Meromyosin, **5:**173
Mescaline
    3,4-methylenedioxyamphetamine and, **5:**449
    characteristics of, **1:**178
    marijuana and, **5:**444
    peyote cactus and, **5:**447
Meselson and Stahl, DNA experiments of, **7:**720
Mesencephalon, **2:**3, 333, 678
    anatomy of, **5:**349–50
    caudal, **2:**334
    and convulsive seizure propagation and control, **6:**782–83
    damage, effects on orienting, **1:**481
    descending projections to caudal brainstem and spinal cord, **4:**714–19

Mesencephalon *(Continued)*
    locomotor region, **4:**772
    rostral, projections to spinal cord, **4:**714
Mesenchymal tissue
    defined, **3:**261
    in volar pads, **3:**85
Mesenchyme
    defined, **5:**161
    myogenesis and, **5:**162
Mesenteric lymph nodes, **5:**384
Mesentery, defined, **3:**57
Mesoderm
    in body pattern formation, **3:**287–88
    bone formation from, **1:**786
    diversity in, **3:**287
    involution of, **3:**285
    local inductions of, **3:**286
    myogenesis from, **5:**161–62
Mesothelioma, **1:**421
Message arguments. *see* Argument processing
Message processing
    motivation in, **5:**803–4
    repetition and, **5:**804–5
Messenger RNA. *see* RNA, messenger
Messengers, intracellular, **2:**316–18. *see also* Second messenger systems
    chemical, **2:**11
    classes listed, **2:**91
    and signal transduction, **7:**20
Mestecky, **5:**138
Mestranol, **5:**599
Meta-analysis
    defined, **5:**599
    of health care technology, **4:**63
Meta-artery, **2:**184
Metabolic channeling, **5:**465
Metabolic clearance rate
    aging and, **3:**893–94
    caloric restriction and, **3:**894
    in dieting, **3:**593
Metabolic controllers
    defined, **3:**377
    role of, **3:**379–80
Metabolic diseases
    disease models, animal, **1:**257
    headaches associated with, **4:**58
    muscle-related, **5:**227–28
    sudden infant death syndrome and, **7:**320
    thyroxine and, **5:**707
Metabolic pump. *see* Biological transport, active
Metabolic tolerance
    of benzodiazepines, **5:**443

Metabolic tolerance *(Continued)*
defined, **5:**439
Metabolism, **2:**240–41, **4:**999–1006
adaptation in protein-energy malnutrition, **4:**894–95
aerobic, via oxidative phosphorylation, **3:**932
aging and, **3:**904
during cardiopulmonary bypass, **4:**94
carrier/receptor activity, **4:**1003–4
catabolism, **2:**240
cofactor, in adrenal medulla, **1:**646
cold stress, **4:**797
defined, **3:**377, **5:**471
of drugs, **7:**103–6
environment and, **4:**1006
enzymatic reactions, **4:**1000–1003
enzyme regulation of, **3:**423–38
feedback inhibition, **2:**241
general importance of, **5:**461–62
genetic errors in proteins, **4:**1003–4
higher thought processes and, **4:**1006
hormones and, **3:**326–27, **4:**1005–6
in hunger, **1:**348–49
levels of control, **4:**1000
marijuana and, **4:**906–7
mitochondrial membrane role, **4:**1004
nutrition and, **4:**1006
pathways, control of, **4:**1001
phosphorylation, **2:**241
of pollutants, **3:**256
polyamine. *see under* Polyamines
purine and pyrimidine, **6:**387–98
rate-limiting enzymes, **4:**1000–1001
regulation of, **2:**240–41
space travel and, **7:**143–45
in starvation, **4:**1005–6
thermoregulation, **4:**794
tissue interactions, **4:**1004
toxicity and, **3:**384
zinc, **7:**879–90
Metabolism, drug. *see under* Drugs
Metabolism, inborn errors
isoenzymes, **4:**578–80
porphyrin, **6:**118–19
treatment of, **5:**708
Metabolites, defined, **3:**919
Metacarpals, **7:**38
Metaclopromide, **2:**149
Metacyelocytes, **4:**171
Metagonimus yokogawai, **2:**389
Metalloenzymes, **7:**885
Metallothionein, **2:**659

Metals
as carcinogens, **7:**552–53
immunotoxicity of, **7:**553
occupational diseases caused by, **7:**551
toxicity, **3:**386–87, **7:**551–53
Metamer, **2:**575, 578–79, 586
Metamerism, **2:**579, 586
Metaphase, **2:**243, 244
chromosome banding in, **3:**839
DNA compaction in, **3:**125, 126
events of, **5:**70–71
Metaplasia
complete, **3:**770
intestinal, **3:**774–75
Metastasis. *see* Neoplasm metastasis
Metchnikoff, Elie, **3:**891
Metencephalon. *see* Pons
Methadone, **2:**786, **3:**212
Methaminodiazepoxide. *see* Chlordiazepoxide
Methamphetamine, **2:**515
Methane, **2:**544
Methanofuran, **2:**543, 544
Methanogen, **2:**545
Methanogenesis, **2:**543
Methanol dehydrogenase, **2:**540
Methemoglobin, **2:**347
Methemoglobinemia, **2:**346, 347
Methionine, **2:**507, 536, 537
degradation of, **3:**421
in phosphatidylcholine synthesis, **5:**887
Methisazone, **2:**399
Methlymercury, **2:**345
Methotrexate, **2:**374, 535, 536
chemoimmunoconjugates with, **4:**370
DNA amplification and, **3:**102
Methoxamine, **2:**22
Methoxatin, **2:**542
3-Methoxy-4-hydroxyphenylglycerol, **2:**848
Methoxyamphetamine, **2:**747
Methoxyhydroxyphenylethyleneglycol, **2:**836
8-Methoxypsoralen, **2:**114, **5:**266
1-Methyl-4-phenyl-1,2,3,6-tetrahydropyridine, **5:**669
1-Methyl-beta-isatin-thiosemicarbazone. *see* Methisazone
N-Methyl-D-aspartate, **2:**313, 339, 523
for Huntington's disease, **4:**273
Methylation
coenzymes and, **3:**438
of DNA, **3:**151–62
gene expression and, **3:**872
in gene imprinting, **3:**795–96
pattern, defined, **3:**151

Methylation, maintenance
    de novo methylation and, **3**:155–56
    DNA replication and, **3**:158
Methylation, sequence-specific. *see* Sequence-specific
    methylation
3-Methylcholanthrene, **2**:119, 121
Methylcobalamine, **2**:537
N⁶-methyldeoxyadenosine, **3**:154–55
5-Methyldeoxycytidine
    5'-CCGG-3' sequences and, **3**:153
    5'-CG-3' dinucleotide sequences and, **3**:152
    chemistry of, **3**:154–55
    detection of, **3**:153
    in eukaryotic DNA, **3**:152
    as fifth DNA nucleotide, **3**:153, 155–56
    late E2A promoter sequence inhibition by, **3**:159
    removal of. *see* Demethylation
Methyldopa, **2**:177
3,4-Methylenedioxyamphetamine, **5**:449
Methylenedioxyphenyl compounds, **5**:288
Methylgloxal, **3**:932
Methylisocyanate, **7**:549
Methylmalonate, **2**:537
L-Methylmalonyl-coenzyme A mutase, **2**:537
Methylmercury compounds
    birth defects caused by, **1**:701
    toxicity of, **3**:387
Methyl(methoxymethyl)nitrosamine, **2**:121
Methylobacterium, **2**:541
Methylphenidate, **1**:310, **2**:515
Methylreductase, **2**:544
Methyltransferases, DNA. *see* DNA methyltransferases
Methyprylon, **2**:787
Metronidazole, **1**:269, **2**:382, 383, 388, 390
Metyrapone, **2**:23
Mevalonic acid, **2**:168, 414, 415
Mevastatin, **2**:168
Meyerhof, Otto, **3**:933
Meynert's nucleus, basal, **2**:686
Mezcalin. *see* Mescaline
Mice, **2**:457. *see also* Murine
    in behavioral genetics research, **3**:836
    chimeric, trisomy 16 and, **3**:186–87
    chromosomes, **2**:457
    experimental metastasis, **4**:1008
    fertilization pathway of, **6**:507–8
    genetic mapping of, **3**:826–27
    immunity, in food-restriction studies, **5**:507
Mice, *mdx*, **5**:203
Mice, nude, and experimental metastasis, **4**:1008
Mice, transgenic
    Alzheimer's disease and, **3**:188

Mice, transgenic *(Continued)*
    angiogenesis model, **4**:1010–11
    defined, **3**:175
    human chromosome 21 and, **3**:187
    metastasis, **4**:1008
    in nutrition/immunity studies, **5**:515
Mice/human chimeric antibodies. *see* Antibodies, chimeric
Micelles, **4**:971
    defined, **3**:47
    formation, **1**:625, **3**:51
    mixed, lipolysis products and, **3**:54
Michaelis constant, **3**:423, 427, 907
Michaelis-Menten assumption, **3**:427, 430
Michaelis-Menten equation, **2**:375
Michaelis-Menten kinetics, **3**:377
Microbial antagonism. *see* Antibiosis
Microbial exotoxins. *see* Exotoxins
Microbial sensitivity tests, **1**:293
Microbiology, scope of, **5**:684
Microbiology laboratory
    antimicrobial susceptibility testing in, **1**:559
    future directions of, **1**:560
    as guide to antimicrobial chemotherapy, **1**:552
    as guide to public health activities, **1**:552
    as guide to understanding organisms, **1**:552
    importance in diagnosis, **1**:552
    modern diagnostic methods in, **1**:552–59
    quality control in, **1**:560
    reporting results in, **1**:560
Microcirculation
    defined, **3**:755
    fluid/solute exchange in, **5**:31–42
    gastric, **3**:757
    villus, **4**:549–50
Micrococcal nuclease, **2**:419, 421, 433
    DNA digestion and, **3**:128
    DNase I and, **3**:128
    mutations of, **3**:420
Micrococcus denitrificans. see Paracoccus denitrificans
Microelectrodes
    pH-sensitive, **5**:873
    use in attention research, **1**:480
Microevolution
    defined, **3**:85
    of dermatoglyphics, **3**:90
Microfibrillar proteins. *see under* Proteins
Microfilaria, **2**:387, 388
Microfluorimetry
    intracellular enzyme studies using, **5**:566
    scanning, **5**:570
Microfluorometry. *see* Cytophotometry
Microfluorometry, flow. *see* Flow cytometry

Microfold cells. *see* M cells
Microglia, **2**:343, **5**:385
Microgravity, **4**:124, 126–27
Microiontophoresis
  advantages of, **5**:6–8
  application of, **5**:6–8
  basic principle of, **5**:6
  current from, **5**:9
  historical background of, **5**:5–6
  major developments in, **5**:6
  and pressure ejection, **5**:5–9
  problems of interpretation using, **5**:8–9
  transport number of drugs used in, **5**:8–9
  variants of, **5**:8
Micronucleus, **2**:488
Micronutrients
  as antioxidants, **5**:509
  appetites for, **1**:347
Microscopic markers, **5**:907
Microscopic rate constants, defined, **3**:426
Microscopy
  in anatomic pathology, **5**:681
  contrast, defined, **3**:271
  in detection of bacterial pathogens, **1**:552
  flow cytometry and, **3**:621, 622
  intravital, **5**:576
  luminescence, **5**:576
  resolution
    defined, **3**:271
    equation for, **3**:271
    in transmission electron microscopy, **3**:273–74
Microscopy, confocal, **5**:570–72, **6**:751–54
  brightfield reflection, **6**:753
  interference, **6**:753–54
  laser scanning, **5**:574
  transmission, **6**:753
Microscopy, electron, **2**:721–30, **3**:271–79
  in anatomic pathology, **5**:681–82
  cryotechnique for, **2**:721–30
  defined, **5**:205
  dehydration, **2**:725
  hybrid techniques, **2**:724–27
  immobilization, essential, **2**:721–22
Microscopy, electron, scanning, **2**:727–29, **6**:735–45
  backscattered electrons, **6**:743
  backscattered images, **6**:742–44
  components of, **6**:737
  electron beam scanning in, **6**:738
  electron pathways in, **6**:738
  electron-solid interactions in, **6**:739–40
  electron sources for, **6**:737
  field emission gun in, **6**:737

Microscopy, electron, scanning *(Continued)*
  of freeze-dried specimen, **2**:727–28
  gun brightness in, **6**:737
  image processing for, **6**:749
  image resolution in, **6**:738
  imaging modes of, **6**:750–51
  at low temperature, **2**:727–28
  metal staining of tissue for, **6**:743
  operation of, **3**:272
  performance limits of, **6**:737
  principles of, **6**:735–37
  at room temperature, **2**:728–29
  secondary electron detector in, **6**:740
  secondary electron imaging in, **6**:740–42
  thermionic guns in, **6**:737
  tissue preparation for, **6**:742
  unwanted beam interactions in, **6**:744
Microscopy, electron, transmission, **3**:272–74
Microscopy, fluorescence, **5**:579. *see also* Microspectro-
      fluorometry
  in anatomic pathology, **5**:682
  confocal, **5**:570–75, **6**:752
  defined, **5**:903
  fluorescent indicators in, **5**:576
  laser scanning, **5**:570–71
  photoelectron microscopy and, **5**:905
  three dimensional, **5**:570
  total internal reflection, **5**:581
  video microscope and, **5**:572–74
Microscopy, immunofluorescence. *see* Microscopy,
      fluorescence
Microscopy, optical, **5**:561–85
  contemporary uses for, **5**:562
  instrumentation in, **5**:562–74
  methods, **5**:561–85
  support of, **5**:584
Microscopy, optical, scanning, **6**:747–54
  advantages of, **6**:747–48
  aligning of, **6**:748–49
  methods of scanning in, **6**:748
  objective lenses used in, **6**:749
  photodetectors in, **6**:749
  scanning system designs in, **6**:748
  schematic diagram of, **6**:748
  spectroscopic imaging, **6**:750
Microscopy, photoelectron, **5**:903–8, 905
Microscopy, polarization, **5**:566
Microsomes, **2**:739, 744
Microspectrofluorometry, **5**:562–64, 564–74
  advantages of, **5**:565
  defined, **5**:561
  grating, **5**:567

Microspectrofluorometry *(Continued)*
  low temperature attachment to, **5**:565
  phase condenser used with, **5**:569
  rapid scanning, **5**:564
  two-dimensional scan by, **5**:569
  ultraviolet absorption, **5**:563
Microspectrophotometry
  defined, **5**:561
  television fluorescence, **5**:567
Microtrabecular lattice. *see* Cytoskeleton
Microtubule-associated proteins
  defined, **5**:11
  microtubule properties and, **5**:12
Microtubule complex, cytoplasmic, **5**:68
Microtubule proteins, phosphorylation of, **5**:893–902,
    6:209
Microtubules, **2**:241, **5**:11–29, 244, 291–92
  astral, **2**:244
  centrosome and, **5**:68, 69–60
  defined, **5**:893
  microtubule protein phosphorylation and, **5**:899–901
  neuronal, isolation of, **5**:16
  of neurons, **5**:894
  in polarized epithelial cells, **6**:61
Microtubules, spindle
  in metaphase, **5**:70
  in prometaphase, **5**:70
Microvascular fluid exchange, **5**:31–42
Microvilli
  in electron microscopic imagery, **3**:274
  small intestine, **3**:76
Micturition control, **5**:43–46
Midbrain. *see* Mesencephalon
Middle ear
  muscle physiology, **3**:231–32
  ventilation, **3**:716
Middle promoters. *see* Promoter regions (genetics)
Middle-Upper Paleolithic transition, **3**:502
Midgut
  anatomy, **3**:64–65
  defined, **3**:57
  fetal, **3**:59
  rotations of, **3**:67
Migraine, **4**:51–53
  kinins and, **4**:599
  oral contraceptives and, **5**:608
  serotonin and, **6**:833
  theories of, **4**:51–52
Migrating motor complexes. *see* Myoelectric complex,
    migrating

Migration, cell. *see* Cell movement
Migration matrix, **4**:435–36
Milk
  aflatoxin M1 in, **5**:265
  antimicrobial systems of, **3**:663
  in "basic four" food group plan, **3**:654
  consumption, **4**:615
  drugs in, **5**:847
  formulas, fatty acid composition of, **5**:540
  in pastoralists' customary diets, **3**:20
Milk, cat, **3**:51
Milk, human
  composition, **4**:605, 612
  fat composition, **3**:48
  digestion of lipids in, **3**:47–56
  fatty acid composition of, **5**:540
  synthesis, **4**:604–7
Milk-alkali syndrome. *see* Hypercalcemia
Milk-ejection reflex, **4**:607–8
Milk fat, defined, **3**:47
Milk sinus, **4**:607
Milk tree, Ndembu society, **4**:701–2
Millipede toxins, **7**:597
Millon Clinical Multiaxial Inventory-II, **1**:610
Minamata Bay (Japan) disaster, **2**:345
Mind
  as collection of adaptations, **6**:500–503
  vs. body, **4**:704–5
Mineral oil, **2**:112
Mineralization
  bone, osteoblast products in, **1**:765
  of cartilage, **2**:212
Mineralocorticoids
  atrial natriuretic factor and, **1**:470
  function of, **1**:73, **7**:270–72
  pathway, in steroid biosynthesis, **7**:260
Minerals. *see also* Elements
  defined, **5**:47–48
  dietary deficiencies, interferons and, **4**:525
  estimated safe dietary intake of, **5**:486
  in hair, **4**:44–45
  in human life, **5**:47–53
  loss, and nutritional quality of food, **5**:494
  space travel and metabolism, **7**:146–47
  turnover of bone, **1**:773–74
Minimal brain dysfunction. *see* Attention deficit disorder
    with hyperactivity
Minimal inhibitory concentration (MIC), **1**:559
Minipill (contraceptive), **5**:600
Minisatellite loci, **3**:875

Minisatellite probes, **3:**137
Minnesota Multiphasic Personality Inventory, **1:**610
Miraculin, as sweetener, **7:**381
Misbehavior, **2:**709
Miscarriage. *see* Abortion
Mites, **2:**390
Mitochondria, **2:**98, 130, 238, 239, 341, 344, 480, 506, 509
    acceptor control of respiration, **1:**459
    in aerobic exercise, **3:**512, 513
    communication with cytosol, **4:**1004
    defined, **2:**480, **3:**317
    DNA, **2:**480
    electron transport chain, **1:**459-61
    inheritance, principles of, **3:**858
    metabolic regulation role, **4:**1004
    optical microscopic methods in, **5:**561-85
    oxidative phosphorylation. *see* Oxidative phosphorylation
    protein transport into, **6:**241-42
    proton translocation, **1:**461
    ribonucleotide transport into, **5:**464
    transducing, **2:**239
Mitochondrial encephalomyopathy, familial, **3:**858
Mitochondrial genes. *see* Genes, mitochondrial
Mitochondrial respiratory chain, **5:**55-63
    complexes in inner membrane, **5:**56-57
    control of, **3:**377-78
    defined, **3:**377
Mitogen response, defined, **5:**373
Mitogenic factors, lymphocyte. *see* Interleukin 2
Mitogens, endogenous. *see* Growth substances
Mitosis, **2:**243, 264, 291, 293, 341, 438, **5:**65-73. *see also*
        Cell division
    and centrosomes, **2:**293
    chromosome transmission in, **3:**849, 850
    and cyclin, **2:**243
    and cytokinesis, **2:**243
    defined, **3:**847, **5:**56
    DNA compaction in, **3:**125
    factor, **2:**243
    interphase, **2:**243
    m-stages, **2:**243
    microtuble protein phosphorylation and, **5:**901
    nuclear envelope in, **5:**451
    phases of, **5:**68-71
    protein phosphatases and, **5:**902
Mitosis-promoting factor. *see* Maturation-promoting
        factor
Mitotic motor, defined, **5:**65
Mitotic spindle apparatus
    aberrations of, **5:**72

Mitotic spindle apparatus *(Continued)*
    microtubules in, **5:**12
    phosphorylation of, **5:**902
Mitral cell layer, olfactory. *see* Olfactory cell layer, mitral
Mitral valve, **2:**180
Mitral valve prolapse, **3:**716
Mixed-function oxidations, ascorbic acid role in, **1:**431
Mixed-lymphocyte reaction, **4:**383
Mobility shift-method (sequence analysis), **3:**172
Models
    of attitude formation, **1:**486
    double ligation, in ganglioside transport study, **3:**746
    for Down's syndrome, **3:**185-86
    dynamic instability of microtubule assembly, **5:**16
    embryological, defined, **3:**281
    of gene amplification, **3:**789
    generalized central capacity, **5:**151-52
    multiple-resource, **5:**153-54
    psychogenetic, of energy intake, **3:**685
Models, animal
    of Down's syndrome, **3:**186-88
    of muscular dystrophy, **5:**203-4
    in Parkinson's disease research, **5:**670-72
Models, biological, in ecotoxicology, **3:**257
Models, genetic, for Down's syndrome, **3:**185-88
Models, mathematical, of pollutant persistence, **3:**256
Models, nerve growth factor
    aged rat, **5:**337-38
    septo-hippocampal, **5:**337
    unlesioned young adult rat, **5:**338
Modularity
    brain damage and, **5:**313
    neocortex, defined, **5:**305
Modus ponens, **2:**621, 625-27
Modus tollens, **2:**621, 625-27
Moenomycins, **1:**281
Molar lamina, **2:**817, 818
"Molecular clock", primate evolution and, **3:**886
Molecular cloning. *see* Cloning, molecular
Molecular diagnostic laboratories, **5:**686
Molecular dipoles
    defined, **3:**10
    polarization, **3:**10
Molecular genetics. *see* Genetics, biochemical
Molecular sequence data, **3:**163-67
Moles (unit of measure), **3:**317
Molybdenum
    biological function of, **7:**607
    deficiency, consequences of, **7:**610
    denaturation of, **3:**418

Molybdenum *(Continued)*
  significance, in body, **5**:49
  toxicity, **7**:610
Moment of force, defined, **5**:173
Monellin, as sweetener, **7**:381
Monilethrix, **4**:48
Monitoring, physiologic, in pediatrics, **5**:703
Monkeys
  New World, fossil record of, **6**:142
  Old World
    characteristics of, **6**:141
    fossil record of, **6**:142-43
    mating, **4**:932-35
Monoacylglycerols. *see* Glycerides
Monoamine oxidase, **2**:12, 366, 529
  Parkinson's disease and, **5**:669
  sensation seeking and levels of, **6**:814-15
Monoamine oxidase inhibitors, **2**:838, 842
  therapeutic use
    depression, **1**:309-10
    of mania and depression, **1**:95
    mood disorders, **5**:95-96
Monoamines, **2**:337, **3**:338, 831
  astrocyte uptake of, **1**:439
  epilepsy and, **3**:460
  hypothesis of depression, **2**:843
  metabolism, suicide and, **7**:333-34
  as neuroregulators, **2**:5, 337
  pituitary hormone secretion and, **5**:396-97
  sensation seeking and levels of, **6**:815
  serotonin, **2**:831
  transporter, pharmacology of, **1**:648-49
Monoamniotic twins. *see* Twins, monozygotic
Monocarboxylic acids, **5**:873
Monochorial twins. *see* Twins, monozygotic
Monocotyledonous plants, **3**:39-40
Monocyte chemotactic and activating factor (MCAF), **1**:28
Monocytes, **2**:613, 751
  characteristics of, **5**:813
  development of, **6**:606
  differentiation, *fos* protein in, **3**:707
  platelet-activating factor and, **6**:48-49
  production and differentiation, **4**:854
  surface markers, **4**:855-56
Monocytopoiesis, inhibition of, **4**:854
Monogamy, hominid, **3**:500-501
Monohybrid cross, defined, **3**:848
Monokines
  immunity and, **5**:510
  in *Plasmodium* infections, **4**:879

Monomorphic loci, defined, **3**:798-99
Mononuclear phagocyte system. *see* Reticuloendothelial system
Mononuclear phagocytes. *see* Macrophages
Monooxygenase, **2**:738
Monooxygenase P-450. *see* Cytochrome P-450
Monosaccharides
  in food, **2**:102-3
  transport, across blood-brain barrier, **1**:725
Monosodium glutamate, **2**:16, **4**:57
Monosomy, defined, **3**:835
Monosomy-13, **2**:441
Monosomy-14, **2**:441
Monosomy-21, **2**:438
Monosomy-22, **2**:439
Monosomy-7 (Turner syndrome), **2**:443
Monosomy, X, **2**:437
Monozygotic twins. *see* Twins, monozygotic
Mons pubis, **6**:564
Montgomery's glands, **4**:607
Mood. *see* Affect
Mood disorders. *see* Affective disorders
MOPP/ABV antineoplastics, **4**:850
Moral reasoning, during adolescence, **1**:66-67
Morantel, **2**:392
Morbidity, inbreeding and, **4**:439-40
Morgan, T. H., **3**:824
Morning sickness
  origins of, **6**:125
  patterns of, **6**:124
  and pregnancy outcome, **6**:125
  prevalence in modern society, **6**:124
Morphine, **2**:26, 338, 340, 593, 746, 747, 786
  characteristics of, **1**:179
  and hyperthermia, **2**:339
Morphine dependence, **1**:43-51
Morphogenesis, **2**:873
  bone, **1**:789
  defined, **3**:281
  inborn errors of, **1**:698-99
  as process in shape of bone, **1**:789
Morphogenesis, neuronal. *see under* Neurons
Morphogens, defined, **3**:281
Morphology
  epithelial tumor cell alterations *in vitro*, **4**:584
  HeLa cells, **4**:108
  leukemic cells, **4**:693
Morrison, **5**:722
Mortality, **5**:99-103. *see also* Death; Fetal death; Infant mortality; Maternal mortality

Mortality *(Continued)*
  central death rates
    defined, **5**:99
    for selected countries, **5**:101
    United States, **5**:100
  inbreeding and, **4**:439
  international, **5**:101
  oral contraceptives and, **5**:608-9
Mosaicism. *see also* Chimera
  chromosomal, **2**:438
  defined, **5**:259
  dispermic, **3**:306
  double, **3**:306
Mosquitoes, as vectors in zoonoses, **7**:895
Mother-infant interaction, **2**:854, 862-63
  imprinting of attachment, **2**:873
Mothers, unmarried, DZ twinning and, **3**:307
Motilin, **2**:27
  erythomycin and, **5**:723
  function of, **6**:89
Motion perception, network modeling of, **5**:389
Motivation, **2**:629
  for altruism, **1**:198, 201, 202-3
  and behavior, **2**:633
  deficits, in learned helplessness, **4**:667-68
  generalized central capacity and, **5**:154
  in human information processing, **5**:803
Motives, altruistic vs. egoistic, **1**:199
Motor, primary, **2**:682
Motor activity. *see also* Movement
  autism and, **1**:506
  development
    birth to 3 months, **1**:574
    3 to 10 months, **1**:576
    10 to 20 months, **1**:577
    20 to 36 months, **1**:578
    3 to 5 years, **1**:579
    5 to 8 years, **1**:580
    8 years to preadolescence, **1**:581
    differentiation vs. integration of processes, **1**:582
    general stages vs. specific cases, **1**:582
    role of shared heredity, **1**:582
  developmental, **2**:873-74
  in infancy, changes in, **6**:820-27
  persuasion and, **5**:807
Motor control, **5**:105-20
  acquiring new patterns of coordination in, **5**:116-17
  central contributions to, **5**:110
  flexibility, goal accomplishment and, **5**:117-20
  general modes of, **5**:108

Motor control *(Continued)*
  handwriting as example of programmed, **5**:117-19
  hemispheric interactions, **4**:133
  hierarchical, **5**:117
  impulse-timing viewpoint of, **5**:111-12
  information programmed in, **5**:111-13
  mass-spring viewpoint as method of, **5**:112-13
  motor equivalence in, evidence for, **5**:119-20
  movement constraints in, **5**:113-17
  reconciliation of peripheral and central viewpoints of,
    5:111
  sensory contributions to, **5**:108-10
  spinal, **7**:215-20
    descending pathways in, **7**:218-20
    interneurons and proprioneurons in, **7**:217-18
    motor neurons and, **7**:215-16
  units of, **5**:115-16
Motor cortex, **2**:7
  dyslexia-dysgraphia and, **3**:218
  monkey, arm movements and, **5**:389
  neurons of, **5**:309
Motor crosstalk. *see* Crosstalk, motor
Motor dysgraphia, **3**:218
Motor dysphasia. *see* Aphasia, Broca
Motor learning, defined, **5**:121
Motor neurons
  activation threshold of, **5**:179
  of bladder, **5**:43-44
  of intrinsic external urethral sphincter, **5**:43-44
  membrane changes and, **3**:698
  size principle of, **3**:694
  somatic, defined, **5**:43
  task group theory of, **3**:696
Motor patterns, in walking and running, **4**:769-71
Motor program, defined, **5**:121
Motor skills
  acquisition of, **5**:121-29
    error detection in, **5**:128
    guidance for, **5**:128
  neocortical lateralization and, **5**:313
Motor units
  defined, **3**:691
  firing rate
    in muscle force production, **5**:179-80
    at recruitment, **3**:698
  of m. triceps, **3**:697
  properties of, **5**:176-78
  recruitment threshold of, **3**:693
Mountain hypothermia, **4**:802
Mountcastle, **5**:309

Mouth, **2:**769-70
    anaerobic bacteria found in, **1:**240-41
    anatomy, **2:**769-81
        embryonic, **2:**818
        primitive, **2:**818
    developmental abnormalities of, **5:**611-12
    as digestive organ, **3:**74
    fibro-osseous conditions of, **5:**615
    fungal infections of, **5:**614
    mucosa, **2:**780-81
    mucous membranes, ascorbic acid absorption, **1:**430
    pathology, **5:**611-16
    salivary glands and, **2:**781
    specific infections of, **5:**613-14
    tumors of, **5:**615-16
Mouth neoplasms, **5:**615-16
Movement, **5:**131-35. *see also* Motor activity
    central commands in organization and order in,
        5:110-11
    constraints, computational simplification of, **5:**113-17
    control of, **5:**105-20. *see also* Motor control
        degrees of freedom and, **5:**113-17
        entrainment as feature of, **5:**115-16
        locus for, **5:**108
        vision receptor system in, **5:**109
    coordination
        amplitude and duration of, **3:**691-92
        direction of, **3:**692-98
    crosstalk and, **5:**155, 157-60
    factors controlling, **5:**105-6
    fast, muscle activation in, **3:**698
    force coupling in, **5:**159
    free fast, muscle spindles in, **3:**723
    impulse-timing viewpoint of, **5:**111-12
    information programmed in, **5:**111-13
    initiation of, **3:**722
    interlimb interactions and, **5:**113-17
    isotonic muscle contraction and, **5:**131
    joint, types of, **5:**133-35
    limb
        control of, cyclical, **5:**113-14
        invariance in, **3:**698
    mass-spring viewpoint of, **5:**112-13
    multi-joint, variety in, **3:**698
    muscle contraction and, **5:**106-7
    periodic. *see* Periodic movement
    phase coupling in, **5:**158
    preparation for, **3:**722
    reconciliation of peripheral and central viewpoints of,
        5:111

Movement *(Continued)*
    role of sensory information after, **5:**110
    sensory contributions to, **5:**108-10
    skeletal, **5:**133-35
        lever activity in, **5:**131-33
        muscles and joints in, **5:**131-33
    voluntary
        factors controlling, **5:**106
        muscle activation in, **3:**697-98
        ventral oral nucleus of thalamus and, **7:**442
MSH, **2:**744
    fetal, **3:**343
    functions of, **5:**727, 729
MSH, alpha. *see* Alpha-MSH
MSH, beta. *see* Beta-MSH
*Msp*I, methylation of, **3:**153
MT, **4:**200-201
mtDNA, defined, **5:**55
Mucilage, seed. *see under* Seeds
Mucins
    of gastric tumors, **3:**769
    of large intestine, **3:**768
    in saliva, **6:**694
Mucociliary apparatus, **5:**691
Mucopolysaccharidoses, **6:**265
Mucosa, **3:**755. *see also* specific types of mucosa
Mucosa-associated lymphoid tissue, **3:**769
Mucosal-associated lymphoreticular tissues
    defined, **5:**137
    IgA induction and, **5:**141
    IgA response induction and, **5:**139
Mucosal diseases, **5:**614-15
Mucosal erosion, defined, **3:**755
Mucous membrane
    defined, **3:**71
    gastrointestinal
        atrophy of, **3:**902
        cancer in, **3:**768
    pemphigoid, **5:**615
    radiation effects, **6:**413
Mucoviscidosis. *see* Cystic fibrosis
Mucus, **2:**34, 37
    in cystic fibrosis, **2:**731
    defined, **3:**71
    gastric, **3:**76
    in upper respiratory defense, **5:**691
Muller, H. J., **5:**251
Muller cells
    characterization, **1:**435-36
    function of, **3:**542

Mullerian ducts, **3**:341

Mullerian inhibiting substance, **2**:246

Mulliparity, **2**:54

Multicellular organisms, **3**:282–83

Multienzyme complexes, **5**:465–66

Multifactorial conditions

    defined, **3**:17

    nutritional deficiencies and, **3**:21

    techno-food and, **3**:22

Multifactorial traits, **3**:855–56

Multigravida, **2**:403

Multilocus probes. *see under* DNA probes

Multimix, defined, **3**:17

Multiple-resource models, defined, **5**:216

Multiple sclerosis, **5**:143–48

    altered specific T-cell populations in, **5**:145–46

    antigen nonspecific suppressor function in, **5**:145

    as autoimmune disease, **5**:143–44

    autologous mixed lymphocyte response in, **5**:144–45

    as demyelinating disease, **5**:280–81

    diagnosis of, **5**:143–44

    experimental models of, **5**:146–47

    genetics of, **5**:146

    immunotherapy for, **5**:1148

    measures of immune function in, **5**:144–45

    pathogenesis of, **5**:147

    pathology, **5**:144

    peripheral blood and central nervous system in, **5**:146

    treatment with interferon, **4**:522

Multiple-task interactions, **5**:149–60

Multiplexing, **2**:475

Mumps vaccine, **4**:466

Munson, **5**:382

Mupriocin, **1**:287–88

Muramidase

    in bacterial cell wall synthesis, **1**:277–78

    catalytic mechanism of, **3**:433–34, 435

    salivary, **2**:781, **6**:695

    thermoinactivation of, **3**:413, 414, 417–18

        ammonia released by, **3**:417

        deamidation in, **3**:416

    thermostabilization of, **3**:420

Muramyl dipeptide, as adjuvant for subunit vaccines, **7**:753

Muramyl tripeptide phosphatidylethanolamine, **1**:684

Murein. *see* Peptidoglycan

Murine/human chimeric antibodies. *see* Antibodies, chimeric

Murine keratinocytes. *see under* Keratinocytes

Murine mammary tumor virus, **2**:63

Murine retroviruses, **6**:644–45

Murrell, K. F. H., **3**:475

Muscarine antagonists, **2**:522

Muscarinic receptors. *see* Receptors, muscarinic

Muscimol, **2**:493

Muscimol receptors. *see* Receptors, GABA-benzodi-azepine

Muscle, skeletal, **2**:125, 126, 180, **7**:23–31

    anatomy, **5**:207–9

    components of, **6**:436–37

    contraction, **7**:26–30

        chemical waste products from, **7**:30

        energy supply for, **7**:29–30

        isometric and isotonic, **7**:31

        physiology of, **7**:28–30

        sliding filament model of, **7**:27

    dystrophin in, **5**:196–98

    effects of training on, **7**:30–31

    elemental composition of, **6**:437

    energy utilization by, **5**:242

    in exercise, **3**:509–10, **5**:189–90

    fast-fiber, **5**:212–13, 215

    fatigue, physiology of, **7**:28–29

    fiber

        slow twitch, **7**:29

        structure of, **7**:24

    filaments, molecular organization of, **7**:25

    force enhancement of, **5**:185–87

    force production in, **5**:182

    force-velocity curve of, **5**:181–82

    genesis of, **5**:163

    glycogen content of, **3**:920–21

    innvervation of, **5**:171

    and joints, movement and relationship between, **5**:131–33

    length-tension curve of, **5**:180–81

    length-tension relationship in, **5**:239

    mass density of, **6**:437

    mechanical properties of, **5**:173–90

    motor end plate and fibers, **7**:24

    motor units, regulation of, **5**:238

    motor units comprising, **7**:24

    myoblast/myotube fusion in, **5**:168

    myofibrillar protein degradation and, **5**:167

    nerves controlling, **7**:24

    nerves supplying, **5**:354

    organization of, **5**:207, **7**:23–26

    pH of, **5**:824

    plasticity of, **3**:512–13

    properties of, **5**:174–76

Muscle, skeletal (*Continued*)
proprioceptors of, **7**:24
regeneration of, **5**:163
regulation of activity, **5**:238–40
relaxation of, **7**:27–28
shapes, **4**:259–60
slow-fiber, **5**:215
staircase effect contraction and, **7**:28
strength and endurance of, **7**:30
stretch-shorten cycle of, **5**:184–85
structure of, **5**:229–30
tension development in, **5**:178–80
Tn-C isoforms of, **5**:220
triacylglycerol in, **3**:583
twitch, **7**:28
type-one myosin light chain production and, **5**:164
Muscle, smooth, **2**:125, 640, **7**:85–87
airways, autonomic innervation, **1**:546–47
bradykinin and, **5**:730
contraction of, Ca^{2+} and, **5**:226
differentiation of, **5**:163
excitation-contraction coupling of, **5**:235–36
genital erectile tissue, autonomic innervation, **1**:548–49
in gut, **5**:723
autonomic innervation, **1**:549–50
histology of, **7**:85
kinins and, **4**:595
macromolecular organization of, **5**:209
mechanism of contraction, **7**:86–87
organization of, **5**:170
phasic, regulation of, **5**:240–41
regulation
of activity, **5**:240
of force in, **5**:241
single and multiunit, **7**:86
special characteristics of, **7**:85
structure, **5**:231
thick filament regulation in, **5**:234
vasoactive intestinal polypeptide and, **5**:718
versus skeletal muscle, **7**:86
Muscle, smooth, vascular, **7**:762
development in lungs, **6**:346–47
endothelium-derived relaxing factor and, **3**:371
nitric oxide and, **3**:374
Muscle, smooth, visceral, **7**:86
Muscle, striated, **2**:640, **5**:229–31
excitation-contraction coupling of, **5**:235–36
structure of, **5**:229–31
thin filament regulation in, **5**:233–34
Muscle contraction, **4**:260–61
activation, in isometric contractions, **3**:692–96
ATP and, **3**:379–80

Muscle contraction (*Continued*)
characterization, **4**:260
electromechanical delay, **5**:184–85
excitation-contraction coupling of, **5**:234–36
glucose consumption in, **3**:512
heat from, **3**:511–12
measurement of, **5**:177–78
molecular organization of, **5**:209–25
movement and isotonic, **5**:131–33
reciprocal inhibition as factor in, **5**:106–7
in reflex-induced contractions, **3**:696–97
regulation of, **5**:225–27
relationship between movement and, **5**:106–7
sliding filament of, **5**:232–33
sliding mechanism of, **5**:209
Muscle relaxation, **5**:236–37
Muscle rigidity, **2**:346
Muscle spasticity
fusimotor overactivity in, **3**:723
reflex compensation and, **3**:723
as spinal cord disorder, **7**:219–20, 230
Muscle spindles
anatomy of, **3**:719–20
innervation of, **3**:721–22
limb-position movement and, **6**:173–74
postural control and, **5**:108
and proprioception, **6**:168–71
structure and innervation of, **6**:169
in voluntary movements, **3**:723
Muscle tone, **7**:230
Muscles. *see also* specific muscles
afferents, reflexes involving, **6**:158–59
aging of, **3**:903–4
agonistic, **4**:260
anatomy, **4**:259–67
antagonistic, **4**:260
atonia, REM sleep and, **7**:72–73
capillaries of, **5**:39
cardiac. *see* Myocardium
cells, **2**:132–33
classes of, **5**:206–7
development, **5**:161–72
of digestive system, **3**:73
dynamics, **5**:173–90
embryonic, formation of, **5**:161–63
energetics, **5**:242
energy demands and metabolic regulation, **4**:1004
in epileptic clonic seizures, **3**:456
fascicles, **5**:174
function, **5**:369–72

Muscles (*Continued*)
  gene expression of nuclei, **5**:168
  glycolysis in, **3**:939
  grading mechanisms of force, **3**:692–93
  insertions, **4**:260–61
  intramuscular electromyographic activity, **3**:694
  invertebrate, contraction of, **5**:226
  laryngeal, **4**:656–57
  length-tension relationship, **5**:238–39
  load-velocity relationship, **5**:239
  loss, and space travel, **7**:150–51
  lower extremity, **4**:265–67
  mechanical advantage
      defined, **3**:695
      joint angle and, **3**:695–96
  membrane, dystrophin and, **5**:203
  metabolism and physiology, space travel and, **7**:145–46
  molecular genetics of, **5**:205–28
  movement. *see* Proprioception; Proprioceptors
  nomenclature, **4**:259
  origins, **4**:260–61
  pectoral girdle, **4**:263–64
  physiology and biochemistry, **5**:229–42
  positive work, defined, **5**:185
  regulation of activity, **5**:238
  repair, growth factors and, **7**:507
  response to cold, **4**:796
  series elastic component, **5**:185
  sheaths, edematous, **5**:41
  stiffness, **5**:241–42
  stimulus frequency, **5**:238
  structure, **5**:229–32
  synergists, **4**:260
  tetanized, stimulus frequency and, **5**:238
  trunk, **4**:263
  upper extremity, **4**:264–65
  zinc metabolism in, **7**:882
Muscular atrophy
  animal studies on, **7**:146
  space travel and, **7**:145
Muscular dystrophy, **2**:481
  Becker's muscular dystrophy, **2**:479
  and cell death, **2**:251
  DNA marker diagnosis of, **3**:146–47
  fathers and, **5**:201
  molecular genetics of, **5**:191–204, 228
  multiplex amplification in diagnosis of, **3**:147
  mutation rate of, **5**:245
  newborn screening test for, **5**:437
  X-linked in female, **2**:481

Muscularis externa, defined, **3**:755
Muscularis mucosa, defined, **3**:755
Musculoskeletal system, fetal, **3**:611
Mushroom poisoning, **7**:591–92
Mustard gas, **2**:110, 112
Mutagenesis, insertional, **2**:265
Mutagens
  chemical, **5**:253
  defined, **5**:243
  detection of, **5**:249–51
  environmental, and cancer, **5**:324
  hyperthermia as, **4**:300
  in lymphocyte chromosomal variations,
      5:248
  metabolism of, **5**:254
  mutagenic risk, **2**:67
  pesticide residues in food as, **7**:561
  trichothecenes as, **7**:665
Mutants, defined, **3**:819
Mutation, **2**:109, **6**:110, 257. *see also* Point mutation
  aflatoxin B1 and, **5**:265
  chromosomal, rates of, **5**:246–48
  consequences of, **3**:872–73
  defined, **3**:791, **5**:243, 244, 819
  in DMD gene, **5**:199–201
  DNA
      change and, **2**:109
      denaturing gradient gel electrophoresis in
          detection of, **6**:527
      diagnosis of cancer-associated, **6**:528
      direct detection of, **6**:526–27
      ribonuclease A cleavage in detection of, **6**:527
      sequencing in detection of, **6**:527–28
  for enhanced enzyme thermostability, **3**:420
  in enhancer regions, **3**:794
  forces controlling, **6**:109–10
  in *fos-jun* heterodimers, **3**:706–7
  in genes, **3**:797, 798–99
  genetic mapping and, **3**:827–28
  genomic
      defined, **5**:243
      nondisjunction and, **5**:244
      in primate evolution, **3**:883
      rate of, **5**:247
  germinal
      age factor in, **5**:254
      defined, **5**:243
      rates of, **5**:244–47
  between groups, **6**:109
  heterogeneity of, **3**:854–55

Mutation (*Continued*)
   historical records in determining, **6:**114
   hybrid-gene synthesis and, **3:**172–73
   kinds of, **3:**882–84
   mating structure and, **6:**112–13
   muscle proteins and, **5:**169
   natural selection and, **6:**110–11
   neutral, **3:**886
   population migration and, **6:**112
   rate, **3:**796, **5:**243–54
      of muscular dystrophy, **5:**192
   recessive, tumor formation and, **7:**706
   resistant, **2:**395
   and retinoblastoma, **6:**637–38
   risk variation in, **5:**253–54
   somatic, **2:**119
      defined, **5:**243
      rates of, **5:**248–49
   synthesis of forces controlling, **6:**113–14
   temperature-sensitive, **2:**257
   virus, and vaccine design, **7:**749
Mutational load, inbreeding, **4:**438
Mutualism, **1:**117, 262
Myasthenia gravis, **2:**301, **5:**255–58
   and acetylcholine receptor deficiency, **2:**330
   and acetylcholine receptors, **2:**301, **5:**257–58
   animal models in, **5:**258
   antibodies and, **4:**408
   as autoimmune disease, **1:**515
   general characteristics of, **5:**255
   pathology, **5:**256–57
   as receptor-specific autoimmune disease, **1:**530
   theories as to cause of, **5:**257–58
   treatment of, **5:**258
MYB genes. *see* Oncogenes
Myclin, **2:**825
Mycobacteria
   antibiotics against, **1:**293
   detection of, in skin infections, **1:**557–58
   resistance to antimycobacterial agents, **1:**328
   synthetic agents specific for, **1:**327, 329
Mycobacterium bovis, **7:**897
Mycobacterium leprae, **4:**449
Mycobacterium tuberculosis, **4:**447
   and AIDS, **1:**156
   laboratory detection of, **1:**557
Mycophenolic acid, **3:**806
Mycoplasma pneumoniae, **4:**450
   in infants and toddlers, **4:**460–61
   in school-age children, **4:**464

Mycotoxicosis
   defined, **5:**260, **7:**659
   food contamination and, **5:**260
   listed, **5:**262
   prevention of, **5:**268–69
   trichothecene, **7:**665–66
Mycotoxins, **5:**259–69. *see also* Toxins, fungal
   chemoimmunoconjugates with, **4:**370
   defined, **3:**671, **7:**659
   in food, **3:**666, 674
   food analysis, **4:**225–26
   trichothecenes as, **7:**659–69
Myelin, **2:**2, 341, 344, 345, 676
   basic proteins, **5:**277
   central nervous system composition, **5:**275
   chemical composition, **5:**274–79
   degradation, **5:**280
   and demyelinating disorders, **5:**271–83
   development, **5:**274
   diseases involving, **5:**280–82
   enzymes associated with, **5:**278
   evolutionary appearance of, **5:**271–72
   formation, **5:**273–74
   function of, **5:**272–73
   glycoproteins associated with, **5:**277–78
   lipids, **5:**274–76
   metabolism, **5:**279–80
   methods of study of, **5:**273
   structure, **5:**273–74
   synthesis, **5:**279–80
Myelin disorders
   genetically determined, **5:**282
   hormonal deficiencies causing, **5:**282
   nutritional deficiencies causing, **5:**282
   toxins causing, **5:**282
Myelin proteins, **5:**276–79
   nerve regeneration inhibition of, **5:**345
Myelination of sensory system, in newborn, **6:**823
Myeloblasts, **5:**427
Myelocytic leukemia. *see* Leukemia, myelocytic
Myelodysplasia, **2:**450, 451
Myeloid cells, progenitors, **4:**169
Myeloma
   characterization, **4:**848
   hypercalcemia of malignancy and, **4:**279
Myeloperoxidase
   activity, during respiratory burst, **5:**429–30
   killing of microorganisms during respiratory burst, **6:**582–83
Myelosuppressive anticancer therapy, **4:**751–52

Myleran, **2:**114
Myo-inositol 1,4,5-trisphosphate. *see* Inositol 1,4,5-tris-
  phosphate
Myoblasts
  cardiac
    development of, **5:**163
    organization of, **5:**170
  clones of, **5:**162
  defined, **5:**161, 191, 205
  differentiation of, **5:**162
  fusion into myotubes, **5:**168
  fusion of, **5:**164
  smooth, development of, **5:**163
  structure of, **5:**168
  transplantation, in *mdx* mice, **5:**203
Myocardial contraction, **2:**125–26, 180–81, **3:**518, **5:**240
  acidosis and, **5:**824
  calcium and, **3:**509
  exercise and, **3:**518
  mechanism, **2:**126–27
  sequence of events, **2:**181
  static exercise and, **3:**520
Myocardial infarction, **2:**160
  defined, **2:**167
  oral contraceptives and, **5:**602–3
  treatment with cardiopulmonary bypass, **4:**97
Myocardial perfusion, in dynamic exercise, **3:**519
Myocardium, **2:**125–236, 157, 180, 181
  action, potential, **2:**134
    depolarization, **2:**134–35
    ion flow, **2:**134
    origin, **2:**134–36
    period, refractory, **2:**134
  activity
    electrical, **2:**134–36
    regulation of, **5:**240
  and calcium, **2:**181–82
  cytoskeleton, **2:**130
  description, **2:**126
  desmosome, **2:**132
  disk, intercalated, **2:**132
  electron micrograph, **2:**131
  energy, **2:**181
  fibers, **2:**181, 253
  function, **2:**126
  in heart rate control, **3:**517
  kinin effects, **4:**594–95
  membrane, **2:**130–32
  membrane potential, resting, **2:**134–35
  metabolism, **2:**181

Myocardium (*Continued*)
  muscle, specialized, **2:**125
  muscle cells of, **5:**162
  myofibril, **2:**181
  nexus, **2:**132
  node, sinoatrial, **2:**132
  nutrient, **2:**182
  organization of, **5:**170
  Tn-C isoforms of, **5:**220
Myoclonic epilepsy, progressive. *see* Epilepsy, myoclonus
Myoclonic seizures
  in epilepsy, **3:**456
  generalized tonic-clonic seizures and, **3:**459
Myocyte, cardiac, **2:**270
  and protein 135-kDa, **2:**270
MyoD1 protein, **5:**168
Myoelectric complex, migrating, **2:**38
  defined, **2:**31
  interdigestive, **2:**35–38
  motilin and, **5:**721
  phase III, **2:**35
Myoepithelial cells, **4:**607
Myofibrils, **2:**125–29, **5:**230, 181
  alignment of, **5:**175
  anatomy, **5:**208
  defined, **5:**161, **7:**24
  in muscle contraction, **5:**209
  organization of, **5:**168–69
  structure of, **5:**206
  thick filaments, structure of, **5:**231–32
  thin filaments of, structure of, **5:**232
  vertebrate skeletal, **5:**210
Myofilaments, anatomy, **5:**208
Myogenesis
  defined, **5:**161
  hormones and, **5:**171–72
  induction of, **5:**170
  innervation and, **5:**171
Myogenic molecules, **5:**167
Myoglobin, **2:**655, **4:**135–36
  alpha-chain hemoglobin and, **3:**165
  as gene family, **3:**798
  and iron, **2:**655
  muscle contraction and, **7:**29
  oxygen equilibrium, **4:**136–37
  tetramer assembly, **4:**140–41
Myomesin
  myosin and, **5:**169
  properties of, **5:**223
Myometrium, **2:**404, **7:**738

Myoneural junction. *see* Neuromuscular junction

Myopathies, genetic aspects of, **5:**227–28

Myopia, defined, **3:**533

Myosin, **2:**126–28, 172, 181, 182, 241, 242
    cardiac contractility and, **3:**518
    crossbridge cycling, regulation of, **5:**233–34
    defined, **5:**173, **7:**25
    genetic analysis of, **5:**169
    in myofibril organization, **5:**168
    polymerization of, **5:**169
    properties of, **5:**210–16
    in smooth muscle, **5:**170
    structure of, **5:**176, 231–32
    ubiquitousness of, **5:**164

Myosin cross-bridges. *see* Cross-bridges, muscle

Myosin head
    actin-activated ATPase activity of, **5:**211
    defined, **5:**211

Myosin heavy chain isoforms, **5:**212
    in *Caenorhabditis elegans*, **5:**164
    genetic analyis of, **5:**169
    isozymes of, **5:**211–12
    skeletal muscle and, **5:**164
    thyroid hormone and, **5:**171

Myosin hinges, defined, **5:**211

Myosin isoforms
    cardiac, **5:**165
    differential location of, **5:**165
    transitions in, **5:**164

Myosin light chains
    alkali, **5:**212–13, 215, 216
    alternative transcription of, **5:**167
    Ca^{2+} and, **5:**206, 226
    cardiac isoforms of, **5:**215–16
    classes of, **5:**212
    phosphorylation of, **5:**218
    synthesis, **5:**166

Myosin power stroke, defined, **5:**226

Myosin tail, defined, **5:**211

Myosin thick filament, defined, **5:**211

Myotatic reflex, **7:**216

Myotubes
    defined, **5:**205
    fusion from myoblasts, **5:**168
    myoblast clones and, **5:**162
    myofibril formation and, **5:**169
    satellite cells and, **5:**163

Myristolation, G proteins and, **4:**21

Myrothecium roridum, **5:**263

Myxedema, **2:**648
    congenital hypothyroidism and, **3:**352
    and thyroid hormone deficiency, **2:**648

Myxovirus infections. *see* Orthomyxovirus infections

N

N-Acetylglucosamine (NAG). *see* Acetylglucosamine

N-CAM. *see* Neural cell adhesion molecule

N1 factor, chromatin assembly and, **3:**130

N-Nitroso compound, carcinogenic, **2:**110

n-Octanol:water partition coefficient
    bioconcentration factor and, **3:**257
    defined, **3:**256

NAD, **2:**530, 655
    chemistry, **2:**530–31
    deamidase, **2:**532
    functions, **2:**532
    glycohydrolase, **2:**532
    metabolism, **2:**531–32
    in oxidative phosphorylation, **5:**473–74
    structure, **2:**531

NAD-linked dehydrogenases, **1:**459–60

NADH, **2:**505–9

NADH-ubiquinone reductase, **5:**58

Nadolol, **2:**171

NADP, **2:**505–9, 530, 532, **3:**377, 913

NADP oxidase, **5:**874
    nature and molecular structure of, **6:**582
    role of, in respiratory burst, **6:**582

Nagana, **7:**684

Nails, **7:**46

Nalbuphine, **2:**786

Nalidixic acid
    antimicrobial activity of, **6:**403
    function of, **6:**402–3
    as quinolone, **6:**400

Naloxone, **2:**26, 28, 338, 787
    alcohol and, **3:**212
    for narcotic overdoses, **5:**723

Naltrexone
    for alcohol intoxication, **3:**212
    in treatment of autism, **1:**510

Naproxen, **2:**785

Napthoquinone, **2:**544

2-Napthylamine, **2:**112

Narcissism, defined, **6:**313

Narcolepsy, **2:**515, **4:**383–84
    amphetamines for, **5:**440
    REM sleep and, **7:**73
    as sleep disorder, **7:**79–80

Narcotic dependence, **1:**44–51. *see also* Substance dependence

Narcotics
    alcohol and, 3:212
    and convulsive seizures, 6:788
    and spinal cord dysfunction, 7:225-26
    and sudden infant death syndrome, 7:321
Nasal cavity
    anatomy of, 6:588-89
    biochemistry, 5:285-89
    cytochrome P-450 enzymes in, 5:285-86
    ester metabolism in, 5:288
    formaldehyde metabolism in, 5:288
    general biochemistry of, 5:288-89
    glucuronidation in, 5:286
    xenobiotic metabolizing enzymes in, 5:285-88
Nasal hair, 4:40
Nasal obstruction, 5:695
Nasal septum, 2:706
Nasalization, 4:635
Nasopharyngeal neoplasms, 4:343
Nasopharynx, 7:578
Nasse's law, 3:852
National Academy of Sciences
    Diet, Nutrition, and Cancer Report, 3:651
    dietary recommendations of, 3:27-28
National Commission on Marijuana and Drug Abuse,
        5:445
National Institute on Aging, 3:896
National Nutrition Monitoring System, U.S., 3:650
National Research Council
    Diet and Health, 3:655
    on ergonomics, 3:476
Native American Indians. see Indians, American
Native conformation, enzyme. see Enzyme conformation,
        native
Natriuresis, 2:143
Natriuretic factor. see Atrial natriuretic factor
Natriuretic hormone, 3:93
Natural killer cells. see Killer cells, natural
Natural resources
    preservation of, 1:146
    restoration of, 1:146
Natural selection. see Selection (genetics)
Nature versus nurture problem, 2:869, 865
Nausea
    marijuana for, 5:444
    vasopressin secretion and, 5:397-98
Ndembu society, 4:701-2
Neanderthals, 2:189, 593
    characteristics of, 3:492
    cold-weather adaptations of, 3:501

Neanderthals (Continued)
    Homo sapiens sapiens evolution and, 3:502
    sex differences in, biocultural, 6:848-49
    speech and, 3:503
Nebulin, and actin, 5:169
Necator, 2:387
Neck, 2:598
    cells (antral gland), 3:768
    and language, 2:598
    muscles of, 4:262
Neck muscles, 4:262
Necrosis, 4:1010, 5:259
nef genes. see Genes, nef
nef viral proteins. see Gene products, nef
Negative work, muscle, 5:185
Neisseria gonorrhoeae, 4:450
    detection of, in genital tract cultures, 1:558
    neonatal infections, 4:458
Neisseria meningitidis, 4:447, 459, 461
Nematodes. see Roundworms
Neoarsphenamine, synthesis and use of, 1:326
Neocortex, 2:3-5, 5:305-15, 673, 681-86
    archicortex and, 5:388
    Brodman's map of, 2:681-85
    fetal, development of, 5:314
    hippocampal formation, 4:231
    volume, 2:679
Neogenome, 4:258
Neolithic face pattern, dyslexia and, 3:225-27
Neomycin phosphotransferase, 3:157
Neonatal intensive care units. see Intensive care units,
        neonatal
Neonatal mortality. see Infant mortality
Neonates. see Infant, newborn
Neonatology
    advances in, 5:704-5
    defined, 5:699
    scope of, 5:701
Neophobia, in food preference and selection, 1:345-46
Neoplasm metastasis, 2:60, 120, 4:1007-15
    altered adhesive properties, 4:1014
    angiogenesis, 4:1010-11
    blood-borne tumor cell arrest, 4:1012-13
    correlation with phenotypic characteristics of cells,
        4:1007-8
    distant, 4:891
    elastin and, 3:268
    entry into circulation, 4:1011-12
    experimental, 4:1008
    extracellular matrix degradation, 4:1014

Neoplasm metastasis (*Continued*)
    fibronectin and, **3:**526
    formation, **4:**1011
    genomic change in tumors, **4:**1009-10
    interaction with immune system, **4:**1012
    invasion of surrounding host tissues, **4:**1013-14
    leukemias, **4:**693
    melanoma, **4:**889
    regional lymph node metastasis, **4:**891
    seed/soil hypothesis, **4:**1014-15
    survival in bloodstream, **4:**1012
    transgenic mice, **4:**1008
    tumor cell motility, **4:**1013
    tumor formation and growth, **4:**1008-9
    tumor heterogeneity, **4:**1010
Neoplasm suppressor genes. *see* Genes, suppressor, tumor
Neoplasms, **2:**53, 445-54, 60-61, 119, 243, 254-55, 448-49, 757, 759. *see also* Carcinoma; Oncogene amplification; specific neoplasms; terms beginning with Tumor
    aberration, acquired, **2:**449
    abnormal cell proliferation and, **3:**326
    aggression, **2:**449
    aging and, **3:**893
    analysis of cells for study of, **5:**317
    angiogenesis, **4:**1010-11
    antigenic modulation, **4:**349
    antigens present in, **5:**319
    ascorbic acid and risk reduction, **1:**432
    biochemistry, **5:**315-20
    birth control devices as risk factor, **1:**695
    bone marrow transplantation for, **1:**796
    and caffeine, **2:**76-77
    carcinogen-induced, study of, **5:**316
    cell type of origin, **5:**329-30
    as cellular disease, **2:**448
    chemically induced, **2:**111-16
    in childhood, **2:**453, **5:**704
    chromose aberrations and, **2:**450-51
    chromosome aberrations and, **2:**448, 449
    chromosome pattern and, **2:**445-54
    clonality of, **5:**319-20
    cross-reactivity, **4:**339, 342-43
    cycle, **2:**255
    Darwinian selection, **2:**449
    defined, **5:**315, 322
    development
        biochemical alterations associated with, **5:**318-19
        biology of, **5:**322-23
        classic pattern of, **5:**322-23

Neoplasms
    development (*Continued*)
        enzyme-altered foci and, **5:**318-19
        nonclassical pattern of, **5:**323
        oncogene hypothesis and, **5:**328-29
    developmental theory of, **5:**327-28
    dietary lipids and, **5:**508-9
    direct biochemical analysis of, **5:**316-17
    disease models, animal, **1:**256
    disorders of DNA repair and chromosome instability as risk for, **5:**324-25
    division, **2:**243, 448
    DNA methylation and, **3:**160
    DNA polymorphism, **2:**449
    dominantly inherited, susceptibility diseases and, **5:**325
    double minutes in, **3:**102
    endometriosis and, **3:**369
    environmental mutagens and, **5:**324
    etiology, **4:**1008-9, **5:**321-31
    flow cytometry and, **3:**624, 625
    formation and growth, **4:**1008-9
    in fruit flies, **7:**706
    gangliosides in, **3:**732-33
    gene mutation and, **7:**706-8
    genetic aspects, **2:**448, **3:**822
    genomic change, **4:**1009-10
    giant cell, **7:**702-3
    granulopoietic effects of lithium, **4:**750-51
    growth
        developmental theory for, **5:**327-28
        rate limiting enzymes and, **5:**317-18
    hematological, listed, **2:**450
    hepatic, in mouse, **2:**117-18
    hereditary, human, **7:**706-8
    heterogeneity, **4:**1010, 1015
    and hormone therapy, **2:**248
    and hyperthermia, **4:**295-301
    immune surveillance, **4:**319-22
    immunogenicity, **4:**338-39
    immunology, **4:**337-49
        antigenic modulation, **4:**349
        foreign cell destruction, **4:**338-39
        immunosurveillance and AIDS, **4:**345-46
        macrophages role, **4:**862
        oncofetal antigens, **4:**341
        oncogene product antigens, **4:**341
        virally induced antigens, **4:**341-43
    in immunosuppressed patients, **4:**320
    immunotherapy with interleukin-2, **4:**532-33
    immunotoxins and, **4:**416, 420, 421-22

Neoplasms *(Continued)*
  induction by radiation, **6:**451–52
  infectious agents and, **4:**450–51
  isoenzyme analysis, **4:**578–79
  isoenzyme pattern, **4:**578
  kinins and, **4:**599–600
  list of, **2:**451
  model for single class of drug-resistant cells in population, **4:**924–27
  monoclonal and polyclonal, **7:**688–89
  monoclonal antibodies for diagnosis, **5:**83–84, 85–86
  monoclonic origin, **2:**449
  mortality, in irradiated populations, **6:**452
  mutation rates in, **5:**248
  necrotic areas, **4:**1010
  neovascularization, **4:**1010
  oncogene hypothesis of, **5:**328–29
  oral contraceptives and, **5:**604–8
  papillomaviruses in progression of, **3:**864–65
  polyclonal
    analysis of, **7:**691
    explanations to account for, **7:**693
  predisposition to, analysis of, **6:**530
  prevention, **4:**788, **7:**851
  psoriasis and, **3:**446
  radiation-induced, **4:**344, **6:**415–18
  radiation therapy for, **6:**448
  recombinant DNA technology and diagnosis, **6:**528–30
  restriction fragment-length polymorphisms and analysis of, **5:**323–24
  selection of neoplasms for study of, **5:**316
  and selenium, **6:**797–98
  sex hormone sensitive, and exercise, **1:**747
  single and composite, **7:**687–88
  small round cell, **7:**700–701
  solid tumor, **2:**450, 451, **6:**529–30
  somatic mutation theory of, **5:**323
  spontaneous, **4:**344
  stages, **2:**449
  theories, **5:**323–30
  therapy
    bispecific monoclonal antibodies in, **5:**86–87
    chimeric antibodies in, **3:**810–11
    glutathione in, **3:**917
    hyperthermia and, **7:**466
  tobacco smoking and, **7:**515–16
  toxins and, **3:**385
  transforming growth factor-alpha and, **7:**628
  transplantable, selection for study of, **5:**316
  treatment with interferons, **4:**523–24

Neoplasms *(Continued)*
  tumor convergence, genetic, **2:**450
  tumor noise, cytogenetic, **2:**449
  viruses and, **7:**781
  vitamin A and, **5:**512
  vitamin C and, **5:**512
  vitamin E and, **5:**512
Neoplasms, connective tissue, **5:**334
Neoplastic transformation
  biology of, **5:**323–23
  classical pattern of, **5:**322–23
  nonclassical pattern of, **5:**323
  papillomaviruses and, **5:**635–543
Neoteny, **2:**872
Neovascularization, **4:**1010
Nephroblastoma. *see* Wilms' tumor
Nephrons
  defined, **3:**93
  physiology of, **3:**94
Nephropathy, Balkan. *see* Balkan nephropathy
Nephrosis, in children, **5:**709
Nephrotic syndrome, in children, **5:**709
Nephrotoxicity, **7:**550
Nernst equation, **5:**872
Nerve cells. *see* Neurons
Nerve endings, **2:**138
Nerve endings, sensory. *see* Receptors, sensory
Nerve fibers, **2:**138, 521–23
  C-type fiber, **2:**138
  defined, **5:**341
Nerve fibers, myelinated, **2:**137
  diagrammatic representation of, **5:**272
  myelinated and unmyelinated, **5:**272
Nerve growth factors, **5:**333–40
  chartin phosphorylation and, **5:**899
  laminin effects, **4:**629
  nerve regeneration and, **7:**506–7
  neuronotrophic factors and, **5:**344
Nerve regeneration, **5:**341–46
Nerve tissue
  defined, **5:**341
  gangliosides in, **3:**730
Nerve transmitter substances. *see* Neuroregulators
Nervous system. *see also* Central nervous system; Peripheral nervous system
  anatomy, **5:**347–60
  behavioral changes in, **6:**40
  in cardiovascular regulation during exercise, **3:**520–21
  chemical transmission and communication, **5:**407–9
  development
    gangliosides in, **3:**737–38

Nervous system
    development *(Continued)*
        serotonin and, **6:**834–35
        zinc and, **7:**887–88
    fetal
        from fourth month, **3:**618–19
        at third month, **3:**611
        thyroid hormones and, **3:**340
    functional changes in, lesion-induced, **6:**39
    in gastric circulation regulation, **3:**762–63
    genesis of, **3:**286
    geriatric, **2:**348
    gingiva, **4:**35
    in heart rate control, **3:**516–17
    hypothermia, **4:**800–801
    immature, **2:**348
    muscle force production and, **5:**179–80
    in neural-immune interactions, **5:**373–86
    neuropeptide Y and, **5:**719–20
    newborn, changes in function of, **6:**823
    organization of, **5:**347–48
    persuasion and, **5:**810
    plasticity, **6:**37–41
    response to toxic agent, **2:**344–46
    serotonin in, **6:**829–35
    structural changes in, **6:**37–39
    toxicology, **2:**1–9, 344–48
    transplantation of, changes associated with, **6:**40–41
    vasoactive intestinal polypeptide and, **5:**718
    vertebrate, regeneration of, **5:**342
Nervous system, autonomic. *see* Autonomic nervous system
Nervous system diseases
    in children, **5:**709–10
    fusimotor system and, **3:**723
    genetic aspects of, **3:**836
    microtubule protein phosphorylation and, **5:**900–901
    micturition control and, **5:**45
    treatment with interferons, **4:**523
NETtalk, capabilities of, **5:**392
Networks
    biologically-based, **5:**388–90
    defined, **5:**387
    stability of, **5:**389
Neuberg, Karl, **3:**938
Neumann computer, **2:**550
Neural cell adhesion molecule (N-CAM), **2:**50
    and mobility, **2:**50
    in myoblasts, **5:**168
    nerve regeneration and, **5:**345

Neural crest, **2:**690, **5:**333
Neural crest cells
    cranial
        bone formation, **1:**786
        migration, **1:**786–87
    defined, **3:**523
    migration of, **3:**524
"Neural Darwinism," defined, **5:**390
Neural firing
    in auditory nerve fibers, **3:**238
    firing frequency, **2:**312
    network modeling of, **5:**388
    in olfactory tract, **5:**525
    in saccadic movements, **3:**558
Neural-immune interactions, **5:**373–86
Neural networks, **2:**137, **5:**387–92
    applied, **5:**390–92
    locomotion, **4:**769–81
    maps and, **5:**388
    memory formation and, **4:**674–76
    metal alloys and, **5:**391
    model, **4:**199, 202
    traveling salesman problem, **5:**392
Neural pathways, **2:**31–33
    afferent, **2:**33
    efferent, **2:**33
    sympathetic, **2:**33, 38, 39
Neural transplantation. *see under* Transplantation
Neural units, defined, **5:**387
Neuralgia
    headaches and, **4:**55–56
    trigeminal, **5:**368
Neuraminic acid
    named, **3:**725
    structural modifications of, **3:**726
Neuraminidase
    influenza viruses, **4:**483–84
    oligosialogangliosides and, **3:**746
Neurites
    defined, **5:**333, 343
    growth, astrocyte role, **1:**441–42
    outgrowth, laminin effects, **4:**626–28
Neuritogenic effect, defined, **3:**737
Neuro-2A neuroblastoma cell line, murine, **3:**738, 740
Neuroblastoma, **2:**451, **5:**547
Neurochemistry, **2:**685–87
    arousel concept, **2:**714
    control of reproduction, **6:**555–57
    of cortex, **2:**685–87
    of learned helplessness, **4:**668–69

Neurochemistry *(Continued)*
 and learning, **2:**714
 and temperament, **2:**714
Neurocrine agents, defined, **3:**318
Neurocrine effectors, defined, **5:**714
Neuroendocrine challenge test, **2:**845
Neuroendocrine system. *see* Neurosecretory systems
Neurofibrillary tangles, in Alzheimer's disease, **1:**211–13,
 5:901, 215–16
Neurofibromatosis
 in children, **5:**710
 DNA marker diagnosis of, **3:**142
Neurofilament proteins. *see* Proteins, neurofilament
Neurohumors. *see* Neuroregulators
Neurohypophysis. *see* Pituitary gland, posterior
Neuroleptics. *see* Tranquilizing agents, major
Neurolinguistics, developmental, **2:**861
 and equipotentiality, hemispheric, **2:**861
 and lateralization of language, **2:**861
 and Lenneberg's tenets, **2:**861
Neuromedin B, **6:**89
Neuromodulators. *see* Neuroregulators
Neuromuscular function, neuropharmacology of, **5:**409
Neuromuscular junction
 development of, **5:**171
 diagram of, **5:**256
 site of action of toxins and antibodies on, **5:**257
Neuron-specific enolase. *see* Phosphopyruvate hydratase
Neuronal plasticity, **6:**37–41
 calpains and, **5:**17
 of central nervous system, **5:**342
 defined, **5:**893
 gangliosides and, **3:**731
 in vivo phosphorylation studies of, **5:**897
 MAP-2 phosphorylation and, **5:**898
 microtubule protein phosphorylation and, **5:**900
Neuronal projection, **2:**677
Neuronotrophic effect, defined, **3:**737
Neuronotrophic factors
 defined, **5:**341
 nerve growth factor and, **5:**334
 nerve regeneration and, **5:**344
 in neuronal death, **5:**333–34
 of Schwann cells, **5:**342
Neurons, **2:**333–35, 673, **4:**198–99, 341, 343. *see also*
 Meganeurites; Microtubules
 activity, continuous, **2:**549
 of animal neocortices, **5:**309
 and catecholamine, **2:**218–19
 categories, **2:**335
  fusiform, **2:**335

Neurons
 categories *(Continued)*
  multipolar, **2:**335
  pyramidal, **2:**335
  stellate, **2:**335
 chromatin, **2:**432–34
 circuitry, **2:**32–33
 cortical, **2:**433–35
  cerebellar, **2:**435
  cultured, **2:**433
  gene, **2:**434
  hypothalmic, **2:**433
  transcription activity, **2:**433
 cultures, nerve growth factor and, **5:**334
 cytokines and, **5:**385–86
 death of, neuronotrophic factors and, **5:**333–34
 defined, **2:**1, **5:**341
 degeneration in Huntington's disease, **4:**271
 differentiation, gangliosides and, **3:**737–42, 745
 endophinergic, **2:**17
 excitability, **2:**311
 ganglioside transport in, **3:**743–48
 gangliosides and, **3:**733–34
 glucose as energy source, **2:**343
 glucose-sensitive, visceral afferent pathway integration
  and, **7:**798
 grandmother neuron, infamous, **2:**549
 hypothalmic, **2:**11, 433
 immunoreactive, **2:**338
 integrative properties of, **7:**389
 laminin effects, **4:**628–30
 lateral geniculate nuclei, and vision, **7:**805–6
 migration, **2:**678
 migration, astrocyte role, **1:**442
 morphogenesis
  microtubule protein phosphorylation and,
   5:899–900
  microtubules in, **5:**897
 morphological response, **2:**344
 motor, role of, in spinal reflex circuits, **6:**158–60
 of neocortex, **5:**309
 neurotransmitter plasticity and, **6:**39
 number of, **2:**675
 oxygen as energy source, **2:**343
 pathways for nine compounds, **2:**18
 pigmented, in Parkinson's disease, **5:**668
 polarization, **2:**5
 population, **2:**549
 potassium channel, **2:**311
 psoriasis and, **3:**452

Neurons *(Continued)*
pyramidal, **2:**677
response to cold, **4:**796
retinal, and response to light, **6:**624–28
sprouting of, spinal cord dysfunction and, **7:**226–27
streams of color in, **2:**588–89
structure, **2:**2
survival, astrocyte role, **1:**441–42
types, **2:**1–2
types of communication between, **7:**225
V1
binocular vision and, **7:**811
color sensitivity of, **7:**810–11
ocular dominance involving, **7:**812
temporal frequency of, **7:**811
and visual information processing, **7:**806–13
visual stimulus selectivity in, **7:**809–11
visceral afferent, **7:**794
voltage potential, **2:**3
Neurons, afferent, **2:**33
Neurons, autonomic
neuropeptides in, **1:**541
neurotransmitter release, **1:**541
neurotransmitters in, **1:**540–41
Neurons, efferent, **2:**33
Neurons, postganglionic
neurotransmitters, **1:**540
sympathetic division of the ANS, **1:**537–38
Neurons, preganglionic
autonomic nervous system, **1:**536–37
neurotransmitters, **1:**540
Neuropathies, **2:**341, 346
Neuropeptide PHI 27. *see* Peptide PHI
Neuropeptide receptors. *see* Receptors, neuropeptide
Neuropeptide Y
-positive nerve fibers, in rat spleen white pulp, **5:**379
in autonomic neurons, **1:**541
functions of, **5:**719–20, **6:**89
as neuroregulator, **2:**314–15, 365
Neuropeptides, **2:**11–17, 26–27
in autonomic neurons, **1:**541
epilepsy and, **3:**460–61
immune responses and, **5:**384–85
listed, **2:**17, 27
in lymphoid organs, **5:**378
opioid, **2:**338
and pituitary hormone release, **2:**27, 36, 38, 365, 493, 494, 685–86
substance P, in allergic response, **1:**189–90
of visceral afferent nerves, **3:**763, **7:**794–96

Neuropeptides *(Continued)*
visceral innervation and, **7:**795
Neuropharmacology, **5:**407–14
of autonomic nervous system, **5:**409
of neuromuscular function, **5:**409
Neuropil, **2:**341, 343, 678
Neuropsychological dysfunction, **2:**719
Neuropsychology, developmental, **2:**859–67
in children, **2:**859–67
disorders, **2:**863–65
test batteries, **2:**865
history, **2:**860
and phrenology, **2:**860
research, **2:**860–63
tests used in, **1:**611
theoretical issues, **2:**860–63
theories, **2:**860
Neuroreceptors. *see* Receptors, sensory
Neuroregulators, **2:**5, 11–17, 217, 219, 323, 333, 337–38, 365, 521, 673, 685, 686, 831, **4:**199, **7:**384–85. *see also* specific neuroregulators
adrenocorticotropic hormone as, **2:**22–23
in Alzheimer's disease, **5:**412
amine and peptide action on, **3:**322
amine blockage by antidepressants, **1:**308–9
amino acid as, **2:**5
in anxiety, **5:**410–11
of auditory nerve, **3:**238
in autonomic neurons, **1:**540–41
in blood, **2:**844–45
brain, **7:**384–85
brain monoamine, relationship to anxiety and depressed mood, **1:**612–13
calcitonin gene-related peptide as, **2:**365
catecholamines as, **2:**12–13, 217
in cerebrospinal fluid, **2:**844–45
criteria, **2:**521
CuZnSOD and, **3:**186
deficiencies in Alzheimer's disease, **1:**214–15
defined, **2:**1, 841, **3:**191, 317, 318, **4:**13, **5:**373, 667
in depression, **1:**132, **5:**410
drug altering function of, **2:**14–15
dyslexia and, **3:**216
effect of drugs on, **5:**408–9
enkephalins as, **5:**375
in epilepsy, **3:**459–61, **5:**412
in fetal hypothalamus, **3:**338
food intake and, **5:**519–20
function altered by drug, **2:**14–15
GABA as, **2:**13–14

Neuroregulators *(Continued)*
  glucocorticoids and, **5:**376
  glutamate as, **2:**521
  glutamate receptor, **2:**312–14
  hormones and, **3:**318
  hypothalamus, **4:**308
  immune system and, **5:**382–85
  information transfer in hippocampus, **4:**231
  levels in Huntington's disease, **4:**271
  mechanism of action, **2:**322
  monoamines as, **2:**5
  motoneurons and, **3:**698
  nerve communication and, **5:**407–9
  neuromodulators, defined, **5:**373
  neuropeptide Y as, **2:**314–15
  neuropharmacology of, **5:**409–13
  in pain, **5:**412–13
  in Parkinson's disease, **5:**411–12
  peptide, brain, **7:**385
  peptide as, **2:**5, 365
  and pituitary hormone secretion, **2:**22–24, 26
  regulation, **2:**22–27
  release, **2:**49, 323, **7:**385–86
      from ANS varicosities, **1:**536–37, 541
      calcium-stimulated, **7:**385
      vesicular exocytosis as mechanism of, **7:**385
  removal, synapse and, **7:**386
  reuptake, **2:**219
  in schizophrenia, **5:**411
  of sleep-wake cycle, **3:**194
  in stroke, **5:**413
  substance P, **2:**365
  in urine, **2:**844–45
  vasoactive intestinal peptide (VIP) as, **2:**365
Neurosecretion, **2:**217
  defined, **5:**394
  of fetal hypothalamus, **3:**337–38
Neurosecretory disorders, **2:**28–29
  diagnosis, **2:**29–29
  treatment, :228–29
Neurosecretory systems, **2:**11, 16, **5:**393–400
  aging and, **3:**897
  hypothalamic outputs, **4:**307
  neurosecretory neurons, **4:**608
  regulation of, **6:**833–34
Neuroses
  defined, **3:**838
  experimental, Pavlovian, **2:**631
Neurotensin, **2:**26, 27, 36, 38, **5:**721
Neurotensin receptors. *see* Receptors, neurotensin

Neuroticism, **2:**711
  biological basis of, **5:**774
  as personality description, **5:**770
Neurotoxins, **2:**15
  pesticide residues in food and, **7:**561
  phosphorylation and, **5:**900–901
  snake, **7:**599–600
  toxicity of, **7:**547–49
Neurotransmitters. *see* Neuroregulators
Neurovestibular system
  changes, countermeasures to space travel, **7:**151–52
  space travel effects on, **7:**149–50
Neutron activation analysis
  calcium measurement and, results from, **7:**540–41
  total body calcium measurement and, **7:**538–41
  in vivo
      basis of, **7:**538
      neutron sources for, **7:**538
      radiation detectors for, **7:**539
Neutropenia, **2:**649–50
  and copper deficiency, **2:**650
  lithium effects, **4:**750–51
Neutrophil-derived relaxant factor. *see* Endothelium-derived relaxing factor
Neutrophils, **2:**613, **5:**427–31
  activation by TNF, **4:**474–75
  in allergic response, **1:**189
  antimicrobial effects of, **5:**430–31
  characteristics of, **4:**166, **5:**812–13
  complement receptors, **1:**299
  contribution to inflammation, **5:**431
  digestive functions of, **5:**429–30
  Fc receptors, **1:**299
  functions, **4:**166, 403, **5:**428–31
  in inflammation, **1:**312–13, **4:**471, **5:**429–30, 819
  life cycle of, **5:**427–28
  microbicidal functions of, **5:**429–30
  migration of chemotactic gradient, **5:**429
  myeloid maturation stages of, **5:**428
  Na^+/H^+ exchange and, **5:**874
  oxygen-related effector mechanisms of, **5:**429–30
  platelet-activating factor and, **6:**47–48
  precursors of, **5:**427
  in psoriatic lesions, **3:**451
  respiratory burst and metabolism of drugs in, **6:**583
  types of, **5:**812–13
Nevus, defined, **7:**49
Newborn. *see* Infants, newborn
Newborn intensive care units. *see* Intensive care units, neonatal

Nexin II, **3:**182

Nexins. *see* Protease inhibitors

Nexus, **2:**132, 274

Niacin, **2:**169, 530
  atherosclerosis therapy with, **1:**454
  food analysis, **4:**221

Nickel, **2:**111, 112
  biological function of, **7:**607
  compounds, carcinogenicity of, **7:**553
  deficiency, **7:**610–11
  significance of, in body, **5:**49
  toxicity, **7:**610–11

Niclosamide, **1:**265, **2:**389, 393

Nicotinamide, **2:**531, 532

Nicotinamide adenine dinucleotide. *see* NAD

Nicotinamide coenzymes, **3:**437

Nicotine, **2:**145
  addictive effect of, **7:**514
  antagonists, **2:**522
  blood vessels and, **7:**513–14
  characteristics of, **1:**178
  nervous system and, **7:**514

Nicotinic acids, **2:**169, 531, 532
  food analysis, **4:**221

Nicotinic receptors. *see* Receptors, nicotinic

Nictitating membrane, **2:**311

Nidation. *see* Ovum implantation

Nidogen, in basement membranes, **4:**624

Nifedipine, **2:**96, 172, 176

Night terror, as sleep disorder, **7:**81–82

Night vision, vitamin A and, **7:**850

Nightmares
  of children, **3:**194–95, 196
  in Freudian psychoanalysis, **3:**192
  incidence of, **3:**195–96
  REM sleep and, **3:**195
  sleep disorders and, **7:**82

Nigrostriatal system, **2:**217, 218

Nitrates, **2:**346
  inhalation-associated headaches, **4:**57
  in well-water, **2:**347

Nitrendipine, **2:**172, 176

Nitric oxide
  chemical properties of, **3:**373–74
  endothelium-dependent vasodilation and, **3:**372–73
  endothelium-derived relaxing factor and, **3:**371
  localized action of, **3:**372–73
  pharmacological properties of, **3:**374
  platelet function inhibition and, **3:**372
  release of, **3:**373

Nitrites, **2:**347
  in cured meats, **3:**679

Nitrofurans
  agricultural uses, **1:**329
  bacterial resistance to, **1:**328
  specificity of action, **1:**327

Nitrogen phosphorus detector, defined, **3:**750

Nitroglycerin, **2:**172, 787
  and angina pain, **2:**172

Nitroimidazole, **2:**382

Nitrosamines
  carcinogenicity of, **3:**679
  in consumer products, **3:**680
  defined, **3:**671
  in food, **3:**678–80
  food analysis, **4:**224–25
  from physiological processes, **3:**680

Nitrosation, in food, **3:**679

Nitroso compounds, gastric carcinoma and, **3:**772

N-Nitroso-N-ethylurea, **2:**121

N-Nitrosodimethylamine. *see* Dimethylnitrosamine

N-Nitrosonornicotine, **2:**110

Nitrosyl chloride, **3:**679

Nitrosyl thiocyanate, **3:**679

Nitrous anhydride, nitrosamines and, **3:**678

Nitrous oxide, **2:**787

Nitrovasodilators, **3:**372

NMDA receptors. *see* Receptors, NMDA

Nociception, **2:**333, 338
  descending mesencephalic projections and, **4:**715
  nociceptive pathway, **5:**365–67
  serotonin in, **6:**831–32

Nociceptors, morphology, visceral afferent systems and, **7:**797

Nodose ganglion, **2:**33, **5:**720

Non-Hodgkin's lymphoma. *see* Lymphoma, non-Hodg-kin's

Nonallelism, evidence of, **3:**792–93

Noncellulosic polymers. *see under* Polymers

Noncellulosic polysaccharides. *see under* Polysaccharides

Nondisjunction
  age factor in, **5:**254
  chromosomal, **2:**438
  defined, **3:**175, **5:**72, 244
  rate of, **5:**247

Nonesterified fatty acids. *see* Fatty acids, nonesterified

Nongonococcal urethritis. *see* Chlamydia trachomatis

Nonionic diffusion. *see* Diffusion, nonionic

Nonlymphoid tissue and organs, **4:**392–93

Nonpenetrant males
  cytogenetic negativity of, **3:**711

Nonpenetrant males *(Continued)*
    defined, **3:**709
    identification of, **3:**
Nonstarch polysaccharides. *see under* Polysaccharides
Nonverbal communication. *see* Communication, nonverbal
Nonvolatile nitrosamines. *see* Nitrosamines
Noradrenaline. *see* Norepinephrine
Noradrenergic innervation of lymphoid organs, **5:**376-81
Norcocaine, alcohol and, **3:**213
Norepinephrine, **2:**5, 9, 12, 18, 24, 38, 141, 147, 152-53, 171, 186, 217-21, 223, 323, 337, 357, 365, 369, 513, 534, 541, 686, 831, 834, 843, 848
    adrenal gland and, **1:**73, 75
    aggression and, **2:**717
    alert states and, **1:**483-84
    in astronauts, **2:**152-53
    in autonomic neurons, **1:**540
    B-lymphocyte proliferation and, **5:**383
    convulsive seizures and, **6:**787-88
    defined, **5:**373
    in gastric circulation regulation, **3:**762
    in vitro studies of, **5:**382
    levels in Huntington's disease, **4:**271
    in mental depression, **2:**223
    metabolites, relationship to anxiety and depression, **1:**612-13
    in mood regulation, **2:**221
    neuropeptide Y and, **5:**720
    perinatal, **3:**350
    pineal body and, **6:**3
    production, from alcohol consumption, **1:**172
    in schizophrenia, **6:**778
    in sensation seeking, **6:**816
    sleep-wake cycle and, **3:**194
    synthesis, in adrenal medulla, **1:**644
    visual plasticity and, **5:**900
Norethindrone
    for bone loss prevention, **3:**482
    in combined oral contraceptives, **5:**599
Norethindrone acetate, coronary disease risk of, **5:**603
Norethindrone enanthate, in minipills, **5:**600
Norethisterone. *see* Norethindrone
Norethynodrel, in Enovid, **5:**599
Norfloxacin, **6:**403-4
Norgestrel, **3:**482
Norm theory, of affective responses, **1:**106
Norpregneninolone. *see* Norethindrone
Northern blot analysis, **2:**431, 432
Nortriptyline, **2:**747

Nose
    airway obstruction of, **6:**358
    anaerobic bacteria found in, **1:**239-40
    anatomy of, **6:**588-89
    lesions in, and noxious chemical inhalation, **5:**694
    removal of inhaled material by, **5:**690
Nosocomial infections, anaerobic bacteria-associated, **1:**243
Notochord, in myogenesis, **5:**161-62
Novel retention tests. *see* Retention tests, novel
Novobiocin, **1:**290(t), 291
Novocaine, **2:**784
Nuclear bag muscle fibers, **3:**720
Nuclear chain muscle fibers, **3:**720
Nuclear envelope. *see* Nuclear membrane
Nuclear family, **7:**236
Nuclear genome, **4:**252
Nuclear lamina, **2:**292, 295-97, 675-77, 817-20
Nuclear magnetic resonance. *see* Magnetic resonance imaging
Nuclear medicine, in radiobiology, **6:**447-48
Nuclear membrane, **2:**238, 291, 295-97, **5:**451-60
    fibrils and, **5:**456
    in prometaphase, **5:**70
    in telophase, **5:**71
Nuclear polyhedrosis viruses. *see* Baculoviridae
Nuclear pore. *see* Nuclear membrane
Nuclear pore complex, **2:**292, **5:**451, 296, 297
Nuclear proteins, **2:**434-36
    localization sequences of, **5:**456
    regulatory, **2:**434-36
    routing signals for, **6:**247
    translocation of, **6:**250
    transport of, **5:**455-57, 458
Nuclear uptake
    conditions for, **5:**456
    inhibition of, **5:**457
Nucleases. *see also* Deoxyribonuclease I
    defined, **3:**125
Nucleic acid hybridization, **2:**431, 432, **3:**669
Nucleic acid probes, **1:**553
Nucleic acid repetitive sequences. *see* Repetitive sequences, nucleic acid
Nucleic acids. *see also* DNA; RNA
    and brain RNA, **2:**431
    metabolism, zinc and, **7:**884
    microspectrography of, **5:**563
    partition coefficients of, **1:**361
    photoelectron imaging of, **5:**908
    protein associations, partitioning for, **1:**362

Nucleic acids *(Continued)*
    separation in aqueous two-phase systems, **1**:361
    structure, **6**:682
    synthesis, and cellular organization, **5**:461–70. *see also*
        Nucleotides
Nucleofilament-100 Angstrom, **2**:421–22
Nucleolus, **2**:244, 291, 292
Nucleophiles, **2**:118, **3**:907
Nucleoplasm, defined, **5**:451
Nucleoplasmin, in *X. laevis* oocytes, **5**:455
Nucleoproteins, **4**:485
Nucleoside analogue. *see* Zidovudine
Nucleoside triphosphates, **6**:251
Nucleosides
    defined, **3**:169
    in prokaryotic DNA, **3**:152
Nucleosomes, **2**:292, 419, 431–34, 476
    assembly of, **3**:130
    chromatin structure, **4**:237
    core particle, **2**:420, 471
    defined, **2**:465
    DNA, folded, **2**:420, 421, 431–34, 471
    DNA in, **3**:125–33
    filament, **2**:423–25
        folding, **2**:423–26
    filament structure, **2**:421–23
        models, **2**:422, 428–30
    folding into 30nm filament, **2**:421–23, 426–29
    function of, **3**:131–33
    histone, **2**:419–23, 432, 434
    histone, folded, **2**:434
    nucleofilament-100 Angstrom, **2**:421–22
    positioned, **3**:130–31
    properties of, **3**:128–29
    structure of, **3**:126
2',3'-Nucleotide-3'-phosphorohydrolase, cyclic, **2**:652
Nucleotide sequence. *see* Base sequence
Nucleotide substitutions
    evolutionary consequences of, **3**:873, 886
    parsimony method and, **3**:887
    as polymorphisms, **3**:875
Nucleotide sugars, isomer configurations of, **3**:729–30
Nucleotides, **2**:317, 467
    biosynthetic pathways, **5**:462–63
    defined, **3**:869
    of elastin, **3**:265
    intracellular compartmentation of, **5**:464
    metabolic imbalance, **5**:470
    as metabolic regulators, **5**:469–70
    purine, in adrenal medulla, **1**:645–46

Nucleotides *(Continued)*
    pyrimidine, synthesis of, **5**:464
    synthesis
        and cellular organization, **5**:461–70
        dynamic protein associations in, **5**:467–69
        intracellular compartmentation and, **5**:464
        mechanisms organizing, **5**:464–70
        multienzyme complexes and, **5**:465–66
        multifunctional proteins in, **5**:465
Nucleotides, cyclic, **5**:417
Nucleus, **2**:31–35, 218, 333
    lens, anatomy, **3**:544
    protein sorting in, **6**:241
    suprachiasmatic, **2**:493–95
    zinc metabolism within, **7**:884
Nucleus basalis, olivary. *see* Olivary nucleus
Nucleus basalis of Meynert. *see* Meynert's nucleus
Nucleus of Onuf. *see* Onuf's nucleus
Nucleus retroambiguus, **4**:718
Nude mouse, **4**:1008
Nuer, conceptualization of time, **4**:700
Nullisomy, **2**:438
Nuremburg Code, **1**:657–58
Nurse cell. *see* Astrocytes
Nurses
    cerebrovascular diseases of, **5**:602, 603
    as emergency surgeons, **4**:919
Nursing homes, nutritional risk factors in, **4**:901
Nurture versus nature problem, **2**:875
Nutrients
    in ADP rephosphorylation, **5**:227
    in customary diets, **3**:18
    deficiencies in protein-energy malnutrition, **4**:894–95
    defined, **3**:649
    essential, defined, **3**:576, 683
    key, in nutrition education, **3**:24–25
    nutritional quality
        addition, **5**:495
        destruction, **5**:491–94
        dilution, **5**:495
    problem, **5**:488–89
    selection
        defined, **3**:589
        food acceptance and, **3**:641
    separation from food, nutritional quality and, **5**:494–95
    supplies, contribution of food groups to, **5**:490–91
    USDA tables of, **3**:656
    utilization
        gastric acid release and, **5**:755
        insulin release and, **5**:756

Nutrients
utilization *(Continued)*
saliva stimulation and, **5:**755
sensory influences on, **5:**755–56
vegetarian diets and, **3:**659
Nutrition. *see also* Child nutrition; Eating; Food; Infant
nutrition
and aging process, **5:**479–81
as factor in prostate cancer, **6:**187–88
feeding behavior and, **3:**592
global fertility and, **1:**749–50
growth rates and size attainment and, **4:**28
immunity and, **5:**503–16
and male reproduction, **1:**748–49
metabolic regulation role, **4:**1006
prepubescent growth and, **3:**351–52
proteins and, **6:**229–35
in puberty onset, **3:**355
selenium in, **6:**789–800
space travel and, **7:**143–45
thermal injury and, **7:**455–56
trace elements in, **7:**603–13
Nutrition disorders, **4:**893–96
deficiencies, causes of, **4:**901
disease models, animal, **1:**257
of the elderly, **4:**897–903
in Federal Republic of Germany, **3:**683
in gatherer-hunters, **3:**20
global fertility and, **1:**749–50
lactose malabsorption and, **4:**615
long-standing, **4:**898
mathematical nutrition education and, **3:**24
milk as treatment, **4:**621
in pregnancy, in developing countries, **3:**26
prevalence, **4:**114
preventative measures, **4:**902–3
protein deficiency and, **6:**235
recurrent, **4:**898
risk factors for, **4:**900–902
sensory functions and, **5:**756–57
specific, **4:**898
sudden malnutrition, **4:**898
techno-food and, **3:**22
Nutrition education
"food group" method of, **3:**24
mathematical method of, **3:**24
Nutrition labeling. *see* Food labeling
Nutrition surveys
in feeding behavior research, **3:**594
methods of, **3:**23–24

Nutrition surveys *(Continued)*
national, **3:**24
United States, **3:**650
Nutritional dietary modification. *see* Nutrition education
Nutritional quality
agricultural practices affecting, **5:**490
of carbohydrates, **5:**501–2
criteria for assessing, **5:**484–85
factors affecting, **5:**489–95
of food components, **5:**499–502
food storage, processing and preparation affecting,
5:490–91
of foods and diets, **5:**483–502
improving, recommendations for, **5:**502
of lipids, **5:**499–500
of major food groups, **5:**495–99
nutrient addition as factor affecting, **5:**495
nutrient destruction affecting, **5:**491–94
nutrient dilution as factor affecting, **5:**495
physical separation of nutrients from foods and,
5:494–95
of proteins, **5:**499
recommended dietary allowances and, **5:**484–85
Nutritional requirements
of children, **5:**708
of the elderly, **4:**899–900
of infants. *see* Infant nutrition
Nutritional status
assessment, **4:**898–99
maternal
DZ twinning and, **3:**307
MZ twinning and, **3:**314
Nuttall, G. H. F., **3:**488
Nyctohemeral rhythm. *see* Circadian rhythm
Nystagmus
optokinetic, **3:**555
physiologic, **3:**555
Nystagmus, vestibular
optokinetic reflex and, **3:**557
smooth eye movements and, **3:**556
waveforms of, **3:**555

O

O-linked *N*-acetylglucosamine. *see under* Acetylglucos-
amine, O-linked
Oatmeal, **3:**656
Oats, dietary fiber from, **3:**40
Obesity, **2:**106, **5:**517–21. *see also* Adipose tissue; Body
weight

Obesity *(Continued)*
in affluent countries, **5:**519
categories of, **3:**247–49
defined, **3:**243
developmental
course of, **3:**248
defined, **3:**243
diet causing, **5:**520
diseases causing, **5:**520
as eating disorder, **1:**351–52
fatty acid composition in, **1:**60
as feeding behavior disorder, **3:**591
genetic factors causing, **5:**520
immunity and, **5:**506
latent, **3:**685–86
measurements of, **1:**671
overnutrition and, **2:**106
pathogenesis of, **5:**518
physical aspects of, **3:**248
psychological aspects of, **3:**248–49
purposeful, **3:**20
research on, **3:**684–85
risks of, **5:**520–21
serotonin and, **6:**832
societal aspects, **4:**84
terms expressed for overweight, **5:**517–18
treatment, **3:**249, **5:**521
types of, **5:**520
versus overweight, **5:**517
Object
color of, **2:**582
and theory of self, **2:**855
Object categories, **7:**89–90
Object-object relations, **2:**853, 855
Objective performance instruments (psychobiology), **1:**611–12
Observational rating scales, **1:**610–11. *see also* specific rating scale
expert's vantage, **1:**611
family ratings, **1:**611
nurses' ratings, **1:**611
Obsessive-compulsive behavior, **4:**997
Occipital junction, **3:**222
Occipital lobe, **2:**351, 683, 684
Occipital sinus, **4:**983
Occlusion, **2:**769
Occult blood, **3:**776
Occupational therapy, **3:**716
Occupational toxicology. *see* Hazardous substances
Ochratoxins, **5:**265–66, **7:**593

Octreotide, **5:**723
Ocular dominance, **5:**311
Ocular flicks, defined, **3:**557
Ocular muscle, **3:**554
Oculocentric map, in space perception, **1:**629
Oculomotor nerve
defined, **3:**553
role of, **3:**557–58
Oculomotor system, **3:**554
Ocytocin. *see* Oxytocin
Oddi's sphincter, defined, **3:**73
Odontoblast, **2:**774, 775, 819–23
Odontogenesis, **2:**821–25
Odontogenic cysts, **5:**614
Odontogenic tumors, **5:**616
Odontometrics, **2:**786, 791
Odorants, **5:**761–62
Odors
aromagram of mixture of, **5:**761
cells having sensitivity to, **5:**766
enhancement, **5:**763
interactions, **5:**763, 764
mixtures
ability to detect and identify constituents of, **5:**761–62
mechanisms of perception of, **5:**765–66
predicting intensity of, **5:**764
predicting smell of, **5:**764–65
total perceived intensities of, **5:**763
perception
neural processing and, **5:**766
receptor cells and, **5:**766
restricted processing capability and, **5:**766
sniffing, complex smell analysis and, **5:**760–61
suppression, **5:**763
evidence for central mechanism of, **5:**765
evidence for peripheral mechanism of, **5:**765
Oedipus complex, **2:**854
Oils
egg yolk, fatty acid composition of, **5:**541
fatty acid and cholesterol content of, **5:**501
fish, fatty acid composition of, **5:**541
Okazaki fragments
defined, **3:**439
in DNA replication, **3:**441, 443, **6:**684
Old people's homes. *see* Nursing homes
Oldowan tools, **3:**497
Oleic acid, in breast milk, **4:**605
Olfaction. *see* Smell
Olfactometers
complex smell perception using, **5:**761
diagram of, **5:**762

Olfactory brain, **4**:232
Olfactory bulb, **2**:679
    anatomy, **5**:524–25
    astrocytes, **1**:437
    physiology, **5**:526–28
Olfactory cell layer, mitral
    anatomy, **5**:524
    order in, **5**:530
    stimulation of, **5**:526
Olfactory cilia. *see* Cilia, olfactory
Olfactory coding
    glomerula in, **5**:530
    topological aspects of, **5**:529–30
Olfactory cortex, primary, **5**:525
Olfactory disorders
    causes of, **5**:529
    G protein defects in, **4**:23
    post-infectious, **5**:696
    post-traumatic, **5**:696
    smell perception and, **5**:744
    as upper respiratory tract condition, **5**:695
Olfactory mucosa, anatomy, **5**:523–24
Olfactory nerve, anatomy, **5**:524
Olfactory tract, anatomy, **5**:525
Olfactory tubercle, **2**:679
(2',5')-Oligoadenylate synthetase, **6**:664
Oligodendroglia, **1**:437–38, **2**:341–46
Oligomers, **4**:972
Oligonucleosomes, **2**:420
Oligonucleotide probes
    allele-specific, polymerase chain reaction and,
        3:139–40
    of peptide hormes, **5**:714
Oligonucleotides, synthesis
    cloning and, **3**:807
    in *in vitro* DNA manipulation, **3**:875
    methods of, **3**:170–72
    objective of, **3**:169
Oligopeptides, **5**:335
Oligosaccharides, **2**:104
    addition to membrane glycoproteins and glycolipids,
        4:977
    and flatulence, **2**:104
    N-linked, glycoproteins and, **5**:454
    O-linked, sperm receptors and, **3**:605
Oligosialogangliosides
    defined, **3**:743
    neuraminidase and, **3**:746
Olivary nucleus, in auditory information flow, **3**:239–40
Olivocochlear bundle, **3**:237

Omega-3 fatty acids. *see* Fatty acids, omega-3
Omentum, **3**:68–69
Omnivores, **2**:593
Onchocerca volvulus, **2**:387, 392–93
Onchocerciasis, **1**:265, **2**:387, **4**:448
    host resistance to, **5**:650
    immune response to, **4**:389
Oncofetal antigens. *see under* Antigens
Oncogene amplification, **3**:102, 788
    chromosomal abnormalities associated with, **5**:545–46
    chromosomal basis of, **5**:545–47
    double minute chromosomes and, **5**:546
    homogeneously staining chromosomal regions and,
        5:546
    in human cancer, **5**:545–50
    in multistage carcinogenesis and tumor progression,
        5:549–50
    mutations and, **5**:549–50
    tumor progression and, **5**:549–50
    tumor specificity of, **5**:547
Oncogenes, **2**:62–64, 121, 243, 256, 263, 265, 441, 451,
    452
    abnormal receptor coding by, **5**:420
    c-erb B-1, glioblastomas and, **5**:548
    cellular, defined, **3**:101
    defined, **2**:53, **5**:415
    ganglioside expression and, **3**:733
    hepadnaviruses and, **4**:763–67
    leukemia, **4**:692
    list of, **2**:121
    myc gene family of, **5**:547–48
    N-*myc*, **5**:547
    neu oncogene, **5**:548–49
    overproduction, **2**:452
    and protein kinases, **6**:205–6
    tumor rejection and, **4**:338–39
    tumorigenic potential, testing of, **5**:328–29
    in "two-hit" theory of cancer genetics, **3**:822
    uterine, estrogen-induced, **7**:743–44
Oncogenesis
    developmental theory of, **5**:327–28
    environmental carcinogens and, **5**:324
    oncogene hypothesis for, **5**:328–29
    recombination theory of, **5**:325–26
    theories of, **5**:323
One-pot ligation techniques, **3**:172
Onions, NSP content of, **3**:39
Ontogeny, **2**:673
    and evolution, **2**:870
    recapitulates phylogeny, **2**:871

Onuf's nucleus
    cortical projections to, 5:46
    L-region and, **5:45**
    motor neurons of, 5:43-44, 45
    pontine tegmentum and, 5:45
    spinal cord, **7:**199
Oocyte maturation inhibitor factor, **3:**633
Oocytes
    annulate lamellae of, 5:553
    Balbiani vitelline body of, 5:553
    defined, **3:**305
    and follicular cells, 5:554
    growth, 5:553
        changes in enzyme activity during, 5:558
        DNA and RNA synthesis during, 5:558-59
        protein synthesis during, 5:559
        regulation of metabolism during, 5:558-59
    maturation, 3:633-34
        induction of, 5:557
        inhibition of, 5:555-56
    meiosis in, 5:552
    primordial germ cells and formation of, 5:551-53
    ultrastructure of mammalian, 5:553-54
Oocytes, *Xenopus*
    gene amplification in, 3:786
    nuclear pores of, 5:452
    nuclear protein transport in, 5:455
Oocytes, *Xenopus laevis*, 3:157
Oogenesis, 3:628-29, 4:950-51, 5:551-59
Oogonia, **2:**438
Oogonia, defined, 5:551
Oophorectomy, **2:**54
Open reading frames
    defined, **3:**463
    in Epstein-Barr virus genome, 3:466
Operant conditioning. *see* Skinner box
Operator regions (genetics)
    DNA binding sites and, 3:122-23
    in prokaryotic gene transcription, 3:109
Operon, defined, **3:**794, 823, 824
Ophthalmia neonatorum, **4:**458
Ophthalmic artery, 3:548-49
Opiate receptors. *see* Receptors, endorphin
Opiates. *see* Narcotics
Opiates, endogenous. *see* Endorphins
Opioid peptides. *see* Endorphins
Opioid receptors. *see* Receptors, opioid
Opioids. *see* Narcotics
Opisthorchiasis, **2:**389
Opisthorchis viverrini, **2:**389

Opisthotonos, defined, **3:**734
Opium
    addiction, 1:43-51
    biosynthesis of, **1:**179
Opponency, **2:**588
Opsonization, 2:614-16, **4:**856
    defined, 2:601, 614
    and phagocytosis, 2:614
Optacon, as aid for blind, **7:**62
Optic chiasm, anatomy, **3:**550
Optic disk, anatomy, **3:**549
Optic dysgraphia, **3:**221
Optic nerve
    anatomy, 3:535, 549-50
    astrocytes, 1:436-37
    regeneration of, **5:**342
Optic system
    anatomy, **3:**550
    axonal migration in, **3:**744
Optic tectum. *see* Superior colliculus
Optical isomers
    separation of, **3:**207
    sweetness and, **7:**364
Optical rotatatory dispersion, 5:567
Optical rotation, 5:567
Optokinetic reflex, **3:**557
Oral cavity. *see* Mouth
Oral contraceptives. *see* Contraceptives, oral
Oral mucosa, 2:770, 780-81
    gustatory, 2:770, 780
    masticatory, 2:780
    specialized, 2:770
Oral rehydration therapy. *see* Fluid therapy
Orangutans
    characteristics of, **6:**141
    evolution of, **3:**490
    mating, **4:**934
    sex differences in, biocultural, **6:**847
Orbit
    anatomy, 3:545, 546
    growth of, 2:706-7
Organ of Corti, 3:232-34
Organ of Zuckerkandi, **1:**72
Organelles
    cross-partitioning, **1:**362
    genomes, **4:**252
    intracellular acid load and, **5:**873
    intraneural transport of, **5:**900
    microtubule transport of, 5:18-20
    pH determination in, **5:**872

Organelles *(Continued)*
  separation and fractionation in aqueous two-phase systems, **1:**361–62
  tubulin and, **5:**897
Organic chemicals, non-ionized, **3:**253
Organic mental disorders, **4:**997
Organisation for Economic Co-operation and Development, **3:**257
Organochlorine insecticides, **7:**561
Organophosphates, **1:**177, **2:**346
Organophosphorate insecticides. *see under* Insecticides
Organophosphorous compounds
  in foods, toxicity of, **7:**560–61
  neurotoxicity, **7:**548
Orgasm
  as biological dimension of sexuality, **6:**880–81
  and sexual response, **6:**866
Orgel, L., **3:**895
Orlowski, M., **3:**913
Ornithine decarboxylase, **2:**119, **6:**68–69
Orofacial region
  autonomic regulation of, **5:**372
  blood flow, autonomic regulation of, **5:**372
  general neural activities of, **5:**361–62
  mechanoreceptors in
    touch and, **5:**362–64
  motor system, features of, **5:**369
  neural basis for pain in, **5:**365–67
  neural basis of function, **5:**361–72
  neuromuscular functions in, **5:**369–72
  pain conditions in, **5:**367–69
  relay of tactile information from, pathways involved in, **5:**364
  sense of touch in, **5:**362–64
  sensory functions in, **5:**362–69
  taste sense in, **5:**365
  thermal sensations in, **5:**364–65
Orofacial responses, in decerebrate organisms, **5:**809
Orotic aciduria, **6:**397
Orthologous, **2:**737
Orthomyxoviridae, **4:**519
Orthomyxovirus infections, **3:**184
Orthopantomography, **2:**809, 810
Orthostatic tolerance, and space travel, **7:**148
Oryctes viruses. *see* Baculoviridae
Os coxa, **7:**38
Osler, William, **5:**700
Osmium tetroxide, **2:**726, 727, **3:**278
Osmophiles, defined, **3:**662
Osmotic balance, defined, **5:**47

Osmotic diuretics. *see* Diuretics, osmotic
Osmotic reflection coefficient, defined, **5:**37
Ossicles, physiology, **3:**230, 231
Ossification, **2:**201, 212, 567, 690, **4:**28–29
  endochondral, **1:**784–85
  intramembranous, **1:**784
  postnatal, **4:**27
Osteoblasts, **1:**782–83, **2:**212
  bone collagens, **1:**763–64
  in bone remodeling, **1:**818–19
  in bone turnover, **1:**773–74
  calcitonin and, **3:**483
  changes in activity, **1:**762
  defined, **3:**481
  developmental pathway, **1:**762
  differentiation, **1:**784
  estrogen and, **3:**485
  functions of bone proteins, **1:**765–66
  gene regulation, **1:**766
  growth hormone and, **3:**484
  noncollagenous proteins, **1:**764
  production of short range protein factors, **1:**762
  products in bone mineralization, **1:**765
  prostaglandin E$_2$ and, **3:**484
  proteins of, **1:**763–64
Osteocalcin, **1:**764–65
Osteoclastic bone loss. *see* Bone resorption
Osteoclasts, **1:**782
  acid production mechanism, **1:**766–67
  activity regulation, **1:**762
  in bone remodeling, **1:**818
  in bone turnover, **1:**773–74
  calcitonin and, **3:**483
  calcitonin receptor, **1:**769
  cell adhesion, **1:**767–69
  defined, **3:**481
  developmental pathway, **1:**762
  enzymes of, **1:**766–67
  extracellular vacuole formation, **1:**767
  gene products, **1:**766–70
  identification, **1:**766
  immunologically defined hemopoietic antigens expressed by, **1:**9–10, 1696
  macrophage colony-stimulating receptor, **1:**769
  major molecules and enzymatic functions of, **1:**767
  mediation of bone resorption, **1:**766–67
  prostaglandin E$_2$ and, **3:**484
  receptor of osteoblast-derived stimulator factor, **1:**769
  sealing zone, **1:**767
  vitronectin receptor, **1:**768

Osteoclasts (*Continued*)
  VLA-2 receptor, **1**:768
Osteocytes, **1**:782-83, **2**:646
Osteogenesis
  direct, **1**:784
  ectopic, **1**:785
  endochondral ossification, **1**:784-85
  growth factors, **1**:787-89
  indirect, **1**:784-85
  intramembranous ossification, **1**:784
  osteoblasts, **1**:782-83
  osteoclasts, **1**:782
  osteocytes, **1**:782-83
  preosteoblasts, **1**:782-83
Osteogenesis imperfecta, **2**:647
Osteoid, **2**:212, 560
Osteolytic metastases, **4**:278-79
Osteoma, osteoid, **7**:698
Osteomalacia, **3**:26
Osteomyelitis
  effect of quinolones on, **6**:409
  and pulmonary tuberculosis, in children, **5**:706
Osteonectin, **1**:764, **2**:91
Osteopenia, **2**:650
Osteopontin, **1**:764-65
Osteoporosis, **1**:775-76, **2**:573
  aging and, **3**:904
  defined, **3**:17
  as diet-related disease, **5**:488
  estrogens and, **3**:481-86
  and fragility, **1**:776-78
  in gonadorelin analogue treatment, **3**:368
  lactose malabsorption effects, **4**:615
  senile, **2**:698
Osteosarcoma
  classification of, **7**:699
  clinical characteristics of, **7**:699
  mice bone tumors and, **3**:702
  retinoblastoma and, **3**:142
Otitis media, **4**:459-60
Ottaway, **5**:381, 384
Ouabain, **2**:284
Outer ear. *see* Ear, external
Outer hair cells. *see* Hair cells, outer
Ovalocytosis, hereditary, **4**:882
Ovarian cysts, **5**:608
Ovarian neoplasms, **5**:606
Ovaries
  anatomy of, **6**:562-63
  embryonic development of, **6**:839

Ovaries (*Continued*)
  fetal
    development of, **3**:340-41
    illustration of, **6**:840
    gonadotrophins and, **3**:629-30
    steroid hormone synthesis in, **7**:252-53
    steroids, biosynthesis of, **7**:263-64
Overeating. *see* Hyperphagia
Overnutrition. *see* Obesity
Overweight. *see* Obesity
Oviducts
  anatomy of, **6**:563
  autonomic innervation of, **1**:543
Oviducts, mammalian. *see* Fallopian tubes
Ovocytes. *see* Oocytes
Ovulation, **2**:498. *see also* Superovulation
  double, causes of, **3**:307
  effect of fat on, **1**:743
  endocrine function in, **7**:739-40
  follicle, and gonadotropins, **7**:350
  food intake and, **1**:748
  LH surge and, **3**:633-34
  timing, in intrauterine insemination, **3**:602
Ovulation induction, **7**:345-51
  clomiphrene citrate and, **7**:347
  follicle stimulating hormone and, **7**:347-48
  gonadotropin-releasing hormone and, **7**:348
  historical perspective of, **7**:346
  human menopausal gonadotropins and, **7**:347-48
  methodology, future directions of, **7**:349-50
  regimens for, **7**:347-49
Ovum, **2**:248
  efined, **3**:599
  fertilization, **4**:954, 956
  maturation, **2**:248
  morphology of, **3**:603
  release, **2**:248
  sperm penetration of, **3**:604-6
Ovum implantation
  defective, in early spontaneous abortion, **3**:302
  defined, **3**:293, 305
  of monochorial twins, **3**:310
Owen's contour lines, **2**:823
Oxaloacetates, , **2**:504-10, **5**:821
Oxamniquine, **2**:393
Oxford Family Planning Association Contraceptive
    Study
  on myocardial infarction, **5**:603
  subjects of, **5**:600
Oxidases, function of, **5**:622

Oxidation
of fatty acids, **3**:575
free radical, and zinc, **7**:883–84
of glutathione, **3**:910, 912, 914
of glyceraldehyde-3-phosphate, **3**:936–37
superoxide formation during, **7**:353–54
Oxidation-reduction
defined, **3**:907, **5**:471
electron transfer in, **5**:61
food spoilage and, **3**:662
pH and, **5**:822
in respiratory chain, **5**:56
Oxidation-reduction potential, defined, **5**:55
Oxidative phosphorylation
acceptor control, **1**:459
ADP/ATP transport system, **1**:462
aerobic metabolism and, **3**:932
ATP synthase complex, **1**:462–63
ATP synthesis, **1**:463–64, **4**:974
cofactors in, **5**:473–74
defined, **3**:931, **5**:55
electrochemical gradient of protons, **1**:458–59
electron transport, **1**:459–61
essential parts of, **1**:458
glycolysis and, **3**:932
H^+/O ratio, **1**:461
H^+/P_i transport system, **1**:462
in heart mitochondrial inner membrane, **5**:57
inhibitors of, **1**:464–65
location of, **1**:458
mitochondrial, **3**:932
P/O ratio, **1**:461
proton translocation, **1**:461
purpose of, **1**:457–58
uncouplers of, **1**:464–65
5-Oxo-L-proline
deficiency of, **3**:916
in gamma-glutamyl cycle, **3**:913
2-Oxoglutarate dehydrogenase, renal, **5**:822
Oxolinic acid, **6**:403
Oxosuccinates. *see* Oxaloacetates
Oxycodone, **2**:786
Oxygen
affinity of hemoglobin, **4**:145
air to blood diffusion of, **5**:619–20
arteriovenous, in exercise, **3**:519
autoregulation, blood flow and, **5**:621
and carbon dioxide exchange, disorders of, **6**:370–74
concentration, intestinal, **4**:556–57
debt, muscle exercise and, **7**:30

Oxygen (*Continued*)
dependence, respiration and, **6**:568
equilibrium, hemoglobin, **4**:136–37
in gastric blood flow regulation, **3**:760
loading and unloading, oxyhemoglobin dissociation
curve and, **5**:620–21
metabolism, zinc in, **7**:883–84
microvessel exchange of, **5**:34, 39
molecular, **2**:507
nitric oxide inactivation and, **3**:373
partial pressures, **4**:137
processing by phagocytes. *see* Respiratory burst
radicals, removal by glutathione oxidase, **6**:793
as radiosensitizer in radiobiology, **6**:446
saturation, defined, **3**:609
tension, blood perfusion of tissue and, **5**:621–22
therapy, for anaerobic infections, **1**:249
Oxygen binding
cooperative binding, **4**:137
heme stereochemistry, **4**:142
hemoglobin S, **7**:7
Oxygen consumption
in aerobic exercise, **3**:505–6
carbonic anhydrase and, **5**:876
in exercise, **3**:519
fetal, **3**:617
gingiva, **4**:37
hypothermia, **4**:800
in nutrient utilization, **5**:472
pH and, **5**:823–24
in physical activity of the aged, **3**:901
relationship of size and metabolism to, **6**:571–72
in respiratory burst, **6**:582
in work, **3**:515–16
Oxygen consumption, maximal
aerobic training and, **3**:512
age and, **3**:507
defined, **3**:505
in dynamic exercise, **3**:518
fatigue and, **3**:507
improvement in, sympathetic discharge and, **3**:509
limits on, **3**:506
stroke volume and, **3**:517
in work, **3**:516
Oxygen deficiency. *see* Anoxia
Oxygen demand
during cardiopulmonary bypass, **4**:94
intestines, **4**:552, 556
Oxygen dissociation curve
for blood, **6**:599
effect of various factors on, **6**:599

Oxygen toxicity
    lung injury and, **6**:352
    superoxide dismutase and, **7**:357
Oxygen transport
    from blood to tissue site, **5**:622
    cascade and, **5**:618
    fetal, **5**:623
    thermal injury and, **7**:453
    to tissues, **5**:617-23
Oxygenases, **5**:622-23
Oxygenators. *see* Heart, mechanical
Oxyhemoglobin dissociation curve, **5**:620, **7**:7
Oxymethalone, **2**:114
Oxymyoglobin, **4**:139-40
Oxyntic cells. *see* Parietal cells, gastric
Oxyntomodulin. *see* Enteroglucagon
Oxytocin, **2**:20, 23, 34, 36, 403, 404
    in fetal hypothalamo-hypophyseal system, **3**:338
    function, **5**:397, 707, **6**:21, 89
    hypothalamus control, **4**:311-12
    milk-ejection reflex, **4**:607-8
    perinatal, **3**:347
    production, **6**:19
Ozone
    exposure routes of, **3**:387
    inhaled, disposition and effect of, **7**:579

P

p97. *see* Melanotransferrin
P cells, in retina, **7**:804
P-positive nerve fiber, **5**:380
P/S ratio
    defined, **3**:567
    use of, **3**:572
PABA. *see* 4-Aminobenzoic acid
Pacemakers, **2**:135-36, 492-93
    defined, **2**:489
Pacific Multifactor test, **3**:838
Pacinian corpuscles
    somatosensory system and, **7**:121
    of visceral sense organs, **7**:796
PAG. *see* Periaqueductal gray
Paget's disease, **1**:823
Pain, **5**:625-33
    acute, defined, **5**:629
    and analgesics, non-narcotic, **2**:785
    ascending system, **5**:626-27
    attitudes toward, **5**:629
    bradykinin and, **5**:730

Pain (*Continued*)
    central nervous system transmitters in, **5**:412-13
    in children, assessment of, **5**:630-32
    chronic, **5**:629
    cold stress to periphery, **4**:796-97
    control, **5**:367
    control, pharmacologic, **2**:785-87
    defined, **5**:625
    dental, mechanism of, **2**:825
    descending system, **5**:627
    developmental aspects of, **5**:630
    gate control theory of, **5**:365, **7**:215, 627-28
    in infants, indices of, **5**:630-31
    in infants and children, management of, **5**:632
    masking, hypnotic induction and, **7**:622
    measurement, **4**:78
    modulation, **2**:338, **5**:626
    in normal adults, testing, **5**:632-33
    in older adults, **5**:633
    in orofacial region, **5**:365-69
    perception, attitudes towards, **5**:628-29
    phasic, defined, **5**:629
    physiology of, **5**:625-27
    and prostaglandins, **2**:785
    psychology of, **5**:628-29
    referred, viscera and, **7**:797-98
    relief, **2**:338
    response, in the spinal-cord injured, **5**:808
    serotonin and, **6**:831-32
    spinal cord
        projections in, **5**:627
        and sensing of pain, **7**:213-15
    temporal forms of, **5**:629
    theories of, **5**:627-28
    thermal injury and, **7**:450
    transduction, **5**:626
    transmission, neural pathways of, **5**:626-27
    transmission fibers, **5**:626-27
    visceral, **7**:797-98
Palaeoanthropology, **6**:877-78
Palate, in fragile X syndrome, **3**:711
Palate, soft, and swallowing, **3**:74
Paleocortex, **2**:673, 679-80
Paleogenome, **4**:258
Pallium, **2**:674, 677, 678
Palmitic acid, **2**:745, **3**:731
Palmitylation, G proteins and, **4**:22
Palmoplantaris pustulosis. *see* Psoriasis
Palsy, progressive supranuclear. *see* Supranuclear palsy,
    progressive

Palytoxin, **7:**589
Pamaquine, **4:**884
Pancreas
  acinar cells of, **7:**568-69
  anatomy, **3:**63-64, 78
  autonomic innervation of, **1:**542
  chemical carcinogens and, **7:**569
  cholecystokinin and, **5:**718, 730
  drugs causing injury to, **7:**568
  endocrine. *see* Islets of Langerhans
  exocrine
    secretion, **2:**38-39
    structure and function in, **7:**567-68
    toxicity in, **7:**567-70
  fetal, development of, **3:**346
  glucagon and, **5:**733
  pancreatic polypeptide and, **5:**719
  secretin and, **5:**718
  secretions, in digestive process, **3:**78
  stress, and exocrine secretion, **7:**288-89
  toxic agents and, **7:**570
  toxicity in, **7:**567-74
"Pancreatic cholera." *see* Cholera, "pancreatic"
Pancreatic cystic fibrosis. *see* Cystic fibrosis
Pancreatic glucagon. *see* Glucagon
Pancreatic hormones
  in children, **5:**708
  cyclic AMP and, **3:**586
  perinatal, **3:**350
Pancreatic insufficiency
  in infants, **3:**52
  phenylalanine deletion and, **3:**144
Pancreatic lipase. *see under* Lipase
Pancreatic neoplasms, **2:**116
  peptide hormones and, **5:**721
  varieties of, **5:**722
Pancreatic polypeptide, **5:**719-20, **6:**89
Pancreatic proteases, **3:**49
Pancreatitis, **2:**651
  and copper toxicity, **2:**651
  drugs causing, **7:**568, 569-70
  kinins and, **4:**598
Pancreatography. *see* Cholangiopancreatography, endoscopic retrograde
Pancreozymin. *see* Cholecystokinin
Pandemic diseases
  defined, **7:**640
  influenza viruses, **4:**487-88
Panencephalitis, subacute sclerosing, **4:**522
Panic
  cannabis and, **5:**445
  LSD use and, **5:**448

Panic disorders, **2:**78-79
  and caffeine, **2:**78-79
  characterization, **4:**997
Pantothenate, **2:**539, 541
Pantothenate kinase, **2:**541
Panum's fusional area (PFA), **1:**636
Papilla, optic. *see* Optic disk
Papilloma
  genital cancer and, **5:**638
  laryngeal, **5:**638
  oral, **5:**615
Papillomaviruses, **4:**450
  cervical cancer and, **4:**343
  epithelial differentiation and, **5:**636-37
  genital cancer and, **3:**859-67
  host specificity of, **5:**637
  HPV-16
    oncogenic functions encoded by, **5:**639-41
    structural similarities of E6 and E7 genes, **5:**641
  HPV-18, oncogenic functions encoded by, **5:**639-41
  HPV E7 protein, **5:**641-42
  human
    characteristics of, **5:**635
    circular genome of, **5:**636
  infections
    clinical manifestations of human, **5:**637-38
    keratinocytes, **4:**585-86
    trends in genital infections, **6:**894-95
  molecular characterization of, **5:**636-37
  natural occurrence of, **5:**636-37
  and neoplastic transformation, **5:**635-43
  open reading frames of genome, **5:**636
  tissue specificity of, **5:**636-37
  in vitro transformation
    assays for human, **5:**638-39
    perspectives on carcinogenesis and, **5:**643
Papoviridae, interferon effects, **4:**518
Paracelsus, **2:**342
Paraclones, **4:**583
Paracoccus denitrificans, **5:**60-61
Paracrine, **2:**321
Paracrine effectors
  defined, **3:**318, **5:**714, 445
  histamine as, **5:**716
  keratinocyte and, **3:**447-48
Paragonimiasis, **2:**389
Paragonimus westermani, **2:**389
Paraldehyde, **2:**787
Parallel-distributed processing, in neural networks, **5:**391
Parallel-distributed processing model, **4:**199, 202

Parallel elastic component, muscle
    of biceps brachii, **5**:184
      identification of, **5**:184
      series elastic component and, **5**:185
Parallel processing of color streams, **2**:588–89
Paralogy, defined, **3**:882
Paralysis, **2**:345
Paralysis agitans. *see* Parkinson disease
Paramyosin. *see* Tropomyosin
Paramyosin muscle, oyster, **5**:209
Paramyxoviridae, **4**:940. *see also* Measles virus
Paraneoplastic syndromes, **4**:275
Paranoid disorders
    cannabis and, **5**:445
    characterization, **4**:997
    in dementia, **2**:766
    in older adults, **1**:133
Paranoid schizophrenia. *see* Schizophrenia, paranoid
Paraoxonase, **2**:732
    as genetic marker for cystic fibrosis, **2**:732
Paraparesis, tropical spastic, **1**:738
Paraphasia, **2**:763
Paraphasias, phonemic. *see* Phonemic paraphasias
Parapithecidae, in Fayum fossils, **3**:489
Paraplegia, **7**:231
Paraquat, **7**:549
Parasites. *see also* specific parasites
    antigenic variation, **1**:300
    defense against, eosinophils and, **5**:812
    drug designs to exploit metabolism, **5**:645–55
    drug resistance of, **5**:650–51
    fatty acid remodeling in, **5**:653
    host modification by, **5**:647–49
    immune evasion by, **5**:647–49
    invasion of host cells, **5**:646–47
    practical applications of molecules, **5**:654
    RNA editing in, **5**:653
    surface, antigenic variation of, **5**:649
    targets for immunoconjugates, **4**:373–74
    taxonomy of, DNA probes in, **5**:654
    tissue penetration by, **5**:647
Parasites, animal, **1**:261–70. *see also* specific parasites
    Ascaris lumbricoides, **1**:270
    Entamoeba histolytica. see Entamoeba histolytica
    Enterobius vermicularis, **1**:269
    Giardia lamblia, **1**:266
    hookworms, **1**:263–64
    pinworms, **1**:269
    Plasmodium, **1**:267
    Schistosoma, **1**:263

Parasites, animal *(Continued)*
    Simulium damnosum, **1**:265
    tapeworms, **1**:264–65
    Toxoplasma gondii, **1**:268
    Trichinella spiralis, **1**:265
    Trichomonas vaginalis, **1**:268–69
    Trichuris, **1**:269–70
Parasitic diseases, **2**:381–94
    by arthropods, **2**:390
    drug design for, **5**:654–55
    by flukes, **2**:390
    immunoassays to diagnose, **5**:654
    immunology, **4**:385–90
      antigenic variation, **4**:387–88
      Bancroft's filariasis, **4**:389
      concomitant immunity, **4**:386, 389–90
      host-protective immunity, **4**:386–87
      immune deviation, **4**:388
      immune evasion, **4**:387–88
      immunodiagnosis, **4**:388–89
      immunopathology, **4**:389–90
      onchoceriasis, **4**:389
      reinfection, **4**:386
      schistosomiasis, **4**:389
      vaccine design, **4**:390
    innate host resistance to, **5**:649–50
    by protozoa, **2**:390
    by roundworms, **2**:386–89
    by tapeworms, **2**:389–90
    vaccination for, **5**:655–56
Parasitic diseases, animal
    ascariasis, **1**:270
    giardiasis, **1**:265–66
    hookworm anemia, **1**:263–64
    malaria, **1**:267
    onchocerciasis, **1**:265
    pinworms, **1**:269
    river blindness, **1**:265
    snail fever, **1**:263
    tapeworms, **1**:264–65
    toxoplasmosis, **1**:268
    trichinosis, **1**:265
    trichomoniasis, **1**:268–69
    trichuriasis, **1**:269–70
Parasitism
    commensalism, **1**:262
    ecology of, **1**:261–62
    molecular mechanism of, **4**:684–85
    mutualism, **1**:262
    symbiosis, **1**:261

Parasitology, molecular, **5:**645–56
Parasomnias, as sleep disorder, **7:**81–82
Parasympathetic nervous system
    anatomy of, **5:**359–60
    as division of autonomic nervous system, **1:**538–39
    efferent fibers, diagram of, **5:**357
    in gastric circulation measurement, **3:**762
    intestinal blood flow regulation, **4:**551
Parasympathetic outflow, **2:**39
Parathion, **2:**347
Parathormone
    in calcium metabolism disorders, **3:**353
    fetal, **3:**346
    perinatal, **3:**349
Parathyroid chromogranin, **5:**660
Parathyroid glands
    anatomy, **5:**657
    embryology, **5:**658
    fetal
        development of, **3:**346
        functions of, **3:**619
    histology, **5:**658
    parathyroid secretory protein of, **5:**660
    species distribution of, **5:**658
Parathyroid hormone, **5:**657–66
    adenylate-cyclase-cyclic AMP system and, **5:**662
    amino acid sequence of, **5:**658
    antagonists, **5:**659–60
    biosynthesis of, **5:**658, 664
    in bone and calcium metabolism, **1:**800
    bone cell function and, **1:**765
    in bone homeostasis, **1:**774
    in bone remodeling, **1:**822
    calcitonin and, **5:**664
    calcitrol and, **5:**664
    in children, **5:**707
    deficiency and excess of, **5:**664–66
    disorders of, vitamin D and, **7:**871
    disposition of, **5:**659
    functions, **5:**660–62, **6:**89
    glucocorticoids and, **5:**664
    hypersecretion of, **1:**807–8
    hyposecretion of, **1:**807–8
    interactions with other hormones, **5:**664
    in plasma, assay of, **5:**660
    plasma calcium concentration and, **5:**663
    plasma Ca^{2+} and, **3:**324
    in regulation of bone and calcium metabolism, **1:**802–3
    regulation of bone resorption, **1:**762
    regulation of secretion, **5:**662–63

Parathyroid hormone *(Continued)*
    structure-activity relationships for, **5:**659–60
    terminal metabolism of, **5:**659
Parathyroid hormone-related protein, **4:**277
Parathyroid hormonelike peptide, **5:**659–60
Parathyroidectomy, **5:**665
Paratope, **4:**314
Paratyphoid fever, **6:**710
Paraventricular hypothalamic nucleus, **4:**719
Paraventricular nuclei, **5:**394
Paraxial mesoderm, **5:**161–62
Parenchyma, defined, **3:**261
Parenchymal cells
    immune system and, **4:**395
    uniquely differentiated, **4:**396
Parenchymatous plant tissues
    dietary fiber from, **3:**36
    noncellulosic polymers of, **3:**38–39
    structure of, **3:**36
Parenting, stress associated with, **7:**293
Paresthesia, **2:**683
Parietal cells, gastric
    defined, **3:**765, **5:**713
    mucosa, types of, **3:**768
Parietal cortex, and space, **2:**10
Parietal lobe, **2:**351, 682, 683
    dyslexia-dysgraphia and, **3:**221–22
Parietal lobe, left inferior, in transcortical sensory dysphasia, **3:**219
Parietal peritoneum, defined, **3:**57
Parieto-occipital system
    dyslexia-dysgraphia and, **3:**219–23
    spatial synthesis and, **3:**221
Parkinson disease, **2:**218–19, 764, 835, **5:**667–73
    amyotrophic lateral sclerosis and, **1:**232
    central nervous system transmitters in, **5:**411–12
    D$_2$ receptors and, **5:**422
    as dopamine deficiency disease, **2:**219
    effects on communication in older adults, **1:**123
    fusimotor overactivity in, **3:**723
    idiopathic, **5:**669
    thalamic nucleus associated with, **7:**442
    therapy, **2:**219, **5:**411–12
Parotid gland, **2:**781
    anatomy of, **6:**690
Paroxetine, **2:**846
Paroxysmal cold hemoglobinuria (PCH), **1:**514–15
Paroxysmal hemoglobinuria, **4:**797
Parthenogenesis, **4:**951
Partial sex linkage. *see* Pseudoautosomal inheritance

Particles
    as airborne toxicants, **7:**577
    common, size distribution of, **7:**554
    inhaled
        clearance and retention of, **7:**580–82
        mechanism of deposition of, **7:**579–80
    pathogenesis of, **7:**554
    toxic effects of, **7:**553–57
Parturition. *see* Labor
Parvalbumin, **2:**89
    crystal structure, **2:**93
Passive electrical properties. *see under* Electrical proper-
    ties
Passive nonionic diffusion. *see* Diffusion, nonionic
Pasteur, Louis, **3:**938
Pasteur effect
    allosteric controls of, **3:**940
    defined, **3:**931
    features of, **3:**932–33
    key regulators of, **3:**940
Pasteurization, **3:**51
Pastoralists, customary diets of, **3:**20–21
Patau syndrome. *see* Trisomy-13
Patch-clamp recordings, defined, **4:**13
Patella, **7:**38
    anatomy of, **1:**376
Patellar ligament, **1:**376
Patellar retinaculum, **1:**376
Patellofemoral joint, **1:**376
Paternity, DNA markers and, **3:**149
Path analysis (statistics)
    in behavioral genetics twin studies, **3:**837
    defined, **3:**835
Pathogens, food-borne. *see* Food-borne pathogens
Pathologists
    functions of, **5:**687–88
    training of, **5:**686
Pathology
    defined, **5:**675, 681
    diagnostic, **5:**681–88
    historical perspective of, **5:**675–76
    molecular, impact of, **5:**679
    oral, **5:**611–16
Pathology, clinical
    defined, **5:**681
    diagnostic pathology and, **5:**685
    scope of, **5:**683–85
Pathology, comparative, **5:**675–80
    divisions of, **5:**676
    future developments in, **5:**679–80

Pathology, comparative *(Continued)*
    impact of genetic engineering on, **5:**679–80
Pathology, forensic. *see* Forensic medicine
Pathology, surgical, **5:**682–83
Pathology department, hospital, **5:**687
Patrick, **3:**777
Pattern inception, primary. *see* Embryonic inductions,
    primary
Pattern recognition, **1:**481–82, **2:**550
Paul-Bunnel-Davidsson test, **3:**465
Pause cells, in saccades, **3:**558
Pavlov, Ivan P., **2:**629–31
Pavlovian conditioning. *see* Conditioning, classical
P34CDC2 protein. *see* Protein P34CDC2
Peanut oil, **3:**572
Peas, flowering, Mendelian studies of, **3:**847, 848–49
Pectic substances
    composition of, **3:**38
    mineral absorption and, **3:**42
Pectins, defined, **3:**35
Pectoral girdle
    bones of, **7:**37
    muscles of, **4:**263–64
Pediatric nutrition. *see* Child nutrition
Pediatrics, **5:**699–712
Pediculus capitis, **2:**390, **4:**47–48
Pedigree, **4:**434
    of cystic fibrosis, **3:**830
    of Huntington's disease, **3:**830
Pedigree analysis
    in autosomal recessive inheritance, **3:**851
    in behavioral genetics, **3:**836–37
    of familial mitochondrial encephalomyopathy, **3:**858
    in genetic counseling, **3:**814–15
    mendelian trait compilation for, **3:**850
    symbols in, **3:**851
Peer groups, **7:**236
Pelage, intevibrissal, **7:**677
Pelvic bones
    archaic *Homo sapiens*, **1:**825–31
        and evolution of, **1:**831
        fossils and, **1:**827–30
        sex differences, **1:**830–31
    human evolutionary studies of, **1:**825–27
    of pelvic girdle, **7:**38
Pelvic inflammatory disease. *see* Adnexitis
Pelvis, **2:**595–96
    as birth canal, **2:**405, 596
    in magnetic resonance imaging, **3:**6, 7
    male and female, **7:**39

Pelvis (*Continued*)
  muscles of pelvic floor, **4**:263
  plexus, **2**:33
  structure, sex differences as factor in, **1**:830-31
  structures and functions of, **1**:826
Pemphigus, **1**:529, **2**:271, **5**:615
Pemphigus vulgaris. *see* Pemphigus
Penetrance (genetics), defined, **3**:835
Penicillin-binding proteins (PBPs)
  peptidoglycan synthesis and, **1**:281-82
  resistance to beta-lactam antibiotics, **1**:283-84
Penicillin G, **1**:288
Penicillins, **2**:372, 784. *see also* beta-Lactams
  cell wall peptidoglycan and, **1**:272-73
  in feed additives, **1**:294
  mode of action, **1**:290(t)
  peptidoglycan transpeptidase and, **1**:282-83
  semisynthetic, **1**:293
  structure of, **1**:273
Penicillium
  aflatoxins of, **5**:264
  food toxicology and, **3**:675
Penicillium citreoviride, **5**:266
Penicillium islandicum, **5**:262
Penicillium notatum, **1**:288
Penile erection, **4**:935-36
Penile neoplasms
  incidence of, **3**:859
  papillomaviruses and, **3**:863
  in partners of women with cervical cancer, **4**:343
Penis, **2**:597, **4**:929
  anatomy of, **6**:562
  flaccid and erect lengths, **4**:935
Pentamidine isethionate, **2**:383, 385
Pentamidine isothionate, in treatment of AIDS, **1**:152-53,
    156
Pentazocine, **2**:786
Pentobarbital, **2**:786
  in dentistry, **2**:786
  withdrawal from, **5**:443
Pentodactyl, **2**:593
Pentylenetetrazol, **6**:783-84
Peppered moth, pollutants and, **3**:254-55
Pepsin, **2**:34, 35
  defined, **2**:31
  delta, **2**:36
  gastric lipase and, **3**:49
  gastrin-releasing, **2**:34
  intestinal vasoactive. *see* Vasoactive intestinal peptide
  list of, **2**:17

Pepsin (*Continued*)
  mapping, **2**:41
  mu, **2**:36
  natriurectic, atrial, **2**:143
  opioid, **2**:26, 35-39
Pepsinogen
  in early gastric cancer, **3**:770
  in gastric juice, **3**:76
  in protein digestion, **3**:80
Peptic ulcer, **5**:723
Peptidase D phenotype in HeLa cells, **4**:107-8
Peptidase inhibitors. *see* Protease inhibitors
Peptidases, signal, in protein targeting, **6**:250
Peptide bond, **4**:183
Peptide chains, **3**:416, 919
Peptide histidine-isoleucine. *see* Peptide PHI
Peptide histidine-methionine. *see* Peptide PHI
Peptide hydrolases
  peptide degradation and, **5**:726
  in protein digestion, **3**:80
Peptide-MHC molecule interactions, **4**:329-30, 335
Peptide peptidohydrolases, **3**:268
Peptide PHI, **5**:718
Peptide PHM. *see* Peptide PHI
Peptide receptors. *see* Receptors, peptide
Peptide YY, **5**:719
Peptidergic nerve fibers
  of gut-associated lymphoid tissue, **5**:381
  in neural-immune interactions, **5**:380-81
Peptides, **5**:725-35, **6**:87-98
  ACTH-related, fetal, **3**:342-44
  active, **6**:88-90
  backbone, **6**:216
  basal levels of, **6**:90
  binding
    and biological activity, **6**:93-94
    receptors and, **6**:92-93
  biological activity of, tissue-specific, **6**:97-98
  biological effects of, **6**:87-95
  biosynthesis of, **6**:95-98, 217
  blood levels of, **6**:91-92
  characteristics of, **5**:725-26
  concentration measurements of, **3**:328-29
  defined, **5**:667, 713
  degradation of, **6**:92
  diversity of, **6**:997-96
  ectopic, **3**:334
  elastin, **3**:268
  endogenous, abolition of pain by, **5**:413
  folding of, **6**:218-20

Peptides *(Continued)*
  gastrointestinal peptide hormones, **4**:555, **5**:713–23
  genetic locus of, **3**:184
  growth factors, wound healing and, **7**:499–500
  hemoglobin, **4**:135–36
  histones, **4**:235–36
  hormone-receptor complexes, internalization of, **6**:996–95
  in immunoglobulin, **4**:832
  listed, **3**:321–22
  mechanisms of action, **3**:331–33
  membrane protein biogenesis, **4**:976
  nature of, **6**:87–95
  as neuroactive substances in retina, **6**:629
  neurosecretion of, regulation, **6**:91
  neurotransmitters and, **3**:322
  as nonclassic hormones, **3**:323
  in oocyte maturation induction, **5**:558
  orphan, functions of, **5**:720–21
  precursors of, **6**:95
  processing of, **6**:95–97
  propeptides, processing of, **5**:726
  proteolysis of, **3**:323
  purification of, **5**:714
  regulation of, **6**:90–91
  secretion, stimulated, **6**:90–91
  signal
      and protein transport, **6**:240
      structure of, **6**:239–40
  in spinal cord function, **7**:205
  subunits, defined, **3**:919
  synthesis, **3**:327
  synthesis, ribosomes and, **6**:672–73
  synthetic, **5**:735
  target cell effects of, regulation and, **6**:92–93
  toxins, plant, **7**:594
  transcription of, **6**:96
  vaccines, synthetic, **7**:746–49
  as visceral sensory markers, **7**:794–96
Peptides, recombinant, **5**:733–34
Peptidoglycan
  antibiotics and biosynthesis of, **1**:278–83
  and bacterial taxonomy, **1**:274
  biosynthesis of, **1**:278–83
  gram-negative
      gram stain response in, **1**:274
      penicillin effects, **1**:272–73, 274–76, 278–83
      structure of, **1**:276–77
  gram-positive
      effect of penicillin on, **1**:272–73, 274–76, 278–83

Peptidoglycan
  gram-positive *(Continued)*
      gram stain response in, **1**:274
      structure of, **1**:275–76
  hydrolases, in bacterial cell wall synthesis, **1**:277–78
Peptidyl transferases
  inhibition by beta-latam antibiotics and beta-lactamases, **1**:282–83
  mechanisms of, **1**:282
  in synthesis of bacterial cell wall, **1**:276
Peptostreptococcus
  habitats, **1**:239, 241
  pathogenicity, **1**:246
Perceived direction
  coordinate systems in, **1**:629
  resolution limits of, **1**:629
  role of eye movements in, **1**:629
Perception, **2**:854, **5**:737–49
  aging and, **3**:901
  of color, **2**:548
  of complex smells, **5**:759–67
  crosstalk and, **5**:156–57
  effect of physiological changes on, **5**:749
  environmental development and, **5**:738–39
  eye and visual pathways in visual perception, **5**:742
  food and nutrition, **5**:751–57
  hearing, **5**:739–40
  and modulation of activity, internal, endogenous, **2**:549
  person, **5**:785–97
  reading, **5**:743–44
  recognition and, **5**:738
  selection process and, **5**:738
  sensory attributes in, **5**:738
  skin senses and, **5**:745–46
  smell, **5**:744
  of sound, **2**:548
  speech, **5**:740–42
  taste, **5**:744–45
  transducers in, **5**:737–38
  vision, **5**:742–43
Perceptron learning rule, **5**:391
Perceptual functions, development of, **5**:747–49
Perceptual inconstancy, and autism, **1**:509
Perceptual selection
  object and subject factors in, **5**:746–47
  reduction of complex data and, **5**:746
  in technical world, **5**:747
Peregrine falcon, **3**:253–54
Perforant pathway, **4**:230
Perforin, **2**:249
  and calcium, **2**:249

Perforin *(Continued)*
  and natural killer cell activity, **5**:299
  purification from cytotoxic T lymphocytes, **4**:814–15
  role in lymphocyte-mediated killing, **4**:815–16
Performance decrements
  analysis of, **5**:151
  cerebral hemispheres and, **5**:152
  crosstalk and, **5**:160
  resource-competition models and, **5**:154
Performance effects, defined, **5**:122
Performance-operating characteristics, **5**:149
  measurement of, **5**:150–51
Performance-resource function, defined, **5**:149
Perfume, production of, **1**:143
Pergolide, **2**:28
Periaqueductal gray, **4**:718
Pericarditis, **2**:157
Pericardium, **2**:201, 202
Peridinium. see Dinoflagellida
Periglomerular cells, olfactory, **5**:524
Perikaryon, **2**:343
Perikymaton, **2**:774
Perilymph, **3**:232
Perimysium, **7**:23
Perinatal endocrinology, **3**:347–50
Perinatal mortality. *see* Infant mortality
Perinatal period, **3**:337, **5**:259
Perineal-bladder reflex, **5**:44
Perineum
  anatomy of, **6**:564
  defined, **5**:43
Period
  defined, **2**:489
  developmental, **2**:869
  free running, **2**:489
Period life tables, **5**:102
Periodic movement
  coupling of, **5**:157–58
  lateral asymmetry and, **5**:158
  phase coupling in, **5**:158–59
Periodic process, **2**:489–501
Periodicity, **2**:489–501
  and biological biological, **2**:489
  circadian, **2**:489–501
  and cortisol secretion in human, **2**:492
  and courtship songs of fruit flies, **2**:491
  development within individual, **2**:495–96
    entrainment, maternal, **2**:495
    in fetus, **2**:495
    sleeping of infant, **2**:495

Periodicity *(Continued)*
  endogenous, **2**:490–91
  entrainment, **2**:491, 495
    light/dark cycle, **2**:491
  in hamster, wheel-running, **2**:490
  hibernation of ground squirrel, **2**:491
  and human health, **2**:499–501
    in jet lag, **2**:499
    in shift work, **2**:499–500
  light entrainment, **2**:494
    and eyes, **2**:494
    and retina, **2**:494
  maternal, **2**:495
  migration of garden warbler, **2**:491
  nonphotic agents, **2**:493
    drug, **2**:493
    feeding schedule, **2**:493
    neurotransmitter, **2**:493
    social factors, **2**:493
  nonphotic effect, **2**:493
  oscillator, endogenous, **2**:491
  phase-response curve, **2**:492
  pseudoperiodic, **2**:491
  and reproduction, **2**:497–99
    and hormones, **2**:497
    and pulsatility, **2**:497–98
  self-sustaining, **2**:490–91
  sleep/wake rhythmicity, **2**:491, 496
    in adult, **2**:496–97
    in infant, **2**:496
  ultradian, **2**:491
  wakefulness/sleep cycle, **2**:491, 496
Periodontal diseases, **2**:805, **4**:33–34, 37, **5**:612–13
Periodontal ligaments, **1**:372
Periodontal membrane, **2**:826
Periodontitis, **4**:37, **5**:612–13
Periodontium, **2**:776–78, 826, 827, **4**:33
Periosteum, **1**:543, **2**:697
Peripheral cues. *see* Cues, peripheral
Peripheral lymph node, **4**:1013
Peripheral nerves
  axonal regeneration and, **5**:344
  nerve growth factor and, **5**:338
  regeneration of, **5**:342
Peripheral nervous system
  anatomy of, **5**:352–60
  autonomic nervous system as part of, **5**:355–60
  axonal transport in, **3**:745–47
  gangliosides and, **3**:739
  kinins and, **4**:595–96

Peripheral nervous system *(Continued)*
  myelin, molecular organization of compact, **5:**278-79
  myelin proteins in, **5:**278-79
  neuronal cells, laminin effects, **4:**628-30
Peripheral temperature, body, **4:**793
Peripheral vascular diseases. *see* Vascular diseases, peripheral
Periplasm, of bacterial cell wall, **1:**275
Peristalsis
  defined, **3:**73, 765
  in small intestine, **3:**77
Peristaltic reflex, of small intestine, **1:**550
Peritoneal dialysis
  definition of, **1:**401
  peritonitis caused by, **1:**413
  recirculating, **1:**411
  regular, **1:**411
Peritoneum, defined, **3:**57
Peritonitis, **1:**413, **2:**38
Peritubular cells, testis, **7:**419-20
Permeability, cell membrane. *see* Cell membrane permeability
Permethrin, **2:**390
Permittivity, relative. *see* Relative permittivity
Permittivity dispersions, in biological tissues, **3:**11
Pernio, **4:**798
Peroxidases
  defined, **3:**318
  in immunocytochemistry, **3:**277
Peroxides
  defined, **3:**907, **5:**667
  detoxification of, glutathione and, **3:**910, 914
  organic, formation of, **3:**914
Peroxisomes
  protein import, **6:**243
  protein transport, **6:**243
Persistent vegetative states, **1:**657
Person. *see* Ego
Person perception, **5:**785-97
  assumptions based on appearance and, **5:**786
  challenge of, **5:**785
  correspondence bias and, **5:**791
  detection of lies and, **5:**787
  discounting principle in, **5:**791
  facial expressions and, **5:**786-89
  impression formation and, **5:**793-94
  impression management in, **5:**785-86
  mental representations formed in, nature of, **5:**794-95
  personality characteristics and, **5:**790-91

Person perception *(Continued)*
  physical attractiveness as factor in, **5:**786
  salience and actor-observer effect in, **5:**791-92
  self-serving biases and, **5:**792-93
  understanding causes of behavior and, **5:**789-93
Personality, **2:**709, **5:**769-75
  abnormal, **2:**709
  altruistic, **1:**205
  antisocial, **2:**709
  behavioral genetics of, **5:**771-73
  biological basis of, **5:**773-75
  criminal, **2:**709
  description of
    factor analysis in, **5:**769-70
    trait theory and, **5:**770-71
  differences, sex differences and, **6:**857-58
  dream recollection and, **3:**196
  illness and, **4:**75
  optimistic versus pessimistic, illness and, **4:**75
  psychology, defined, **5:**769
  trait analysis
    family studies and, **5:**771-73
    twin studies and, **5:**771-73
  traits
    environmental influences on, **5:**772-73
    genetic influences on, **5:**772-73
    relationship of personality disorders to, **5:**778
    theory, **5:**770-71
Personality development, **1:**573-83, **2:**853-58
  birth to 3 months, **1:**573-75
  3 to 10 months, **1:**575-76
  10 to 20 months, **1:**576-77
  20 to 36 months, **1:**577-78
  3 to 5 years, **1:**578-79
  5 to 8 years, **1:**579-81
  8 years to preadolescence, **1:**581-82
  cognitive, **2:**855-58
    four levels of self, **2:**857-58
    models, **2:**855-58
  differentiation vs. integration of processes, **1:**582
  general stages vs. specific cases, **1:**582
  obesity and, **3:**248-49
  object relations theory, **2:**855, 857-58
  psychoanalytic
    Erikson's theory, **2:**854-55
    Freudian, **2:**853-54
    Loevinger's theory, **2:**855
    non-Freudian, **2:**854-55
    Sullivan's theory, **2:**854
  psychobiology, **2:**869-79
    adaptations, **2:**872

Personality development
  psychobiology *(Continued)*
    behavior, **2**:873
    change, environmental, **2**:873
    dynamism, **2**:873
    Edelman's classifications for, **2**:873
    emotions, **2**:874
    experiences, **2**:874
    learning, **2**:874
    maturation, **2**:873
    organization levels, **2**:874
    periods, **2**:869
    precursors, **2**:872
    regulatory processes, **2**:873-74
    of self. *see* Self
    stages, **2**:869, 862
    transformations, **2**:872-73
  psychological
    anal phase in, **6**:306
    phallic-oedipal period in, **6**:306
    pleasure-unpleasure principle in, **6**:306
  psychosexual, **2**:854
    anal, **2**:854
    genital, **2**:854
    latent, **2**:854
    oral, **2**:854
    phallic, **2**:854
  schema, **2**:853
  Selman's theory, **2**:857-58
  shared heredity and, **1**:582
Personality disorders, **5**:777-83
  antisocial, **5**:779
  avoidant, **5**:779
  biological parameters associated with, **5**:781
  borderline, **5**:779
  characterization, **4**:998
  dependent, **5**:779
  diagnosis of, **5**:778-80
  epidemiology of, **5**:778
  etiological models for, **5**:780-81
  historical background of, **5**:777
  histrionic, **5**:779
  narcissistic, **5**:779
  obsessive-compulsive, **5**:779
  and other conditions, relationship between, **5**:781-82
  paranoid, **5**:778-79
  passive-aggressive, **5**:779-80
  personality development and, **5**:777-78
  schizoid, **5**:778-79
  schizotypal, **5**:778
  treatment of, **5**:782

Personality Research Form, **1**:610
Personality scale, **6**:810-11
Persuasion
  and distraction, **5**:803-4
  health psychology, **4**:81
  mood and, **1**:110-11
Persuasive communication, **5**:799-810
Perturbations, in locomotion, **4**:771
Pertussigen. *see* Pertussis toxins
Pertussis. *see* Whooping cough
Pertussis toxins
  in ADP ribosylation, **4**:14, 22
  cDNA of, **4**:18
  G_{i2}-alpha- mRNA and, **4**:21
  intracellular activity of, **5**:1-3
  laboratory detection of, **1**:556
Pertussis vaccine, **4**:446
Pes hippocampi, **4**:227
Pes planus. *see* Flatfoot
Pesticides
  in foods
    analysis, **4**:225
    behavioral toxicity of, **7**:562
    carcinogenicity of, **7**:562
      and dose-response studies involving, **7**:564
    enzyme induction from, **7**:562
    estimation of human risk from, **7**:564-65
    FDA tolerance settings for, **7**:563
    importance of knowledge of, **7**:563
    mutagenicity of, **7**:561
    neurotoxicity of, **7**:561
    reproductive toxicity of, **7**:562
    risk assessment for toxicity of, **7**:564
    sources of, **7**:563
    teratogenicity of, **7**:562
    toxicity of organophosphorous and carbomate, **7**:560-61
    types of toxicity from, **7**:560-62
  metabolism, species differences in, **7**:560
  organophosphorous, **7**:548, 560-61
  resistance to, genetic aspects of, **3**:254-55
  selective exposure of, **7**:559-60
  selective toxicity of, **7**:560
PET scan. *see* Tomography, emission-computed
Petit mal status. *see* Status epilepticus
Petrosal sinuses, inferior, **4**:983-84
Petrosal sinuses, superior, **4**:983-84
Peyer's patches, **4**:399
  defined, **5**:137
  disassociated cells from, **5**:140

Peyer's patches *(Continued)*
    functional anatomy of, **5:**139–40
    lymphocyte homing receptors, **4:**1013
    solitary lymphoid nodules and, **5:**139
    vasoactive intestinal peptide and, **5:**384
Peyote. *see* Mescaline
PGG2, in inflammation, **4:**472
pH. *see* Hydrogen-ion concentration
Phagocytes
    definition of, **5:**811
    description of, **5:**811–13
    development of, **6:**606
    elastin peptides and, **3:**268
    immune response, **4:**830
    in inflammation, **5:**818–19
    membrane, **2:**614
    mononuclear
        characteristics of, **5:**813
        classes of, **6:**607
    movement, inflammation and, **5:**815
    origin, **5:**811–13
    receptor-ligand interaction of, **5:**814
    receptors for, **5:**814
    respiratory burst in. *see* Respiratory burst
Phagocytosis, **2:**239, 242, 614. *see also* Endocytosis
    adrenergic agents and, **5:**382
    antibody-antigen complexes in, **1:**299
    of axonal debris, **5:**338
    defined, **2:**601
    flow cytometry of, **3:**624
    as function of polymorphonuclear neutrophils,
        5:428–29
    inflammation and, **4:**471
    macrophages, **4:**855–57
    nitric oxide and, **3:**375
    oxygen-dependent pathway of, **5:**814–15
    oxygen-independent pathways of, **5:**815
    phagocytes and, **5:**813–15
    and pinocytosis, differences between, **5:**814
    process of, **6:**607
Phagosomes
    formation, **4:**856–57, **5:**814
    fusion with lysosome, **4:**857
    macrophages and, **5:**813
Phalanges, **7:**38
Phalloidin, fluorescence visualization and, **5:**907
Phallotoxins, **7:**592
Pharmacodynamic tolerance, defined, **5:**439
Pharmacogenetics, **5:**829–30, 829–33
    biochemical basis for, **5:**832

Pharmacogenetics *(Continued)*
    drug metabolism and, **5:**830–32
    objectives of, **5:**830
    underlying causes of, **5:**829–32
Pharmacokinetics. *see under* Drugs
Pharmacology, **2:**783–88
    dental, **2:**783–88
    oral, **2:**783–88
Pharmacotherapy. *see* Chemotherapy
Pharyngitis, in school-age children, **4:**464
Pharynx
    airway obstruction of, **6:**358
    anatomy of, **6:**589
    swallowing and, **3:**74
Pharynx, calf, **3:**49
Phase angle, **2:**492
Phase coupling. *see* Coupling, phase
Phase shift, **2:**492–93
Phenacetin, **2:**114
Phencyclidine abuse, **5:**446–47
Phenelzine, **1:**309–10
Phenetics, dental, **2:**791–98
Phenformin, **2:**747
    lactic acidosis and, **5:**827
Phenocopies, in behavioral abnormality classification,
    3:838
m-Phenolatepyridium, **2:**533
Phenothiazines, **2:**219
    antipsychotic drug, **2:**219
    in treatment of autism, **1:**510
Phenotype, **2:**475, 689
    defined, **3:**135, 175, 847, 848, **5:**243
    expression, astrocyte role, **1:**441–42
    primate, **3:**885
    sentinel, **5:**245–46
Phenylalanine
    deletion of, in cystic fibrosis, **3:**144
    hydroxylating system, **5:**864
Phenylalanine hydroxylase, **2:**536
    deficiency, in phenylketonuria, **3:**144
    gene
        chromosome location of, **5:**865
        molecular structure of, **5:**865
    haplotype 1 and 4 chromosomes, **5:**867–68
    haplotype 2 and 3 mutant chromosomes, **5:**866–67
    haplotypes, **5:**866–68
    molecular cloning of human, **5:**864–65
    mutations
        allele-specific oligonucleotide hybridization
            analysis of, **5:**867, 869
        polymerase chain reaction in detecting, **5:**869

Phenylalanine hydroxylase (*Continued*)
   phenylketonuria and activity of, **5**:864
   primary structure determination of, **5**:864-65
Phenylalkylamines, **2**:84-85
   characteristics of, **1**:178
Phenylethanolamine N-methyltransferase, **1**:645,
    2:337
Phenylketonuria, **4**:1001-3
   biochemical basis of, **5**:863-64
   carrier screening of, **5**:868-69
   diagnosis of, DNA probe in, **6**:525
   direct mutation analysis of, **5**:868-69
   DNA marker diagnosis of, **3**:144
   genetic aspects of, **3**:839
   haplotype analysis of, **5**:866
   hyperphenylalaninemia, **3**:144
   inbreeding and, **4**:440
   mathematical nutrition education and, **3**:24
   as model for newborn screening, **5**:435-36
   molecular genetics, **5**:863-70
   population genetics of, **5**:868
   prenatal diagnosis of, **5**:865-66
   prevention of, **5**:868
   screening for, ethical considerations in, **5**:869
   treatment programs for, **5**:435-36
Phenylthiocarbamide, **2**:811, **5**:751
Pheochromoblasts, **3**:346
Pheochromocytoma, **1**:78
   hypertension caused by, **4**:290
   nerve growth factor and, **5**:334
   signs of, **3**:353
Pheomelanins, in hair, **4**:45
Pherome receptors. *see* Chemoreceptors
PHI peptide. *see* Peptide PHI
Philadelphia chromosome, **2**:441, 452
   and leukemia, **2**:453
   and study of tumor clonality, **7**:690
Phlebotomy. *see* Bloodletting
PHO5 gene, transcription rate, **3**:132
Phobic disorders, **4**:997
Phocomelia, **1**:701
Phonation
   disorders of, **7**:157
   laryngeal functions in, **4**:661
   laryngeal reflexes, **4**:650-51
Phonemes, defined, **3**:215
Phorbol esters, **2**:272, 309, 317, 318
   Na^+/H^+ exchange and, **5**:874
   nerve growth factor and, **5**:336
Phorias. *see* Strabismus

Phosphatases, **2**:534
   defined, **3**:318
   in muscle physiology, **6**:206
   nasal cavity, **5**:288-89
   phosphoprotein, **6**:206-8
   and protein phosphorylation, **6**:207-8
Phosphates
   free energy of hydrolysis of, **1**:54
   function of, **5**:50
   high-energy
     coenzyme transfer of, **3**:435-37
     defined, **3**:931
   hydrolysis, mechanism of, **1**:54
   hypercalcemia of malignancy treatment, **4**:280-81
   inorganic, **2**:528
     in glycogen degradation, **3**:923
     as metabolic controller, **3**:379
     in mitochondrial respiratory chain, **3**:377-78
   in nuclear magnetic resonance assessment of pH,
     5:873
   parathyroid hormone metabolism in intestine, **5**:662
   parathyroid hormone regulation
     of bone, **5**:662
     of kidney, **5**:662
   P_i/H^+ transport system, **1**:462
Phosphatidate phosphohydrolase, **3**:583
Phosphatidic acids, **5**:885-86
Phosphatidyl inositol polyphosphates, functions of,
    3:584
Phosphatidylcholines, **5**:886
Phosphatidylethanolamine, **5**:883-85
Phosphatidylglycerol, **5**:883-85
Phosphatidylinositide diphosphate, **5**:336
Phosphatidylinositides, **5**:336
Phosphatidylinositol glycans, **4**:682-83
Phosphatidylinositols, **5**:890
Phosphatidylserines, **5**:886
Phosphite-triester oligonucleotide synthesis method,
    3:170-71
Phosphocholine cytidyltransferase, **3**:583
Phosphocreatine, **2**:182, **5**:227
Phosphodiesterases
   in adenylate cyclase regulation, **3**:331
   defined, **3**:318
   nucleases and, **3**:128
   in phototransduction, **6**:619
Phosphoenolpyruvate, **2**:505-8
   carboxykinase, **2**:508
   in fourth-stage glycolysis, **3**:938
   pH and, **5**:821

Phosphoenolpyruvate carboxykinases
    in gluconeogenesis, 5:828
    pH and, 5:822
    renal lactate disposal and, 5:827
Phosphofructokinase, 2:509
    adenosine triphosphate and, 5:821
    deficiency, in glycogen storage disease type VII,
        5:227
    Down's syndrome and, 3:179
    fructose 2,6-biphosphate and, 3:940
    glycolysis and, 3:936
    GSH/GSSG ratio and, 3:914
    inhibition of
        by lowered pH, 5:825
        by pH$_i$ fall, 5:874
    as rate-limiting enzyme, 4:1000
    stimulation of, by pH$_i$ rise, 5:874
Phosphoglucomutase
    in glycogen formation, 3:921
    phenotype in HeLa cells, 4:108
6-Phosphogluconate dehydrogenase
    NADPH and, 3:913
    phenotype in HeLa cells, 4:108
3-Phosphoglycerate
    in fourth-stage glycolysis, 3:937-38
    in third-stage glycolysis, 3:936-37
Phosphoglycerate kinase
    in fourth-stage glycolysis, 3:938
    heat shock-stimulated, 4:103
    in third-stage glycolysis, 3:936-37
Phosphoinositide, 2:13
Phosphoinositide cycle, defined, 5:415
Phosphoketolase, 2:528
Phospholamban, 2:97-98
    cardiac muscle activity and, 5:240
    substrate for protein kinase, 2:97
Phospholipase C, 2:44, 238, 304, 308, 327, 329, 4:16-17
    antibody conjugates with, 4:372-73
    in bacterial virulence, 1:242
    in beta cell triggering, 1:567
    in Ca^{2+} messenger system, 3:333
    Clostridium perfringens and, 5:3
    and G protein-linked receptors, 1:84
    G proteins and, 5:417
    inflammation, 4:472
    phospholipid catabolism and, 5:889
Phospholipase D, 5:889
Phospholipases
    phospholipid catabolism and, 5:888-89
    reactions in human cells, 5:888

Phospholipases (*Continued*)
    snake venom, 7:600
    sources of, 5:888
Phospholipases A, 4:17
    association with events in pregnancy, 6:585
    G proteins and, 5:417
    in inflammation, 1:313, 4:472
    phospholipid catabolism and, 5:888-89
Phospholipid bilayer
    defined, 3:581
    fluidity of, 3:582
Phospholipids, 2:662, 663
    in activated lymphocytes, 5:508
    catabolism, phospholipases and, 5:888-89
    chemistry of, 3:567
    deacylation-reacylation of, 5:886
    functional aspects of, 5:890-91
    in gingiva, 4:36
    human milk fat and, 3:48
    human synthesis of, 5:885
    hydrolysis, arachidonic acid and, 3:584
    in membrane, 2:413-14
        anchoring of proteins to, 5:890-91
    metabolism, 5:883-91
    red cell membrane, 6:539
    as sources of second messengers, 5:890
    synthesis, 3:583, 5:885-88
        procaryotic, 5:883-85
        regulation of animal, 5:887-88
        regulation of procaryotic, 5:884-85
    transfer, in cells, 5:889-90
Phosphonomycin. *see* Fosfomycin
5-Phosphopantetheine, 2:13, 15, 539, 541
Phosphopantetheinyl coenzymes, 2:539
4'-Phosphopantothenate, 2:539, 541
4'-Phosphopantothenyl-L-cysteine, 2:541
    decarboxylase, 2:541
    synthetase, 2:541
Phosphoprotein phosphatases, 3:331
    defined, 5:903
Phosphoproteins
    cAMP and, 3:331
    dephosphorylation of, 3:331
    as second messengers, 3:333
Phosphopyruvate hydratase
    heat shock-stimulated, 4:103
    sodium fluoride and, 3:939
Phosphoribosylamine, 6:390-91
Phosphoribosyltransferase, 2:532
Phosphorolysis, 3:936-37

Phosphorus
fetal, **3**:346
in oligonucleotide synthesis, **3**:170
perinatal, **3**:349
Phosphorylase
crystalline structure of, **3**:927–28
discovery of, **3**:923
in glycogen metabolism regulation, **3**:924–25
pyridoxal phosphate and, **3**:923
regulation of, **3**:928
Phosphorylase A, **2**:238, 241, **3**:929
Phosphorylase B, **2**:238, 241, **3**:929
Phosphorylase kinase, **2**:330
cAMP-dependent protein kinases and, **3**:331
in glycogen metabolism regulation, **3**:925
phosphorylation of, **3**:925
Phosphorylation, **2**:241
by cAMP-dependent protein kinase, **3**:331
defined, **2**:306, **5**:893
of fructose-6-phosphate, **3**:933
ion pumps, **4**:561–62
of microtubule protein, **5**:893–902
oxidative, **2**:507, 509
in phosphotriester method, **3**:170
protein, **6**:201–11. *see also* Protein phosphorylation
in regulating G protein-linked receptors, **1**:91–92
of sperm membrane proteins, **3**:602
of sugar, **3**:933
transacting factor modification and, **3**:115
Phosphoryltransferase, **2**:528
Phosphotriester oligonucleotide synthesis method, **3**:170
Phosphotriose isomerase. *see* Triosephosphate isomerase
Photobiology, principles of, **5**:588, **7**:66
Photochemistry, **7**:66
Photoelectric effect, **5**:903–8
Photoelectron spectroscopy. *see* Spectroscopy, photo-electron
Photolyases, in DNA repair, **6**:549
Photon absorptiometry
selection of regions for, **7**:541
total body, calcium measurements and, **7**:541–42
Photons. *see under* Radiation
Photophosphorylation, defined, **3**:931
Photopic sensitivity, spectral, **2**:585
Photopigments, **2**:583–85
Photoreceptors, **2**:584–86
anatomy, **3**:542
cone
response to light, **6**:626
structure of, **6**:616

Photoreceptors *(Continued)*
responses, and light absorption, **6**:618–19, 626
retinal, **7**:836–37
rod
and response to light, **6**:626
structure of, **6**:616
spectral sensitivity, **2**:584
three-dimensional space, **2**:586
Photosensitivity disorders, **5**:266
Photosynthesis
photoelectric effect and, **5**:904
in spinach plants, **3**:257–58
Phototaxis, **2**:306
Phototransduction
events occurring in, **6**:619
mechanics of, **4**:16
Phrenology, **2**:860
Phrobol esters, **5**:905
Phthirus pubis, **2**:390
Phylogeny
defined, **3**:881
DNA sequence data for, **3**:798
genomes and, **3**:882
larynx, **4**:659–60
Physical activity. *see* Exertion
Physical exercise. *see* Exercise
Physiologic availability. *see* Biological availability
Physiological aging. *see* Aging
"Physiological" heart hypertrophy. *see* Heart enlargement, "physiological"
Physiology
comparative, respiration, **6**:567–79
fetal, **3**:616–19
of hunger, **1**:348–49
Physique, **2**:713–14
Physostigmine, **1**:180, **2**:23
Phytates, in dietary fiber, **3**:43
Pia mater, **5**:351
cranial, **4**:986–87
spinal, **4**:989–90, **7**:195
Pica
behavioral aspects of, **3**:592
defined, **3**:31
in pregnancy, **3**:31
Pick's disease, **2**:763, 764
Picornaviridae, **4**:519
Pierce-Donachy pump, **1**:387
Pigment epithelium of eye, **3**:540–41
Pigmentation, skin. *see under* Skin
Pigments
hair, **4**:45–46

Pigments *(Continued)*
  skin, **7:**44
  transport of granules, **5:**900
Pigs. *see* Swine
Pili annulati, **4:**48
Pili torti, **4:**48
Piloerection, and regulating body temperature, **1:**756
Pilus, in conjugation, **3:**824
Pimephales promelas. see Fathead minnows
Pineal body, **6:**1–7
  anatomical location of, **6:**1–2
  astrocytes, **1:**436
  and eyes, neuroanatomical connections between, **6:**2
  innervation of, **6:**2–3
  melatonin metabolism and excretion in, **6:**5
  melatonin synthesis in, **6:**3
  morphology of, **6:**1–2
  and norepinephrine release, **6:**3
  in puberty, **3:**355
  serotonin synthesis in, **6:**3
Pineal gland. *see* Pineal body
Pinocytosis, **5:**814
Pinworm infestation, **2:**387
Piperazine, **2:**84–85
  ascariasis treatment with, **1:**270
  as calcium antagonist, **2:**84
  citrate, pinworm treatment with, **1:**269
Piperonylbutoxide, **2:**390
Piriform cortex, rat, **5:**388
Pisum sativum. see Peas, flowering
Pit, coated, **2:**238, 239, 299
Pitch perception, **5:**157
Pithecanthropus erectus. see Homo erectus
Pithomyces chartarum, **7:**593
Pituicytes, **1:**436, **6:**19
Pituitary adenoma
  oral contraceptives and, **5:**608
  prolactin secretion and, **5:**399
Pituitary diseases, in children, **3:**352
Pituitary gland, **2:**11–29, 143, 834, **6:**9–21
  aging of, **3:**902
  anatomy, **5:**394, **6:**556
  ascorbic acid concentrations, **1:**430
  blood supply, **5:**395
  development of, **6:**9–11
  diagrammatic illustration of, **6:**10
  fetal
    gonadotropin-releasing hormone and, **3:**342
    at midgestation, **3:**619
  hypothalamus, inputs, **4:**306

Pituitary gland *(Continued)*
  hypothalamus, outputs, **4:**307
  immunity and, **5:**374–75
  metabolic regulation role, **4:**1005–6
  in mothers of twins, **3:**307
  pars distalis functions, **5:**394
  pars intermedia functions, **5:**394
  pars tuberalis, anatomy, **5:**394
  structure of, **6:**9–11
Pituitary gland, anterior
  at birth, **3:**347
  cell types, **6:**11, 13–19
  fetal, development of, **3:**338–45
  function, control of, **6:**11–13
  hypothalamic hormone regulation of, **5:**395
  structure and function of, **6:**11–13
Pituitary gland, posterior
  anatomy, **5:**394
  at birth, **3:**347
  defined, **5:**393
  functions of neurohypophysial hormones, **5:**397
  milk-ejection reflex, **4:**608
  structure and function, **6:**19–21
Pituitary gonadotropin, **3:**305
Pituitary growth hormone. *see* Somatotropin
Pituitary hormones, **2:**31
  neuropharmacology of, **5:**396–97
  from pars distalis, **5:**394
  regulation of, **5:**397–400
  squamous cell tumors and, **3:**335
Pituitary hormones, anterior
  function of, **6:**12
  hormone production by cell types, **6:**13–19
  regulation of, **6:**833–34
Pituitary hormones, posterior, **6:**19–21
  in children, **5:**707
  regulation of, **6:**834
Pituitary neoplasms
  G protein defects and, **4:**23
  thyroid-stimulating hormone and, **5:**400
Pivot joints, **1:**374
Placenta, **2:**408, 566
  in cortisol transfer, **3:**344–45
  with damaged contents, **3:**303
  defined, **3:**293, 609
  effect of maternal smoking on, **7:**518
  examination of, **3:**302
  functions of, **3:**612–13
  hormone production in, **5:**708
  monochorial, structure of, **3:**308

Placenta (Continued)
in phosphorus transfer, 3:346
steroid hormone synthesis by, 7:253-54
in thyroid hormone transfer, 3:339-40
Placental lactogen
function of, 6:91
mammogenesis, 4:604
Placentation
of MZ twins, 3:314
of twins, 3:305-15
Plague, as flea-borne zoonosis, 7:896
Plain film radiography. see Radiography, plain film
Plakoglobin, 2:269, 270
Plank's constant, defined, 3:423
Plankton, omega-3 fatty acids of, 5:534
Plant cells
flow cytometry of, 3:625
microfibrils in walls, 3:37
Plant flavonoids, 3:678
Plant proteins. see Proteins, plant
Plant toxins, 7:594-95. see also Mycotoxins
gut-associated lymphoreticular tissues and, 5:137
immunotoxins, 4:417-18
Plantago ovata Forsk, 3:41
Plants
breeding, and crop sciences, 1:144
dietary fiber from, 3:36
fatty acid content of, 5:533, 535
in hominid diet, 3:500
resistance to pollutants, 3:254
Planum temporale, 2:8
Plaque, hardened, in artery wall, 2:167, 811
Plasma. see Blood
Plasma clearing factor. see Lipoprotein lipase
Plasma membrane, 2:236-38
Ca^{2+} pump, 4:563
Golgi apparatus role in transport to, 4:11
lipoprotein receptors of, 6:32
proteins
restriction to diffusion of, in polarized epithelial cells, 6:65
role in epithelial cell polarity, 6:64-65
Plasmalemma, 2:127, 130
and laminar coat, 2:130
vesicles
defined, 5:31
in macromolecule exchange, 5:34
Plasmapheresis
in blood treatment, 1:414
cascade membrane, 1:414-15

Plasmids, 2:462
bacterial resistance, 1:292, 293
defined, 2:439, 701, 801, 3:101, 169
of epidermal growth factor, 3:173
in gene amplification, 3:789
resistance plasmids, 1:292
Plasmin. see Fibrinolysin
Plasminogen, 2:170, 668
kringle domains, 4:184
as streptokinase activator, 2:170
Plasminogen activators, 4:182
fibrin and, 3:810
Plasmodesmatum, 2:273-74
in plant cell, 2:273
stucture, 2:274
Plasmodium
defense mechanisms, 4:879-80
evolution of, 4:880
integrated evolution with humans, 4:880-81
intracellular invasion by, 5:646
life cycles, 4:873-74
merozoite-erythrocyte interface, 4:875
poaching of host cell enzymes, 4:875
proteins in erythrocyte envelopes, 4:875-76
resistance to, missing host factors and, 5:650
subgenera infecting humans, 4:873-74
Plasmodium falciparum, 2:384, 392
Burkitt's lymphoma and, 3:464
P. malariae, 2:984
P. ovale, 2:384
P. vivax, 2:384
vaccine for, 5:655-56
Plasmodium infections. see Malaria
Plasticity, 2:859, 875. see also Adaptation
Plasticity, neuronal. see Neuronal plasticity
Plastoquinones, 2:544
Plate mesoderm, lateral, in myogenesis, 5:162
Platelet activating factor, 6:43-52
antagonists of, 6:51
and asthma, 6:50
biosynthesis of, 6:46-47
derivation of, 3:584
effect cells, 6:45, 47-49
effect on tissues and organs, 6:50-51
function of, 6:57
and immune response, 6:48-49
in inflammation, 1:313, 314, 4:470-71, 473
lung injury and, 6:352
mechanisms of cell activation by, 6:49
receptors for, 6:49

Platelet activating factor (*Continued*)
    in salmonella infection, **6**:708
    and shock, **6**:51
    as source of second messenger, **5**:890
    sources of, **6**:45–46
    structure of, **6**:44–45
Platelet aggregation, **2**:848
    nitric oxide and, **3**:372
    omega-3 fatty acids and, **3**:579
Platelet-derived growth factor, **2**:62, 258–59, 324
    in bone and calcium metabolism, **1**:806
    in bone cell differentiation, **1**:788
    in reversing healing deficits, **7**:505–6
    in tissue repair, **7**:503–4, 506–8
Platelet GPIV, as collagen receptor, **6**:55
Platelet receptors, **6**:53–58
    adenosine compounds as, **6**:53–54
    collagen as, **6**:54–55
    epinephrine as, **6**:55–56
    fibrinogen as, **6**:56
    laminin as, **6**:56–57
    non Willebrand factor as, **6**:57–58
    platelet-activating factor as, **6**:537
    structure of, **6**:55
    thrombin as, **6**:57
Platelet transforming growth factor. *see* Transforming
        growth factor beta
Platelets, **2**:167
    clumping in artery, **2**:169
    imipramine and, **2**:846
    inhibition of function by nitric oxide, **3**:374
    microtubules of, **5**:901
    platelet-activating factor and, **6**:47
    serotonin and, **2**:845
Pleiotropy, defined, **3**:835
Pleistocene (Ice Age), **2**:189
Plesiadapis, **6**:135–36
Pleura, organization of, **6**:591
Plexiform layer, olfactory external. *see* External plexiform
        layer, olfactory
Plexus of Raschkow, **2**:775, 825
Plicamycin
    hypercalcemia of malignancy treatment, **4**:280–81
    side effects, **4**:280
Pliocene, **2**:189
Pneumococcal infections, vaccines for, **4**:446
Pneumocystis carinii, **4**:450, **6**:280
Pneumocystis carinii pneumonia. *see* Pneumonia,
        Pneumocystis carinii
Pneumocystosis, **6**:280

Pneumocytes, type I, **6**:364–65
Pneumocytes type II, **6**:365–66
Pneumogastric nerve. *see* Vagus nerve
Pneumonia, **4**:450
    bacterial, and influenza, **4**:490
    in children, **5**:706
    in infants and toddlers, **4**:461
    muscular dystrophy and, **5**:192
    neonatal, **4**:459
Pneumonia, *Pneumocystis carinii*, **2**:383–86
    and AIDS, **1**:156, **2**:385, 160
    and eflornithine, **2**:386
    fungal, **2**:385
    gamma-difluoromethylornithine therapy for, **6**:72
    and pentamidine isethionate, **2**:383
    and trimethoprim-sulfamethoxasole, **2**:383, 385
Pneumotachrography, **6**:363–64
Poikilotherms, **1**:38–40
Point mutations
    age factor in, **5**:254
    defined, **5**:243
    in Gaucher disease, **3**:779–80
    in primate evolution, **3**:882–83
Poisoning, of children, **5**:710
Poisons. *see also* Toxicology
    in food supply, **3**:18, 27
pol genes. *see* Genes, *pol*
pol viral proteins. *see* Gene products, *pol*
Polarized fluorescence, **5**:580
Polarography, in drug analysis, **3**:201–2
Poliomyelitis, **1**:255
Poliovirus vaccine, **4**:444–46
Poliovirus vaccine, oral, trivalent, **4**:466
Pollutants. *see also* Air pollutants
    chemical interactions among, **3**:257, 258
    chemical persistence, **3**:255–56
    defined, **3**:251
    doses
        defined, **3**:256
        safety factor for, **3**:258
    effect prediction of, **3**:255–56
    habitat alteration by, **3**:252–53
    intake route of, **3**:255
    measurement of, **3**:258
    social aspects of, **3**:251
    toxicity of, **3**:253–54
Poly A, defined, **5**:205
Poly-ADPR synthetase, **2**:533
Polyacrylamide gel electrophoresis. *see* Electrophoresis,
        polyacrylamide gel

Polyamines
  incorporation into cellular molecules, **6**:68
  metabolism, **6**:67–73
    in bacteria, **6**:72
    in fungi, **6**:73
    general outline of, **6**:67–68
    in mammalian cells, **6**:68–71
    in protozoa, **6**:71–72
    in viruses, **6**:73
  transport of, **6**:70–71
Polyarteritis nodosa, **4**:410
Polybrominated biphenyls, effect on interferons, **4**:525
Polychlorinated biphenyl compounds. *see* Biphenyl compounds, polychlorinated
Polycystic kidney disease. *see* Kidney, polycystic
Polycystic ovary syndrome, high LH secretion in, **3**:630
Polydactyly, in twins, **3**:313
Polyembryony. *see* Twins, monozygotic
Polyergin. *see* Transforming growth factor beta
Polyethylene glycol-salt partitioning systems, **1**:364
Polygamy, hominid, **3**:501
Polygenic diseases, **3**:148
Polygenic inheritance
  in behavioral genetics, **3**:836
  of common diseases, **3**:821
  defined, **3**:85, 819
  of depression, **3**:844
  of schizophrenia, **3**:843
Polymenorrhea. *see* Menstruation disorders
Polymer-polymer partitioning, **1**:356. *see also* Aqueous two-phase systems
Polymer-salt partitioning, **1**:356, 364
Polymerase acidic protein, **4**:485
Polymerase basic protein 1, **4**:485
Polymerase basic protein 2, **4**:485
Polymerase chain reaction, **2**:475, 481
  allele-specific oligonucleotide probes and, **3**:139–40
  defined, **2**:465, 475
  diagnosis and, **3**:136
  diagnosis of genetic disease and, **6**:526
  DNA synthesis and, **3**:173
  gene mapping and, **2**:481
  in *in vitro* DNA manipulation, **3**:875–76
  inverse, **2**:475
  Mendelian inheritance and, **2**:463
  in muscular dystrophy diagnosis, **3**:147, **5**:201
  in papillomavirus infection diagnosis, **3**:862
  polymorphism restriction fragment length, **2**:463, 480, 737
  as test for human immunodeficiency virus, **1**:738

Polymerases, RNA. *see* RNA polymerases
Polymeric IgA. *see* IgA, polymeric
Polymers
  defined, **3**:318
  noncellulosic, in parenchymatous plant tissues, **3**:38–39
Polymorphism (genetics), **6**:79–85
  defined, **3**:135
  diagnosis and, **3**:136
  in Gaucher disease, **3**:780–82
  herpesvirus genomes, **4**:191
  nature of, **3**:875
  origins of human, **6**:81
  polymorphic loci, defined, **3**:798–99
  and prehistory of modern humans, **6**:83–85
  quantitative, **6**:79–80
  quantitative versus genetic, **6**:85
  risk assessment and, **3**:138
Polymorphonuclear leukocytes. *see* Neutrophils
Polymyositis
  antigens associated with, **1**:336
  autoantibodies associated with, **1**:336
  characteristics of, **1**:335
Polymyxin E, **1**:290(t)
Polynesian origin, based on teeth, **2**:797
Polynuclear aromatic hydrocarbons, **4**:224
Polyomavirus transforming antigens. *see* Antigens, polyomavirus transforming
Polypeptides. *see* Peptides; Vasoactive intestinal peptide
Polyploidy, **2**:437
Polyposis syndrome, familial, gastrointestinal cancer and, **3**:771
Polyps, adenomatous. *see* Adenomatous polyps
Polyps, colonic. *see* Colonic polyps
Polyps, hamartomatous. *see* Hamartomatous polyps
Polyps, inflammatory oral, **5**:615
Polyrhythms, coupling of, **5**:157, 158
Polysaccharides
  anaerobic bacterial flora and, **3**:43
  noncellulosic, in parenchyma cell walls, **3**:37
  nonstarch
    in cereals, **3**:40
    defined, **3**:35
    in the diet, **3**:36
  pectic. *see* Pectic substances
  viscous, **3**:41–42
Polysomes, **2**:235
  defined, **5**:259
Polyspermy
  causes of, **3**:606

Polyspermy *(Continued)*
  vitelline block to, **3:**606
  zona pellucida and, **3:**605
Polystyrene microspheres, **4:**373
Pongidae. see Apes
Pongo pygmaeus. see Orangutans
Pons, **2:**3
  anatomy of, **5:**350
  REM sleep and, **3:**194
Pontine micturition center, **5:**44–45
Pontine reticular formation, paramedian, **3:**558
Popliteal ligament, oblique, **1:**376
Population
  crop production and, **1:**144
  defined, **3:**251
  environmental change and, **3:**252
  growth, prehistoric, **3:**492
Population differentiation, **6:**99–105
  culture as measure of, **6:**103–4
  evolutionary aspects of, **6:**101–3
  geography as measure of, **6:**103–4
  population history and, **6:**103–4
  scientific concerns about, **6:**99
  terminology of, **6:**100
Population genetics, **6:**107–15
  genetic diversity of, and mating structure, **6:**112–13
  human, polymorphism and origins of, **6:**83–84
  migration, and genetic variation, **6:**112
  mutation and, **6:**110
  polymorphism and, **6:**79–85
  quantitative polymorphism in, **6:**79–80
Population history, worldwide, **2:**794–96
  American native, **2:**797
  dental morphology, **2:**796
  oceanic, **2:**797
  teeth characteristics, **2:**794–96
Pore formation, in cellular membrane, **2:**612
Pores, microvessel
  defined, **5:**33
  in macromolecule exchange, **5:**35
Porins, **4:**973
  in bacterial cell wall, **1:**275
Pork
  tapeworm and, **2:**389
  uncooked, and trichinosis, **7:**655–56
Pornography, and violence against women, **1:**4–5, 1058
Porphyria
  acute intermittent, **6:**118–19
  congenital erythropoietic, **6:**118
Porphyrins
  biosynthesis of, **6:**117–18

Porphyrins *(Continued)*
  catabolism of, **6:**120–21
  metabolism, **6:**117–21
  structure of, **6:**118
Portal vein, defined, **3:**755
Portwine stain, **5:**594
Positioned nucleosomes. *see under* Nucleosomes
Positron-emission tomography. *see* Tomography, emission-computed
Post-column derivatization, defined, **3:**206
Postconcussion syndrome, **4:**56
Postgastrectomy dumping syndrome, **4:**601
Postherpetic neuralgia. *see* Shingles
Postpartum hemorrhage, and maternal mortality, **4:**916, 919
Postpartum infection, **4:**919–20
Postsynaptic potentials
  inhibitory, signal transmission and, **7:**386–87
  neurotransmitters and, **7:**386–88
Posttranslational modification, defined, **5:**893
Posture
  abnormalities, and autism, **1:**506
  bipedality, **2:**593
  control of
    sensory information and, **5:**109
    vision receptor system in, **5:**109
  digestive system orientation and, **3:**68
  human, **2:**593–94
  space travel effect on, **7:**149–50
  ventricular filling and, **3:**517
Postviral fatigue syndrome. *see* Fatigue syndrome, chronic
Potassium, **2:**175, 237, 311–12, 314, 317–18, 366
  binding to transport proteins, **4:**560–61
  excretion, aldosterone and, **3:**99
  function, as essential element, **5:**48
  ion pumps, **4:**568
  sodium transport and, **3:**95
  in sperm capacitation, **3:**602
  transport, in choroid plexus, **1:**718
  in vegetables, processing and, **5:**494
Potassium channels, **2:**311, **7:**19
Potassium deficiency, in thiazide therapy, **3:**98
Potassium-sparing diuretics. *see* Diuretics, potassium-sparing
Potatoes, **2:**106, **3:**655
Potentiation, chemotherapeutic, **2:**371–79
  long-term, **2:**312, 314
Pott, Percival, **3:**383
Poultry
  in "basic four" food groups plan, **3:**654
  nutrient composition of, **5:**497

Poverty, food choice and, **3**:650

Poxviridae, interferon effects, **4**:520

Practice (psychology)

blocked vs. random, **5**:123-25

role of, **5**:123

scheduled variations in, **5**:125-26

Prader-Willi syndrome, **2**:439

Prausnitz-Kustner reaction, **4**:411

Pravastatin, **2**:168

Praziquantel, **1**:263, **2**:389, 390, 393

Prazosin, **2**:177

Prealbumin

fetal, **3**:339

thyroid hormone transport and, **7**:488-89

vitamin A transport and, **7**:855-56

Prebeta lipoproteins. *see* Lipoproteins, VLDL

Precipitation methods, of enzyme isolation, **3**:406

Precorneal tear film, **3**:536

Predation

aggression and, **1**:114

in primate behavioral ecology, **6**:128-29

Predators, **6**:128-29

Predentine, **2**:775

Predictive food microbiology, defined, **3**:661

Prednisone, **2**:787

Prefrontal cortex, **4**:723

Prefrontal lobe system, **3**:217, 218

Pregnancy, **2**:875. *see also* Maternal mortality

alcohol consumption and, **1**:175-76

alcohol/marijuana interactions in, **3**:212-13

animal models, **1**:257

birth-related risks

in caffeine, **2**:75-76

drug and chemical exposure, **1**:700-701

identification of, **1**:703

breast development, **4**:603-4

complications, oral contraceptives and, **5**:608

congenital defects during, **7**:412-13

cytomegalovirus infection, **4**:189-90

diet in, **3**:31-33, 659

drug kinetics during, **5**:854

essential fatty acids for, **5**:539

fetal alcohol syndrome, **1**:701

hair loss during, **4**:42

induced hypertension, **4**:919

length of, **2**:403

marijuana use and, **4**:908-9

multiple, **3**:307

nausea and vomiting, **6**:123-25. *see also* Morning sickness

Pregnancy (*Continued*)

nutrition disorders and, **3**:26

parity effect, **4**:28

respiratory burst activity and, **6**:584-85

steroid biosynthesis during, **7**:275

and uterine contraction, **2**:403

zinc metabolism in, **7**:888

Pregnenolone

function of, in adrenal gland, **1**:74

structure of, **7**:246

delta^5-Pregnolone, **3**:345

delta^5-Pregnenolone-sulfate, **3**:345

Prejudice

defined, **7**:233

stereotypes and, **7**:233-41. *see also* Stereotypes and prejudice

Prejudice reduction

changing stereotypes and, **7**:238-41

education and, **7**:238-39

institutional support as factor in, **7**:240-41

intergroup contact in, **7**:239-41

intergroup cooperation and competition as factor in, **7**:240

intimacy of intergroup interaction and, **7**:240

status of groups as factor in, **7**:239-40

Preleukemia, **2**:450

Premenstrual syndrome, **2**:716

characterization, **4**:245-46

and criminality, **2**:717

symptoms, **2**:717

Premotor neurons

defined, **3**:553

eye movement and, **3**:558

Premunition, **4**:386

Prenatal care, treatments of birth defects, **1**:704

Prenatal diagnosis

of autosomal recessive diseases, **3**:820

birth defects, **1**:704

clinical genetics and, **5**:703

DNA analysis in, **3**:139

of Down's syndrome, **3**:176

embryofetopathology and, **3**:303-4

of Gaucher disease, **3**:782

in genetic counseling, **3**:815-16

Prenatal endocrinology, 337-47

Prenatal learning. *see* Learning, prenatal

Preneoplastic cells, **4**:1009

Prentice, R. L., **5**:600

Prenylation, microsomal, **2**:544

Preoptic area, **4**:304

Preoptic area, cat, **5:**46

Preprohormone, **3:**327

Prepropeptides, **5:**726

Prepuberty

    defined, **3:**337

    endocrinology of, **3:**350-53

Presbycusis, **1:**122, **5:**749

Presbyopia, **5:**749

Preservatives, food analysis, **4:**224

Presoteoblasts, **1:**782-83

Presyncope, **2:**147

Presyngamy, **2:**437

Previable fetuses. *see* Fetuses, previable

Prilocaine (Citanest), **2:**784

    in dentistry, **2:**784

Primagravida, **2:**403

Primaquine, **4:**884

Primary adrenal hyposecretion. *see* Addison's disease

Primary hormonal hypersecretion. *see* Hormones, abnormal secretion of

Primary olfactory cortex. *see* Olfactory cortex, primary

Primary pattern inception. *see* Embryonic inductions, primary

Primary visual cortex. *see* Visual cortex, primary

Primate behavioral ecology, **6:**127-33

    food distribution and availability as factor in, **6:**129-30

    infant care as factor in, **6:**130-31

    kinship as factor in, **6:**130

    mating patterns and, **6:**11-132

    predation as factor in, **6:**128-29

Primates, **2:**189, 593, **6:**135-43

    archaic, **6:**135-36

    bipedal, hominids as, **3:**495-97

    catarrhine, fossil record of, **6:**142

    classification of, **6:**135, 137

    defined, **6:**135

    dermatoglyphics of, **3:**91-92

    evolutionary relationships, **6:**135-36

    extinct

        skeletons of, **6:**139

        skulls of, **6:**139

    families, temporal distribution of, **6:**138

    fossils, **3:**490-91

    genetic relations among, **3:**881-89

    gut anatomy of, **3:**60-61

    higher

        fossil record of, **6:**141-43

        living haplorhines as, **6:**140-41

        representative living, **6:**140

        and tarsiers, **6:**139

Primates (*Continued*)

    lower

        fossil record of, **6:**136, 137-39

        living, **6:**136

        living representative, **6:**139

    mating, **4:**932-35

    Miocene, **3:**489

    as models for human disease, **5:**678

    nonhuman, anthropological studies of, **6:**878

    phylogeny of, **6:**136

    sex differences, biocultural, **6:**846-48

    social awareness of, **3:**503

    species, list of, **6:**128

    taxonomy of, **3:**487-88

Primitive ectoderm. *see* Blastoderm

Primitive people, Neolithic face patterns and, **3:**226

Primitive streak. *see* Gastrula

Primordium

    bone, **1:**787

    defined, **3:**609

Primosome, defined, **5:**467

Prions, in Alzheimer's disease, **1:**214

Proalbumin. *see* Prealbumin

Pro-opiocortin. *see* Pro-opiomelanocortin

Pro-opiomelanocortin

    endorphin and, **5:**731-32

    fetal, **3:**342

    functions, **1:**73-74, **5:**727, 729

Probes, DNA. *see* DNA probes

Problem solving, **6:**145-48

    common blocks to, **6:**147

    defining, **6:**145

    expertise in, **6:**148

    functional fixity and, **6:**147

    general characterization of, **6:**145-46

    main stages in, **6:**146

    random practice and, **5:**124-25

    sets and, **6:**147

Problems

    adversary and nonadversary, **6:**146

    dimension of, **6:**145-46

Probucol, **1:**454

Procainamide, **2:**173

Procaine, **2:**784

Processivity, defined, **3:**439

Procollagen, **2:**204, 559-64, 641, 566

Procrastination, **6:**149-55

    behavioral theory of, **6:**152

    clinical research on, **6:**153-54

    cognitive behavior modification theory of, **6:**152-53

Procrastination *(Continued)*
definition of, criteria for, **6:**150–52
emotional upset and, **6:**151–52
empirical research on, **6:**153–54
as inefficient behavior, **6:**151
integration of concepts relevant to, **6:**154–55
personality types and, **6:**153–54
psychoanalytic theory of, **6:**152
task demand and, legitimacy of, **6:**151
task quality affected by, **6:**151
time frames associated with, **6:**150
Product labeling. *see* Food labeling
Product of conception. *see* Conceptus
Proenzymes. *see* Enzyme precursors
"Profile of Mood States," **1:**610
Progabide, **2:**850
Progenitor cells. *see* Stem cells
Progestational hormones
biosynthesis of, **7:**263–64
for endometriosis, **3:**368
functions of, **7:**275–76
in minipills, **5:**600
sperm capacitation and, **3:**603
in steroid contraceptives, **5:**599
stroke risk of, **5:**602
Progesterone, **2:**54, 59, 247, 248, 497, 744
antral follicle growth and, **3:**631
in birth control, **1:**691–92, 693
in corpus luteum formation, **3:**634
effect on cognitive/motor tasks, **4:**248
effect on fertilization, **1:**4
effects on female sexual behavior, **4:**244–45
in estrous cycle, **7:**739–40
estrus and, **4:**932
gonadotrophins and, **3:**629
initiation of lactation, **4:**605–6
LH surge and, **3:**633
metabolic transformation of, **7:**280
in oocyte maturation, **5:**556
perinatal, **3:**349
in pregnancy, biosynthesis of, **7:**275
receptor retention in breast tumor cells, **2:**59
sperm capacitation and, **3:**601, 603
and survival of breast cancer patients, **2:**59
synthesis of, **7:**252–53
Progestin. *see* Progestational hormones
Progestogen
defined, **3:**481
in estrogen replacement therapy, **3:**482, 483
Prognathism, in fragile X syndrome, **3:**711

Programming, linear, in neural networks, **5:**392
Prohormone, in hormone synthesis, **3:**327
Proinsulin
in insulin synthesis, **3:**327
structure of, **5:**732
Prokaryotes, **2:**236
DNA binding sites in, **3:**118
eukaryote genes and, **3:**794
flow cytometry of, **3:**625
gene transcription in, **3:**109
genomes, **4:**252
promoter regions of, **3:**123
ribosomal proteins and
organization of genes, **6:**675–77
evolution of, **6:**674–75
Prolactin, **2:**19–21, 25–27, 54, 58, 248, 259, 846, 847
drug interference with, **5:**397
in fetal pituitary gland, **3:**339
function of, **6:**88
immunity and, **5:**375
initiation of lactation, **4:**605–6
lactational amenorrhea, **4:**608–9
maintenance of lactation, **4:**606
mammogenesis, **4:**604
perinatal, **3:**347
production and function of, **6:**14–16
regulation of, **5:**399
secretion
serotonin and, **6:**834
sleep and, **7:**75
somatotropin and, **5:**733
Prolactin release inhibiting hormone, **2:**20, 22
Prolactin releasing hormone, **2:**22
Prolactinoma, **2:**29, **5:**399
Proliferating cell nuclear antigen (PCNA), **1:**334–35
Proline, **2:**506
in elastin, **3:**264
residues, in elastin, **3:**265
Proline hydroxylase, **2:**563
Prolonged QT hypothesis, **7:**319
Prolylhydroxylase, **1:**431, **2:**562
ProMACE antineoplastics, **4:**850
Prometaphase, **2:**243, 244, **5:**70
Promoter regions (genetics), **2:**479. *see also* Glucocorticoid response element
in antibody production, **3:**806
defined, **3:**151, **5:**205, 261, 701
DNA methylation and, **3:**156
of elastin, **3:**265
eukaryotic, **3:**123–24

Promoter regions (genetics) *(Continued)*
  in eukaryotic gene transcription, **3:**110–11
  of glucocerebrosidase genes, **3:**778
  inactivation by DNA methylation, **3:**160–61
  inactivation reversal, **3:**158
  methylation of, **3:**153
  of myosin light chains, **5:**214
  prokaryotic, **3:**123, 166
Promyelocytes, **4:**171
Pronation, **3:**691, **5:**133
Prone position. *see* Pronation
Prontosil rubrum, synthesis and use of, **1:**326
Pronuclei, in meiosis resumption, **3:**606
Proof, conditional, **2:**619
  schema for, **2:**619
Proofreading activity (genetics), **3:**441
Propafenone, **2:**173, 747
10-Propargylestr-4-ene-3,17-one, **2:**744
Properdin, **2:**606, 608, 610
Prophase, **2:**243, 244, **5:**68–70
Propidium iodide, **2:**458, 468, **3:**623
Propionibacterium, **1:**239, 241, 246
Propionic acid, food analysis, **4:**224
Propliopithecus, in Fayum fossils, **3:**489
Proposition. *see* Conditional
Propranolol, **2:**171, 325, 326, 747
Proprioception
  mechanisms of, **6:**173–75
  perceived motor commands and, **6:**174–75
  and proprioceptors, **5:**133–35, **6:**167–76
  reflexes, **6:**157–65
  touch and limb, spinal cord and, **7:**213
Proprioceptors
  background information on, **6:**167
  cutaneous, **6:**172–73
  Golgi tendon organ, **6:**171–72
  joint, **6:**172
  muscle spindle, **6:**168–71
  properties of, **6:**168–73
  and proprioception, **5:**133–35, **6:**167–76
  sensations from, **6:**173–74
  skeletal muscle, **7:**24
Propulsion and locomotion, **4:**771
Propyl gallate, food analysis, **4:**223
N-Propylajmaline, **2:**747
Prosencephalon, **2:**3
Prosody, **2:**7
Prosopagnesia
  bilateral occipital lesions and, **3:**222
  right parieto-occipital system lesions and, **3:**223

Prospective parents. *see* Parents, prospective
Prostacyclin, **2:**170, 366
Prostaglandin receptors. *see* Receptors, prostaglandin
Prostaglandins, **2:**754, 784
  in acute phase response, **1:**26, 27
  in bone resorption, **1:**807
  bone resorption inhibition by, **1:**762
  defined, **3:**371
  effects on bone cell function, **1:**765
  excess dietary lipids and, **5:**507, 509
  in fetal circulation, **6:**345–46
  functions of, **3:**578–79
  in gastric mucosal blood flow regulation, **3:**760
  linoleic acid and, **3:**577, 578
  as mediator of hypercalcemia of malignancy, **4:**278
  in postprandial intestinal hyperemia, **4:**556
  soft analog design of, **7:**107–8
  uterine production of, estrogen-induced, **7:**744
Prostaglandins D, **1:**186–89
Prostaglandins E, **2:**848
  estrogen and, **3:**483–84
  in fever, **1:**759–60
  monocytopoiesis inhibition, **4:**854
  nutrition and, **5:**511
Prostaglandins I, **3:**376
Prostate, **2:**250, 251, 745
  anatomy of, **6:**178–79, 561
  autonomic innervation of, **1:**543
  biology of, **6:**178–79
  function of, **6:**178–79
  hyperplasia, benign, **2:**251
  message-2 gene, **2:**250
  role of hormones in, **6:**179
Prostatic hypertrophy, **2:**251
Prostatic neoplasms, **2:**745, **6:**177–90
  age-adjusted incidence, **6:**183
  androgen-dependent, **2:**745
  animal and tissue culture models of, **6:**184–86
  clinical aspects of, **6:**180–82
  diet and nutrition as cause of, **6:**187–88
  environmental factors as cause of, **6:**187
  epidemiology, **6:**182–84
  estrogen for, **3:**369
  etiology, **6:**186–90
  genetic factors and, **6:**184
  geographic variation in, **6:**182
  hormones
    effects of, **6:**180
    as endogenous factor in, **6:**189–90
  influence of life style on, **6:**183

Prostatic neoplasms *(Continued)*
  metastases involving, **6:**180
  methods of staging, **6:**180–81
  mortality rates, **6:**183
  occupational risk for, **6:**189
  pathology of, **6:**179–80
  sexual behavior and, **6:**188
  socioeconomic factors and, **6:**183–84
  techniques for detecting, **6:**180–81
  treatment of, **6:**180–81
Prosthetic groups, **2:**527
  electron transfer between, **5:**61
  respiratory chains and, **5:**61
Protamine gene, **2:**479
Protanomaly, **2:**590
Protease inhibitors
  in acute phase response, **1:**31
  hormone actions and, **5:**715
  neuronotrophic effects of, **5:**345
Proteases, **2:**646
  cascade, coagulation, **7:**493–94
  neutral, calcium-dependent, **2:**44
  parasite penetration of tissue and, **5:**647
  in processing routing signals, **6:**250
  transforming growth factor-beta and, **7:**636–37
alpha-2-Protein, nucleosome positioning and, **3:**132
beta-Protein, in Alzheimer's disease, **1:**215–16
Protein, M-line, **5:**223
Protein binding
  DNA and, **3:**117
  DNA methylation and, **3:**158
  effect on brain, **1:**720
Protein C
  activated, termination of clotting cascade, **4:**182
  myosin and, **5:**168
  properties of, **5:**223
Protein-calorie malnutrition, **4:**893–94, **6:**235
  defined, **3:**17
  in early industrial cities, **3:**21
  immune function and, **4:**895–96
  immunity impairment and, **5:**504, 505, 510, 511
  in infants of developing countries, **3:**26
  infection and, **4:**896
  macronutrient deficiencies in, **4:**894–95
Protein conformation, **6:**218–20
  defined, **3:**919
  stability of, **6:**223
  ultracentrifugation for studying, **7:**717
Protein-DNA specificity, defined, **3:**117
Protein engineering, enzyme thermostability and, **3:**420–21

Protein extraction/separation
  affinity systems, **1:**357, 361, 364
  in aqueous two-phase systems, **1:**361
  dextran-polyethylene glycol systems, **1:**363–64
  nucleic acids, **1:**361
  polyethylene glycol-salt systems, **1:**364
  product recovery from polymer phase, **1:**364
"Protein folding problem," **3:**164
Protein G, **7:**388
Protein isoforms. *see* Isoforms
Protein 85K, **5:**223
Protein kinase A, **1:**92
Protein kinase C, **2:**48, 119, 271, 272, 276, 302, 308, 317, 329, 330
  activation of, **6:**49
  in Ca^{2+} messenger system, **3:**333
  in *fos* induction, **3:**705
  functions of, **5:**896
  G protein phosphorylation and, **4:**22
  nerve growth factor and, **5:**336
  in protein phosphorylation, **6:**204
  psoriasis and, **3:**452
  in regulation of G protein-linked receptors, **1:**92
  sphingolipids and, **7:**185
  sphingosine inhibition of, **7:**188
  substrates of, **4:**17
Protein kinase C Type II, **6:**203–4
  functions of, **5:**895
  tau protein phosphorylation and, **5:**898
  tubulin phosphorylation and, **5:**897
Protein kinases
  activation, and glycogen metabolism, **7:**20–21
  in brain microtubule protein preparations, **5:**894
  in Ca^{2+} messenger system, **3:**333
  calcium-calmodulin-dependent, **6:**203–4
  in cell-cycle control, **5:**901
  cellular signal transduction and, **7:**20
  chartin phosphorylation and, **5:**899
  consensus sequences for, **6:**206
  cyclic AMP-dependent, **6:**202
  in cyclic AMP messenger systems, **3:**331
  defined, **2:**321, **5:**893
  in glycogen metabolism regulation, **3:**925
  and protein phosphorylation, **6:**201–6
  in sperm capacitation, **3:**601–2
  sphingolipids and, **7:**187–88
  visual plasticity and, **5:**900
Protein P34CDC2, **5:**901
Protein phosphatases. *see* Phosphoprotein phosphatases
Protein phosphorylation, **6:**201–11
  and cancer, **6:**205–6
  cell-surface receptors as substrates for, **6:**208–9

Protein phosphorylation (*Continued*)
    general characteristics of, **6:**201
    ion channels as substrates for, **6:**209–10
    mechanism of, **6:**202, 203
    metabolic enzymes as substrates for, **6:**210–11
    multisite, **6:**211
    role of protein kinases in, **6:**202–6
    sphingolipids and, **7:**187–88
    structural proteins as substrates for, **6:**209
    substrates for, **6:**208–11
beta-Protein precursor, **1:**215–16
Protein S-100, **2:**94
Protein synthesis, **5:**559
    aflatoxin B1 inhibition of, **5:**265
    bacterial, antibiotic inhibitors of, **1:**291
    characteristics of, **6:**685–86
    genetic disease and, **6:**216–17
    regulation of, **6:**216
    ribosomal involvement in, **6:**673–74
    in salivary gland cells, **6:**691
    sense and antisense strands in, **6:**685
    steps in, **6:**672–74
    termination, mechanism of, **6:**673
    trichothecene inhibition of, **5:**264
    trichothecenes and, **7:**667
Protein targeting
    basic concepts of, **6:**237–43
    general principles of, **6:**245–46
    molecular mechanisms, **6:**245–55
    routing as stage in, **6:**246–51
    routing signal receptors in, **6:**247–48
    routing signals in, **6:**246–47
    translocation as stage in, **6:**246–51
Protein toxins, **4:**416–18
    antibody conjugates with, **4:**367–69
    microbial, **5:**1–4
Protein transport
    in endoplasmic reticulum-Golgi complex, **6:**252–55
    functions of, **6:**214
    ion interaction, **4:**560–61
    into mitochondria, **6:**241–42
    into nucleus, **6:**241
    into peroxisomes, **6:**243
    structure, **4:**560
    target pathways in, **6:**251–55
Protein tyrosine kinase, **2:**44, 58, 61, 62, 227, 231, 232, 329, 453
    and cancer, **6:**205–6
    growth factor receptor, **6:**210
    messenger systems, **3:**333
    in microtubule protein phosphorylation, **5:**897

Proteinase inhibitors. *see* Protease inhibitors
Proteins, **2:**30, 47, 148, 270, 272, **6:**213–27. *see also*
    specific proteins
    aging and catabolism, **3:**895
    allosteric, ligand binding in, **6:**224
    amino acid patterns in, **6:**230–31
    antibody fusion, **3:**809–10
    biosynthesis, **6:**216–17
    in body pattern formation, **3:**283
    calcium-binding. *see* Calcium-binding proteins
    carrier, metabolic regulation role, **4:**1003–4
    catalytic, function of, **6:**214
    cellular, interaction of viral gene products with, **5:**641–42
    clotting, structural domains, **4:**182–85
    cofactor structural and functional domains, **4:**185
    contractile
        assembly of, **5:**169–70
        muscle tissue and, **5:**163
    cotransporters, **4:**558
    defined, **3:**649, 791
    degradation of, **3:**419, **6:**226–27
    detection, **6:**191–99
        by antibody binding, **6:**196–97
        Folin method for, **6:**192
        historical perspective of, **6:**192
        Kjeldahl method for, **6:**192
        Lowry method of, **6:**192
        with metallic stains, **6:**194–96
        quantitation standards for, **6:**198–99
        quantitative, **6:**198–99
        quantitative integral comparisons in, **6:**199
        ultraviolet light in, **6:**192
        underlying principles of, **6:**191–92
    detection of specific proteins, **6:**196–97
    dielectric properties of, **3:**10
    dietary. *see* Dietary proteins
    DNA-binding sites. *see also* DNA-binding proteins
        protein-protein interactions and, **3:**120–21
    dystrophin-associated, **5:**198
    EBV-encoded, **3:**467–70
    elastic, of muscle, **5:**223–24
    electromorphic variants of, **5:**246
    energy transduction by, **6:**225–26
    enzymatic modifications in Golgi apparatus, **4:**4
    as enzymes, detection of, **6:**196
    of Epstein-Barr virus particle, **3:**472
    evolutionary origins of, **6:**218
    extracellular matrix, **7:**636–37
    finger domains, **4:**184–85

Proteins *(Continued)*
    folding of, **3:**419, **6:**251
    fos, **3:**702, 704-5
        in *fos* gene transcription, **3:**706
        function of, **3:**707
        transactivation by, **3:**706-7
    functions, **6:**213-14
    ganglioside transfer, **3:**747-48
    genes and, **3:**792
    genetic errors and metabolic regulation, **4:**1003-4
    in Golgi apparatus, **4:**7
    growth-associated, in nerve regeneration, **5:**344
    growth factor domains, **4:**184
    guanine nucleotide-binding, **4:**14-15
    guanosinetriphosphate-binding, **7:**17-19
    in hair, **4:**46-47
    heat-shock. *see* Heat-shock proteins
    histone. *see* Histones
    homologous, genetic basis of, **6:**218
    immunochemical, functions of, **6:**213-14
    immunogenic early, in Epstein-Barr virus, **3:**471
    informational, functions of, **6:**213-14
    interaction with molecules
        small molecules, **7:**720
        other molecules, **6:**223
    intermediate filament, **5:**224
    intracellular
        calcium-binding, **2:**92-95
        exchange proreins, **2:**92
    and kidney, **2:**194
    latent infection membrane, in Epstein-Barr virus, **3:**469-70
    ligand-binding, **6:**224
    and liver, **2:**194
    loss, and hookworm infection, **4:**116
    membrane. *see* Membrane proteins
    metabolism, respiratory quotient of, **5:**475
    microfibrillar, in elastic fiber, **3:**262
    microtubule-associated. *see* Microtubule-associated proteins
    mitochondrial
        routing signals for, **6:**247
        translocation of, **6:**250
    modification, cellular signal transduction and, **7:**21
    morphogenic, bone, structure and function, **1:**764
    multifunctional, and nucleotide synthesis, **5:**464
    muscle, **5:**163-64
        function of, **6:**214
        varieties of, **5:**164
    myofibrillar, degradation of, **5:**167

Proteins *(Continued)*
    myosin, **5:**211
    of nerve growth factor, **5:**335
    neurofilament, **5:**224
    non-collagenous, **2:**201, 207-8
    non-histone, **2:**434-35, 470
    as nonclassic hormones, **3:**323
    nonstructural, influenza viruses, **4:**485
    nuclear. *see* Nuclear proteins
    odor-binding, **5:**525
    oligomerization of, in sorting and transport of proteins, **6:**251
    orthologous, **2:**737
    of osteoblasts and bone, **1:**763-64
    in parenchyma cell walls, **3:**37
    peroxisomal, routing signals for, **6:**247
    plant, **3:**655
    of plasma, **2:**P611
    posttranslational modification of, **6:**217-18
    in pregnancy, **2:**194
    pumps
        functions of, **6:**214
        and ion transport, **6:**226
    quaternary structure of, **6:**222
    red cell membrane, **6:**542-45
    reexport of, during sorting and transport, **6:**251-52
    regulatory. *see* Regulatory proteins
    repressor. *see* Repressor proteins
    of respiratory chain complexes, **5:**57
    ribosomal
        characteristics of cDNA and their, **6:**672
        eukaryotic and prokaryotic similarities in, **6:**674-75
        internal duplications in, **6:**674
    routing signals, proteolytic processing of, **6:**250
    salivary, **6:**693-95
    satiating power of, **6:**729
    secondary structures of, **6:**220-21
    secretion, by salivary gland cells, **6:**691
    secretory
        model for sorting and transport of, **6:**253
        pathway for, **6:**238-41
        receptor-mediated sorting of, **6:**254
        routing signals for, **6:**246-47
        sorting and transport of, **6:**252-55
        stages in translocation of, **6:**248
    sequence analysis of, **3:**164, 166
    single cell, **5:**259
    sorting
        in endoplasmic reticulum-Golgi complex, **6:**252-55
        in polarized cells, **6:**241

Proteins *(Continued)*
   sorting-pathways, **6:**238, 243
   in sperm capacitation, **3:**602
   in sperm/egg binding, **3:**605
   stress. *see* Heat-shock proteins
   structural
      functions of, **6:**213
      levels of, relationship between, **6:**222–23
      as substrate for phosphorylation, **6:**209
      of vertebrate skeletal myofibrils, **5:**221–24
   structure of, **6:**14–216, 221–22
   sweet sensitive, **7:**363–64
   sweeteners, **7:**380–81
   tandem repeat domains, **4:**185
   translocation
      in endoplasmic reticulum, **6:**239–40
      insertion of routing signal in, **6:**248–49
   turnover rate, gingiva, **4:**37
   ultracentrifugation of different species of, **7:**717
   and urea, **2:**194
   vegetable, **3:**654
   of vertebrate skeletal myofibrils, **5:**210
   vimentin, **5:**224
   viral, **4:**191–92
   vitamin K-dependent, **4:**184
   of zona pellucida, **3:**605
Proteins, dietary. *see* Dietary proteins
Proteins, radioactive
   autoradiography to detect, **6:**197
   detection of, **6:**197–98
   fluorography to detect, **6:**197–98
Proteoglycans, **2:**201, 202, 563, 566, 573, 639–42, 645, **6:**257–26
   biosynthesis of, **6:**263–64
   from cartilage, **6:**259–60
   in cartilaginous tissue, **1:**365–67
   chondroitin sulfate. *see* Chondroitin sulfate
      proteo-glycans
   in cytotoxic T lymphocytes, **4:**816
   in development, **6:**264–65
   and disease, **6:**264–65
   distribution of, **6:**260–62
   electron micrographs of, **6:**263
   families, diversity of, **6:**264
   function of, **6:**260–62
   general characteristics of, **6:**257–58
   gingiva, **4:**36
   interstitial, small, **6:**261
   of interstitium, **5:**35
   intracellular, **6:**263

Proteoglycans *(Continued)*
   large
      aggregating, **6:**260–61
      nonaggregating, **6:**261
   matrix, **2:**202–3, 206–8
   osmotic pressure, in cartilaginous tissue, **1:**368
   in parenchymatous plant tissues, **3:**38
   plasma membrane, **6:**262
   structure and function, **1:**764–65
   structure of, **6:**258–60
   sugar chain formation in Golgi apparatus, **4:**8
Proteolysis, defined, **3:**318
Proteolytic cleavages, in nerve growth factor, **5:**335
Proteolytic enzymes. *see* Peptide hydrolases
Proteosomes, as adjuvant for subunit vaccines, **7:**754
Prothrombin, **2:**170, 171
   activator, **4:**180–81
   kringle domains, **4:**184
Prothrombin time test, **7:**495
Proto-oncogene proteins c-*fox*, **5:**336
Proto-oncogenes, **2:**120, 250, 265, 318, 451, 872
   activation, **2:**265
   in autoimmune diseases, **5:**515
   carcinogenesis and, **3:**674
   defined, **2:**53, **3:**701, 859
   DNA-binding proteins and, **5:**167
   growth-regulated, **2:**253, 256–58
   locus, **2:**452
   nerve growth factor and, **5:**336
   in psoriasis, **3:**446
   in uterus, estrogen-induced, **7:**743–44
Protofilaments, **5:**12
Protohominids. *see* Australopithecus
Proton gradient, **2:**238, 239
Protons
   astrocyte, **1:**438–39
   Bohr effect, **4:**138
   in carbohydrate metabolism, **5:**824
   electrochemical gradient of, **1:**459
   ion pumps, **4:**569
   motive force
      in chromaffin-vesicle membrane, **1:**647
      defined, **5:**871
   proton-dependent antiporters, **4:**565
   pumping, by respiratory chain complexes, **5:**62
   as source of space radiation, **6:**428
   translocating pump, in chromaffin vesicles, **1:**646–47
   translocation, in mitochondrial bioenergetics, **1:**461
Protoplast fusion, in gene transfection, **3:**805

Protoporphyrins, **2**:655, 737
Protozoa
    host cell invasion by, **5**:646
    immune evasion and, **5**:649
    intramacrophage, **4**:387
    polyamine metabolism in, **6**:71–72
Protozoan infections, **2**:382–86, **6**:267–81
    general characteristics of, **6**:267–68
    human, **6**:267–81
    listed, **2**:383
    as scalar-borne zoonoses, **7**:897–98
Protraction, as joint movement, **5**:133
Provider-patient interaction, **4**:76
Psalterium, **4**:228
Pseudoautosomal inheritance, **3**:851–52
Pseudoautosomal region of the sex chromosomes. *see* Sex
        chromosomes
Pseudodementia, **1**:132, 210
Pseudogenes, **2**:480
    defined, **2**:469, **3**:792
    of glucocerebrosidase, **3**:778, 779, 781
Pseudohermaphroditism, **2**:251
    and cell death, **2**:251
    hormonal causes of, **3**:342
Pseudohydroparathyroidism, and receptor defect, **2**:301
Pseudohyperparathyroidism, **4**:277
Pseudohypoparathyroidism, **2**:301, **5**:665
    G protein defects in, **4**:23
    parathormone and, **3**:353
Pseudohyposteroidism, **7**:266
Pseudomonas aeruginosa, **1**:287–88
Pseudomonas exotoxin, **4**:368–69
Pseudomonas fluorescens, **1**:287–88
Pseudomonas infections, **5**:706
Pseudopodia, defined, **3**:318
Pseudoprecocious puberty, defined, **3**:357
Pseudopromoter (genetics). *see* Promoter regions
        (genetics)
Pseudosites
    defined, **3**:117
    DNA binding sites and, **3**:120
Pseudoxanthoma elasticum
    defined, **3**:261
    RNA level of, **3**:269
Psilocybine
    marijuana and, **5**:444
    sources of, **5**:447
Psoralens, **5**:266
Psoriasis, **2**:329
    antigen-presenting cells in, **3**:450–51
    cancer and, **3**:446, 447

Psoriasis (*Continued*)
    in children, **5**:710
    dermal dendrocytes in, **3**:451
    dielectric properties of, **3**:13
    EGF receptor expression in, **3**:447
    endothelial cells and, **3**:451–52
    interleukins and, **3**:450
    neurons and, **3**:452
    omega-3 fatty acids and, **3**:580
    protein kinase C and, **3**:452
    T cell proliferation and, **3**:449, 450
    TGF-beta in, **3**:448
    ultraviolet radiation and, **7**:68
    wound healing and, **3**:445–46
Psyche
    ego as part of, **6**:308
    id as part of, **6**:307
    restructuring of, after childhood, **6**:312
    structure of, **6**:308
    superego as part of, **6**:308
Psychedelic agents. *see* Hallucinogens
Psychiatric disorders. *see* Mental disorders
Psychiatry, **5**:777–83
Psychic satiation. *see* Satiation
Psychoanalysis, **6**:293–303
    classical treatment in, **6**:294–95
    clinical interpretation in, **6**:311
    clinical observation in, **6**:311
    concept of determinism in, **6**:306–7
    concept of sexual and aggressive drives in,
        6:307
    dream research in, **3**:192–93
    as dynamic psychology, **6**:307
    and early childhood experiences, **6**:309
    empathy as, factor in, **6**:311
    fantasy transference in, **6**:309
    fundamentals of, **6**:306–10
    genetic principle in, **6**:309
    information transference, **6**:295
    memory as topic for, **5**:401
    methodology of, **6**:310–12
    nature of, **6**:293–94
    object relations in, **6**:313
    scope of, **6**:296
    social applications of, **6**:314–15
    social value of, **6**:296
    techniques in therapy, **6**:296
    therapy, **6**:310–12
    topographic principle in, **6**:308
    vis-à-vis psychotherapy, **6**:297–98

Psychoanalytic theory
    new developments in, **6:**315–17
    personality disorders and, **5:**780
Psychobiology
    componential model, **1:**608–9
    developmental, **2:**869–79
    drug effects on behavior/emotions, **1:**613–14
    measurement of emotion, behavior, and cognition,
        1:609–12
    medical model, **1:**608
    multivantaged approach, **1:**609, 612–14
    neurotransmitter function and behavior/emotions,
        1:612–13
    objective performance instruments, **1:**611–12
    observational rating scales, **1:**610–11
    self-report tests, **1:**609–10
Psychodynamic therapy for autism, **1:**509
Psychogenetic models. *see* Models, psychogenetic
Psychological criteria for beginning of human life, **1:**656
Psychological development. *see under* Personality devel-
    opment
Psychological refractory period. *see* Refractory Period,
    psychological
Psychological risk factors, **4:**78
Psychology. *see also* Health psychology
Psychology, clinical, **4:**70
Psychology, comparative, **2:**870
Psychology, developmental, **2:**870
Psychology, industrial
    branches of, **3:**475
    intelligence testing and, **3:**474
Psychology, medical, **4:**70
Psychomotor, defined, **2:**511
Psychoneuroimmunology, **4:**74–75, **6:**319–27
Psychopathy, **2:**709–19
    "born criminal concept," **2:**711
    criteria for, **2:**709
    deficiency in anticipating punishment, **2:**715
    and nervous system, autonomous, **2:**715–16
    and superego, **2:**710
Psychophysics, **4:**197
Psychophysiological disorders, **6:**329–38
    alimentary tract ulcers as, **6:**335–36
    causes of, **6:**331–32
    concept physiological systems and, **6:**330–32
    conditioning as explanation for, **6:**335
    deficiency in anticipating punishment, **2:**715
    electroencephalogram and, **2:**714
    evidence for
        animal studies in, **6:**334–35
        current evidence, **6:**335–37
        prevalence of, **6:**332–35

Psychophysiological disorders (*Continued*)
    evolution as explanation for, **6:**332
    general characteristics of, **6:**329–30
    hypertension as, **6:**336–37
    personality as explanation for, **6:**333
    physiological systems and, concept of, **6:**330–32
    psychodynamic explanations for, **6:**332–33
    respiratory system and, **6:**337
Psychophysiology, **2:**715–16
Psychoses, senile. *see* Dementia, senile
Psychoses, substance-induced
    amphetamines and, **5:**440
    cannabis and, **5:**445
    drug treatment, **2:**219
    phencyclidine and, **5:**447
Psychoses, toxic. *see* Psychoses, substance-induced
Psychosocial deprivation, **3:**352
Psychosomatic medicine, **4:**70–71
Psychotechnology, and ergonomics, **3:**474
Psychotherapy. *see also* Psychoanalysis
    corrective emotional experience in, **6:**301–2
    dynamic, nature of, **6:**293–94
    educational activities as mechanism of, **6:**301
    expressive, **6:**298–99
    maintenance supportive, **6:**300–301
    personality disorders and, **5:**782
    psychoanalytic, **6:**293–303
    psychoanalytically based, modes of, **6:**297
    supportive, **6:**299–303
    transference cure as mechanism of, **6:**301
    in treatment of mood disorders, **5:**96
    vis-à-vis psychoanalysis, **6:**297–98
Psychotic disorders, **2:**711
    and androgen, **2:**711
    cruelty, **2:**711
    defined, **3:**838
    hostility, **2:**711
    insensitivity, **2:**711
Psychotomimetic agents. *see* Hallucinogens
Pterin coenzymes, **2:**535–36
Pteroyl polyglutamate hydrolase, **2:**536
Pteroylglutamic acid. *see* Folic acid
Puberty, **2:**497–98, **4:**29–30, **6:**339–44. *see also* Adoles-
    cence; Prepuberty
    abnormal, hormonal aspects of, **3:**357–58
    biological changes during, **1:**65–66
    breast development, **4:**603–4
    defined, **3:**337
    delayed, **6:**343
    development, sequences in, **6:**340

Puberty *(Continued)*
  endocrinology of, **3**:353–58
  environmental factors affecting, **6**:341
  in girls, sequence of, **6**:340
  gonadotropin regulation in, **5**:398
  grading of, nomenclature for, **6**:340
  growth patterns and sexuality during, **6**:884
  hormonal changes at, **6**:342
  induction of, **6**:342–43
  initiation rituals at, **6**:884
  and mammary gland, **2**:54–55
  onset of, **3**:355–57
  precocious, **6**:343–44
  progression of, evaluation of, **6**:340
  sexual characteristics at, **7**:273
  sexual signs during, **6**:341–42
  weight, and first menstrual cycle, **1**:744–45
Puberty, delayed, **3**:358
Puberty, precocious
  21-hydroxylase deficiency in, **3**:353
  types of, **3**:357–58
Public health interventions, **4**:82–83
Pufferfish toxin, **2**:345, **3**:673
Pulmonary circulation, **2**:171–62
  blood flow, **6**:597–98
  clinical applications from knowledge of, **6**:353
  control of pulmonary perfusion in, **6**:375
  effect of low oxygen content on, **6**:598–99
  fetal, **6**:345–46
  lung injury and, **6**:351–53
  metabolic functions of, **6**:598
  neurohumoral aspects of, **6**:346–47
  pressure fall in, **6**:374
  pressures within, **6**:597–98
  role of eicosanoids in, **6**:350–51
  vascular flow resistance in, **6**:374–75
  vasoactive agents and, isolated vessel studies, **6**:348
  vasomotor tone, factors affecting, **6**:347
  ventilation-perfusion ratio of, **6**:370
Pulmonary diseases. *see* Lung diseases
Pulmonary edema, **5**:41
  causes of, **6**:598
  furosemide for, **3**:98
Pulmonary embolism, **5**:604
Pulmonary gas exchange, **6**:373–74, 598–99
Pulmonary hypertension. *see* Hypertension, pulmonary
Pulmonary perfusion
  control of, alveolar oxygen concentration and, **6**:375–76
  sympathetic control of, **6**:375
Pulmonary region, characteristics of, **7**:578

Pulmonary surfactants
  alveolus volume range and, **5**:618
  defined, **3**:609
  fetal
    cortisol and, **3**:345
    formation of, **3**:614
Pulmonary toxicology, **7**:575–85
  airborne toxicants and, **7**:577
  assessing risks and controlling exposures in, **7**:584–85
  disposition of inhaled toxicants and, **7**:577–78
  historical aspects of, **7**:575–77
  organic chemicals and, **7**:549
  responses to inhaled toxicants and, **7**:582–84
Pulmonary tree, **6**:591–92
  conducting division, **6**:591–92
  respiratory division, **6**:592
Pulmonary tuberculosis. *see* Tuberculosis, pulmonary
Pulmonary ventilators. *see* Ventilators, mechanical
Pulpitis, dental caries and, **5**:612
Pulsed field gel electrophoresis. *see* Electrophoresis, pulsed field gel
Pulvinar lesions, effects on covert orienting, **1**:481
Pumiliotoxins, **7**:599
Purdue Pegboard Test, **1**:611
Pure Food and Drug Act, **2**:512, **3**:672
Pure water. *see* Water
Purging, defined, **3**:243
Purine-pyrimidine metabolism, **6**:387–98
Purinergic receptors. *see* Receptors, purinergic
Purines, **2**:506
  bases, biosynthesis and degradation of, **6**:390
  biosynthesis. *see also* Nucleotides
    genetic locus of, **3**:184
    multienzyme catalysis in, **5**:466
    regulation of, **6**:390–91
  metabolism
    in adrenal medulla, **1**:645–46
    disorders of, **6**:392–93
    metabolic pathways, **6**:387–90
    overview of human, **6**:388
  ribonucleosides and, **6**:390
  ribonucleotides and, **6**:390, 391
  ring, biosynthesis of, **6**:389
  synthesis
    de novo, **6**:387–88
    by *Leishmania*, **4**:682
    nucleotides and, **6**:388–89
    salvage pathways in of, **6**:389
Purkinje cells, **2**:313
Purkinje fibers, **2**:157, **5**:162

Pursestring sutures, **4**:93

Pustular psoriasis of palms and soles. *see* Psoriasis

Putamen, **2**:359

Putrescine, **6**:68

Puzzle-solving, **7**:471

Pycocyanase, **1**:287-88

Pyloric antrum

    defined, **3**:75, **5**:713

    in fat digestion, **3**:81

Pyramidal cells, defined, **5**:307

Pyrantel pamoate, **2**:387, 392

Pyrethroid, **2**:390, 393

Pyrethrum, **4**:883

Pyridine alkaloids, **1**:178

Pyridine nucleotide coenzymes, **2**:530-33

Pyridinoline, **2**:510

Pyridoxal, **2**:533, **4**:221

Pyridoxal kinase, **2**:533, 534

Pyridoxal phosphate, **2**:533-34

    catalysis rate and, **3**:437

    in phosphorylase, **3**:923

Pyridoxal phosphate oxidase, **2**:533, 534

Pyridoxamine, **2**:533

Pyridoxaminephosphate, **2**:533, 534

Pyridoxine, **2**:533, 534

    defined, **2**:527

    food analysis, **4**:221

    pyridoxal phosphate and, **3**:923

Pyridoxine deficiency, and immunity, **5**:513

Pyrimethamine, **1**:268, **2**:392

Pyrimethamine-sulfadiazine, **2**:384

Pyrimethamine-sulfadioxine, **2**:384

Pyrimidines

    bases, biosynthesis of, **6**:396

    biosynthesis, **6**:394-95. *see also* Nucleotides

        of nucleosides, **6**:396

        of nucleotides, **6**:396

        scheme for *de novo*, **6**:395

    degradation, **6**:397

    dimers, removal of, in DNA repair, **6**:549

    interconversions, **6**:395-97

    metabolism

        disorders of, **6**:397-98

        metabolic pathways, **6**:394-97

        overview of, **6**:394

        regulation of, **6**:397

Pyrogens, endogenous, **1**:30

Pyrolitic compounds, **3**:774

Pyrophosphatase, inorganic, **2**:528

Pyrophosphate, **3**:921

Pyrophosphorylase, **2**:541

Pyropoikilocytosis, **6**:545

Pyrroloquinoline quinone, **2**:540, 542

Pyruvate carboxylase, **2**:507-10, **5**:821

Pyruvate dehydrogenase, **2**:509

Pyruvate kinase, **2**:506

    in fourth-stage glycolysis, **3**:938

    SV40 T-antigen sequence in, **5**:455

Pyruvates, **2**:504, 509, 510

    in fifth-stage glycolysis, **3**:938-40

    in fourth-stage glycolysis, **3**:937-38

    lactates and, **5**:822

    oxidation of, **3**:932

Pyrvinium pamoate, **1**:269

Q

Q cycle hypothesis

    defined, **5**:55

    tenets of, **5**:61-62

Quackery, in health food movement, **3**:25

Quadruplets, Weinberg's rule and, **3**:306

Quality assurance, in food handling, **3**:668-69

Quality control, in drug analysis, **3**:209

Quality of life, **4**:78-79

Quantal release, defined, **2**:306

Quantitative genetics, defined, **3**:835

Quantum theory, **5**:904

Quaternary salts, in drug designs, **7**:106-9

Questionnaires

    in feeding behavior research, **3**:596

    in food acceptance research, **3**:641

Quinacrine hydrochloride, **2**:382, 383

Quinidine, **1**:181, **2**:173

Quinine, **2**:382-83, 391

    antimalarial, **4**:884

    applications of, **1**:181

    resistance, **2**:384

Quinol, **2**:544

Quinolate, **2**:532

Quinolinic acid model of Huntington's disease, **4**:272

Quinolones, **6**:399-410

    adverse reactions from, **6**:406-7

    antimicrobial activity of, **6**:403-4

    bacterial resistance to, **1**:328, **6**:404-5

    chemistry of, **6**:399-402

    classification of, **6**:399-402

    clinical uses of, **6**:407-10

    drug interactions, **6**:407

    mechanism of action, **6**:402-3

Quinolones *(Continued)*
  anucleus, structure of derivatives of, **6**:400–410
  penetration of, into tissues, **6**:406
  pharmacology of, **6**:405–6
  specificity of action, **1**:327
  toxicity of, **6**:406–7
Quinonoid dihydrobiopterin reductase, **5**:864
Quipazine, **2**:22
Quisqualate, **2**:313, 339, 523, 525
Quisqualate receptors. *see* Receptors, quisqualate

R

R repetitive sequences. *see* Repetitive sequences, nucleic
    acid
Rabbit, conditioning, classical, **2**:311
Rabies, **7**:896–97
Races. *see* Racial stocks
Rachitis. *see* Rickets
Racial stocks
  autosomal recessive diseases of, **3**:820
  classification of, and population differentiation, **6**:100
  death rates and, **5**:100–101
  dermatoglyphics of, **3**:90–91
  food preferences of, **3**:650–51
  gene distribution and, **6**:102
  genetic disorders of, counseling for, **3**:814
  genetic similarity of, **3**:885
  geographical, defined, **6**:100
  local, defined, **6**:100
  micro, defined, **6**:100
  mitochondrial DNA of, **3**:889
  population differentiation due to, **6**:102–3
  thalassemia risk and, **3**:145–46
  twinning in, **3**:307
Radial loops, dermatoglyphic, **3**:88
Radiation
  and aplastic anemia, **6**:416
  average dose of, in United States, **6**:442
  of body heat. *see* Body temperature regulation
  cyclooxygenase inhibitors and, **6**:424
  detectors, for neutron activation analysis, **7**:539
  diagnostic, and *in utero* exposure, **6**:416
  electromagnetic, **6**:443, **7**:65–66
  fractionated, tissue tolerance and, **6**:473–75
  genetics
      chromosome mapping and, **3**:828
      hybrid mapping, in genome study, **3**:878
  hazards of, **6**:448–53
  linear transfer energy and, **6**:412

Radiation *(Continued)*
  medical uses of, **6**:447–48
  natural, and risk of cancer, **6**:412
  particulate, characteristics of, **6**:443
  photon interaction with tissue, **5**:591, **6**:439
  quality, effect on cells, **6**:446–47
  sources and kinds of, **6**:441–43
  tobacco products as source of, **6**:443
  tumors induced by, **4**:344
Radiation, cosmic. *see* Cosmic radiation
Radiation, ionizing
  as cause of DNA damage, **6**:548
  effect on skin, **6**:463–64
  exposure to, **2**:347
  free radicals and, **3**:914–15
  and leukemia, **4**:691
Radiation, whole body. *see* Whole-body irradiation
Radiation dosage, **6**:446
Radiation effects, **6**:411–25, 441–53
  on aorta and major blood vessels, **6**:414
  on biological systems, **6**:441–53
  on blood vessels, **6**:465
  on body tissue, **6**:435–40
  on bone marrow, **6**:413
  carcinogenesis, **6**:415–18, 451–52
      age at exposure and, **6**:418
      dose response, **6**:417
      induction periods in, **6**:417
  on cardiovascular system, **6**:414, 424
  cell death, **2**:251
  cellular, **6**:445–47
  on central nervous system, **6**:412–13
  electromorphic mutations as, **5**:246
  on epidermal cells, **6**:464
  on fetus, **6**:414, 452–53
  on gastrointestinal system, **6**:424, 479
  genetic, **6**:418–20, 450–51
      biochemical variants and, **6**:419–20
      sentinel phenotypes and, **6**:419–20
  on hair, **6**:466
  immunological, **6**:420–23
  induced menopause, **6**:416
  of ionizing vs. nonionizing radiation, **6**:444
  on Langerhans cells, **6**:467
  late
      degenerative effects of, **6**:449–50
      hazards of, **6**:449–52
      on skin, **6**:465–66
  on liver, **6**:414
  on lungs, **6**:414–15

Radiation (*Continued*)
    on mucous membranes, **6:**413
    mutation, **5:**252
        estimate of risks of, **6:**420
    on organ systems, **6:**412–15
    on pigmentation, **6:**466–67
    prenatal, **6:**452–53
    on radiologists, **6:**416
    on respiratory system, **6:**414–15
    risk of breast cancer, **6:**415
    on skin, **6:**413
    on skin pigmentation, **6:**466–67
    on small intestine, **6:**472–73
    theories on, **6:**412
    on urogential system, **6:**4415–14
    on vascular endothelium, **6:**424
Radiation injuries
    acute gastrointestinal syndrome, **6:**471–72
    characteristics of, **6:**411–12
    endothelial cells and, **6:**424
    nature of, **6:**443–45
    repair of, **6:**45–446
Radiation protection, **6:**418, 446, 465
Radiation-sensitizing agents, **6:**446–47
    effect on skin, **6:**464–65
Radiation therapy
    for cervical cancer, **6:**416
    effects of, **6:**416–18
    for leukemia, **4:**695
    for lymphoma, **4:**849–50
    meningiomas, **4:**991
    for prostate cancer, **6:**181
    total lymphoid irradiation, **4:**364
Radiation tolerance
    of the small and large intestine, **6:**469–80
    tolerance dose and, **6:**478
Radicals
    oxygen-derived, **7:**358–59
    oxygen-free, **7:**556
Radicular arteries, of spinal cord, **7:**193–94
Radicular cysts, **5:**614
Radio frequency pulse, **4:**864–65
Radioactive proteins. *see* Proteins, radioactive
Radioautography. *see* Autoradiography
Radiography
    diagnostic, **3:**1–8
        effects of, **6:**416–18
    imaging of body tissue by, **6:**439–40
    in laboratory evaluation of allergy, **1:**194
    lactose malabsorption diagnosis, **4:**619
    plain film, in diagnosis, **3:**1–3

Radiography, interventional
    defined, **3:**1
    techniques of, **3:**7–8
Radioimmunoassay, **6:**455–61
    antigen heterogeneity as problem with, **6:**459–60
    antisera preparation, **6:**457
    in drug analysis, **3:**208, **5:**838
    in hormone concentration measurement, **3:**329–30
    hormone detection by, **5:**713–14
    kinins, **4:**593
    labeled antigen preparation, **6:**457–58
    nonspecific interference as problem with, **6:**458–59
    parasite detection, **4:**388
    principle of, **6:**455–56
    problems associated with, **6:**458–60
    reactions that form basis of, **6:**456
    reagents preparation, **6:**456–60
    sensitivity of, **6:**456
    for separation of antigen-antibody complexes, **6:**458
Radioimmunoconjugates, **4:**371–72
Radioimmunoimaging, **4:**371–72
Radioimmunotherapy, **4:**371–72
Radioisotopes. *see also* specific radioisotopes
    antibody conjugates with, **4:**371–72
    carcinogenesis, **4:**344
    counting, in drug analysis, **3:**201
Radioligand assay, drug analysis using, **5:**838–39
Radiological equipment
    for diagnosis, **3:**1–2
    portable, **3:**2
Radiologists, and radiation effects, **6:**416
Radiomimetic damage
    defined, **5:**259
    in toxic syndromes, **5:**264
Radionuclide tomography, computed. *see* Tomography, emission-computed
Radionuclides. *see* Radioisotopes
Radioreceptor assay. *see* Radioligand assay
Radiosensitivity. *see* Radiation tolerance
Radiotherapy, **6:**439–40, 447–48
    and hyperthermia, **4:**297–98
    treatment of ankylosing spondylitis, **6:**416–17
Radius, **7:**38
Radon
    average dose in United States, **6:**442
    cigarette smoking and, **3:**388
    effect on population, **6:**412
Rage, **2:**718
Rainbow trout. *see* Trout, rainbow
Rainier, **3:**843

Ralgro, toxicity of, **5**:267

Ramazzini, **3**:383

Ramus, **2**:697, 705–6

Random practice. *see* Practice (Psychology)

Randomized clincial trials, **4**:62–63

Ranvier's nodes, **1**:437

Raphe nuclei, **2**:34, 35, 686

    magnus, **4**:712–15

        projections to spinal cord, **4**:712–14

    oscurus, **4**:712–14

    pallidus, **4**:712–15

        projections to spinal cord, **4**:712–14

Rapid eye movements. *see* Sleep, REM

Raschkow's plexus, **2**:775, 825

Rash. *see* Dermatitis

Rashevsky, Nicholas, **5**:388

Rate coding

    defined, **5**:173

    muscle force production and, **5**:180

Rate-limiting step, chemical, **3**:931

Rationing, health care. *see* Health care rationing

Rats

    aged, in nerve growth factor model, **5**:337–38

    amino acid pyrolysis products and, **3**:678

    immunity of, in food-restriction studies, **5**:507

    young adult, in nerve growth factor model, **5**:338

Rauwolscine. *see* Yohimbine

Rayleigh scattering, **5**:590

Raynaud's disease, **2**:87

RB gene. *see* Genes, retinoblastoma

RB protein. *see* Retinoblastoma protein

Reaction range, genetic. *see* Genetic reaction range

Reactionary obesity, defined, **3**:248

Reactive hyperemia. *see* Hyperemia, reactive

Reactive oxygen species, **6**:582–83

Reading

    development of, perception and, **5**:748

    and eye movement, **6**:482–87

    and language, perception and, **5**:743–44

    by NETtalk, **5**:392

    process of, **6**:481–87

    processes and comprehension, **6**:481–92

    silent, inner voice and, **6**:487

    in technical work, perception and, **5**:743–44

    temporal integration of, **5**:743

    visual processes, **4**:201–2

    word identification and, **6**:484–86

Reading comprehension

    comprehension process, **6**:487–91

    and dyslexia, **6**:491–92

Reading comprehension *(Continued)*

    in older adults, **1**:120

    sentence parsing and, **6**:488–89

    and speedreading, **6**:491

    text understanding and, **6**:489–91

Reading disability. *see* Dyslexia

Reading frame (genetics), **5**:199–201

Realistic conflict theory, **7**:235–36

Reasoning

    deductive, **6**:495–98

    defined, **6**:493

    evolutionary approaches to, **6**:500–503

    inductive, characteristics of, **6**:498–99

    mind versus brain in, **6**:494

    and natural selection, **6**:493–503

    and social conduct, **6**:501–3

    and thinking, **7**:470

Reasoning, conditional, **2**:619–28

    aim of, **2**:619–20

    Carol's motorcycle, **2**:619

    complex, **2**:626–27

    development, **2**:619–28

    inference schema, **2**:619–21, 623, 627

    inferences under a supposition, **2**:623

    and knowledge, **2**:620

    and logic, **2**:622–23

    reductio ad absurdum, **2**:619, 621

    and Sherlock Holmes, **2**:619

    from a simple, **2**:624–26

    from a supposition, **2**:619–28

    and THOG task, **2**:627

    from a universally quantified, **2**:626

Rebaudioside, as sweetener, **7**:375–76

Receptive field, visual. *see* Visual fields

Receptor-mediated signal transduction. *see* Signal trans-
    duction

Receptor surface maps, in network modeling, **5**:389

Receptors, acetylcholine. *see* Receptors, cholinergic

Receptors, adrenergic, **2**:325, 368, 369, 848–49

    cell membrane, **6**:516–17

    characteristics of, **1**:81

    definition of, **1**:81

    effect on adenyl cyclase, **1**:81, 84

    mammalian, cloning of, **1**:84

    properties of, **5**:421–22

    structure and function, **1**:81–92

Receptors, adrenergic, beta, **2**:36, 217, 337

    beta-adrenergic receptor kinase (betaARK), **1**:92

    heart rate and, **3**:516

    as substrate for protein phosphorylation, **6**:208–9

Receptors, angiotensin, **5:**424

Receptors, antigen, **6:**519–22

Receptors, antigen, T-cell, **7:**478–79

Receptors, benzodiazepine. *see* Receptors, GABA-benzodiazepine

Receptors, beta adrenergic. *see* Receptors, adrenergic, beta

Receptors, biological. *see* Receptors, endogenous substances

Receptors, bradykinin, **5:**730

Receptors, calcitonin, **1:**769, **5:**424

Receptors, CCK$_B$, **5:**424

Receptors, cell. *see* Receptors, endogenous substances

Receptors, cholecystokinin, **5:**731

Receptors, cholinergic, **2:**287–88, 301, 314, 337, 522

    cell membrane, **6:**517

    function, **2:**288

    in gastric circulation regulation, **3:**762

    in myasthenia gravis, **5:**257–58

    properties of, **5:**421

    structure, **2:**287–88

    topography diagram, **2:**288

    type muscarinic, **2:**522

    type nicotinic, **2:**522

Receptors, colony-stimulating factor, **5:**816

Receptors, complement, **2:**57, 614–15

    defined, **2:**614–16

    in phagocytosis, **6:**610

    polymorphonuclear neutrophils, **1:**299

Receptors, corticosteroid. *see* Receptors, glucocorticoid

Receptors, cytoadhesin, **3:**528

Receptors, diazepam. *see* Receptors, GABA-benzodiazepine

Receptors, dopamine

    in Parkinson's disease, **5:**672

    properties of, **5:**422

    topographical model of, **1:**86–87

Receptors, endogenous substances, **2:**137, 138, 199, 299–304, 304, 333, 338–40

    -binding, **2:**337

    and adenylate cyclase activity, **6:**514

    agonist, **2:**15, 841

    antagonist, **2:**15, 841

    antibody against, **2:**301

    arteriolar, in gastric circulation regulation, **3:**762

    assay, **2:**301–2

    and autoimmune diseases, **2:**301

    binding sites for, **3:**124

    biochemistry, **6:**513–22

    biological criteria for identifying, **2:**300–301

    biological response, **2:**300–301

Receptors, endogenous substances (*Continued*)

    blockade, **2:**15

    with catalytic activity, **5:**420

    cell membrane, **6:**516–19

    characteristics of, **3:**325, 326

    and chemical modification, **2:**302

    cortical, **2:**686

    defective

        in pseudohydroparathyroidism, **2:**301

        in testicular feminization, **2:**301

    defined, **2:**57, **3:**371, 455, **4:**13, **5:**415–16, 713

    design, **2:**325

    disposal of, **2:**302

    and endococytosis, **2:**302

    versus enzymes, **2:**301

    excitatory amino acid, **5:**423–24

    families of, **5:**416–21

    G-protein-linked, **5:**416–17, 420

    and intracellular messenger, **2:**302

    joint, motor control and, **3:**699

    kinin, **4:**596–97

    L-aminophosphonobutyric acid, **5:**424

    ligand-gated ion channel, **5:**416

    ligand interaction and, **2:**299–300, 303

    ligand-receptor association process, **3:**179

    ligand-receptor complexes, **5:**875

    location and models for activation of, **6:**514–16

    mannose-6-phosphate, **5:**875

    metabolic conversion of, **2:**301

    metabolic regulation role, **4:**1004

    muscle activation and, **3:**698–99

    and myasthenia gravis, **2:**301

    number of, **3:**330

    osteoblast-derived stimulatory factor, **1:**769

    and proteolytic lysosome, **2:**302

    saturability, finite, **2:**300

    and Scatchard plot, explained, **2:**302

    and signal recognition, **7:**15–16

    as signal selector, **2:**325–26

    signal transduction and, **7:**15–21

    signal transmission and, **2:**302

    specificity, **2:**300

    sphingolipids as, **7:**187

    stimulation, **2:**14–15

    structure, **2:**302

    synthesis, **2:**302

    theory, **2:**57

    types of, **6:**514–16

    zinc finger, **2:**57

Receptors, endorphin, **2:**338, **5:**424

Receptors, epidermal growth factor-urogastrone, **2**:61
transforming growth factor alpha, **7**:627
transforming growth factor beta, **7**:636
Receptors, estrogen
antral follicles and, **3**:632
in bone cells, **3**:484
in breast tumor cells, **2**:59
in uterus, **7**:741–42
zearalenone and, **5**:267
Receptors, extracellular matrix glycoprotein. *see* Receptors, cytoadhesin
Receptors, FC
classification, **1**:298
Ig molecule, **1**:677
for IgE, **1**:299
macrophage, **5**:815
in phagocytosis, **5**:814, **6**:610
polymorphonuclear neutrophils, **1**:299
Receptors, FSH, **3**:633
Receptors, GABA-benzodiazepine, **2**:34
cellular signal transduction and, **7**:16–17
epilepsy and, **3**:460
functions of, **5**:416
properties of, **5**:423
Receptors, gamma-aminobutyric acid. *see* Receptors, GABA-benzodiazepine
Receptors, gastrointestinal hormones
gastrin
antagonists for, **5**:723
cholecystokinin and, **5**:717
peptide binding by, **5**:717
peptide hormone, **2**:58
vasoactive intestinal peptide, **5**:384
Receptors, glucagon, **5**:733
Receptors, glucocorticoid
aging and, **3**:896–97
nuclear transport of, **5**:456
Receptors, glutamate
epilepsy and, **3**:460
kainate subtype of, **5**:416, 423–24
Receptors, glycine
functions of, **5**:416
properties of, **5**:423
Receptors, GQ$_{1b}$, **3**:733
Receptors, growth substances, **6**:514
gangliosides and, **3**:734
glioblastomas and genes, **5**:548
interferon-gamma and, **3**:448–49
melanoma, **4**:965
squamous cell carcinomas and genes, **5**:548
TGF-alpha and, **3**:447

Receptors, guanine nucleotide regulatory protein
alignment of, **1**:85
amino acid sequence of, **1**:82–83
characteristics of, **1**:81
coupling of, mutations affecting, **1**:89–91
palmitoylation of, **1**:91
regulation of, by phosphorylation, **1**:91–92
transmembrane organization of, **1**:86–89
Receptors, histamine, **5**:423
Receptors, hormone. *see* Receptors, endogenous substances
Receptors, immunologic, **6**:515
Receptors, insulin, **2**:301, **5**:732, **6**:94, 517–18
Receptors, integrin, **3**:529–31
Receptors, interleukin-2, **4**:527–33
antibody against, and graft rejection, **4**:363
interaction with interleukin-2, **4**:530–31
signal transduction, **4**:530–31
structure, **4**:529–30
Receptors, interleukin-3, **5**:816
Receptors, interstitial cell-stimulating. *see* Receptors, LH
Receptors, LDL, **1**:452
Receptors, lectin. *see* Receptors, mitogen
Receptors, LH, **3**:633
Receptors, luteinizing hormone. *see* Receptors, LH
Receptors, lymphocyte homing, **4**:1013
Receptors, macrophage colony-stimulating, **1**:769, **5**:816
Receptors, mitogen, **6**:519
Receptors, muscarinic
defined, **2**:321
properties of, **5**:421
Receptors, muscimol. *see* Receptors, GABA-benzodiazepine
Receptors, N-Methyl-D-aspartate, **4**:231
Receptors, nerve growth factor
cholinergic neurons and, **5**:338
culture studies of, **5**:334
melanoma, **4**:965
on Schwann cells, **5**:338
types of, **5**:335–36
Receptors, neural. *see* Receptors, sensory
Receptors, neurohumor, **6**:514
astrocytes, **1**:439
defined, **2**:841
neuropeptide, **5**:415–25
neuroregulators, **2**:312–16
neurotensin, **5**:424
reduced in number, **2**:844
Receptors, neurotransmitter. *see* Receptors, synaptic
Receptors, nicotinic
acetylcholine, functions of, **5**:416
and cellular signal transduction, **7**:16–17

Receptors, nicotinic (*Continued*)
    phosphorylation of, **6**:209–10
    properties of, **5**:421
Receptors, NMDA, **5**:423
Receptors, peptide
    in brain, **5**:424–25
    functions of, **5**:726
    opioids and, **5**:719
    structure of, **6**:94–95
Receptors, pherome. *see* Chemoreceptors
Receptors, prostaglandin, **1**:769
Receptors, protein, **2**:521, 522
Receptors, protein-tyrosine kinase, **5**:548–49
Receptors, purinergic, **5**:424–25
Receptors, quisqualate, **5**:423
Receptors, sensory
    atrial, **2**:138
    classes of, **7**:675
    deficiencies, in Alzheimer's disease, **1**:214–15
    G proteins and, **5**:416
    mucosal, **4**:555
    of olfactory mucosa, **5**:523–24, 525–26, 530
    retinal, **7**:803
    role at nerve terminal, **2**:323
    somatosensory system
    and muscle spindle, **7**:121
    and skin, **7**:120–22
    of trigeminal system, **7**:671–79
    taste
        properties of, **7**:532–33
        theory of, **7**:363–65
    touch and, **7**:52–53
    transduction, vision and, **7**:803
Receptors, serotonin
    properties of, **5**:422–23
    subtypes, signal transduction mechanisms of, **6**:830–31
Receptors, somatostatin
    adenylate cyclase and, **5**:733
    properties of, **5**:424
Receptors, sperm, **3**:605
Receptors, steroid, **2**:53, 58, 302
    in G_{i1} protein, **4**:21
    steroid action and, **7**:265–66
    steroid binding to, **6**:516
    zinc-rich, **2**:53
Receptors, stretch. *see* Mechanoreceptors
Receptors, synaptic, **5**:415–25
    chloride-containing, inhibitory postsynaptic potential
        and, **7**:386–87
    postsynaptic, **2**:337

Receptors, synaptic (*Continued*)
    presynaptic, **2**:323, 337
Receptors, T-cell, **2**:1, 2, 225, 226, **4**:832–33, **7**:395–400
    antigen binding to, **6**:520–22
    biochemistry of, **7**:395–96
    CD-2 as, **7**:398
    CD-4 as, **2**:3, 6–8, 227, 230–32
    CD-8 as, **2**:3, 6–8, 227, 230–32
    cDNA of, **7**:396
    development of, **7**:397–98
    in experimental allergic encephalomyelitis, **5**:147
    as gene family, **3**:798
    germline organization of, **7**:396
    idiotypes, **4**:316
    immune system, **4**:329
    interleukin-2 and, **6**:518–19
    receptor antigen for, **7**:478–79
    receptor-CD3 complex, schematic of, **7**:396
    receptor gene segments, **7**:397
    receptor genes, chromosome location of, **7**:396
    receptor V gene products, **7**:397
    repertoire and diversity of, **7**:396
    soluble, **7**:399
    V domains, **4**:314
Receptors, tachykinin, **5**:720
Receptors, taste
    mechanisms for stimulus activation of, **7**:534
    mechanisms of, **7**:531–32
    modern receptor theory of, **7**:363–65
    nature of, **7**:363–65
    signal transduction mechanisms for, **7**:533–35
    sweet, **7**:533
Receptors, thyroid hormone, **5**:171, **7**:489–90
Receptors, thyrotropin, antibodies, ELISA, **4**:354–55
Receptors, transferrin, **4**:416
Receptors, transforming growth factor. *see* Receptors,
    epidermal growth factor-urogastrone
Receptors, tryptamine. *see* Receptors, serotonin
Receptors thyroid-stimulating hormone, **2**:301
    antibody against, **2**:301
Recessiveness (genetics)
    defined, **3**:847
    molecular basis of, **3**:854
Recklinghausen's disease of nerve. *see* Neurofibromatosis
Recognition, defined, **3**:215
Recombinant DNA. *see* DNA, recombinant
Recombination, genetic, **4**:952–53
    cancer theory, **5**:325–27
    chromosome mapping and, **3**:824–25
    defined, **3**:135, 403, 823

Recombination, genetic (Continued)
    in enzyme isolation, 3:409
    genetic variation and, 6:110
    principles of, 3:849
    X-linked recessive disorders and, 3:829-30
Recommended dietary allowances
    cautions concerning, 3:23
    defined, 5:485
    for the elderly, 4:899-900
    of National Academy of Sciences, 3:27-28
    nutritional quality assessment and, 5:484-85
    table of, 5:486
Recovery, spontaneous, in classical conditioning,
        2:629, 631
Recruitment, follicular
    course of, 3:631-32
    defined, 3:627
Recruitment, motor unit
    defined, 5:173
    sequence of, 5:178-79
Recruitment, muscle, defined, 3:691
Recruitment threshold, muscle
    defined, 3:691
    formula for, 3:694
    isometric recruitment and, 3:698
    of m. biceps brachii motor unit, 3:696
    of motor units, 3:693
Rectifier, delayed, 2:305
Rectum, 2:33, 3:66, 82
Recurrent laryngeal nerve, 4:657
Red cell mass. see Erythrocyte volume
Red cell membrane. see Erythrocyte membrane
Red cells. see Erythrocytes
Red-green color deficiency, 2:590
Red-mold disease, cereal, 5:263
Red tide. see Dinoflagellida
Redox-driven transport, 4:974
Reduced bonds, counting of, 5:474-75, 476
Reducing agents, ascorbic acid, 1:431
Reductio ad absurdum argument, 2:619, 621, 625-28
Reductionism, defined, 3:17
Reduviids, trypanosomiasis and, 7:682
Reflectance
    color of natural abjects, 2:589
    spectral, 2:582
Reflex
    autonomic, 2:32
    and baroreceptor, 2:186
    of cutaneous and joint afferents, 6:159-60
    and eye-movement, 2:312-13

Reflex (Continued)
    and muscle afferents, 6:158-59
    proprioceptive, 6:157-65
        circuits for, 6:158-61
        H reflex as, 6:1664-63
        muscle vibration and, 6:164
        stretch reflex as, 6:163-64
        tendon jerk as, 6:161
        unloading reflex as, 6:162
    recurrent inhibition of circuits, 6:160
    "Reflex reinforcement," 3:723
    skin, spinal cord dysfunction and, 7:230-31
    spinal cord injury and, 5:808-9
    tendon, spinal cord dysfunction and, 7:230
    transcortical circuits and, 6:160-61
    vagovagal, 2:36
Reflex, Babinski, 7:219
Reflex, cardiogenic, 2:137-46
    atrial, 2:139-45
    and hormones, 2:143-44
    from receptor, 2:144
    and sympathetic afferent nerves, 2:145
Reflex, conditioned. see Conditioning, classical
Reflex, monosynaptic, vagovagal, 2:31
Reflex, stretch
    defined, 3:719
    muscle activation in, 3:696-97
    regulation of, 3:723
    spinal reflex pathway of, 7:216
Reflex, vestibulo-ocular, 2:312-13
    space travel and, 7:149-50
    visual stability and, 3:556-57
Reflex pathways
    in "isometric" muscle contractions, 3:723
    in movement preparation, 3:722
Refractive index, 5:590
Refractive index detection
    HPLC, 4:211
    sugar separation, 4:218
Refractory period, psychological, 5:152
Regeneration, 1:444-45
Regeneration, nerve. see Nerve regeneration
Regional immunology. see Immunology, regional
Regression, 2:245
Regulation of gene expression. see Gene expression
        regulation
Regulatory proteins
    binding strength of, 3:119
    DNA and, 3:117-18
    functional specificity of, 3:119-20

Regulatory proteins *(Continued)*
  nonspecific DNA binding and, **3:**120
  sequence variability of, **3:**121-22
  site size of, **3:**121
Regulatory standards, toxicological
  defined, **3:**383
  development of, **3:**384
Regurgitation, defined, **3:**243
Rehydration therapy. *see* Fluid therapy
Reinforcement
  defined, **2:**511
  schedule, **2:**629
  theory of interpersonal attraction, **4:**539-40
Reitan-Indiana test battery, **2:**865-66
REL genes. *see* Oncogenes
Relative body weight. *see* Body weight, relative
Relative permittivity
  of biological tissues, **3:**11
  defined, **3:**9
  of polarization, **3:**10
  skin conductivity and, **3:**13
  of whole blood, **3:**15
Relative rate test (genetics), **3:**887
Relative risk. *see* Risk
Relaxation time
  of polarizable systems, **3:**11
  of skin dielectric dispersion, **3:**13
Relaxing factor, endothelium-derived, **2:**366
Religious criteria for beginning of human life, **1:**656
REM sleep. *see* Sleep, REM
Remnant particle, **2:**411
Renal agenesis, **1:**702(t)
Renal ammoniagenesis. *see under* Ammonium
Renal function, impaired. *see* Kidney failure
Renal neoplasms. *see* Kidney neoplasms
Renal transplantation. *see* Kidney transplantation
Renin, **2:**144, 174
Renin-angiotensin system, **2:**145
  at birth, **3:**349
  features of, **5:**730
  and hypertension, **4:**291-92
Rennin secretion, **1:**474
Renovascular hypertension, **4:**289
Renshaw cell, **7:**217
Reoviridae, **4:**519-20
Reperfusion
  free radical damage during, **7:**358-59
  of muscle, in dynamic exercise, **3:**519
Reperfusion injury, vitamin E in treatment of, **7:**877-78
Repetition of messages, **5:**804, 805, 806

Repetitive sequences, nucleic acid, **2:**431, 435, 469-70, 477, 486-87, **3:**163-67, 796, **4:**185
  characteristics of, **3:**874
  computer analysis of, **3:**163-67
  defined, **3:**261, 777
  detection of mutations by sequencing, **6:**527-28
  Dideoxy analysis method, **3:**173
  DNA probes and, **3:**138
  in *fos* gene transformation, **3:**705
  genomic mutations and, **3:**883
  repetitive elements of human genome, **2:**477
  retrotransposition and, **2:**478, **3:**884
  sequence analysis of, **3:**164
  tandem repeats, variable number, **2:**669
Replication, virus. *see* Virus replication
Replicon, **2:**293
Report on Nutrition and Health, **3:**652
Reporter genes. *see* Genes, reporter
Representation
  defined, **3:**215
  nature of, **2:**555-56
Repressor proteins
  gene regulation by, **3:**117
  operator sites and, **3:**122
Reproduction, **2:**497-99, 599, **4:**929-30. *see also* Meiosis
  anatomy, **4:**935-36, **6:**559-65
  asexual, **4:**929
  bisexual, **4:**930
  breeding rhythm, seasonal, **2:**497-99
  cycle, seasonal, **2:**498
  disease models, animal, **1:**257-58
  female
    anatomy of, **6:**562-65
    role of body fat and menarche in, **1:**741-50
  genetic variation and, **6:**110
  gonadotropin secretion and, **5:**398-99
  history of neuroendocrinology of, **6:**555-56
  hormone regulation of, **3:**326
  in human evolutionary studies of pelvis, **1:**826
  in humans, **2:**599
  male
    anatomy of, **6:**559-62
    and nutrition, **1:**748-49
  and mother-child dependency, **2:**599
  neurochemical control of, **6:**555-57
  neuroendocrinology of, **6:**555-56
  platelet-activating factor and, **6:**51
  sexual, **4:**929
  and sexuality, **6:**887-88
Reptiles, venomous, **7:**599-600

Research
on birth defects, **1:**704–5
diagnostic pathology and, **5:**685–86
double-blind experiments, **2:**67
Research Diagnostic Criteria and Diagnostic and Statistical Manual, **4:**665, 668, 993, 995
Research on Aging Act, **3:**896
Reserpine, **2:**12, 219, 832, 843
antipsychotic drug, **2:**219
and depression, **2:**832
for Huntington's disease, **4:**272
and hypertension, **2:**219
as inhibitor of monoamine transporter, **1:**648
Resistance, natural. *see* Immunity, natural
Resistance training, defined, **3:**505
Resistance transfer factor. *see* F factor
Resistant starch. *see under* Starch
Resolution, microscopic. *see under* Microscopy
Resonance, normal, **7:**158–59
Resonance disorders, **7:**158–59
Resource allocations, medical, **1:**659
Resource-competition models. *see* Models, resource-competition
Resources, defined, **5:**149
Respiration
adaptation to body size, **6:**572–76
in burrowing mammals, **6:**578
chemoreceptors and, **5:**878
comparative physiology, **6:**567–79
control of, **5:**618–19
by ADP, **3:**377
defined, **3:**940
disordered, as cause of sudden infant death syndrome, **7:**318–19
elements associated with, **6:**602–3
dependence of oxygen for, **6:**568
in diving mammals, **6:**577
at high altitudes, **6:**578
mammalian, newborn, **6:**576–77
in mammals adapted to special environments, **6:**577–78
perinatal, **3:**617, 618
pH homeostasis and, **5:**877
rate, gingiva, **4:**36–37
Respiration, artificial, **5:**705, 825
Respiration disorders, **5:**880
Respirators. *see* Ventilators, mechanical
Respiratory alkalosis. *see* Alkalosis, respiratory
Respiratory burst, **6:**581–85
activity, and association with human pregnancy, **6:**585–85

Respiratory burst *(Continued)*
biochemical basis of, **6:**582–83
and effect on physiological states, **6:**584–85
effect on tissues, **6:**583
enzymatic basis of, **6:**582–83
in eosinophils, **6:**583
general characteristics of, **6:**581–82
inflammation and, **4:**471–72
killing of microorganisms during, **6:**582–83
in lymphocytes, **6:**583
and macrophages, **6:**610–11
metabolism of drugs and, **6:**583
in polymorphonuclear neutrophils, **5:**429–30
Respiratory burst oxidase, **5:**429–30
Respiratory chain, mitochondrial. *see* Mitochondrial respiratory chain
Respiratory chain complex, defined, **5:**55
Respiratory distress syndrome, **6:**601
acute
following cardiopulmonary bypass, **4:**95
treatment with pulmonary bypass, **4:**97
animal models, **1:**258
Respiratory quotient
calculation of, **5:**475
defined, **5:**471
Respiratory syncytial virus, **4:**460, 461
Respiratory system
aging of, **3:**902
anatomy, **6:**587–93
cold stress response, **4:**797
deposition of particles in, **7:**581
effect of radiation on, **6:**414–15
function of, **6:**595, **7:**578
gas exchange in, **6:**568
inhaled toxicants in, **7:**578–82
lower, origin of, **4:**658
mammalian, design of, **6:**571
mucosal immunology of, **5:**139
organization of, **6:**587–90
physiology and biochemistry, **6:**595–603
and psychophysiological disorders, **6:**337
response to inhaled chemicals, **7:**582–84
as target for intoxicants, **7:**577–78
Respiratory tract
antimicrobial defenses in, **5:**692
defenses, **5:**690–92
function, **5:**690
immune defenses in, **5:**692
lesions and tissue susceptibility in, **5:**695
metabolism of xenobiotics in, **5:**691

Respiratory tract *(Continued)*
  mucociliary clearance as defense in, **5:**691
  noxious chemical and particulate inhalation and,
      5:694–95
  olfactory dysfunction in, **5:**695
  pathophysiology of, **5:**689–97
  protective reflex responses of, **5:**690–91
  structure, **5:**690
Respiratory tract diseases, **4:**450, **5:**705
  air pollutants in, **5:**694–95
  allergic rhinitis as, **5:**693–94
  epithelium role, **5:**690
  selected, **5:**692–96
  toxic chemicals in, **5:**694–95
Respiratory tract infections
  air pollutants and, **3:**387
  causes of bacterial, **1:**555–57
  in day-care centers, **4:**463
  detection of bacterial pathogens in, **1:**555–56
  effect of quinolones on, **6:**408
  in infants and toddlers, **4:**460–61
Respiratory tract neoplasms, **7:**583–84
Response
  conditional, **2:**630
  unconditional, **2:**630
Response-produced feedback. *see* Feedback, intrinsic
Restraint eating
  experiment in, **3:**685–86
  survey of, **3:**688
Restriction endonucleases. *see* DNA restriction enzymes
Restriction fragment length polymorphisms, **2:**463, 483,
      661, 668–71
  in alpha-1-antitrypsin deficiency, **3:**146
  applications of, **3:**138–39
  atherosclerosis-linked genes, **1:**453
  characterization of, **3:**136–39
  in cystic fibrosis gene identification, **3:**832
  defined, **3:**135, 175, 835
  in diagnosis, **3:**136
  family studies, **2:**670–71
  genetic disease analysis using, **6:**524–25
  genetic mapping and, **3:**180, 182
  interspecies backcross and, **3:**827
  karyotype detection and, **3:**149
  linkage maps and, **3:**877
  and Mendelian inheritance, **2:**463
  in pedigree analysis, **3:**836–37
  and prenatal diagnosis of phenylketonuria, **5:**865–66
  as technique for discrimination of monoclonal and
      polyclonal tumors, **7:**689–90
  in thalassemia diagnosis, **3:**146

Restriction mapping, **2:**483–86, **3:**878, **6:**524–25
Retardation, mental. *see* Mental retardation
Retention period, intrauterine. *see* Intrauterine retention
      period
Retention (psychology), **1:**493, **5:**122
Retention tests
  confounded effects and, **5:**122
  defined, **5:**121
  novel, **5:**125, 126
  practice and, **5:**123
Reticular activating system, **2:**217
Reticular formation
  caudal pontine, projections to spinal cord, **4:**712–14
  medullary medial, projections to spinal cord, **4:**712–14
Reticulocytes, **4:**126
Reticuloendothelial system, **1:**299, **6:**605–13
  autism and, **1:**508–9
  biological role of, **6:**606–7
  development (ontogeny) of, **6:**606–7
  disorders of, **6:**612–13
  historical background of, **6:**605–6
  homeostasis and, **6:**611–12
  interrelationship with tissue injury, **6:**611–12
  molecular basis of biologic activities of, **6:**609–11
  organization of, **6:**606–7
  other systems of host defense and, **6:**611–12
Reticulum, **2:**89, 97–98
Reticulum, endoplasmic, **2:**97, 238–40, 341, 344, 563
Retina, **6:**615–31, **7:**802–5
  alpha-linolenic acid deficiency and, **3:**577
  amacrine cells of, **6:**620
  anatomy, **3:**540–43, **6:**616
  bipolar cells of, **6:**619–20, **7:**837
  cell connections in, **7:**837
  cell types found in, **6:**620
  cellular and synaptic organization of, **6:**619–24
  characteristics of, **6:**615–17, **7:**836–38
  disparity in, **7:**811
  function of, **6:**618
  functional organization of, **6:**627–28
  ganglion cells of, **6:**621–22, **7:**803–4, 837–38
  glial cells of, **7:**838
  images on, **7:**822–24
  interplexiform cells of, **6:**621–22
  network modeling of, **5:**389
  neuroactive substances found in, **6:**628
  neuronal responses in, **6:**624–28
  neurons of, **5:**311
  omega-3 fatty acids in, **5:**537–38
  pharmacology of, **6:**628–30

Retina *(Continued)*
  photoreceptors of, **6:**617–19
  receptive field maps, **6:**622
  receptive field of, **6:**622–23
  receptors of, **7:**803
  synaptic organization of, **6:**623–24
  tumors of, infant. *see* Retinoblastoma
  visual information processing by, **6:**622–23
  visual pigments of, **6:**617–18
Retinal pigment epithelium. *see* Pigment epithelium of eye
Retinoblastoma, **2:**64, 65, 451–54, **4:**1009–10, **6:**424
  and cancers accompanying, **6:**636
  clinical management of, **6:**635
  disease characteristics of, **6:**633–35
  DNA marker diagnosis of, **3:**142–43
  as hereditary tumor, **7:**706–7
  Knudson's hypothesis for, **6:**636
  molecular genetics of, **6:**633–39
  mutation rates in, **5:**248
  mutations resulting in, **6:**637–38
  phenotype, events causing expression of, **7:**707
  risk factors for, **6:**638
Retinoblastoma protein
  as cell cycle regulator, **6:**636–37
  interaction with DNA tumor virus, **6:**636–37
Retinoblastoma tumor suppressor gene, **5:**641–42
Retinohypothalmic tract, **2:**494
Retinoic acid
  embryonic body pattern diversity and, **3:**291
  nuclear receptors for, **7:**858
Retinoids
  in cancer regression, **3:**775
  chemistry, **4:**819–21
  humeral immunity and, **4:**826
  lymphocyte responses, **4:**819–28
  mechanism of action, **4:**826–28
  physiology and metabolism, **4:**821
  structure, **4:**819–21
  vitamin A and, **5:**512
Retinol. *see* Vitamin A
Retinol-binding proteins
  plasma, **7:**855–56
  three-dimensional structure of, **7:**856
  vitamin A metabolism and, **7:**848, 854
Retirement, longevity and, **4:**784
Retraction, as joint movement, **5:**133
Retrobulbar region, **2:**679
Retrograde axonal transport. *see* Axonal transport, retrograde
Retrograde menstruation. *see* Menstruation, retrograde

Retrograded starch. *see* Starch, resistant
Retroperitoneal organ, defined, **3:**57
Retrotarsal fold. *see* Fornix conjunctivae
Retrotransposition (genetics)
  of DNA element, **2:**478
  genomic mutations and, **3:**884
Retroviridae, **2:**395
  cancer causing, **6:**656–57
  endogenous, **6:**656
  genome, organization of, **6:**655
  interferon effects, **4:**520, 521
  life cycle of, **6:**653–56
  murine leukemia, characteristics of, **6:**644–45
  oncogenic, **6:**656–57
  replication, **6:**654–56, 668, 688
  RNA, reverse transcription of, **6:**655
  structure of, **6:**653–54
  vectors
    experiments with, **6:**657–58
    in gene therapy, **6:**659
    for gene transfer, **6:**653–59
    and helper cell, **6:**657
    natural, **6:**656–57
    problems and limitations with, **6:**658
Retrovirus infections, **5:**677
Retrovirus vaccines, **6:**641–52
  for bovine leukemia, **6:**646
  feline, **6:**645–46
  feline leukemia, **4:**342
  feline retrovirus, **6:**645–46
  human, **6:**647–51
  mammalian leukemia, **6:**644–46
  murine, **6:**644–45
  principles of vaccination, **6:**643–44
Re-uptake, defined, **3:**455
rev genes. *see* Genes, *rev*
rev viral proteins. *see* Gene products, *rev*
Revenge, media protrayal of, **1:**3, 1058
Reverse genetics
  defined, **3:**830
  in DMD gene cloning, **5:**193
Reverse transcriptase, **2:**395, 478
  clone libraries and, **3:**876
  and retroviruses, **6:**654
  virus replication and, **7:**777
Reverse transcription, Alu element and, **3:**874
Reversed phase, HPLC, **4:**208
Rewarming
  frostbite, **4:**808
  hypothermia, **4:**803–6

Reye's syndrome
    aflatoxins and, **5**:265
    in school-age children, **4**:465
Rh-Hr blood-group system, **4**:381
Rhabdomyosarcoma, **2**:451
Rhabdoviridae, interferon effects, **4**:519
Rhamnogalacturonans, **3**:38
Rhesus monkey, **2**:339, **4**:932–33, 935
Rhetorical messages, persuasion and, **5**:805
Rheumatoid arthritis
    autoimmunity and, **1**:529–30
    IL-2 and, **4**:474
    mechanisms of, **4**:410
Rheumatoid factor, **1**:302
Rhinencephalon. *see* Limbic system
Rhinitis, **5**:608
Rhinitis, allergic, **5**:693–94
Rhizobium, and soil biology, **1**:147
Rhodanase, detoxication activity of, **5**:287
Rhodopsin
    functions of, **5**:417, **6**:617
    photopigment, **2**:585
    sensitivity curve, spectral, **2**:585
    sequence of events in light absorption by, **6**:618
    vitamin A and, **7**:850
Rhombencephalon. *see* Hindbrain
Rhythm, biological. *see* Circadian rhythm; Periodicity
Rhythmicity. *see* Periodic movement
Ribavirin, **2**:399
Riboflavin, **2**:529, **4**:221
1-beta-D-Ribofuranosyl-1,2,4-triazole-3-carboxamide.
    see Ribavirin
Ribonuclease, pancreatic
    bovine, **3**:413
    cleavage, in detection of genetic diseases, **6**:527
    irreversible inactivation of, **3**:418
Ribonuclease A. *see* Ribonuclease, pancreatic
Ribonuclease H, **2**:400
    and DNA replication, **6**:667
    and retrovirus replication, **6**:668
    and RNA/DNA hybrids, **6**:667
    and transcription, **6**:668
Ribonuclease L, **6**:664
Ribonuclease P, **6**:667
Ribonuclease S, **3**:420
Ribonucleases
    cyclizing, **6**:661
    general characteristics of, **6**:661–62
    inhibitor, **6**:664
    inhibitor, human placental, **6**:662

Ribonucleases *(Continued)*
    metabolic involvement, **6**:661–68
    pancreatic-like, **6**:662–63
    phosphodiesterase-type, **6**:662
    phosphotransferase-type, **6**:662
    pyrimidine-specific, **6**:662–63
    and ribosomal RNA processing, **6**:665–66
    RNA-processing, **6**:665–67
    and RNA stability, **6**:663–64
    specificity for RNA/DNA hybrids, **6**:667–68
Ribonucleic acid. *see* RNA
Ribonucleoproteins, defined, **5**:451
Ribonucleotide reductases, **2**:655
    bacterial, **2**:537
    in DNA nucleotide precursor synthesis, **3**:471
    in nucleotide synthesis, **5**:462, 464, 465
Ribonucleotides
    biosynthesis pathways, **5**:462
    degradation, purine biosynthesis and, **6**:391
Ribosomal proteins
    amino acid sequence in, **6**:672
    characteristics of cDNA and, **6**:672
    internal duplications in, **6**:674
Ribosomal RNA. *see* RNA, ribosomal
Ribosome-inhibiting proteins, **4**:417
Ribosomes, **2**:235, **6**:671–79
    biogenesis of, **6**:675–77
    domains, function of, **6**:677–78
    eukaryotic and prokaryotic, **6**:674–75
    evolution of, **6**:674–75
    function, **6**:672–74
    inactivation, **4**:418
    and RNA, **2**:291
    in RNA transport, **5**:459
    structure of, **6**:672
Ribs. *see* thorax
Rice, **2**:104, 106
    in "basic four" food groups plan, **3**:656
    dietary fiber from, **3**:40
Rice Act (Japan), **5**:266
Ricin
    antibody conjugates with, **4**:368–69
    immunotoxins, **4**:417–18
Rickets
    calcium deficiency and, **5**:708
    in early industrial cities, **3**:21
    history of, **7**:860
    vitamin D-dependent, **7**:870–71
    vitamin D-resistant, **7**:870
Ridge-and-furrow system. *see* Derm

Ridge details, illustrated, **3**:90

Ridge dissociations, defined, **3**:89

Rifampicins, **1**:290(t)

Rifampin, **5**:706

Rifamycins, **1**:291

Right to die, **1**:136

Rigor mortis, defined, **7**:28

Rimantadine, **2**:399

Ring chromosomes, **2**:439, **5**:259

Ringed hair, **4**:48

Risk, defined, **5**:599

Risk assessment, toxicological, **3**:383, 388–90

Risk factors

    in Alzheimer's disease, **1**:209

    amoebiasis, **1**:218–19

    asbestos exposure, **1**:422–23

    atherosclerosis, **1**:451–52

    birth defects, **1**:701

    cancers, ascorbic acid role in reduction of, **1**:432

    Down's syndrome, **1**:699

    in genital cancer, **3**:859–60, 865–66

    malnutrition in the elderly, **4**:900–902

    in mutation, **5**:253–54

    of oral contraceptives, **5**:599–609

    in papillomavirus infection, **3**:866

    psychological, **4**:78

    of techno-food, **3**:22

Rivalry (behavior), **1**:574

River blindness, **1**:265

RNA

    analysis by nucleic acid hybridization, **2**:431–32

    in brain, **2**:431, 432

    characteristics of, **3**:871

    degradation, **2**:294

    editing

        and genetic information storing, **5**:653

        trypanosome, **5**:653

    folding of, **3**:164

    genes, **2**:478

    hnRNA, **2**:294

    hRNA, **2**:479

    influenza viruses, **4**:485

    in precellular world, **6**:675

    replication, **6**:681–88

    sequence analysis of, **3**:164

    stability, and ribonuclease, **6**:663–64

    structure, **6**:682

    synthesis

        chemistry of, **3**:109

        and oocyte growth, **5**:558

RNA (*Continued*)

    transcribed, **3**:792, **4**:251

    transport, through nuclear pores, **5**:457–59

RNA, cap, **2**:294, 295

RNA, heterogeneous nuclear, **6**:686

RNA, messenger, **2**:43, 235, 238, 258, 291, 294, 295, 318, 431, **4**:252

    chromosome 21 and, **3**:177

    decay, ribonuclease involvement with, **6**:663–64

    decoding, transfer RNA recognition and, **5**:76

    of DMD gene, **5**:194

    in EBV genome transcription, **3**:466

    estrogen-inducible, **3**:182

    formation of, **6**:684–86

    of *fos* gene, **3**:705

    of G proteins, **4**:21, 22–23

    gene expression and, **3**:871–72

    inhibition, in muscle, **5**:167

    lactase, **4**:617

    of light-chain myosin, **5**:213

    in mammalian cardiac ventricles, **5**:171

    maturation, ribonucleases and, **6**:666

    modification and processing of, **6**:685–86

    muscle differentiation and, **5**:165

    of muscle phosphorylase, **5**:228

    myoblast synthesis of, **5**:168

    nerve growth factor and, **5**:335

    potassium-sparing diuretics and, **3**:99

    and protein synthesis, **6**:685–86

    regulated by interferons, **4**:517

    splicing. *see* RNA splicing

    translation of, **6**:216

        to protein, **6**:685–86

        ribosomes and, **6**:672–73

    transport of, **5**:459

RNA, polyadenylated, **5**:459

RNA, ribosomal

    eukaryotic and prokaryotic characteristics of, **6**:674–75

    expression of, **3**:874

    maturation, ribonucleases and, **6**:665–66

    processing, and ribonucleases, **6**:665–66

    secondary structure of, **6**:672

RNA, small nuclear, **1**:334

RNA, transfer, **2**:235, 294, **4**:251–52, **5**:75–80, 459, **6**:673–74

    alanine, effect of base pair change and identity of, **5**:78–79

    amino acid recognition of, **5**:79

    and aminoacyl tRNA synthetases, **5**:77–78

    in autoimmune disease, **1**:335

RNA, transfer (*Continued*)
  enzymes that recognize, **5**:77
  evolution of molecular recognition of, **5**:80
  maturation, and ribonucleases, **6**:667
  minihelices, attachment of alanine to, **5**:79
  molecular recognition of, **5**:75–80
  and protein synthesis, **6**:685
  recognition of, rules of genetic code and, **5**:76
  sequence variants of, **5**:77–78
5S RNA gene, histone H1 and, **3**:132
RNA ligase, **2**:294
RNA polymerase I, **5**:558
RNA polymerase II, **2**:258, 478
  promoter regions and, **3**:123
  transcription factor IID and, **3**:131
RNA polymerase III, **3**:132
RNA polymerases, **2**:294, 478, 479
  binding sites of, **3**:794
  in DNA binding, **3**:117, 121
  DNA transcription and, **5**:468–69
  eukaryotic promoters and, **3**:123–24
  prokaryotic promoters and, **3**:123
  in transcription initiation, **3**:109, 110–11
RNA processing, ribonucleases and, **6**:665–67
RNA splicing, **2**:294, **6**:685–86
  alternative, in muscle development, **5**:166
  in chimeric antibody production, **3**:808
  defined, **4**:13, **5**:205
  of DMD gene, **5**:199
  in elastin, **3**:265
RNase. *see* Ribonucleases
Robertsonian translocations
  defined, **3**:175
  in mice, **3**:186
Robotics, **2**:552–53
  antirepresentationalism, **2**:553, 555
  empty symbol problem, **2**:552
"Robusts, the." *see* Australopithecines, "robust"
Rodbell, **4**:14
Rodent/human hybrid cell, **2**:456
  chromosome isolation, **2**:456
Rodents
  denervation of, **5**:383
  fibroblast transformation, **4**:583–84
  and trichinosis, **7**:648
Rods and cones
  anatomy, **3**:541–42
  defined, **2**:575
  as eye receptors, **7**:803
  photopigment, **2**:591

Rods and cones (*Continued*)
  retinal, characteristics of, **7**:836–37
  as retinal photoreceptors, **6**:617–18
  rods, **2**:583, 585
  terminals, in retina, **6**:624
Roentgen, Wilhelm Conrad, **3**:1
Roentgenography. *see also* Radiography
  defined, **3**:1
Roller pump
  heart-lung machine, **4**:90, 92
  in open heart surgery, **1**:388
Rolling circle replication, defined, **3**:101
Romaña's sign, **7**:685
Root caries index, **2**:806
Roseman, S., **3**:726
Rosenbloom, Joel, **3**:265
Roseola infantum, **4**:462
Rotation, as joint movement, **5**:133
Rotavirus infections, **4**:618
Rotaviruses, in children, **4**:462
Rotch, Thomas Morgan, **5**:700
Roundworm chemotherapy, **2**:386
  diseases, **2**:386–88
  drugs, listed, **2**:386
Roundworms, intestinal, **1**:269–70
Royal College of General Practitioners' Oral Contraception Study
  on cerebrovascular disease risk, **5**:601–2
  on ischemic heart disease, **5**:602
  on myocardial infarction, **5**:603
  subjects of, **5**:600
  on vascular diseases, **5**:604
Royal free disease. *see* Fatigue syndrome, chronic
RRNA, **2**:291, 294, 469
Rubber, synthetic, **3**:938
Rubber workers, **2**:347
Rubella
  birth defects caused by, **1**:701
  fetal infections, **4**:456–57
Rubella syndrome, **7**:780
Rubella vaccines, **4**:445, 466
Rubella virus, **1**:8
Ruffini terminals, **7**:676
Ruminants, **3**:60
Rumination
  defined, **3**:243
  as eating disorder, **3**:247
Running, motor pattern in, **4**:769–71
RVMLC1, **5**:215–16
Ryanodine, **2**:97

Rye, in "basic four" food groups plan, 3:656

S

S-100 proteins
    calcium-binding, 4:962
    in microtubule regulation, 5:17
S phase
    of cell growth, 2:262-64
    defined, 5:65
    histone protein synthesis, 4:237-38
S-substituted-N-acetyl-L-cysteine. *see* Mercapturic acid
Sac, pericardial. *see* Pericardium
Saccades
    characteristics of, 3:555
    defined, 3:553
    in fixation, 3:557
    horizontal, 3:558
    neural aspects of, 3:557-58
    neuronal activity for, 3:554
    premotor mechanisms of, 3:558
    visual cortex and, 3:559
Saccadic eye movements. *see* Saccades
Saccharin, as sweetener, 7:377-78
Saccharolytic organisms, in colon, 3:43
Saccharomyces cerevisiae, 2:321, 323-24, 3:132
Sacral cord, and micturition control, 5:44
Sacral vertebrae, 7:768
Sacrum, 2:595, 7:768
SAD. *see* Seasonal Affective Disorder
Saddle joints, 1:374, 376
Safety assessment, toxicological
    defined, 3:383
    development of, 3:384
Safflower oil, 3:578
Safrole, 2:111
Sagittal sinus, inferior, 4:983
Sagittal sinus, superior, 4:982
Saimiri, 2:593
Salamanders, 5:528
Saliva, 2:781, 811-13
    and bacteria, 2:811
    composition of, 2:781, 6:693-95, 812
    dental film, 2:811
    in Epstein-Barr virus infection, 3:463
    families of, 6:694
    flow, 2:813
    formation, and acinar cells, 6:692
    general characteristics of, 6:689
    pH, 2:813
    stimulation, nutrient utilization and, 5:755

Saliva glands, 2:813
Salivary gland mixed tumor, 2:451
Salivary gland neoplasms, 5:616
Salivary gland viruses. *see* Cytomegaloviruses
Salivary glands
    anatomy of, 6:690
    autonomic innervation of, 1:542, 547-48
    ducts, 6:690
    dysfunction, 6:697-99
    fluid and electrolyte transport in, 6:691-92
    functions, 2:813, 6:697
    infections involving, 6:698
    pharmacologic agents and, 6:698
    physiology, 6:690-93
    protein synthesis and secretion by, 6:691
    and saliva, 6:689-700
    secretion
        neural control of, 6:692-93
        stress and, 7:286-87
Salivary proteins, 6:691, 693-95
    antimicrobial activity of, 6:696
    buffering capacity of, 6:696
    as digestive enzymes, 6:696
    functions of, 6:695-97
    posttranslational processing by, 6:696
    tissue coating of, 6:695-96
    tooth mineralization by, 6:696
Salivation, in digestive process, 3:74
Salivation of dog. *see* Conditioning, classical
Salmonella, 6:701-14
    bacteriophage typing of, 6:704
    biochemical characteristics of, 6:703
    carrier status, 6:713
    characteristics of, 6:701-2
    classification of, 6:702-3
    common, antigenic grouping of, 6:705
    entigenic differentiation of, 6:703
    in foods, 3:664-65, in foods, 6:711
    habitat of, 6:705-6
    negatively stained micrograph of, 6:705
    pathogenicity
        for humans, 6:706-12
        mechanisms of, 6:706-8
            scheme showing, 6:707
    physiological properties of, 6:703
    serotypes, most frequently reported, 6:713
Salmonella cholerasuis, 6:711-12
Salmonella enteritidis, 6:710-11
Salmonella infections
    bowel infections, 4:462
    clinical diagnosis of, 6:712

Salmonella infections *(Continued)*
    diarrhea, types of, **6:**707
    diseases, cominant serovars in, **6:**708–9
    enteritis caused by, **4:**450
    enterotoxin and diarrhea, **6:**707
    epidemiology of, **6:**713
    reported cases of, **6:**714
    septicemia, **6:**711–12
Salmonella infections, animal, **6:**712
Salmonella paratyphi, **6:**710
Salmonella typhi
    carrier state of, **6:**713
    infection, typhoid and, **6:**709–10
Salmonella typhimurium, **6:**710–11
Salsolinol, **1:**179–80
Salt depletion, experimental studies of, **6:**716–18
Salt, dietary. *see* Sodium, dietary
Salts, in aqueous two-phase systems, **1:**356
Salty taste
    compounds eliciting, **7:**530
    mechanisms of signal transduction for, **7:**534
Sanders, **5:**382
Sanger sequencing, defined, **3:**877
gamma-Sarcin, **6:**677–78
Sarcocystis species. *see under* Sarcosporidiosis
Sarcolemma, **2:**129–30, **5:**229
    calcium transport mechanisms in, **5:**237
    defined, **5:**174, **7:**24, 191
    electrical properties of, **5:**207
    structures, **2:**130
Sarcomeres, **2:**129–30, **5:**230
    anatomy, **5:**174–75, 208
    bands, **2:**126, 127
    cardiac contractility and, **3:**518
    differential movement of, **5:**182
    fine structure of, **5:**231
    Frank-Starling mechanism and, **3:**519
    in myofibril organization, **5:**169
    skeletal muscle, **7:**24–26
Sarcoplasm
    anatomy, **5:**208–9
    defined, **5:**174
Sarcoplasmic reticulum, **2:**97–98, 125, 127, 129, **5:**230
    anatomy, **5:**209
    calcium pump, **4:**563, **5:**236
    defined, **5:**161, 230
    in muscle contraction, **5:**225
Sarcoptes scabiei, **2:**390
Sarcosporidiosis, **6:**278–79
Satellite cells
    defined, **5:**191
    myotube formation and, **5:**163

Satiation
    compared with satiety, **6:**724
    disorder of, **3:**686–87
    experimental strategies for investigating, **6:**728
    food characteristics and, **6:**723–33
    food habits, eating patterns and, **6:**723–24
    hunger and, **6:**724–25
    measurements of, **6:**726–28
    obesity and, **3:**684–85
    physiological mechanisms mediating, **6:**7731
    sensory factors and, **6:**730
Satiety
    cholecystokinin and, **5:**718
    cholecystokinin 8 and, **5:**731
    defined, **3:**589
    fat and, **3:**576
    obesity and, **3:**591
    palatability and, **3:**641
    physiological events associated with, **6:**731–32
    processes identified with, **6:**725–26
    thirst and, **7:**475
Saturation, **2:**575
    defined, **2:**575
    variation in, **2:**578
Savory taste, compounds eliciting, **7:**530
Saxitoxin, sources of, **3:**673
Scabby wheat poisoning, **7:**666
Scabies, **2:**390
Scalds, as thermal injuries, **7:**447
Scanning electron microscopy. *see* Microscopy, electron, scanning
Scapula, **7:**37–38
    muscles of, **4:**264
Scars. *see* Cicatrix
Scatchard equation for receptor-binding, **2:**375
Scatchard plot, explained, **2:**302
Scavenging, by early hominids, **2:**196, **3:**500
Schedule for Affective Disorders and Schizophrenia, **1:**611
Schema congruity theory of emotions, **1:**105
Schema-triggered affect, **1:**106
Schiff's base, defined, **3:**437
Schistosoma, **1:**263
Schistosoma haematobium, **2:**389
Schistosoma japonicum, **2:**389
Schistosoma mansoni, **2:**389
    antigenic disguises of, **5:**649
    protease expression by, **5:**647
Schistosoma mekongi, **2:**389
Schistosomiasis, **2:**389
    control of, **4:**448

Schistosomiasis (*Continued*)
  distribution, 4:447–48
  immune responses to, 4:389
Schizocortex, 2:681
Schizophrenia, 2:9, 218, 219, 222, 832, 871, 877, 6:769–78
  acute, treatment of, 6:775–76
  autism and, 1:507
  behavioral family management of, 6:765
  biological theories of, 6:777–78
  cannabis and, 5:445
  case management as factor in treatment of, 6:757–58
  central nervous system transmitters in, 5:411
  characteristics of, 4:997, 6:755–56
  characterization, 4:997
  classification of, 6:772–73
  clinical diagnosis of, 6:770–71
  clinical features of, 6:769–72
  community therapy for, 6:760
  and confused lateralization, 2:865
  course and long-term outcome of, 6:771–72
  defined, 6:769
  dopamine and, 2:9, 218, 222, 5:672
  dopaminergic neuron and, 2:9
  drug therapy, 6:774–77
  early description and terminology of, 6:769–70
  epidemiology of, 6:773
  family involvement in treatment, 6:764–66
  genetic aspects of, 3:841, 843, 6:773
  group therapy, 6:761–64
  individual treatment for, 6:757–59
  kidney dialysis and, 1:410
  left-hemispheric dysfunction and, 2:865
  long-term treatment of, 6:776
  metapsychological theories of, 6:773–74
  neuroleptic therapy, 2:832
  neurotransmitter dysfunction as cause of, 6:777
  outreach programs, 6:758–59
  parental influences and, 6:774
  and Parkinsonian syndrome, 2:219
  perceptual disturbances in, 6:771
  phencyclidine and, 5:447
  psychomotor disturbances in, 6:771
  psychosocial and pharmacological treatment, 6:766
  psychosocial interventions, 6:757
  psychosocial treatment, 6:755–67
  psychotherapy, 6:757
  psychotic symptoms in, 6:770
  relapse rates, 6:765
  simple, defined, 6:772
  social learning-token economy therapy, 6:760

Schizophrenia (*Continued*)
  social skills training, 6:762–64
  social support and case management, 6:758
  stress-vulnerability-coping skills model of, 6:764
  therapeutic communities for, 6:760–61
  thought disturbances in, 6:770
  Training in Community Living Program, 6:758–59
  treatment of, 6:774–77
  treatment-related side effects, 6:776
Schizophrenia, borderline. *see* Schizotypal personality disorder
Schizophrenia, catatonic, defined, 6:772
Schizophrenia, chronic
  community therapy for, 6:759–60
  treatment of, 6:775–76
Schizophrenia, disorganized, defined, 6:772
Schizophrenia, hebephrenic. *see* Schizophrenia, disorganized
Schizophrenia, paranoid
  amphetamine psychosis and, 5:440
  defined, 6:772
Schizophrenic disorders. *see* Schizophrenia
Schizosaccharomyces pombe, 5:901
Schizotypal personality disorder, defined, 6:772
Schlemm, canal of. *see* Canal of Schlemm
Scholastic aptitude test (SAT), 6:859
Schumuffner's dots, 4:875
Schwann cells
  defined, 5:333
  glial cells and, 5:342
  laminin and, 5:345
  nerve growth factor and, 5:335–36, 338
Sciatic nerve
  implantation of, 5:339
  regeneration of, 5:341
Sciatic nerve, rat, 3:745–46
Scintigraphy, 1:449
Scintigraphy, computed tomographic. *see* Tomography, emission-computed
Scintillation proximity assay, 5:838
Sclera, 3:536, 550
Scleral foramina, 3:536
Scleroderma, 2:574, 5:615
  antigens associated with, 1:337
  autoantibodies associated with, 1:337
  characteristics of, 1:335–38
Scoliosis
  in children, 5:709
  muscular dystrophy and, 5:192
Scopolamine, 2:149

Scorbutus. *see* Scurvy
Scorpion venom, **7:**570, 596
Scripts for social behavior, **1:**8–9, 1058
Scrotum
    anatomy of, **6:**560
    edema of, **5:**41
Scurvy, **1:**428, 431, **2:**648, **4:**37, 431
    in early industrial cities, **3:**21
    history of, **1:**428
    and vitamin C, **2:**648
Sea urchins, **6:**506–7
Seasonal affective disorder, **2:**500
Sebaceous glands, **7:**45–46
Sebum, **3:**903, **7:**46
Second messenger systems, **2:**321. *see also* Messengers,
    intracellular; Signal transduction
    of cyclic GMP, **3:**375
    defined, **5:**725
    of hormones, **3:**331–33
    for nerve growth factor, **5:**336
    olfactory, **5:**526
Secondary hormonal hypersecretion. *see* Hormones, ab-
    normal secretion of
Secretagogue, defined, **5:**725
Secretin, **2:**322
    discovery of, **5:**713
    function of, **5:**718, **6:**89
    pancreas and, **3:**78
Secretions, defined, **3:**71
Secretogranin II, in adrenal medulla, **1:**642
Secretory enzymes. *see* Enzymes, secretory
Secretory granules. *see* Cytoplasmic granules
Secretory IgA. *see* IgA, secretory
Sedimentation coefficients, **7:**715–16
Sedimentation equilibrium
    density gradient, **7:**719–20
    experiments
        high-speed, **7:**718–19
        low-speed, **7:**718
    ultracentrifugation, **7:**718–21
Sedimentation velocity, **7:**713, 715–18
Seed/soil hypothesis, **4:**1014–15
Seeds
    of dicotyledonous plants, **3:**39
    mucilage from, **3:**41
Segregation, genetic. *see* Genetic segregation
Seizures, **2:**345. *see also* Convulsions; Epilepsy
    in children, **5:**709
    defined, **3:**455
    in fragile X syndrome, **3:**716

Seizures *(Continued)*
    generalized
        in Lennox-Gastaut syndrome, **3:**458
        tonic-clonic seizures, on awakening, **3:**459
    neocortex and, **5:**314
    subcortical mechanisms of generation, **6:**779–88
Selection, **2:**872
Selection, follicular, defined, **3:**627
Selection (genetics)
    and adaption, **2:**872
    defined, **3:**85, 881
    in design of brain, **6:**494
    genetic variation and, **6:**110–11
    inbreeding and, **4:**440
    point mutations and, **3:**882–83
    reasoning and, **6:**493–503
    role of, in reasoning, **6:**500
Selectivity index, **2:**371, 395
Selegiline, **5:**672
Selenium
    biochemical functions of, **6:**792–94
    biological function of, **7:**607
    blood concentrations of, **6:**794
    and cancer, **6:**797–98
    content, or major food classes, **6:**791
    daily intake of, estimated, **6:**791
    deficiency, **7:**611
        consequences of, **6:**795–96
    dietary
        allowances for, **6:**797
        core foods for, **6:**791
        factors affecting bioavailability of, **6:**791–92
    glutathione peroxidase deficiency and, **3:**916
    in human diets, **6:**790–91
    imbalance, and immunity, **5:**514
    metabolism of, **6:**791–92
    in nutrition, **6:**789–800
    "selenophilia" and, **3:**23
    significance in body, **5:**49
    status of humans, evaluating, **6:**794–95
    toxicity, **6:**798, **7:**611
Selenosis, **6:**798
Self. *see also* Ego
    schema, and depression, **2:**855–56
    and speech development, **2:**857
Self-concept, **2:**856, **4:**704–5, **6:**801–2. *see also* Identity
    crises
    during adolescence, **1:**67–68
    recognition, **2:**853
    sense of, **6:**803–4

Self-concept *(Continued)*
theory of, **2:**856
understanding, **2:**853
Self-determination, for older adults, **1:**136
Self development. *see* Personality development
Self-disclosure, and interpersonal attraction, **4:**544
Self-esteem. *see* Self-concept
Self-fulfilling prophecies
person perception and, **5:**786
stereotypes and prejudice and, **7:**238
Self-image. *see* Self-concept
Self-interest motivations, **7:**235
Self-monitoring, **6:**801-8
advertising and, **6:**807
attitude-action relationship and, **6:**803
consumer behavior and, **6:**807
development of, **6:**804-5
friendships and, **6:**805
high, characteristics of, **6:**802
jobs and careers and, **6:**806-7
low, characteristics of, **6:**802
romantic relationships and, **6:**805-6
sense of self and, **6:**803-4
situational specificity and conformity and, **6:**802-3
Self-nonself discrimination, **1:**571-72, **4:**324
Self-perception. *see* Self-concept
Self-regulation, cybernetic theory of, **1:**107
Self-report tests, **1:**609-10. *see also* Specific tests
for abnormal behavior or psychopathology, **1:**610
types of, **1:**610
Self-sacrifice
altruism and, **1:**199
logic of inclusive fitness, **1:**204
Self-tolerance, immunological, **4:**837
Semantic memory, **4:**673, 676
Semen
characteristics of, **6:**178-79
constituents of, **3:**600
emission of, **7:**174
sperm number in, **7:**174
Semiconservative replication. *see* DNA replication, semiconservative
Seminal vesicles
anatomy of, **6:**561
autonomic innervation of, **1:**543
function of, **6:**178-79
Seminiferous tubules, **7:**419-32
anatomy of, **6:**560
diagram of, **7:**422
epithelium
cycle of, **7:**428-29
germ cell associations of, **7:**430

Seminiferous tubules *(Continued)*
fluid, characteristics of, **7:**423
peritubular cells of, **7:**419-21
Sertoli cells of, **7:**422-32
of testes, **7:**419-20
Semistarvation. *see* Starvation
Semitendinosus muscle, cat, **5:**182
Sen, Amartya, **3:**562-63
Senate Select Committee on Nutrition and Human Needs,
U.S., **3:**651
Sendai virus, **2:**481
Senescence. *see also* Aging
defined, **3:**891
Senescence, cellular. *see* Cell survival
Senile dementia of the Alzheimer's type. *see* Alzheimer's
disease
Senile osteoporosis. *see* Osteoporosis
Sensation
in feeding behavior research, **3:**596
in food acceptance, **3:**643-45
motoneuron activation and, **5:**179
muscle activation and, **3:**699
Sensation seeking
biological basis for, **6:**812-15
effect of environmental factors on, **6:**816
heart rate during, **6:**812
psychobiological model for, **6:**815
psychopathology of, **6:**811-12
sensation seeking scale, **6:**809-12
and stimulus intensity, **6:**812-13
theory and train description, **6:**809-10
trait for, **6:**809-17
Sensemilla, **4:**905, 909
Sensitization, **2:**306
defined, **2:**306
histocompatibility antigens, **4:**360-61
Sensory activity, in infancy, **6:**819-27
Sensory dysphasia. *see* Dysphasia, sensory
Sensory functions, and nutrition, **5:**756-57
Sensory-motor development
Baldwin's theory of, **6:**825
course of, data on, **6:**826-27
courses and processes of newborn, **6:**824-27
Gibson's theory of, **6:**825-26
Piagetian theory of, **6:**825
theories of newborn, **6:**825-27
Sensory receptors. *see* Receptors, sensory
Sensory system, **2:**6-7, **7:**119-26
basic components of, **7:**120
food intake and, **5:**753-55

Sensory system *(Continued)*
information processing in, **7**:119–20
peripheral pathways in, **7**:120–22
receptors in, **7**:120–22
spinal cord dysfunction and, **7**:231
Sentence comprehension
in older adults, **1**:120
parsing, and reading comprehension, **6**:488–89
Sentence production, in older adults, **1**:121
Sentinel phenotypes. *see* Phenotypes, sentinel
Sepsis
hospital-related, history of, **4**:914
lung injury from, **6**:352
Sepsis neonatorum, **4**:458–59
Septicemia, **6**:711–12
Septo-hippocampal test model. *see under* Models, nerve
growth factor
Septum, paleocortical, **2**:680
Sequence motifs (genetics), **3**:151
Sequence-specific methylation
defined, **3**:151
DNA-protein interactions and, **3**:158–59
gene activity and, **3**:160
gene expression and, **3**:153
gene inactivation and, **3**:156–59
inactivating effect of, **3**:152
late E2A promoter inactivation and, **3**:157
restriction endonucleases and, **3**:153
Sequence-tagged site, DNA, **2**:486
and physical map of genome, **2**:486
of restriction, **2**:486
Serial homology. *see* Homology, serial
Serine, **2**:10
beta-tubulin and, **5**:898
hydroxymethyltransferase, **2**:534
MAP phosphorylation and, **5**:898
in nuclear pore glycoproteins, **5**:454
phosphorylation of, **3**:926, 928, 929
Serine esterases, **4**:816
Serine proteases, **3**:798
Serine-threonine protein kinases, **5**:896
Serotonin, **2**:5, 18, 22–25, 33, 34, 218, 317, 323, 326, 337,
340, 368, 369, 494, 534, 686, 716, 834, 843
in aging, **6**:834–35
anterior pituitary hormones and, **6**:833–34
in anxiety, **6**:832
in autonomic neurons, **1**:540
and blood platelets, **2**:845, 846
in carcinoid syndrome, **3**:769
carcinoids and, **3**:769

Serotonin *(Continued)*
cardiovascular function and, **6**:833
central nervous system stimulants and, **6**:832–33
conversion to melatonin, **6**:3
defined, **2**:321
in depression, **2**:9
in depressive disorders, **6**:832
distribution of, **6**:829–30
in eating disorders, **6**:832
in hallucinogen abuse, **6**:832
and 5-hydroxyindoleacetic acid, **2**:845
levels in Huntington's disease, **4**:271
in limbic motor system, **4**:712
metabolites, and relationship to anxiety and depression,
1:612–13
migraine and, **6**:833
in the nervous system, **6**:829–35
in nervous system development, **6**:834–35
neurons and, **2**:337
in nociception, **6**:831
in obesity, **6**:832
in pain, **6**:831
in personality disorders, **5**:781
in pineal body, **6**:3
in postprandial intestinal hyperemia, **4**:555–56
receptor subtypes
neurochemical profile of, **6**:830
pharmacological profile of, **6**:830
physiological functions, **6**:831
signal transduction mechanisms of, **6**:830–31
role in diseases, **6**:831–33
in sensation seeking, **6**:816
in sleep, **6**:831–32
sleep-wake cycle and, **3**:194
in spinal cord function, **7**:205
in transgenic-CuZnSOD mice, **3**:188
Serotonin receptors. *see* Receptors, serotonin
Serovars, salmonella, **6**:702–3
Sertoli cells, **6**:839
secreted products of, **7**:423–25
secretion of sulfated glycoprotein-2, **7**:425
testis, **7**:420, 422–27
Serum sickness, **4**:410
Sesamoid cartilages, laryngeal, **4**:655–56
Set point, temperature, **4**:793
Set point theory of weight control, obesity and, **3**:248
Severe combined immunodeficiency, **4**:837
bone marrow transplantation for, **1**:793–94
gene therapy for, **1**:797
Sewage, environmental damage of, **3**:253

Sex, **2:**8
bisexual potential animals, **4:**930
fetal development and, **6:**881–82
longevity and, **4:**784
menopause and, **6:**888–89
menstrual cycle and, **6:**886
and Mullerian inhibiting substance, **2:**246
origin and uniqueness of, **6:**877–80
palaeoanthropology and, **6:**877–78
stages critical to development, **6:**842
Sex behavior, **2:**340, **4:**934–36, **6:**865–74. *see also* Sexual
response
aging and, **6:**868–69
anthropological and evolutionary perspectives,
6:877–89
attitudes, and self-monitors, **6:**805–6
biological dimensions of, **6:**880–82
cohabitation and, **6:**872
cultural patterns and, **6:**882–83
extramarital, **6:**872, 885–86
fantasy as pattern of, **6:**870
female, hormonal influences, **4:**244–46
incest and, **6:**886–87
life cycle and, **6:**871–72
male, hormonal influences, **4:**246
marriage and, **6:**872, 885
masturbation as pattern of, **6:**870
nocturnal penile tumescence, **4:**936
in older adults, **1:**136
olfaction and, **5:**529
in oral contraceptive studies, **5:**607
orgasm, **4:**936
patterns of, **6:**869–71
patterns of growth and, **6:**882
petting and foreplay as patterns of, **6:**870
premarital, **6:**871–72
reproduction and, **6:**887–88
sexual climax, **4:**932–33
sexual desire, **6:**865–66
sexual dysfunctions and, **6:**873–74
sexual excitement, **6:**866
sexual orientation and, **6:**868–69
sexual positions as patterns of, **6:**870–71
sexual response and, **6:**888–89
within social context, **6:**882–89
societal aspects, **4:**84
stages of arousal, **4:**936 .
of young
initiation rituals and, **6:**884
rules of behavior in, **6:**884–85
socialization and sanctions over, **6:**883–85

Sex behavior, animal, **4:**929–37
comparative aspects, **4:**930–32
exteroceptive factors, **4:**930–31
great apes, **4:**932–33
mammal matings, **4:**931–32
Old World monkeys, **4:**932–35
olfaction and, **5:**528–29
postural reflexes, **4:**931–32
primate
body size and, **6:**879
mate selection and, **6:**878
sexual receptivity and, **6:**879
Sex characteristics
anthropological studies of, **6:**878
biocultural, **6:**845–53
animal insights into, **6:**846
behavioral patterns and, **6:**850–51
body composition and, **6:**849–50
brain anatomy and, **6:**851
cultural variation and, **6:**850
environmental variation and, **6:**850
hormonal effects and, **6:**851
individual variability and, **6:**852
mortality and, **6:**851–52
in prehistoric humans, **6:**848–49
primates and, **6:**846–48
sex ratio and, **6:**851–52
biological, **6:**837–43
defined, **3:**85
dermatoglyphic, **3:**90
in dietary protein needs, **6:**234
future of, **6:**862–63
and gender classification, **6:**883
hormonal effects, **4:**242–46
in motivation for sexual activity, **6:**871
orientation, variations on, **6:**869
psychological, **6:**855–63
in children, **6:**856–57
general characteristics of, **6:**855–56
and interest differences, **6:**857–58
and personality differences, **6:**858–60
response to observed violence, **1:**3–4, 1058
sexual orientation and, **6:**860–61
in psychopathology, **6:**861–62
in puberty, **3:**353–55
role of sex and gender classification in, **6:**883
Sex chromosomes
abnormal interchanges involving, **6:**287–88
ancestry of pseudoautosomal sequences of, **6:**291–92
boundaries of pseudoautosomal region of, **6:**290–91

Sex chromosomes *(Continued)*
conservation of pseudoautosomal sequences of, **6**:291
defined, **3**:847
expression of pseudoautosomal genes of, **6**:288–90
genetic properties of pseudoautosomal region of, **6**:285–87
inactivation of, **6**:289–90
numerical aberrations in, **3**:821, 840
physical and genetic maps of pseudoautosomal region of, **6**:286–87
pseudoautosomal region of, **6**:283–92
pseudoautosomal sequences of, **6**:284–85
recombination in pseudoautosomal region of, **6**:284–85
Sex determination
gene action, **3**:340
human, during embryogenesis, **2**:246
Sex differences. *see* Sex characteristics
Sex differentiation
genetic differences, **6**:837–39
hormonal aspects of, **3**:340–42
hormonal effects, **4**:242–46
Sex differentiation disorders, **3**:353
Sex dimorphism. *see* Sex characteristics
Sex disorders. *see also* Sex behavior
characterization, **4**:998
therapies for, **6**:873
Sex factors
bacterial. *see* F factor
in death rates, **5**:100–101
in high density lipoprotein level, **3**:574
in mutational risk, **5**:254
Sex hormones
alcohol consumption and, **1**:174
bone cell function and, **1**:765
functions of, **3**:320
lymphocyte proliferation and, **5**:507
male, effect of alcohol consumption on, **1**:174
mammogenesis, **4**:604
prepubescent growth and, **3**:350, 351
sensation seeking and, **6**:814
steroids, **2**:247
Sex linkage, partial. *see* Pseudoautosomal inheritance
Sex-linked diseases, **3**:146–48
Sex-linked dominant inheritance. *see* X-linked inheritance
Sex maturation, in adolescence, **1**:65, **3**:356
Sex ratio
in anorexia nervosa, **3**:246
in bone loss, **3**:904
in bulimia nervosa, **3**:247
defined, **3**:305

Sex ratio *(Continued)*
in eating disorders, **3**:246
in fragile X syndrome, **3**:709
Sexes
cognitive differences between, **6**:858–60
dimorphism, in childhood, **6**:856–57
future behavior of, sex differences and, **6**:862–63
general intelligence of, tests to determine, **6**:858
orientation
disorders affecting, **6**:860–61
factors contributing to, **6**:860–61
scholastic achievement of, tests to determine, **6**:858–60
spatial ability differences among, **6**:859
specific abilities of, tests to determine, **6**:858–60
Sexual arousal, **5**:808, 809
Sexual dimorphism. *see* Sex characteristics
Sexual intercourse, **6**:880–81
Sexual partners, multiple
as cancer risk factor, **3**:859
as papillomavirus infection risk factor, **3**:866
Sexual response
of adult female, **6**:865–67
of adult male, **6**:865–67
cultural influences on, **6**:868
desire and, **6**:865–66
erotic stimuli and, **6**:868
excitement and, **6**:866
mechanisms of, **6**:867–68
myotonia as mechanism of, **6**:867
neural mechanisms associated with, **6**:867
orgasm and, **6**:866
Pavlovian conditioning and, **6**:867–68
plateau phase of, **6**:866
reflex mechanisms of, **6**:867
resolution and, **6**:866–67
social influences on, **6**:868
vasocongestion as mechanism of, **6**:867
Sexually transmitted diseases, **4**:450, **6**:891–903
bacterial infections and, **6**:896–99
clinical care for, **6**:900–901
control of, **6**:899–903
governmental responsibilities for, **6**:892
resources for, **6**:892
current importance of, **6**:891–92
detection of bacterial pathogens in, **1**:558–59
effect of quinolones on, **6**:409–10
facilities providing services, **6**:893
as factor in prostate cancer, **6**:188
future directions in, **6**:903
health education messages on, **6**:899–900

Sexually transmitted diseases (*Continued*)
  intervention
      early detection in, **6:**901
      partner notification in, **6:**902-3
      patient counseling in, **6:**902
      treatment in, **6:**901-2
  oncogenic, **4:**343
  population at risk of, **6:**892
  prevention of, **6:**899-903
  public interest in, **6:**893
  trends in, **6:**893-99
  trichomoniasis, **1:**268-69
Shale oil, **2:**112
Sham feeding, **2:**35-39
Shape-space homeostasis, immune system, **4:**325-26
Sharpey's fiber, **2:**776, 777, 779, 824-27
Sheer waves, **7:**725-26
  in tissue, **7:**725-26
Shelf life of food. *see* Food preservation
Shellfish toxins, **3:**673, **7:**590-91
Shift work, **2:**499
Shifts, chromosomal. *see* Insertional translocations,
      chromosomal
Shigella infections, **4:**462
Shikimate pathway, **2:**544
Shinbone, **7:**39
Shingles
  mouth, **5:**613
  as orofacial pain condition, **5:**369
Shock
  and cardiac failure, **2:**145
  kinins and, **4:**597-98
  lactic acidosis in, **5:**826
  and renal failure, **2:**145
  vascular, systemic, **4:**410-11
Shock, anaphylactic. *see* Anaphylaxis
Shock, Nathan, **3:**897
Shock, septic, **5:**4
Short, **3:**630
Short-latency reflex. *see* Reflex, stretch
Short-term memory, **4:**673, 676-77
Shoshin-Kakke. *see* Beriberi, cardiac
Shoulder, **2:**597
Shoveling tooth, **2:**789, 792, 793, 797
Shunting inhibition
  defined, **5:**389
  in ganglion cells, **5:**389
Sialic acids
  defined, **3:**725
  ganglioside degradation and, **3:**732
  named, **3:**725
  structures of, **3:**726

Sialoglyclolipids, **3:**738
Sialomucins, **3:**770
Sialylparagloboside, **3:**734
Sialytransferase, **3:**600
Siamese twins. *see* Twins, conjoined
Sib correction method, in genetic segregation, **3:**853-54
Sicca syndrome, **6:**697
Sickle-cell anemia. *see* Anemia, sickle-cell
Sickle-cell hemoglobin. *see* Hemoglobin, sickle
Sickle cell trait, and malaria, **4:**881-82
Side effects, defined, **3:**93
Sideroblast, **2:**654
Sigmoid, anatomy, **3:**82
Sigmoid sinus, **4:**983
Signal peptidase, **4:**976
Signal peptides
  defined, **5:**725
  functions of, **5:**726
Signal recognition particle, **6:**247-48
Signal recognition particle cycle, **6:**239
Signal sequences, peptides. *see* Signal peptides
Signal transduction, **2:**227, 321-31, 230. *see also* Cell
      communication; Second messenger systems
  aberrant, **2:**329
  adaptation, **2:**328
  amplification, **2:**328
  amplification effect and, **4:**15
  in cancer, **2:**328
  cell-cell, **2:**322-23
  cell surface, **7:**15-21
      multicomponent systems and, **7:**17-19
      self-transducing system as type of, **7:**16-17
  and channel opening, **2:**327
  by contact, **2:**322
  crosstalk, **2:**329-30
  defined, **4:**13
  in diabetes, **2:**329
  by drug, **2:**330
  and enzyme activation, **2:**327
  and enzyme inhibition, **2:**327
  features, basic, tabulated, **2:**327
  by gap junction, **2:**322
  by growth factor, **2:**324
  by hormone, **2:**323-24
  intracellular, **2:**321, 326-27
  and ion flux, **2:**92
  of keratinocytes, **3:**452
  by local mediator, **2:**325-26
  mechanisms, **4:**975
  modification by drug, **2:**330

Signal transduction (*Continued*)
  by molecules, **2:**323–25, 328
  by neurotransmitter, **2:**323
  pathways, **2:**92–93
  principles for, **2:**91–92, 327
  in psoriasis, **2:**329
  receptor, **2:**325–27
  sensitivity, **2:**328
  transient, **2:**328
  transmembrane, **2:**321
Signaling, chemical, **2:**273
  autocrine, **2:**324
  endocrine, **2:**324
  humoral, **2:**273
  paracrine, **2:**324
  by receptor, **2:**273
  by transduction, **2:**326–27
  by transmitter, **2:**273
Significance, statistical, **2:**67
Silencers (genetics), defined, **3:**112
Silica, toxicity of, **7:**555
Silicon
  biological function of, **7:**607
  deficiency, **7:**611
  toxicity, **7:**611
Silicone, in gas chromatography, **3:**753
Silver
  in photoelectron microscopy, **5:**907
  stains
    color effects of, in protein detection, **6:**195–96
    for protein detection, **6:**194–96
    sensitivity of protein detection with, **6:**196
Simian B virus, in humans, **4:**188
Simian EBV-like viruses, **3:**469
Simian immunodeficiency virus, **5:**677
Simian immunodeficiency virus vaccine, **6:**647
Simian sarcoma virus proteins. *see* Antigens, polyomavirus
    transforming
Simian sarcoma viruses, **5:**874
Simian virus. *see* SV40
Similarity, interpersonal attraction, **4:**541
Similarity principle, **2:**630
Simple inheritance
  defined, **3:**819
  of diseases, **3:**819–21
Simple repetitive sequences. *see* Repetitive sequences,
    nucleic acid
Simulated annealing, in neural networks, **5:**391
Simulium damnosum, **1:**265
Simvastatin, **2:**168

SINES, **4:**252, 255
Singing
  laryngeal functions in, **4:**661–62
  voice and, **6:**590
Single cell protein. *see* Proteins, single cell
Single-channel model. *see* Models, single-channel
Single-gene defects, **1:**698–99
Single-ovum twins. *see* Twins, monozygotic
Single-stranded DNA binding proteins. *see* DNA-binding
    proteins
Sinoatrial node, **2:**136, 183, 186
  in aerobic training, **3:**508–9
  regulation of heart activity and, **3:**517, **5:**240
Sinomenium acutum, **1:**179
Sinuses, **2:**179, 182, 704, 706
  coronary, **2:**179, 182
  dural venous, **4:**982–84
  paired, **4:**983–84
  posterior intercavernous, **4:**984
  sagittal
    inferior, **4:**983
    superior, **4:**982
  transverse, **4:**983
  unpaired, **4:**982–83
  venous, **2:**182
    blood flow in, **4:**984
Sinusitis, headaches caused by, **4:**55
Sinusoidal grating, **7:**802
SIS genes. *see* Oncogenes
Sister chromatid exchange. *see also* Chromatids
  in Bloom syndrome, **5:**251
  defined, **3:**101, **5:**68
  in DNA amplification, **3:**104–5
  mutagens and, **5:**250
Sivapithecus, fossils of, **3:**490
Size principle, motor unit
  defined, **3:**694, **5:**173
  type S motor units and, **5:**179
Sjögren's disease, **1:**332–35, 530, **2:**813
  and autoimmunity, **1:**530
  and dental caries, **2:**813
  characteristics of, **1:**332–35
Skeletal age, **4:**27
Skeletal age measurement. *see* Age determination by
    skeleton
Skeletal muscles. *see* Muscle, skeletal
Skeleto-muscular activity. *see* Motor activity
Skeleton, **7:**33–39
  aging of, **3:**904
  anatomy of, **7:**33–39

Skeleton *(Continued)*
    appendicular, **7:**37–39
    axial, **7:**34–37
    functions of, **7:**33
    movements of, **5:**133–35
    neoplasms of, **7:**698–703. *see also* Bone neoplasms
    subdivisions of, **1:**772
Skin, **2:**641, 645, **4:**187, **7:**41–49
    afferents, reflexes involving, **6:**159–60
    aging of, **3:**903, **7:**48–49
    anaerobic bacteria found on, **1:**239
    autonomic innervation of, **1:**543
    dermis and subcutaneous tissue of, **7:**46–48
    development of, **7:**41–42, 42
    dielectric properties of, **3:**13
    diverse properties, **4:**392
    fetal, at third month, **3:**611
    of fingertip
        schematic vertical section through, **7:**53
        tactile units on, **7:**52
    functions of, **7:**42–43
    general features of, **7:**41–42
    hair, **7:**45
    light absorption, **5:**593
    light penetration and interactions in, **5:**593
    light scattering in, **5:**590, 594
    maps, in somatosensory cortex, **5:**388
    as means of photoprotection, **7:**68–69
    melanin content
        Caucasian, **5:**594
        Negroid, **5:**594
    nerve supply, **7:**48
    optical properties of, **5:**593
    photoprotection by, **7:**68–69
    photorepair of, **7:**68–69
    platelet-activating factor and, **6:**51
    proprioceptors, **6:**172–73
    radiation effects, **6:**413, 448
        factors controlling, **6:**463–64
        late radiation effects, **6:**465–66
    radiobiology of, **6:**463–64
    radioprotectors for, **6:**465
    radiosensitivity of, **6:**463–67
    reflexes, and spinal cord dysfunction, **7:**230–31
    ridged, histology of, **3:**87
    sebaceous glands of, **7:**45–46
    section, schematic of, **7:**42
    structure of, **7:**43–48, 46–48
    sweat glands, **7:**44–45
    thermal injuries and, **7:**448–49
Skin *(Continued)*
    trichothecenes and, **7:**663
    trigeminal system innervation of, **7:**673–75
    ultraviolet radiation and, **7:**65–69
    vitamin D synthesis in, **7:**861–62
Skin diseases, **7:**47–49
    associated with autoimmunity, **1:**529
    in children, **5:**710
    elastin-related, **3:**268
    vitamin A treatments for, **7:**851
Skin diseases, infectious, **6:**409
Skin neoplasms, **7:**48–49. *see also* Melanoma
    in immunosuppressed patients, **4:**320–21
    papillomavirus infection and, **5:**637–38
    psoriasis and, **3:**446, 447
    ultraviolet radiation and, **7:**67
Skin pigmentation, **7:**44
    radiation effects, **6:**466–67
    ultraviolet radiation and, **7:**67
Skin sensory system, **7:**51–63
    classes of sensory receptors on, **7:**675
    facial sensory receptors, **7:**677–78
    glabrous, sensory receptors of, **7:**678
    haptic system and, **7:**60–62
    manual exploration recognition and, **7:**60–62
    object motion across, and touch, **7:**59–60
    perception and, **5:**746
        cognitive attributes of, **5:**746
        technical uses of, **5:**746
        transduction and, **5:**745
    physical stimuli and, **5:**745–46
    point localization task and, **7:**54
    sensory aids for disorders involving, **7:**62–63
    sensory inputs to hypothalamus, **4:**306
    somatosensory system and receptors, **7:**120–22
    spatial, **7:**54–55
    tactile character recognition and, **7:**56–58
    temporal, **7:**55–56
    thalamic response to stimuli of, **7:**441
    three-dimensional object recognition and, **7:**60–62
    two-dimensional stimuli and, **7:**60
    two-point test for, **7:**54
Skin temperature
    circulation, as mechanism for regulation of, **1:**754–55
    effect of heat on, **6:**465
    and regulation of body temperature, **1:**757–58
Skinner, B. F., **2:**634
Skinner box, **2:**634
Skipper-Schabel model, **4:**922–23
Skull, **2:**702–7
    anterior view of, **7:**35

Skull (*Continued*)
  bones of, **7**:34
  and face, **2**:705-6
  foramina, **7**:34
  of infant, **2**:707
  inferior view of, **7**:36
  lateral view of, **7**:35
  plain film radiography of, **3**:2
  sinuses of, **6**:588-89
  superior view of, **7**:36
Skylab Medical Experiment Altitude Test, **4**:124
Sleep, **7**:71-76
  bedwetting and, **7**:82
  and biological rhythms, **2**:496-97, **7**:74-75
  and breathing, **7**:73-74
  and chronobiology, **7**:74-75
  delta, defined, **7**:82
  endocrine hormone secretion during, **7**:75
  and fatigue, **2**:496
  and hormones, **2**:497
  non-REM, features of, **7**:71-73
  orthodox, **7**:72
  pattern of, **2**:496
  periodic leg movements during, **7**:79
  serotonin and, **6**:831-32
  structure, **2**:496
  ultradian rhythm, **2**:496
Sleep, REM, **1**:133, **2**:339, 491, 496-97, 835
  of animals, **3**:195
  brain physiology and, **3**:194
  of children, **3**:194
  dreaming and, **3**:193-94
  features of, **7**:71-73
  lucid dreams, **3**:196
  phasic events during, **7**:72
  REM and non-REM
      functions of, **7**:73
      physiological characteristics of, **7**:72
Sleep apnea syndromes, **7**:73, 79, 318-19
Sleep disorders, **7**:77-83. *see also* Insomnia
  autism and, **1**:506
  characterization, **4**:998
  hypersomnia as, **7**:79-80
  insomnia as, **7**:77-79
  in older adults, **1**:133-34
  parasomnias as, **7**:81-82
  sleep terror, **7**:81-82
  treatment of, **7**:82-83
Sleep stages, **1**:133, **2**:496
Sleep-wake cycle, **2**:496-97
  ACTH secretion and, **5**:398

Sleep-wake cycle (*Continued*)
  delayed, **7**:81
  disorders of, **7**:80-81
  factors controlling, **7**:74-75
  generalized tonic-clonic seizures and, **3**:459
  neurotransmitters of, **3**:194
  and sleep disorders, **7**:81
  time of waking, **2**:496
  waking up, **2**:500
Sleeping sickness. *see* Trypanosomiasis
Sliding filament theory of isometric tension, **5**:180
Sling, fibrous, **2**:769
Slow-fiber skeletal muscles. *see* Muscle, skeletal, slow-fiber
Slow twitch motor unit, defined, **5**:177
Slow virus diseases, **7**:778-80
Slump famine, **3**:562
Small intestine. *see* Intestine, small
Smallpox vaccine, **4**:444, 464
Smallpox virus, **2**:399
Smell, **5**:523-31. *see also* Olfactory disorders
  food acceptance and, **3**:644
  kidney dialysis and, **1**:410
  neonatal, mammal, **5**:528
  network modeling of, **5**:388
  olfactory information processing, **5**:523-31
  olfactory tissue, biochemistry unique to, **5**:289
  perception and, **5**:744
      chemical stimuli and, **5**:744
      disorders involving, **5**:744
      sensory and cognitive attributes of, **5**:744
      transduction, **5**:744
  role in appetite, **1**:345
  sense of, conditions affecting, **5**:695-96
  sensory inputs to hypothalamus, **4**:306
Smells, complex
  human behavior towards, **5**:759
  instrumental analysis of, **5**:760
  perception of, **5**:759-67
Smith-Birnsteil experiment, **2**:484
Smoke inhalation injury, **5**:711
Smoking, **2**:114, 116, 177
  aging and, **3**:894
  and arteriosclerosis, **7**:517
  blood interferon levels and, **4**:525
  and cancer, **7**:515-16
  and cardiovascular disease, **7**:517-18
  chronic obstructive pulmonary disease and, **7**:516-17
  constituents of, **7**:513
  deaths associated with, **7**:512

Smoking (Continued)
  design and manufacture of cigarettes, 7:512-13
  diseases associated with, 7:512
  hypnosis for stopping, 7:623
  impact on health, 7:511-20
  intrauterine growth retardation and, 3:347
  marijuana, 4:906-8
  maternal, and child health, 7:518-19
  nasal cavity enzymes and, 5:288
  oral contraceptives and, 5:602, 603, 604
  pollution-induced lung diseases and, 3:388
  resolution of addiction without treatment, 4:86
  societal aspects, 4:84
  and weight loss, 7:514
Smoking, passive, 7:516-17
Smooth muscle. see Muscles, smooth
Smooth pursuit eye movements. see under Eye movements
Snail fever, 1:263
Snakes, poisonous, 7:599-600
Social behavior
  conduct and reasoning, 6:501-3
  development
    birth to 3 months, 1:574-75
    3 to 10 months, 1:576
    10 to 20 months, 1:577
    20 to 36 months, 1:578
    3 to 5 years, 1:579
    5 to 8 years, 1:580-81
    8 years to preadolescence, 1:581
    differentiation vs. integration of processes, 1:582
    general stages vs. specific cases, 1:582
    role of shared heredity, 1:582
  dilemma and, 1:575-77
  learning, and treatment of autism, 1:511
  moods and, 1:108-9
  personality disorders and learning theory, 5:780
  role of aggression in, 1:116
  scripts, 1:8-9, 1058
Social categories, 7:89-99
  accumulation of exemplars in formation of, 7:91
  altering consequences of, 7:96-98
    change in social contact and, 7:97-98
    changing category boundaries and, 7:97
    individuation and, 7:98
    overlapping categories and, 7:97
  assumptions about category members, 7:93-94
  characteristics of, 7:89-90
  criteria for membership into, 7:90
  formation of, 7:90-91

Social categories (Continued)
  implications of
    for behavior, 7:93-96
    encoding effects and, 7:94
    memory effects and, 7:94-95
    self-identity and, 7:95-96
  and object categories, differences between, 7:89-90
  patterns of similarity in formation of, 7:91
  prototypes in formation of, 7:91
  selection of, 7:91-93
    category salience in, 7:92
    ease of access in, 7:92
    information value in, 7:92
    motivation in, 7:92
    perceiver's set in, 7:92
    social behavior in, 7:92-93
Social class
  in endometriosis epidemiology, 3:361-62
  overweight status and, 3:684
Social conditions
  as affective response to food, 1:501
  in famines, 3:563-64
  in primate behavioral ecology, 6:132
  support, 4:76
Social conformity
  eating disorders and, 3:245
  food acceptance and, 3:646
  versus individuals, 4:704-5
  obesity and, 3:248
Social criteria, for beginning of human life, 1:656
Social environment
  hominid evolution and, 3:503-4
  persuasion and, 5:806
Social exchange theories, 4:539-40
Social intellect
  defined, 3:495
  hominid evolution and, 3:503-4
Social relativism, 4:703
Socialization, 1:116-17, 2:709. see also Personality development
Societies
  aggression against external threats, 1:117
  ergonomic, memberships of, 3:479
Society for Developmental Neuroscience, 2:870
Society for Neuroscience, 2:870
Socioeconomic status, and longevity, 4:784
Sodium
  absorption
    in colon, 3:82
    in small intestine, 3:81
  as essential element, 5:48

Sodium *(Continued)*
  excretion
    osmotic diuretics and, **3:**100
    urinary, **2:**143, 145
  function, **2:**289
  glucose absorption and, **3:**79
  and local anesthetic, **2:**783–84
  in photoreceptor responses, **6:**619
  renal reabsorption of, **3:**95–96
  replacement, thermal injury and, **7:**452–53
  retention, in essential hypertension, **4:**291–92
  secretion, in adrenal medulla, **1:**651–53
  in vegetables, effect of processing on, **5:**494
Sodium, dietary, **6:**716–18. *see also* Sodium chloride
  appetite and
    clinical reports of, **6:**717–18
    extracellular thirst and, **7:**474
  salt preference
    developmental factors, **6:**719–20
    genetic factors, **6:**718–19
    in humans, **6:**715–21
    physiological determinants, **6:**716–18
    psychological factors, **6:**719
Sodium bicarbonate, pancreatic, **3:**78
Sodium channels, **2:**237, 288–89, 314, 523
  structure, **2:**288–89
  taste cell, **7:**533
  voltage-sensitive channels, **5:**421
Sodium chloride
  atrial natriuretic factor and, **1:**470
  gastric carcinoma and, **3:**772
Sodium metabolism. *see also* Sodium metabolism; Sodium
    transport system
  ion pumps, **4:**567–68
  sodium-calcium exchanger, **2:**98
  sodium ion/hydrogen ion exchange, **5:**874
  sodium potassium pump, **2:**283–85
    function, **2:**284
    mechanism of action, **2:**285
  sodium pump, **2:**175, 283–85
Sodium potassium ATPase, **2:**237
Sodium transport system. *see also* Sodium channels
  atrial natriuretic factor and, **1:**473
  in choroid plexus, **1:**718
  Na^+, glucose symporters, **4:**566
  Na^+, HCO_3/H^+, Cl^- transporter, **4:**567
  Na^+, K^+, 2 Cl^- symporters, **4:**566
  Na^+, K^+ pump, ATP-dependent, **4:**559, 562
  Na^+, serotonin, Cl^-/K^+ transporter, **4:**567
  Na^+/Ca^{2+} antiporters, **4:**564

Sodium transport system *(Continued)*
  Na^+ Cl^-, GABA symporter, **4:**567
  Na^+ glutamate/K^+ transporter, **4:**567
  Na^+/H^+ antiporters, **4:**564–65
  in salivary gland cells, **6:**692
  sodium glucose cotransporter, **2:**186–287
    structure, **2:**286
    topography, diagram of, **2:**286
Sodium valproate, **2:**29
Soft drugs, **7:**101–17
Soil
  cartography, **1:**147
  characteristics of, **1:**145
  chemistry, **1:**147
  classification, **1:**147
  fertility, **1:**148
  function of, **1:**146
  mineralogy, **1:**147
  physics, **1:**147
  preservation of, **1:**146
  restoration of, **1:**146
  technology, **1:**147–48
Soil sciences. *see also* Agronomy
  characteristics of soil and, **1:**145
  disciplines within, **1:**146–48
    soil biology, **1:**147
    soil chemistry, **1:**147
    soil fertility, **1:**148
    soil genesis, classification and cartography, **1:**147
    soil mineralogy, **1:**147
    soil physics, **1:**147
    soil technology, **1:**147–48
Solar flares, **6:**431–32
Solar particle events, **6:**432
Solenoid, **2:**434
Soleus muscle
  gastrocnemius and, **5:**188
  in vertical jump, **5:**187
Soleus muscle, cat
  fiber angulation in, **5:**183
  force-velocity pattern of, **5:**185–86
Solitary lymphoid nodules. *see* Lymphoid nodules, solitary
Solute exchange
  microvascular, **5:**31–42
  physics of, **5:**37–38
  physiology of, **5:**38
Solutions
  concentration in cartilaginous tissue, **1:**367
  lipid, **4:**728–31
  movement through matrix, **1:**367
  partition in cartilaginous tissue, **1:**367

Solvent abuse, **5:**443–44

Solvent exchange, **5:**36–37

Solvent reservoir, HPLC, **4:**209

Somatic mediation, of operantly conditioned autonomic responding, **1:**662–64

Somatoform disorders, **4:**997–98

Somatomedins, **2:**22, 62, 259, **4:**503–11

    biosynthesis in fetal tissues, **4:**504–6

    biosynthesis regulation, **4:**507–8

    in bone and calcium metabolism, **1:**805, 806

    in bone cell differentiation, **1:**789

    characterization, **4:**504

    circulating forms, **4:**507

    fetal growth and, **4:**508–10

    gene expression, **4:**504–6

    receptors, in fetal tissues, **4:**508

    somatotropin and, **5:**733

Somatosensory cortex, **5:**388

Somatosensory system. *see* Sensory system

Somatostatin, **2:**18–22, 28, 34–38, 300, 338, 494

    action mechanisms of, **5:**715

    analogues of, **5:**723

    in autonomic neurons, **1:**541

    defined, **5:**393

    functions of, **5:**720, **6:**89

    gastrin secretion and, **5:**716

    growth hormone and, **5:**399, **6:**14

    insulin and, **5:**732

    intestinal smooth muscle and, **5:**723

    levels in Huntington's disease, **4:**271

    pancreatic tumors and, **5:**722

    purification of, **5:**714

    radioimmunoassay of, **5:**714

    substance P and, **5:**384

Somatostatin receptors. *see* Receptors, somatostatin

Somatotopy, **2:**682

Somatotrophs of anterior pituitary, **6:**13–14

Somatotropin

    function of, **5:**733, **6:**88

    imbalances of, **5:**707

    sources of, **5:**734

Somatotropin-releasing hormone

    defined, **5:**393

    function of, **6:**89

    growth hormone and, **6:**14

    in growth hormone regulation, **5:**399

Somatotypes, and criminality, **2:**713–14

Somites

    differentiation of, **5:**162

    migration of, **1:**786–87

Somitic mesoderm, **5:**162

Sommer's sector, **4:**232

Sonography, **7:**305

Soot, **2:**112

Sorbic acid, in food analysis, **4:**224

Sorbitol, **2:**103, 105

Sotalol, **2:**171, 173

Sound frequency, encoding, **2:**548

Sour taste

    compounds eliciting, **7:**530

    signal transduction for, **7:**535

South Atlantic anomaly, as source of space radiation, **6:**429–30

Southern hybridization, **6:**383–84

Space flight, **2:**10. *see also* Space stations

    biochemistry and physiology of, **7:**143–52

    bone and mineral metabolism during, **7:**146–47

    cardiovascular changes during, **7:**147–48

    endocrine changes in, **7:**144–45

    energy metabolism and nutrition in, **7:**143–45

    hematology and immunology in, **4:**123–28

        anemia, **4:**123–24

        functional capacity of space-born red cells, **4:**128

        magnitude of red cell mass loss, **4:**124–25

        mechanisms of red cell mass deficit, **4:**126–27

        parameters, **4:**127–28

        pattern of red cell mass deficit, **4:**125–26

        recovery, **4:**127–28

        red cell mass detection method, **4:**124

    illnesses and injuries in, **7:**133

    lower body negative pressure, **7:**151

    motion sickness and, **7:**150

    muscle metabolism and physiology during, **7:**145–46

    neurovestibular system and, **7:**149–50

    problems caused by exposure to and return from, **7:**134

    space radiation and, **6:**432–33

    spatial orientation and, **7:**149–50

Space flight health care

    anesthesia control in, **7:**136–37

    fluid containment and collection in, **7:**137–38

    health care systems, **7:**127–42

    new space medicine and, **7:**130

    resources required for, **7:**134–35

    surgery and critical care transport in, **7:**138–39

    systems testing for, **7:**140–41

Space medicine, and exploration of inner solar system, **7:**128–30

Space perception

    binocular motion in depth as factor in, **1:**632

    and binocular vision, **1:**627–40

Space perception (*Continued*)
  characteristics of, **1:**628
  disorders of, **2:**10
  eye movements and, **1:**629
  frame of reference as parameter of 3-D perception in, **1:**629
  global cues as factors in, **1:**630–31
  kinetic cues for direction in, **1:**632
  local analysis of direction and orientation, **1:**631
  looming and optic flow as factors in, **1:**632–33
  motion parallax as factor in, **1:**634
  optical transformation of 3-D to 2-D space in, **1:**628
  orientation sense in, **1:**630
  perceived direction in, **1:**629
  perceived distortion of 3-D space in 2-D projections in, **1:**631–32
  primary visual cortex and, **5:**311
  stimulus covariance as parameter of 3-D perception in, **1:**629
  3-D, parameters of, **1:**628–29
  unconscious inference as parameter of 3-D perception in, **1:**628
Space radiation. *see* Cosmic radiation
Space-related coupling. *see* Coupling, space-related
Space Station Freedom, **7:**130–31
  configuration and dimensions of, **7:**128–29
  environmental health system of, **7:**140
  exercise countermeasure facility in, **7:**139–40
  health care system of, **7:**130–31
  health maintenance facility of, **7:**131–39
Space Station Mir, **7:**131
Space stations
  American and Soviet approaches to health care, **7:**130–31
  crew return vehicles of, **7:**129–30
  health maintenance facility
    clinical laboratory and imaging diagnostics of, **7:**135–36
    costs of being unprepared in, **7:**132–34
    costs of preparation of, **7:**132
    role of telemedicine in, **7:**135
    surgical procedures in, **7:**138–39
  health maintenance facility subsystems, **7:**135
  medical equipment and pharmaceuticals, desirable characteristics of, **7:**136
  mission duration of, **7:**128–29
  presence in earth orbit, **7:**128–29
  spaceflight and, **7:**127–42
Sparteine, **2:**747
Spasmogenicity, **2:**612, 613

Spasticity. *see also* Muscle spasticity
  as spinal cord disorder, **7:**219–20, 230
Spatial coding by neuron, **2:**587–88
  receptive field, **2:**587, 588
  and retina, **2:**588
  and visual neuron, **2:**588
Spatial filters
  sensitivity of, three-dimensional and contour plots of, **7:**818
  in vision psychophysics, **7:**817–19
Spatial sampling, and peripheral vision, **7:**819–21
Specialization, cerebral. *see* Cerebral specialization
Spectinomycin, **1:**290(t)
Spectrin, **2:**41–51, 317
  in axon, **2:**45
  in brain, **2:**41–51
  defined, **2:**41
  in early experiments, **2:**42
  in erythrocytes, **2:**41, 47
  function, **2:**48–50
  isoforms, **2:**46–48
  location at synapse, **2:**46
  in membrane skeleton, **2:**41
  model of molecule, **2:**44
  myofibril organization and, **5:**169
  in neural process, **2:**42
  in non-erythrocytes, **2:**42
  protein interactions, **2:**46–48
  in red cell membrane structure, **6:**544–45
  structure, **2:**43–44
  subunits, **2:**43–44
Spectrometry, fluorescence, time resolved, **5:**578
Spectrometry, photoelectron, **5:**904
Spectroscopy, photoelectron
  defined, **5:**903
  photoelectron imaging, defined, **5:**903
  UV, **5:**904
  X-ray, **5:**904
Spectrum analysis, mass
  complex smell analysis using, **5:**761
  coupled to HPLC, **4:**212
  drug analysis using, **5:**839
Speech, **4:**633–35
  as biological process, **7:**154
  development of, **2:**857, **7:**154–55
  fluent, word processing of, **5:**741
  as form of communication, **7:**154
  formant frequency encoding, **4:**634–35
  laryngeal functions in, **4:**661–62
  nasalization, **4:**635

Speech *(Continued)*
  neural mechanisms for, **4**:636
  normalization, **4**:635
  of older adults, **1**:120
  production, **5**:740, **7**:154–55
  recognition, perception and, **5**:741
  resonance disorders, **7**:158–59
  source-filter theory, **4**:634–35
  technology, perception and, **5**:742
  vowel and consonant sounds in, **5**:740–41
Speech acoustics, **5**:740–41
Speech disorders. *see also* Communication disorders
  fluency disorders, **7**:159–60
  in older adults, **1**:122–23
  sound production disorders, **7**:158
Speech-language pathology, **4**:202, **7**:153–63
Speech Sounds Perception Test, **1**:611
Speedballs, defined, **5**:441
Speedreading. *see under* Reading
Sperm, **2**:875, **7**:165–77
  abnormalities, **7**:175
  biochemistry of, **7**:175
  cross-section of, **7**:172
  destruction by macrophages, **4**:858
  ejaculation, **7**:174
  electron micrograph of, **7**:167
  enzymes released from, **7**:166
  formation, **7**:170–73. *see also* Spermatogenesis
  mammalian, **7**:429
  membranes, biochemistry of, **7**:175
  at moment of fertilization, **7**:176
  morphology of, **7**:166–70
  nucleus, **7**:166
  purpose of, **7**:165
  storage, in male, **7**:173–74
  surface antigens, **7**:175
  survival, **7**:174–75
Sperm capacitation, **7**:166
  antiestrogens and, **3**:603
  defined, **3**:599
  discovered, **3**:600–601
  in vitro fertilization and, **3**:599, 602–3
  intracellular ions in, **3**:602
  progestation agents and, **3**:603
  requirements for, **3**:601
  sequence of, **3**:601–2
Sperm head, **7**:168, 169
Sperm receptor
  carbohydrate involvement in, **6**:510–11
  functional characteristics of, **6**:510
  mechanism of action, **6**:510–11

Sperm receptor *(Continued)*
  molecular genetics of, **6**:510
  nature of, **6**:509–10
  structural characteristics of, **6**:509–10
Sperm transport, **7**:173–74
Spermadine, **2**:423
Spermatocytes, **7**:427–28
Spermatogenesis, **4**:950, **7**:170–73
  clonal nature of, **7**:428
  hormonal control of, **7**:173
  regulation of, **7**:429–32
  stages of, **7**:427–32
Spermatogonia, **2**:438, **7**:171–73
Spermidine, **6**:68
Spermine, **2**:382, **6**:68
Sperry, Roger, **5**:313
Sphenoparietal sinuses, **4**:984
Spherocytosis, defined, **6**:545
Sphingolipids
  behavior and occurrences, **4**:745–47
  biology of, **7**:183–84
  biosynthesis of, **7**:181
  breakdown products, **7**:185, 186–88
  cellular organization of, **7**:181
  characterization, **4**:744
  degradation of, **7**:181–82
  function of, **7**:184
  mechanism of action of, **7**:186–88
  in membranes, **4**:970
  metabolism, **4**:744–45, **7**:181–86
    and biology, **7**:179–89
  structure of, **7**:180–82
Sphingomyelin
  biosynthesis, **4**:746, **5**:886–87
  formation in Golgi apparatus, **4**:8
Sphingosine, **7**:187
Sphygmomanometers, **4**:289
Spice, production of, **1**:143
Spider venoms, **7**:596
Spin-echo sequence, **4**:868–69
  clinical imaging, **4**:868–69
  human head, **4**:868, 870
Spina bifida
  polygenic/multifactorial causes, **1**:700
  prevalence of, **1**:702(t)
Spinal arteries, **7**:194–95
Spinal column. *see* Spine
Spinal cord, **7**:191–220
  amygdaloid projections to, **4**:721–23
  anatomy of, **5**:351–52

Spinal cord *(Continued)*
    ascending pathways in, **7:**212–13
    autonomic afferents in, **7:**212
    ben nucleus of the stria terminalis projections to,
        4:721–23
    blood vessels of, **7:**193
    central pattern generator for locomotion, **4:**773–74
    cerebrospinal fluid of, **7:**195–96
    chemoarchitecture of, **7:**201–5
    clinical implications of, **7:**213–20
    connections, **7:**205–13
        afferent pathways in, **7:**207–11
        propriospinal pathways in, **7:**205–9
    corticospinal system in, **7:**209–11
    cytoarchitecture of, **7:**197–200
    dendroarchitecture of, **7:**200–201
    descending brain stem pathways to, **7:**211
    descending systems to, **7:**219
    dorsal horn
        major components of upper laminae of, **5:**628
        peptide fibers in, **7:**202
    dorsal root ganglia of, **7:**193, 208–9
    dysfunction. *see* Spinal cord dysfunction
    efferent pathways in, **7:**211–13
    external morphology of, **7:**192–96
    functional aspects of, **7:**213–20
    general characteristics of, **7:**192
    gray matter of, **5:**352, **7:**196–205, 207–11. *see also* Gray
        matter, spinal cord
    hypothalamic projections to, **4:**719–21
    internal neuronal organization of, **7:**196–213
    interneuron motor control in, **7:**217–18
    locomotor network, **4:**773–76
    lumbar
        dendroarchitectonic scheme of, **7:**200
        transverse section of, **7:**196
    meninges, **5:**351, **7:**195
    mesencephalon projections to, **4:**714–19
    micturition control and, **5:**46
    motor centers-cerebellum interaction, **4:**776–77
    motor control in, **7:**215–20
    motor neuron organization in, **7:**215–17
    pain and temperature sense and, **7:**213–15
    pathways in, somatosensory system and, **7:**122
    periqueductal gray projections to, **4:**719
    posterior view of, **5:**352
    prefrontal cortex projections to, **4:**723
    propioneuron motor control in, **7:**217–18
    recurrent inhibition and, **7:**217
    reflex circuits of, **6:**158–60

Spinal cord *(Continued)*
    sensory control in, **7:**213
    shock and, **7:**227–29
    somatosensory pathways from receptors to, **7:**122
    spinal nerves of, **7:**193
    sympathetic outflow in, **7:**212
    touch and limb proprioreception and, **7:**213
    ventral horn of, **7:**203
    white matter of, **5:**352, **7:**196
Spinal cord dysfunction, **7:**221–32
    autonomic dysreflexia and, **7:**232
    basic concepts in, **7:**221
    bowel and bladder function with, **7:**231–32
    central nervous system integration in, **7:**223–25
    central pattern generator cells and, **7:**222–23
    clinical features of, **7:**229–32
    defined, **7:**222
    nature of, **7:**223–27
    nature of recovery from, **7:**223–27
    paraplegia as, **7:**231
    pharmacological changes during, **7:**225–26
    sensory system disturbance and, **7:**231
    spinal shock and, **7:**227–29
    sprouting nerve cells and, **7:**226–27
    synapse transmission alteration and, **7:**227
    synapse unmasking after, **7:**227
    understanding, research problem illustrating,
        7:222–23
    various states of, **7:**222–23
Spinal cord injuries, reflex survival after, **5:**808–9
Spinal fractures, **1:**779
Spinal ganglia. *see* Ganglia, spinal
Spinal nerves
    anatomy, **5:**353–54
    peripheral regions innervated by, **5:**354–55
Spinal shock, **7:**227–29
Spinal veins, **7:**195
Spindle
    chromosomal, **2:**244
    meiotic, **2:**291
    mitotic, **2:**291
Spindle microtubules. *see* Microtubules, spindle
Spindles, muscle. *see* Muscle spindles
Spine, **2:**597. *see also* specific areas, such as Cervical
    vertebrae
    bones of, **7:**36–37
    lower, schematic outline of, **7:**192
    muscles of, **4:**263
    and spinal cord relationships, **7:**192–93
Spiramycin, **2:**382

Spirolactone. *see* Spironolactone
Spirometry
    and measurement of lung volumes, **6**:596–97
    as test of airway obstruction, **6**:363
Spironolactone, **3**:99
Splanchnic mesoderm, in myogenesis, **5**:162
Splanchnic nerves, **2**:32, 37
Spleen
    anatomy, **3**:64
    autonomic innervation of, **1**:543
    beta cell development in, **1**:565–66
    noradrenergic innervation of, **5**:377, 381
    regional immunology, **4**:399
    role in *Plasmodium* infections, **4**:879
Spleen, rat, **5**:379
Splenic artery, anatomy, **3**:756
Splenium, posterior, **3**:220–21
Splenomegaly, hyperreactive malarious, **4**:389
Spliced leader peptides. *see* Signal peptides
Splicesome, **2**:294, **6**:666
Splicing, RNA. *see* RNA splicing
Split-brain surgery, **2**:352–53
Split genes. *see* RNA splicing
Spodoptera frugiperda, **3**:157
Spoilage associations
    defined, **3**:661
    of foods at ambient temperature, **3**:662
Spondylitis, ankylosing, **4**:384
Spontaneous abortion. *see* Abortion, spontaneous
Spontaneous recurrence. *see* Flashbacks, drug
Sporidesmins, **7**:593
Squalene, **2**:415
Squalene synthetase, **2**:415
Squamous cell carcinoma, **4**:581, **5**:548, 615–16
Squint. *see* Strabismus
Squirrel monkey. *see* Saimiri
Stadler, L. J., **5**:251
Stage-specific embryonic antigens. *see* Antigens, stage-
    specific embryonic
Stages, developmental. *see* Growth
Stains and staining. *see also* Immunostaining
    in cytogenetic maps, **3**:877–78
    for electron microscopy, **3**:278–79
    in flow cytometry, **3**:622–23, 625
    fluorescent, in protein detection, **6**:194
    history of, **5**:682
    of odorants, **5**:527
    organic, in protein detection, **6**:193–94
    for scanning electron microscopy, **6**:743
    silver, for protein detection, **6**:194–96

Stanford-Binet test, **3**:838
Stanolone, **2**:301
    male genital organogenesis and, **3**:342
Staphylococcal nuclease. *see* Micrococcal nuclease
Staphylococcus aureus, **4**:450
    in food, **3**:666
    neonatal infection, **4**:459
    toxins, intracellular activity of, **5**:4
Staphylococcus cell wall, **1**:275–76
Starch, **2**:101, 103–4
    defined, **3**:35
    in dicotyledonous seeds, **3**:39
    digestion of, **3**:43
    in food, **2**:101, 103, 104
    level of consumption, **2**:101
    resistant
        defined, **3**:35
        in plant food products, **3**:37
    types of, **2**:101
Starch, laundry, cravings for, **3**:31–32
Starling, **5**:713
Starling forces, defined, **5**:36
Starling's curve, as test for artificial heart, **1**:393
Starling's law of the heart. *see* Frank-Starling mechanism
Starvation. *see also* Famine
    in anorexia nervosa, **3**:246
    defined, **3**:243
    in famines, **3**:563
    immunity and, **5**:506
    mobilization of energy and proteins in body, **4**:1005–6
State-dependent learning, **2**:67
    defined, **2**:67
    inotropic state, **2**:179
Statherin, in saliva, **6**:694
Static exercise. *see* Isometric contraction
Stationary phase, HPLC, **4**:208–9
Statistics
    in inductive reasoning, **6**:499
    of regulatory protein base pair choice, **3**:121–22
Statocyst, **2**:307
Status epilepticus, **3**:458
Stead, **5**:381
Stearic acid
    in food, **3**:571
    gangliosides and, **3**:731
Steatorrhea, **3**:573
Steely hair syndrome, **2**:850–51, **3**:269
Steering control, in locomotion, **4**:771
Stein-Leventhal syndrome. *see* Polycystic ovary syndrome
Stellaris joints, **1**:374, 376

Stellate cells, defined, **5**:307

Stem cells

  characterization, **4**:168–69

  immune system, HIV infection, **1**:20

  lymphoid, **4**:169–71

    in bone marrow transplantation, **1**:791–92

  myeloid, **2**:259, **4**:169

  pluripotent hemopoietic, **4**:167–68

  restricted, **4**:168

    characterization, **4**:168

    functions, **4**:168

Step cycle, in locomotion, **4**:774–76

Stereo-depth scaling, and binocular vision, **1**:638–39

Stereocilia

  anatomy, **3**:234

  deflection of, **3**:236, 237

  hair cell intracellular voltage change and, **3**:235

Stereopsis. *see* Depth perception

Stereoscopic procession, **7**:825–26

Stereoscopic vision. *see* Depth perception

Stereotypes and prejudice, **7**:233–41

  acquiring and maintaining, **7**:234–38

    cognitive orientation as theory for, **7**:236–38

    psychodynamic orientation as theory for, **7**:234

    sociocultural orientation as theory for, **7**:235–36

  categorization process involving, **7**:237–38

  changing

    reducing prejudice and, **7**:238–41

    role of education in, **7**:238–39

  defined, **7**:233

  distinctions between, **7**:233–34

  influence of society on, **7**:235–36

  origin of, **7**:234

  Stereotype behavior, defined, **5**:523

  understanding, **7**:233–34

Stereotypy, apomorphine-induced, **2**:223

Stern, **3**:838

Sternum, **2**:597

Steroid contraceptives. *see* Contraceptives

Steroid-deficiency syndrome, **1**:77

Steroid-excess syndromes, **1**:78

Steroid 21-hydroxylase, **2**:744

Steroids, **2**:58, 326, 414, 744, **7**:257–83. *see also* Adrenal cortex hormones

  actin, model of, **7**:265

  adrenal, **1**:72–74

    biosynthesis of, **7**:258–64

    in protein-energy malnutrition, **4**:894

  in arachidonic acid inhibition, **3**:579

  atrial natriuretic factor and, **1**:470

Steroids (*Continued*)

  binding

    allosteric model showing, **7**:266

    to intracellular receptors, **6**:516

  biosynthesis of, **7**:258–64

  and birth control, **1**:691–93

  characterization, **4**:747–48

  circulating, measurement of, **7**:278

  classification of, **7**:243–44

  defined, **2**:321

  degradative metabolism of, **7**:276–80

  DNA binding sites and, **3**:124

  drug design strategies for, **7**:112–13

  as drugs, **7**:281–82

  follicular genesis of, **3**:630

  formation, **4**:747

  functions of, **7**:258, 267–76, 280–81

  in host defense, **1**:311

  immunosuppressive effects, **1**:684–85

  inflammation therapy, **1**:312–13, **4**:477

  levels, estimation of, **7**:276–80

  listed, **3**:320

  mechanism of action, **7**:265–66

  nomenclature, **7**:258

  in oocyte maturation, **5**:555–56

  ovarian

    biosynthesis of, **7**:263–64

    endometriotic implants and, **3**:367

  receptors. *see* Receptors, steroid

  secretion

    abnormalities in, **7**:280–81

    control of, **7**:258–64

  soft, therapeutic indices of, **7**:114

  structure of, **7**:258, 267

  synthesis, **3**:327, 328, **7**:243–55,

    adrenal, **7**:245, 246–52

    in antral follicles, **3**:632

    congenital adrenal hyperplasia as disorder in, **7**:249

    early steps in, **7**:244

    in Graafian follicle, **3**:633

    hormonal regulation of, **7**:254

    ovarian, **7**:252–53

    placental, **7**:253–54

    testicular, **7**:253

  testicular

    biosynthesis of, **7**:262

    metabolism, **7**:425

  thyronines and, **3**:330–31

  transport, in blood, **7**:264–65

Steroids, anabolic. *see* Anabolic steroids

Steroids, gonadal. *see* Gonadal steroids
Sterols, **2**:415
    in membranes, **4**:970
    plant, **4**:747
Stethoscope, in pediatrics, **5**:700
Stevioside, as sweetener, **7**:375-76
Stilbestrol, **2**:111
Stimulants, addiction to, **2**:511-20
    clinical characteristics, **2**:516
    epidemiology, **2**:515-16
Stimulation. *see also* Arousal
    central, **2**:511
    central nervous system, effect of serotonin on, **6**:832-33
    of cortical cells, **5**:309
    of isolated spinal cord, **5**:808
    isoluminant, **2**:580-81
        defined, **2**:575
Stimulation, electric, **2**:833
Stimulus, generalization, **2**:629
    conditional, **2**:630
    unconditional, **2**:630
Stomach, **2**:32-34
    alcohol consumption and, **1**:175
    anaerobic bacteria found in, **1**:240-41
    anatomy, **3**:61-62, 74-75, 756-57
    blood circulation in, **3**:755-63
    blood flow
        after eating, **4**:554
        measurement of, **3**:758-59
        regulation of, **3**:760
    fat digestion in, **3**:572-73
    function, **2**:34
    hormonal activity in, **5**:716
    medullary pathways, **2**:32
    microvascular anatomy, **3**:757
    mucosa, protein expression by, **3**:182
    nervous pathways, **2**:32, 33
    perfusion pressure, **3**:761
    physiology, **3**:75-76, 755-56
    precancerous state of, **3**:770
    receptive relaxation, **1**:549
        defined, **3**:755
Stomach neoplasms
    diet and, **3**:674
    genotoxic factors, **3**:772
    in Japan, **3**:767, 775
    prevention of, **3**:775-76
    types of, **3**:769
Stomatitis, primary herpetic, **5**:613
Stool. *see* Feces

Storage temperature of food. *see* Temperature, of food storage
Strabismus, in fragile X syndrome, **3**:716
Straight sinus, **4**:983
Strain-toward-symmetry theory, **4**:539
Strand slippage, genomic mutations and, **3**:883
Stratum corneum
    defined, **3**:445
    light penetration by, **5**:593
    structure of, **7**:43-44
Street drugs, **3**:211-14. *see also* Narcotics; Substance abuse
Strepshirine primates. *see* Primates, lower
Streptococcal infections
    neonatal, **4**:457-58
    in school-age children, **4**:464
Streptococcus mitis, **2**:372
Streptococcus pneumoniae, **4**:450, 459, 461
Streptococcus pyogenes, **1**:555
Streptococcus thermophilus, **3**:667
Streptokinase, **2**:170
Streptomycin, **2**:372
    biosynthesis, **1**:290
    mode of action, **1**:290(t)
Streptozotocin, as cause of diabetes, **7**:571-72
Stress
    adaptive behavior, **2**:221
    aging and, **3**:900, 903
    alteration of gastrointestinal function, **7**:285-92
    appraisal, **4**:74
    chronic
        and the handicapped child, **7**:297
        and hypertension development, **4**:292
    corticotropin-releasing hormone and, **5**:398
    defined, **7**:285
    disease and gastrointestinal responses, **7**:291
    eating disorders and, **3**:244
    families with a handicapped child and, **7**:293-98
    gastrointestinal function and, **7**:291
    gastrointestinal motor function and, **7**:289-90
    gastrointestinal secretory function and, **7**:286-89
    headache associated with, **4**:55
    immunity and, **5**:374, **6**:320-22
    metabolic regulation role, **4**:1006
    normal parenting and, **7**:293
    opioid peptides and, **5**:375
    pain stimuli and, **5**:808
    as predictor of suicide, **7**:333
    psychoneuroimmunology, **4**:74-75
    vasopressin secretion and, **5**:397-98

Stress proteins. *see* Heat-shock proteins

Stressed volume, defined, **7**:758

Stretch, muscle, energy storage and, **5**:185

Stretch reflex. *see* Reflex, stretch

Stretch-shorten cycle. *see under* Muscle, skeletal

Stria of Retzius, **2**:822

Stria vascularis, **3**:232

Striate cortex. *see* Visual cortex

Striated border. *see* Microvilli

"String of pearls" (dermatoglyphics), defined, **3**:89

Stroke. *see* Cerebrovascular disorders

Stroke volume

    in dynamic exercise, **3**:518

    in exercise, **3**:517-19

    oxygen consumption and, **3**:517

Stromal cells

    characterization, **4**:171-72

    functions, **4**:171-72

    of gastrointestinal tract, **3**:770

Strongyloides stercoralis, **2**:387

    autoinfection, **4**:115, 118

    transmission, **4**:114

Strongyloidiasis, **4**:118-19, 121

    hyperinfection syndrome, **4**:119

    malabsorption syndrome, **4**:118-19

Strontium, in drinking water, **2**:813

Stroop phenomenon, **5**:156, 160

Strychnine, **2**:522, **5**:416, 423

Studying and learning strategies, **7**:309-14

    cognitively active learner in, **7**:310

    comprehension monitoring for, **7**:311

    creating supportive personal climates for, **7**:313-14

    knowledge required to use, **7**:310-11

    metacognition and, **7**:310-11

    role of goal setting and using in, **7**:312

    types of, **7**:312

    what we need to know about, **7**:309

Studying and learning tasks

    elaboration strategies

        for basic, **7**:313

        for complex, **7**:313

    organization strategies

        for basic, **7**:313

        for complex, **7**:313

    rehearsal strategies

        for basic, **7**:312-13

        for complex, **7**:313

Stunkard, Albert, **3**:684

Sturtevant, A. H., **3**:824

Stuttering, as speech fluency disorder, **7**:159-60

Styryl dyes, **5**:577

Subarachnoid cisterns, **4**:985

Subarachnoid hemorrhage, **5**:601

Subarachnoid space, spinal, **4**:989

Subcortical nuclei (gray matter), **3**:224

Subcutaneous tissue, **7**:46-48

Subdural hematoma, **4**:987-88

Subdural spaces

    cranial, **4**:980

    spinal, **4**:988

Suberin, dietary fiber and, **3**:39

Subglottis, **4**:654-55

Subicular complex, **4**:227, 229

Subiculum, **2**:690

Sublingual gland, **6**:690

Submandibular gland, **6**:690

Submandibular gland, mouse, B335, **5**:334

Submaxillary gland. *see* Submandibular gland

Submersion hypothermia, **4**:802

Submissive behavior, **1**:115

Submucosa, defined, **3**:755

Submucosal arteriolar plexus, **3**:757

Submucosal arterioles

    in gastric mucosal blood flow regulation, **3**:760-61

    vasodilatation of, **3**:762

Subnucleus caudalis, **5**:366

Substance abuse

    characterization, **4**:997

    by children, **5**:710

    headaches associated with, **4**:56-57

Substance dependence, **1**:43-51, **5**:439-49. *see also* Alcoholism; Substance withdrawal syndrome

    active phase, **4**:85-86

    in America, **1**:45-46

    characteristics of, **1**:43-44, **4**:997

    costs to society, **4**:84

    defined, **5**:439

    determinants of, **1**:46-49

        biological, **1**:47

        integrating, **1**:48-49

        psychological, **1**:47

        sociocultural, **1**:48

    headaches associated with, **4**:56-57

    insomnia and, **7**:78-79

    marijuana and cannabinoids, **4**:909

    motivational model for, **1**:43, 50

    postrecovery condition, **4**:87

    predisposing factors, **4**:84-85

    psychology, **4**:83-87

    relapse, **4**:87

Substance dependence *(Continued)*
    resolution without treatment, **4:**86–87
    societal aspects, **4:**83–84
    symptoms of, **1:**44
    systematic motivational counseling for, **1:**43
    taking up the behavior, **4:**85
    treatment of, **1:**49–50, **4:**87
Substance P, **2:**5, 18, 26, 27, 323, 338, 365
    in autonomic neurons, **1:**541
    function of, **6:**89
    in gut-associated lymphoid tissue, **5:**381
    in immune response, **5:**384
    levels in Huntington's disease, **4:**271
    muscle fatigue and, **7:**29
    in spleen, **5:**381
    in visceral sensory system, **7:**795
Substance withdrawal syndrome, **2:**67
    in barbiturate abuse, **5:**443
    defined, **5:**439
    headaches associated with, **4:**56–57
Substantia nigra, **2:**217
    and convulsive seizures, **6:**782
    fetal, transplantation of, **5:**671
    neurotranmsitters in Huntington's disease, **4:**271
    in Parkinson's disease, **5:**668
Substrate activation, **3:**437–38
Substrate cycling, **5:**469
Substrates, defined, **3:**71, 393
Subtilisins, **3:**421
Subtraction angiography, digital. *see* Angiography, digital
    subtraction
Subunit vaccines, **7:**749–51
Succinate-ubiquinone reductase, **5:**58
N-Succinimidyl-3-(2-pyridyldithic)propionate, **4:**418–19
Succinyl coenzyme A, **2:**506–8
Succinylcholine, **5:**832
Sucking behavior
    human babies, **4:**607
    prenatal, **3:**618
Sucrase, **2:**105
Sucrose, **2:**101, 103, 104
Sudden infant death syndrome, **4:**635, 652, **7:**315–22
    cardiac arrhythmia as cause of, **7:**319
    clinical care of those at risk for, **7:**320
    defined, **7:**315
    developmental vulnerability and, **7:**320
    disordered respiratory control as cause of, **7:**318–19
    epidemiology of, **7:**317–18
    historical perspective of, **7:**316
    impact on survivors, **7:**317

Sudden infant death syndrome *(Continued)*
    maternal smoking and, **7:**518–19
    metabolic dysfunction as cause of, **7:**320
    neurobiology of, **7:**320
    pathological findings in, **7:**318
    prevention of, **7:**321
    research strategies for, **7:**318–19
    risk of, **5:**706
    societal attitudes to, **7:**316–17
    theories for, **7:**318–19
Suffering
    healing, and the human condition, **7:**323–25
    meaning of, **7:**324
    society's response to, **7:**324–25
Sugar, nucleotide. *see* Nucleotide sugars
Sugar phosphate esters, **3:**933
Sugar residues, from pectic substance extraction, **3:**38
Sugar substitutes. *see* Sweeteners
Sugars, **2:**102–3. *see also* Carbohydrates; Sweeteners
    food analysis, **4:**217–18
    food spoilage and, **3:**663
    intake in prepared food, **2:**102
    passive diffusion, **3:**80
    from plant tissues, **3:**36
    relative sweetness, **2:**103
    simple, as sweeteners, **7:**373–74
    sweetness of. *see* Sweetness
Suicide, **7:**327–35
    altruistic, **7:**329
    anger, irritation and dissatisfaction as predictors of,
        7:333
    anomic, **7:**329
    biological factors in, **7:**333–34
    cellular, programmed, **2:**250
    common traits of, **7:**328
    concept of, **7:**328
    defined, **7:**327
    and depression, **5:**94
    egoistic, **7:**329
    epidemiology of, **7:**330–31
    family, as predictor of suicide, **7:**333
    fatalistic, **7:**329
    human, **2:**845–50
    indirect self-destructive behaviors as, **7:**328
    lethal object availability as predictor of, **7:**332
    media effects on, **1:**7–8, 1058
    negative life events as predictor of, **7:**333
    and paroxysmal abnormalities, **7:**334
    physical illness as predictor of, **7:**333
    predictors of, **7:**331

Suicide *(Continued)*
    predispositions, electroconvulsive therapy and, **7**:334
    prevalence of, **7**:330–31
    prior suicide attempts as predictor of, **7**:332
    as problem-solving behavior, **7**:328
    social isolation as predictor of, **7**:332–33
    talk, as predictor of, **7**:332
    types of, **7**:329–30
Sulbactam, **2**:372
Sulfa drugs, synthesis and use of, **1**:326
Sulfamethoxazole-trimethoprim, combined, **2**:372
Sulfanilamide, **1**:273
Sulfation factor. *see* Somatomedins
Sulfhydryl compounds
    chemistry of, **3**:910
    defined, **3**:907
    glutathione reductase and, **3**:913
    in potentially reversible enzyme thermoinactivation
        mechanisms, **3**:415
Sulfite dehydrogenases. *see* Sulfite oxidases
Sulfite oxidases
    deficiency, as mechanism of sulfite sensitivity, **7**:339
    molybdenum denaturation and, **3**:418
Sulfites
    asthma induced by, **7**:339
    drug uses for, **7**:341
    in food, **7**:337–43
    in food analysis, **4**:224
    nonasthmatic response to, **7**:338
    sensitivity
        clinical manifestations of, **7**:338
        diagnosis of, **7**:339–40
        prevalence of, **7**:338
        treatment for, **7**:340
Sulfonamides
    agricultural uses, **1**:329
    mechanism of action, **1**:326–27
    medical uses of, **1**:328–29
    synthesis and use of, **1**:326
    toxoplasmosis treatment with, **1**:268
Sulfur dioxide
    exposure routes of, **3**:385
    inhaled, disposition and effect of, **7**:579
    in nitrosamine inhibition, **3**:679
    spinach photosynthesis and, and **3**:257–58
"Summer diarrhea." *see* Diarrhea, "summer"
Sunburn, **2**:648. *see also* Ultraviolet rays
Sunscreens, **7**:68–69
Supercritical fluid chromatography. *see* Chromatography,
    supercritical fluid

Superego, **2**:710
    defined, **6**:308
    in psychology, **2**:710
Superfecundation, defined, **3**:306
Superfetation, defined, **3**:306
Superior colliculus, **5**:389
Superior laryngeal nerve, **4**:657
Superiority (behavior), **1**:574
Superovulation
    exogenous gonadotrophins in, **3**:633
    historical perspective, **7**:346
    in laboratory species, **7**:346
    LH and, **3**:633
    and ovulation induction, **7**:345–51
Superoxide
    biological production of, **7**:357
    defined, **5**:667
    dismution, by superoxide dismutase, **7**:353
    formation, in polymorphonuclear neutrophil, **5**:429–30
    and generation of hydroxyl radicals, **7**:353
    pathophysiological importance of, **7**:358–59
    production of
        platelet-activating factor and, **6**:48
        during respiratory burst, **6**:583
Superoxide dismutase, **7**:353–59
    bovine erythrocyte, **7**:355
    copper-zinc-containing, **7**:355–56
    cytosolic, **2**:652
    free radicals and, **3**:894, 914
    function of, **7**:357, 885
    history and discovery of, **7**:354
    iron-containing, **7**:356
    manganese-containing, **7**:356
    measurement of, **7**:357–58
        chemical assays, **7**:358
        chemiluminescent assays, **7**:358
        electrochemical assays, **7**:358
        enzymatic assays, **7**:357
        nuclear magnetic resonance assays, **7**:358
        photochemical assays, **7**:357
    nitric oxide and, **3**:373–74
    oxygen toxicity and, **7**:354
    parasite, host modification and, **5**:648
    rate constants for, **7**:357
    structure of, **7**:355–56
    use of, **7**:358–59
Supersensitivity, **2**:217
Supination, defined, **3**:691, **5**:133
Supposition, **2**:619–28
    and conditional reasoning, **2**:619–28

Supposition *(Continued)*
and empirical findings, **2:**623-27
in preschool children, **2:**623-24
and truth, **2:**621-22
Suppression, genetic phenomenon of, **5:**77
Suppressor cells, **4:**333
and autoimmune disease, **1:**522
characteristics of, **4:**811-12, **7:**481
functions, **4:**403
parasite invasion and, **5:**649
radiation effects, **6:**422
tumor rejection and, **4:**340
Suppressor factor, T-cell, glioblastoma-derived. *see* Transforming growth factor beta
Suppressor factors, immunologic, **4:**340
Suprachiasmatic nuclei
defined, **5:**396
visual system and, **7:**843
Suprachoroidea, anatomy, **3:**538
Supraciliaris, anatomy, **3:**539
Supraglottis, anatomy, **4:**654
Supralaryngeal airway, **4:**635-36
Supranucleur palsy, progressive, **2:**763
Supraoptic nuclei, defined, **5:**394
Suprapontine micturition control, **5:**45-46
Suramin, **2:**387, 392
Surface markers
for leukemia, **4:**693
monocytes and macrophages, **4:**855-56
Surfactants, pulmonary. *see* Pulmonary surfactants
Surgeon General, U.S.
Healthy People, **3:**651
Report on Nutrition and Health, **3:**652
Surgery
for anaerobic infections, **1:**247
animal models, **1:**258-59
for endometriosis, **3:**366-67
for leukemia, **4:**695
in space flight care systems, **7:**138-39
stroke treatment by, **7:**306-8
Surveillance, immune, **2:**248, 249
Sustentecular cells
defined, **5:**523
functions of, **5:**523
Sutton, W. S., **3:**849
Sutures
joint, and skull bones, **7:**34
between skull bones, **1:**371, 372
Sutures, pursestring, **4:**93
SV6-1 *Bam*/HFK, **4:**585

SV40 T antigens. *see* Antigens, polyomavirus transforming
SV40 virus, **2:**263
and DNA replication, **2:**263
interferon effects, **4:**518
keratinocytes infection, **4:**585
Svedberg equation, for ultracentrifugation, **7:**713
Svennerholm, L., **3:**726
Swainsonine, **4:**12
Swallowing. *see* Deglutition
Sweat glands, **2:**593
in regulation of body temperature, **1:**755-56
structure of, **7:**44-45
Sweden, and maternal mortality, **4:**914
Sweet foods. *see* Foods, sweet
Sweet taste
AH, B, gamma hypothesis to explain, **7:**365-70
compounds eliciting, **7:**530
mechanism of, **7:**362
nature of, **7:**362-63
signal transduction for, **7:**534
theories to explain, **7:**363-70
Sweeteners
absorption of, **3:**80
acesulfam-k as, **7:**378
aspartame as, **7:**378-80
in beverages, **2:**101
carbohydrate
halogenated sucrose derivatives as, **7:**374-75
simple sugars as, **7:**373-74
terpenoid glycosides as, **7:**375-76
dihydrochalcone, **7:**377
flavanoid, **7:**377
in foods, **2:**101
labeling of, **3:**646
nonsugar, **2:**101, 102
sorbitol, **2:**102
xylitol, **2:**102
protein, **7:**380-81
rating of, **3:**644
saccharin as, **7:**377-78
synthetic and natural, **7:**372
Sweetness
amino acids, **7:**368
chemistry, **7:**361-81
compounds, classes of, **7:**367
intensity, of different compounds, **7:**372
of natural and synthetic sweeteners, **7:**372
nature of, **7:**362-63
sensory system associated with, **7:**362
stereochemical basis of, **7:**363-70

Sweetness *(Continued)*
  temporal aspects and, **7:**370-72
  theories to explain, **7:**363-70
Sweetness modality
  in food acceptance, **3:**645
  prenatal, **3:**619
Swine
  aflatoxicoses in, **5:**260-61
  infection, and trichinosis, **7:**648-49
  mandible, **2:**706
  in scalar-borne zoonoses, **7:**898
  zearalenone and, **5:**267
Sydenham, Thomas, **5:**699
Sylvatic trichinosis, **7:**649
Symbiosis, **1:**261
Symbolism (psychology), **3:**646
Symmetry, in vertebrate body pattern formation, **3:**290-91
Sympathectomy, **2:**38
  chemical, **5:**382-83
  and diarrhea, **2:**38
Sympathetic nervous system, **2:**33, 38, 39, 141
  in aerobic exercise, **3:**508-9, 510-13
  anatomy of, **5:**356-59
  blood pressure and, **3:**510
  defined, **5:**373
  in dynamic exercise, **3:**519
  elevation in essential hypertension, **4:**290-92
  in gastric circulation regulation, **3:**762
  intestinal blood flow regulation, **4:**551
  motor routes of, **5:**359
Sympathins. *see* Catecholamines
Symphisis, **2:**778
Symphysiotomy, **4:**918
Symphysis joints, **1:**373
Symporters, **4:**558, 566-67. *see also* Biological transport,
    active
  Cl⁻-coupled, **4:**567
  Na⁺, Cl⁻, GABA, **4:**567
  Na⁺, glucose, **4:**566
  Na⁺, K⁺, **4:**2, 566
  Na⁺-dependent, **4:**566
Symptom Checklist 90, **1:**610
Synapses, **2:**337, 341, 344, 521, 831, **4:**199
  bipolar ribbon, in retina, **6:**623-24
  chemical, types of, **7:**383-84
  circuits, **7:**389-90
  conventional, in retina, **6:**623
  defined, **2:**673, **3:**455
  enhancement, neurotransmitter release and, **7:**390-93
  function, astrocyte role, **1:**443-44

Synapses *(Continued)*
  kindling and, **7:**392
  latent, and functional plasticity, **6:**39-40
  membranes, gangliosides in, **3:**733
  morphology, **7:**383-84
  negative feedback at, **7:**391
  neuromodulation, **7:**388-89
  number of, **2:**675
  photoreceptor cell, in retina, **6:**623
  physiological response at, **7:**387
  physiology of brain, **7:**383-93
  plasticity, **7:**390-93
  structural modification of, **6:**38-39
  structural plasticity of, **6:**38-39
  transmission, **2:**49, 316, **5:**408, **7:**227
  turnover, **7:**393
  unmasking, spinal cord dysfunction and, **7:**227
Synapsin I, **2:**44, 46, 48, 50
Synaptic cleft, **2:**11, 17, 49, 344
Synaptic receptors. *see* Receptors, synaptic
Synaptic rule
  defined, **5:**387
  in network modeling, **5:**389
Synaptic vesicles, **2:**344, **3:**186, 522, 678
Synaptogenesis, **5:**346
Synaptonemal complex, **2:**419, **4:**951
Synarthroses, **1:**371, 372
Synchondroses, **1:**372
Syncope, **2:**147
Syncytia, HIV-infected T cells, **1:**18-19
Syncytiotrophoblastic cells, **3:**338
Syncytium, **2:**275, **4:**425, 427, **5:**72
Syndactylism, **2:**251
Syndecan, defined, **6:**262
Syndesmoses, **1:**372
Syndrome, defined, **7:**406
Synergism, defined, **3:**211
Synergism-antagonism, quantitation of, **2:**377
  chemotherapeutic, **2:**371-74
  equation of drug combination, **2:**377
  methods for, **2:**377
Synesthesia, and LSD use, **5:**447-48
Syngamy, **2:**437
Synostosis, **1:**372
Synovial fluid, **1:**373
Synovial joints
  anatomy of, **1:**374, 376
  arthrodial, **1:**374
  ball-and-socket, **1:**374
  biaxial, **1:**374, 376

Synovial joints *(Continued)*
  bursae, **1**:373
  carticular cartilage, **1**:373
  condyloid, **1**:374
  ellipsoidal, **1**:374
  ginglymus, **1**:374
  gliding, **1**:374
  hinge, **1**:374
  hyaline cartilage, **1**:373
  ligaments, **1**:373
  pivot, **1**:374
  saddle, **1**:374, 376
  sellaris, **1**:374, 376
  synovial fluid, **1**:373
  tendon sheaths, **1**:373
  trochoid, **1**:374
Synovial membrane, **1**:373
Synoviosarcoma, **2**:451
Synovium, **2**:640
Syntax, neural mechanisms for, **4**:636
Syphilis, **2**:813, **4**:450. *see also Treponema pallidum*
  as cause of spontaneous abortion, **1**:8
  and dental caries, **2**:813
  fetal infections, **4**:456-57
  oral, **5**:614
  serologic test for, **1**:739
  trends in, **6**:896-97
Syringomyelia, **7**:214-15
Systemic circulation. *see* Blood circulation
Systems analysis, and health psychology, **4**:81-82
Systolic blood pressure, **4**:285

T

T1 and T2 relaxation, **4**:865-66
T-cell growth factor. *see* Interleukin 2
T-cells. *see* T-lymphocytes
T effector cells. *see* Helper cells
T-lymphocyte receptors. *see* Receptors, T-cell
T-lymphocytes, **2**:225-30, 751-56, **4**:830-31
  activation, **2**:752-56, **4**:333, 361
  activity, **4**:334
  aging and, **3**:903
  alloreactive, **4**:363
  antigen activation of, **7**:480
  antigen receptor on, **2**:753
  antigen recognition, **4**:834
  in autoimmune disease, **1**:520-22, 531
  beta-carotene and, **5**:512-13
  CD3 cell surface molecule, **1**:677

T-lymphocytes *(Continued)*
  CD4-positive. *see* CD4-positive T cells
  cell distribution of, **7**:397
  cell surface molecules of, **7**:479-80
  characterization, **4**:165-66, 811, 821-22
  compared with B lymphocytes, **4**:335
  defects, **4**:837
  deficiency, tumor incidence and, **4**:320-21
  development, **2**:752
  dietary protein deficiency and, **5**:505
  dietary restriction and, **5**:514-15
  effect of platelet-activating factor on, **6**:48-49
  effector molecules, **4**:334
  epitopes, synthetic vaccines design and, **7**:747-48
  expression of cell surface proteins, **1**:677
  Fc-dependent inactivation, **1**:301
  functional types of, **7**:481
  functions, **4**:165-66, 402-3
  in gastrointestinal lymphoma, **3**:770
  graft rejection, **4**:362
  growth factor. *see* Interleukin-2
  in gut-associated lymphoid tissue, **5**:381
  helper. *see* Helper cells
  herpesviridae, **4**:190
  HIV depletion of, **1**:18-20
  idiotype-specific, **4**:317
  IgA and, **5**:140
  immune destruction of, **1**:19-20
  immunological self-tolerance, **4**:837
  immunoparasitology, **4**:386
  inducer. *see* Helper cells
  inflammatory, **4**:333-34
  interactions with MHC, **4**:358
  macrophage activation and, **5**:817
  maturation, **7**:480-81
  memory, characteristics of, **7**:481
  MHC-coded surface antigens and, **5**:506
  migration into lymph nodes, **4**:360
  migration patterns, **4**:396-97
  in multiple sclerosis
    altered receptor, **5**:146
    altered specific, **5**:145-46
  in neural-immune interactions, **5**:380-81
  oligoclonal, defined, **5**:146
  positive selection, **4**:330
  precursor, **2**:752
  prevention of Fc signaling, **1**:303
  progenitors, **4**:169-71
  proliferation and differentiation, role of cytokines in, **7**:481-82

T-lymphocytes *(Continued)*
  psoriasis and, **3:**449, 450
  radiation effects, **6:**421–22
  recognition of antigen-presenting cells, **1:**677
  in regulation of immune response, **5:**144
  sensitization, **4:**360–61
  subsets and cellular interactions, **4:**835–36
  suppressor. *see* Suppressor cells
  surface markers, **2:**752
  synthetic vaccines and requirement for help, **7:**747
  targets for immunoconjugates, **4:**373–74
  in thymus, development of, **7:**478–79
  tumor cell destruction, **4:**1012
  tumor rejection and, **4:**340
  vasoactive intestinal peptide and, **5:**384
  and zinc, **2:**659
T-lymphocytes, cytotoxic, **1:**677, **2:**225, **4:**333, 361–62
  cell killing without lymphocyte exocytosis, **4:**816–17
  characteristics of, **4:**812, **7:**481
  defined, **2:**225
  granule exocytosis and lethality, **4:**812–13
  granulolytic properties, **4:**813–14
  homologies with complement membrane attack complex, **4:**815
  macrophages and, **4:**859
  perforin purification from, **4:**814–15
  perforin role in, **4:**815–16
  pore-forming activity, **4:**814–15
  proteoglycans in, **4:**816
  resistance to self-lysis, **4:**815
  serine esterases in, **4:**816
  target cell lysis, **4:**813
  target cell recognition and binding, **4:**812
T3 receptors. *see* Receptors, thyroid hormone
T4 receptors. *see* Receptors, thyroid hormone
T tubules. *see* Transverse tubules
Tachistoscopy, **2:**353, 862
Tachycardia, **2:**146
  marijuana and, **4:**908
  ventricular, **2:**172
Tachycardia, paroxysmal, atrial, **2:**146, 174
Tachykinins
  functions of, **5:**720
  peptides, and visceral innervation, **7:**795–96
  psoriasis and, **3:**452
Tachyphylaxis, **2:**511, 514
Tadoma method for deaf-blind, **7:**62–63
Taenia ovis, **5:**656
Taenia saginata, **2:**389
Taenia solium, **2:**389

Taeniaestatin, **5:**648
Talin, **2:**270, 271
Tamofixen, **2:**60
Tandem repeats. *see* Repetitive sequences, nucleic acid
Tangier disease, **2:**667
Tanning, ultraviolet absorption and, **5:**594
Tanshinone, **2:**81
Tanycytes, **1:**435
Tapeworm diseases, **2:**386–90
  chemotherapy for, **2:**386
  cystic, **2:**390
  extraintestinal, **2:**396, 390
  intestinal, **2:**386, 389
Tapeworms, **1:**264–65
Taphonomy, **2:**189, **3:**495
Tardive dyskinesia, **6:**776
Tarichatoxin. *see* Tetrodotoxin
Tarsiers
  fossil record of, **6:**141–43
  and higher primates, **6:**139–43
Tarui disease. *see* Glycogen storage disease type VII
Task performance. *see also* Multiple-task interactions
  capacity and, **5:**150–51
  crosstalk and, **5:**155
  ordering in, **5:**123–25
  practice and, **5:**123
Task performance decrements. *see* Performance decrements
Tasks, defined, **5:**150
Taste
  aversions, conditioned, **1:**346–47
  bitter. *see* Bitter taste
  effect of kidney dialysis on, **1:**410
  food acceptance and, **3:**644
  molecules acting as modifiers, **7:**531
  perception, **5:**744–45
  peripheral taste system, **7:**527–31
  qualities, **7:**529–30
  role in food selection, **1:**345–49
  salty. *see* Salty taste
  savory. *see* Savory taste
  sense, in orofacial region, **5:**365
  sensitivity, factors affecting, **7:**530–31
  sour. *see* Sour taste
  stimuli and sensory attributes of, **5:**745
  sweet. *see* Sweet taste
  temporal aspects of, **7:**370–72
  tongue and, **7:**527–35
  transduction, **5:**744–45
Taste buds, **2:**780
  electrophysiological properties of, **7:**532–33

Taste buds *(Continued)*
    organization of, **7**:527-29
    schematic representation of, **7**:529
Taste disorders, **5**:745
Taste receptors. *see* Receptors, taste
tat genes. *see* Genes, *tat*
tat viral proteins. *see* Gene products, *tat*
TATA box, **4**:256
    defined, **5**:205
    RNA polymerase II and, **3**:123
    in RNA polymerase II system, **3**:111
    transcription factor IID and, **3**:131
Tau proteins
    CaM-kinase II and, **5**:895-96
    characteristics of, **5**:14
    phosphorylation of, **5**:896, 897, 898
Taxol, **5**:899
Tay-Sachs disease, **3**:732, **4**:579-80
Taylor Manifest Anxiety Scale, **1**:610
TCA cycle. *see* Citrate cycle
Tea, **2**:68
Tear film, precorneal. *see* Precorneal tear film
Teats, **4**:607
Techno-food. *see* Food, formulated
Technology
    hominid, **3**:497-99
    hominid brain development and, **3**:503
    Middle-Upper Paleolithic, **3**:502
Tectum, **2**:3
Teeth. *see* Tooth
Tegmental field, **4**:714-15
Tegmentum, **2**:3
    defined, **5**:43
    function, **2**:3
Tegument. *see* Skin
Telencephalon, **2**:3, 679
Telencephalon, dorsal, bird. *see* Dorsal telencephalon
Television, effects on aggression. *see* Media effects on aggression
Telogen, **4**:41
Telogen effluvium, **4**:42
Telomerase, schematic function of, **7**:403
Telomeres, **2**:470, 472, 473, 477, 484, **7**:401-4
    defined, **2**:465
    function of, **3**:872, **7**:401
    human
        characteristics of, **7**:403
        cloned in yeast, **7**:404
    listed, **2**:477
    repeat unit (TTAGGG), **2**:473
    replication of, **7**:402-3

Telomeres *(Continued)*
    terminal repeated sequences of, **7**:402
Telopeptide, **2**:54
Telophase, **2**:243, 244, **5**:71
Temperament, **2**:709
    personality disorders and, **5**:780
    social behavior and, **5**:771
Temperature. *see also* Body temperature; Body temperature regulation; Hyperthermia; Hypothermia
    adaptation to, **1**:38-39
    changes, in orofacial region, **5**:364-65
    crop production and, **1**:143
    spinal cord and sense of, **7**:213-15
    variation, monitoring of, **7**:459-60
    in velocity of ultrasound in tissue, **7**:725
Temporal coding, **3**:238
Temporal cortex, inferior, **3**:220
Temporal coupling. *see* Coupling, temporal
Temporal lobe, **2**:1, 8, 9, 351, 683, 684, 871, 874
    damage, **2**:8
    speech perception and, **3**:241
Temporomandibular joint, **2**:779
    and acromegaly, **2**:779
    receptors in, jaw position perception and, **5**:362
Temporomandibular joint syndrome, **4**:55, **5**:367-68
Ten-State Nutrition Survey, **3**:650
Tendon reflex. *see* Reflex, stretch
Tendon sheaths, **1**:373
    of synovial joints, **1**:373
Tendons, **2**:560, 572
    defined, **7**:23
    in elastic energy storage, **5**:187
    and fibroblast, **2**:567
    muscle performance and, **5**:174, 183-84
    patellar retinacula, **1**:376
    reflexes
        spinal cord dysfunction and, **7**:230
        spinal cord motor neurons and, **7**:216
    repair, growth factors and, **7**:507
    in stretch-shorten cycle, **5**:185
Tension curve, total. *see* Total tension curve
Tentorium cerebelli, **4**:981
Teratogens, **2**:67. *see also* Abnormalities, drug-induced
    alcohol/marijuana interactions as, **3**:212-13
    consequences from action of, **7**:409-10
    defined, **2**:67, **3**:383, **7**:412
    human, **7**:411-18
        identification of, **7**:415-16
        list of, **7**:416
        possible, **7**:417

Teratogens
    human *(Continued)*
        unlikely, **7**:417
    hyperthermia as, **4**:300
    maternal dietary cravings/aversions and, **3**:32
    as mutagens, **7**:409
    pesticide residues in food and, **7**:562
    principles of teratology, **7**:413–15
Terminal repeats. *see* Repetitive sequences, nucleic
    acid
Terrestrial animals. *see under* Animals
Terrorism. *see* Violence
Tertiary care centers, neonatal. *see* Intensive care
    units, neonatal
Test-tube fertilization. *see* Fertilization *in vitro*
Testicular feminization syndrome, **6**:842, 860
Testicular steroids, **7**:262–63
Testing, animal, alternatives to. *see* Animal testing
    alternatives
Testis, **2**:597. *see also* Macroorchidism
    aging in, **7**:436
    anatomy of, **6**:560
    blood supply of, **7**:434
    compartmentalization of, **7**:419
    development, **6**:342, **7**:435–36
    diagram of, **7**:420
    embryonic
        development of, **6**:839
        illustration of, **6**:840
    function, **7**:419–36
    germ cell development, **7**:427–28
    innervation, **7**:435
    interstitial area of, **7**:434–35
    at puberty, **7**:435
    seminiferous tubules of, **7**:419–20
    steroid hormone synthesis in, **7**:253
    zinc metabolism in, **7**:883
Testis determining gene, **6**:838
Testosterone, **2**:246, 248, 301, 493, 497, 498, 864. *see*
    also Feminization, testicular
    biosynthesis by testis, **7**:262
    and criminality, **2**:716
    defective receptor, **2**:301
    effects of
        on aggressive behavior, **1**:115, **4**:246–47
        on female sexual behavior, **4**:244–46
        on male sexual behavior, **4**:246
        mechanism of action and, **7**:272
        on spatial ability, **6**:859
    fetal
        abnormal secretion of, **3**:342
        chorionic gonadotropin and, **3**:340, 342

Testosterone
    fetal *(Continued)*
        in genital duct development, **3**:341
        in genital organ development, **6**:839–40
        and hair loss, **4**:42
        levels in seasonal breeding, **4**:931
        metabolism by prostate gland, **6**:179
        and monoamine oxidase, **2**:716
        perinatal, **3**:348
        in puberty, plasma concentrations of, **6**:343
        secretion
            control of, **7**:263
            Leydig cells and, **7**:432
        synthesis, in testes, **7**:253
        in treatment of prostate cancer, **6**:181
Testosterone 5a reductase, and synthesis, **7**:253
Tetanus toxin
    gangliosides and, **3**:734
    intracellular activity of, **5**:2–3
Tetanus vaccines, **4**:445
Tetany, **7**:28. *see also* Muscle contraction
    in calcium metabolism disorders, **3**:353
    neonatal, **3**:349
Tetrabenazine, **4**:272
Tetrachlorodibenzodioxin, **3**:389
Tetrachloroethylene, **1**:264
Tetracyclines, **2**:384
    biosynthesis, **1**:290
    in dentistry, **2**:785
    in feed additives, **1**:294
    mode of action, **1**:290(t), 291
Tetraethyl lead inhalation, **2**:342
Tetrahydrobiopterin, **1**:646
Tetrahydrocannabinol, **4**:905–6
    as antiemetic, **4**:909–10
    ethanol and, **3**:212
    from marijuana, **5**:444
Tetrahydrofolate, **2**:536
    functions of, **3**:437–38
Tetrahydrofolate coenzymes, **5**:465
Tetrahydrofolate reductase, **2**:535, 536
    gene amplification in, **3**:788
Tetrahydrofolylpolyglutamate, **2**:535, 536
Tetrahydrolipopterin, **2**:535, 536
Tetrahydromethanopterin, **2**:543, 544
Tetrahydropteroyl-L-glutamate derivatives, **2**:535
Tetrandrine, **2**:81
Tetraploidy, lethal, **2**:437
Tetrapolar bioelectrical impedance analysis, **1**:667–73,
    667–74, **5**:517. *see also* Body composition
    body composition assessment analysis, **1**:669–72

Tetrapolar bioelectrical impedance analysis (*Continued*)
  effect of physical factors on, 1:672
  factors affecting measurements in, 1:672-73
  instrumentation in, 1:668
  measurements in, 1:668
  model for, 1:668
  physical principles of, 1:668
Tetrazolides, arylsulfonyl. *see* Arylsulfonyl tetrazolides
Tetrodotoxin, 2:345
  as fish poison, 7:598
  pufferfish and, 2:345, 3:673
  structure of, 7:589
Texture perception, and touch, 7:58-59
TH cells. *see* Helper cells
Thalamus, 2:3, 5, 673, 5:349, 7:437-46
  anatomy of, 7:438-40
  embryonic development of, 7:437-38
  histological section of, 7:440
  language and, 3:224
  lesions in
    effects on orienting, 1:481
    motor symptoms from, 7:445
    neuropsychological symptoms from, 7:445-46
    sensory deficity from, 7:444-45
  neocortex and, 5:307
  neural circuits in, 7:443
  nuclei, naming of, 7:439
  Parkinson's disease and, 7:442
  pathological changes in, 7:444-46
  photographic identification of, 7:439
  physiology of, 7:440-44
  primary olfactory cortex and, 5:525
  response to cutaneous stimuli, 7:441
  schematic illustration of, 7:437-38
  somatosensory pathways from receptors to, 7:122-23
  syndromes of, 7:444-45
  ventral caudal nucleus of, 7:440-41
  ventral intermediate nucleus of, 7:442-44
  ventral oral nucleus of, 7:441-42
Thalassemia, 2:655, 4:143, 149
  bone marrow transplantation and, 1:796
  defined, 5:259
  DNA marker diagnosis of, 3:144-46
  and malaria, 4:143-44, 882
Thalidomide, birth defects caused by, 1:701, 7:405, 414
Thames River, pollution of, 3:253
Thaumatin, as sweetener, 7:380-81
THC. *see* Tetrahydrocannabinol
Thebaine, 1:179
Thecal cells, defined, 3:627

Thelarche, premature, defined, 3:357
Theobromine, 2:67, 69
Theophylline, 2:67, 69-71
Therapeutic index of drugs, 2:371, 7:102
Thermal injuries, 7:447-58, 448
  acids causing, 7:451
  burns as, 7:448
  causes of, 7:447-48
  in children, 5:711
  circulatory failure from, 7:452-53
  complex burn, 7:448
  early surgery for, 7:453
  fluid therapy for, 7:452-53
  incidence of, 7:447-48
  infection control of, 7:453-54
  mechanisms of, 7:448-49
  metabolic changes from, 7:454-56
  nutritional support for, 7:453
  physical problems from, 7:457
  process
    duration of, 7:449-58
    events within 30 seconds of exposure, 7:449-50
    events 30 seconds and 3 minutes after exposure, 7:450
    events 3 minutes to 3 hours after exposure, 7:450-52
    events 3 hours to 3 days after exposure, 7:452
    events 3 days to 3 weeks after exposure, 7:454-57
    events 3 weeks and 3 months after exposure, 7:457
    events 3 months and 3 years after exposure, 7:457-58
  psychological and social problems from, 7:456-57
  scalds as, 7:447
  severe, nutritional input in, 7:455-56
  water for treatment of, 7:450
  wound healing from, 7:454
Thermal sensation, in orofacial region, 5:364-65
Thermistor, core temperature measurement, 4:793
Thermogravimetry, in drug analysis, 3:201
Thermophilic enzymes. *see* Enzymes, thermophilic
Thermoreceptors, 7:797
Thermoregulation. *see* Body temperature regulation
Thermotolerance, 4:105
  cell cycle regulation and, 7:464
  clinical applications of, 7:466
  heat shock response and, 7:461
  induction of heat shock proteins and, 7:463
  in mammalian development, 7:459-66
Theta rhythm
  brain mapping of, 3:216-17
  defined, 3:215
  of olfactory cortex, 5:388

Thiabendazole, **2:**387
  in treatment of trichinosis, **7:**655
Thiamine, **2:**527
  food analysis, **4:**220–21
  interconversions, **2:**528
Thiamine monophosphatase, **2:**528
Thiamine pyrophosphate, **2:**527–29, **3:**437
Thiamine triphosphatase, **2:**528
Thiaminokinase, **2:**528
Thiazide diuretics. *see* Diuretics, thiazide
Thigh muscles, **4:**265–66
  spindles of, **3:**720
Thin-layer chromatography
  drug analysis using, **5:**837
  lipids, **4:**734
Thinking, **7:**467–72. *see also* Thought
  connectionist approach to, **7:**469–70
  creative thought and, **7:**471–72
  current research regarding, **7:**470–72
  current theory of, **7:**469–70
  daydreaming and, **7:**471
  decision-making and, **7:**472
  defined, **7:**467
  dimensions of, **7:**468
  information-processing approach and, **7:**469
  issue-relevant, **5:**803, 804
  limiting factors in, **7:**468–69
  as mental modeling, **7:**467–68
  mental models vs computer model in, **7:**467–68
  metabolic regulation role, **4:**1006
  modal model for, **7:**469
  puzzle-solving and, **7:**471
  reasoning and, **7:**470
  role of imagery in, **7:**468
  and scientific creativity, **7:**471–72
Thinking, associative, **2:**218
  and dopamine, **2:**218
  and schizophrenia, **2:**218
Thioctic acid, **2:**540
Thiocysteine, **3:**417
Thioester, **2:**539
Thiols. *see* Sulfhydryl compounds
Thirst, **7:**473–76
  angiotensin II as activator of, **7:**475
  brain activation sites and, **7:**474
  cellular dehydration as signal for, **7:**474
  dual depletion theory of, **7:**474
  hypothalamus role, **4:**309–10
  satiety and, **7:**475
  signals for, **7:**473–74

Thirst *(Continued)*
  vascular hypovolemia as signal for, **7:**474
THOG task, **2:**627
Thomas, D. B., **5:**600
Thoracic vertebrae, **7:**768
Thorax, **2:**596
  bones of, **7:**37
  endometriosis in, **3:**369
  muscles of, **4:**263
  plain film radiography of, **3:**2
  ribs, **2:**596–97
Thorndike's Law of Effect, **2:**633, **5:**127
Thought, **2:**853. *see also* Thinking
  automatic, **2:**853
  conscious, **2:**853
  preconscious, **2:**853
  unconscious, **2:**853
Threats
  aggressive nature of, **1:**114
  external, aggression and, **1:**117
Threonine, **2:**507
  MAP phosphorylation and, **5:**898
  in nuclear pore glycoproteins, **5:**454
Threshold, **2:**137
Thrill-seeking behavior, **2:**711, **6:**810
Thrombin, **2:**170, 171
  action on fibrinogen, **4:**180, 182
  enzyme inhibitors of, **3:**398
  in platelet activation, **6:**57
  Protein C activation, **4:**182
Thrombocytes, **4:**167
Thromboembolism, **1:**693–94, **5:**604
Thrombolytic agents, **2:**170
Thrombomodulin, **4:**182
Thromboplastin. *see* Tissue factor
Thromboplastinogen. *see* Factor VIII
Thrombosis, **2:**144, 167, 171
  oral contraceptives and, **5:**602
  tissue factor and, **7:**498
Thrombospondin, **1:**764
Thromboxane A2, **2:**169, 170, 369
  inflammation, **4:**473
  and platelet clumping, **2:**169
Thromboxanes, **1:**27
Thylakoid vesicle, **2:**238
Thymic cortex, **2:**752
Thymidine
  BrdU and, **3:**624
  DNA and, **2:**255
  in DNA synthesis, **5:**250

Thymidine (*Continued*)
　　5-methyldeoxycytidine and, **3:**154
Thymidine dimers, DNA damage and, **3:**894
Thymidine kinase, **2:**395-98
　　of Epstein-Barr virus, **3:**471
　　gene, **2:**257
Thymidylate, **2:**536
Thymidylate synthetase, **2:**372, 535
Thymine, **2:**117
Thymocyte mitogenic factor. *see* Interleukin 2
Thymocytes. *see* T-lymphocytes
Thymulin, **2:**649-59
Thymus, **2:**225, 227, 752, 753, **7:**477-82
　　in autoimmunity, **1:**525
　　development of, **7:**478-79
　　general characteristics of, **7:**477-78
　　in immune system aging, **6:**76
　　location of, **4:**830, **7:**477
　　lymphocytes from, **4:**830-31
　　microenvironment, and T cell maturation, **7:**481
　　noradrenergic innervation of, **5:**377, 380-81
　　organization of, during fetal development, **7:**478
　　removal of, in treatment of myasthenia gravis, **5:**258
　　structure of, **7:**478-79
Thymus, rat, **5:**380
Thymus-dependent lymphocytes. *see* T-lymphocytes
Thymus hormones, **1:**684
Thyroarytenoid muscle, **4:**656-57
Thyroglobulin, **3:**328, **7:**485
Thyrohyoid membrane, laryngeal, **4:**656
Thyroid cartilage, laryngeal, **4:**655
Thyroid diseases
　　of children, **3:**352
　　iodine deficiency and, **7:**609
　　oral contraceptives and, **5:**608
　　thyroid-stimulating hormone imbalances in, **5:**400
Thyroid gland
　　aging of, **3:**902-3
　　autonomic innervation of, **1:**542
　　development of, **7:**483-85
　　in electron microscopic imagery, **3:**276
　　fetal
　　　　development landmarks in, **7:**484
　　　　development of, **3:**339-40
　　　　at midgestation, **3:**619
　　follicular cells
　　　　animal, organization of, **7:**484
　　　　diagram of, **7:**485
　　　　hormone production by, **7:**485-87
　　　　iodine metabolism in, **7:**485-87

Thyroid gland (*Continued*)
　　functions, regulation of, **7:**487
　　hormone content of, **7:**486
　　hormone synthesis in, **3:**327-28
　　iodine content of, **7:**486
　　iodine metabolism defects, **7:**487
　　metabolism, agents interfering with, **7:**487
　　parafollicular cells of, **7:**490-91
　　structure of, **7:**483-85
Thyroid hormone deficiency. *see* Myxedema
Thyroid hormone receptors. *see* Receptors, thyroid hormone
Thyroid hormones, **7:**483-91. *see also* specific hormones
　　action of, **7:**488-90
　　in bone and calcium metabolism, **1:**805
　　bone cell function and, **1:**765
　　effects of, **7:**489
　　G protein mRNA and, **4:**21
　　in gene activation control, **5:**171
　　iodine removal from, **7:**490
　　lactase activity effects, **4:**617
　　levels in protein-energy malnutrition, **4:**894
　　maintenance of lactation, **4:**606
　　metabolism of, **7:**488-90
　　perinatal, **3:**348
　　precursors of, **7:**486
　　prepubescent, **3:**351
　　structure of, **7:**486
Thyroid neoplasms, oral contraceptives and, **5:**608
Thyroid-releasing hormone. *see* Thyrotropin-releasing hormone
Thyroid-stimulating hormone. *see* Thyrotropin
Thyroid-stimulating hormone receptors. *see* Receptors, thyrotropin
Thyroiditis, **1:**527-28, **3:**352
Thyroliberin. *see* Thyrotropin-releasing hormone
Thyronines, steroidal hormones and, **3:**330-31
Thyrostimulating hormone. *see* Thyrotropin
Thyrotoxicosis
　　maternal autoimmune, neonatal hyperthyroidism and, **3:**340
　　perinatal, **3:**348
Thyrotrophs, of anterior pituitary, **6:**18-19
Thyrotropin, **2:**19, 20, 23-24, 27, 301, 834
　　and autoimmune disease, **1:**515
　　fetal, **3:**339
　　function of, **6:**88
　　metabolic regulation role, **4:**1006
　　perinatal, **3:**348
　　production and activity of, **6:**18-19

Thyrotropin (*Continued*)
  regulation of, **5:**400
  secretion and sleep, **7:**75
  in thyroid function, **7:**487–88
  in thyroid-stimulating hormone regulation, **5:**400
  thyrotropin-releasing hormone and, **3:**322, 324–25
Thyrotropin-releasing hormone, **2:**19, 20, 35, 37, 39
  defined, **5:**393
  fetal, **3:**339
  functions of, **3:**322, **6:**89, 324–25
  perinatal, **3:**347
  production and activity of, **6:**18–19
  spinal cord dysfunction and, **7:**226
  in treatment of amyotrophic lateral sclerosis, **1:**235
Thyroxine, **2:**415
  fetal, **3:**339, 340
  imbalances of, **5:**707
  metabolic regulation role, **4:**1006
  perinatal, **3:**348
  production, **7:**485
  thyroid-stimulating hormone and, **5:**400
Thyroxine-binding globulin, fetal, **3:**339
Thyroxine receptors. *see* Receptors, thyroid hormone
Tibial collateral ligament, **1:**376
Tibialis anterior muscle
  sole of foot and, **3:**724
  spindles of, **3:**722
Tibiofemoral joint, **1:**376
Tic douloureux, **5:**368, **7:**678
Ticks, **2:**390
  toxins produced by, **7:**597
  as vectors in zoonoses, **7:**893–95
"Tight coupling", bioenergetic, **3:**931
Tight junction, **2:**271–73
Time, cultural construction, **4:**698–700
Time factors, in behavioral structure, **3:**590
Time-resolved fluorescence spectroscopy, **4:**355–56, **5:**578–80
Time-sharing, verbal-manual, **2:**862
Timolol, **2:**171, 172, 247
Tinidazole, **2:**382, 383
Tinnitus, **5:**740
Tissue, connective. *see* Connective tissue
Tissue biopsy. *see* Biopsy
Tissue culture. *see also* Cells, cultured
  fecal, **1:**557
  genital tract, **1:**558–59
  methods, **1:**553
  in Parkinson's disease treatment, **5:**672
  skin, **1:**557–58
  wound, **1:**557–58

Tissue edema safety factor. *see under* Edema, interstitial
Tissue factor, **7:**493–98
  blood coagulation cascade and, **7:**493–95
  clinical use of, **7:**495
  cofactor function of, **7:**494–95
  in disease, **7:**498
  expression
    by cells, **7:**496–97
    inducers of, **7:**497
  function inhibitor, **7:**497
  homeostasis and, **7:**497
  molecular biology of, **7:**496
  protein structure of, **7:**495–96
  regulation of, **7:**496–97
Tissue homogenization
  defined, **3:**403
  technique of, **3:**405
Tissue repair
  biological activities in, **7:**501
  cellular events in, **7:**501–2
  deficient, **7:**504–6
  epithelial, growth factor therapy and, **7:**506
  extracellular events in, **7:**501–2
  growth factor therapy in, **7:**506–8
  growth factors and, **7:**499–509, 500, 502
  model of PDGF and TGF-beta-mediated, **7:**504
  normal, **7:**501–4
  physiologic and pathologic processes, **7:**508
  subcellar events in, **7:**502–4
Tissue stiffness
  defined, **5:**173
  of muscle, **5:**183–84
  of tendon, **5:**183
Tissue typing. *see* Histocompatibility testing
Tissues. *see also* Cells; specific types of cells
  autofluorescence spectra, **5:**596
  blood perfusion of, **5:**621–22
  body
    mass density data for, **6:**439
    nutritional factors influencing composition of, **6:**436–38
    radiation interaction
      data for, **6:**439
      properties of, **6:**435–40
  combined absorption and scattering of, **5:**591–92
  dielectric properties of. *see* Dielectric properties of tissue
  fluorophores, **5:**596
  infrared absorption by, **5:**589
  light absorption, **5:**588, 592

Tissues *(Continued)*
　light effects, **5:**596–97
　light scattering in, **5:**590–91, 592
　optical properties, **5:**587–97, 592
　optical window and photon absorption by, **5:**591
　oxygen transport to, **5:**617–23
　parasite invasion of, **5:**646–47
　plant. *see* Plant cells
　radiative de-excitation of, **5:**596
　soft
　　and hard tissue repair, **7:**507–8
　　hydrogel of, **2:**645
　　infections, effect of quinolones on, **6:**409
　　water, lipid, and protein content of, **6:**438
　spectroscopy, **5:**592
　total absorption spectrum in, **5:**589–90
　transforming growth factor-beta and development, **7:**637–38
　ultraviolet absorption by, **5:**589
Titin. *see* Connectin
TNF. *see* Tumor necrosis factor
Tobacco, **2:**114. *see also* Smoking
　characteristics of, **7:**512–13
　curing and aging of, **7:**512–13
　marijuana and, **5:**446
　nitrosamines in, **3:**680
　products, as source of radiation, **6:**443
　smokeless, **2:**114
Tocopherol. *see* Vitamin E
Tocotrienols, in food analysis, **4:**219–20
Toes, **2:**594
　muscles of, **4:**266
Togaviridae, **7:**521–25
　classification of, **7:**521
　control measures of, **7:**525
　diseases associated with, **7:**522–24
　epidemiology of, **7:**524
　molecular biology of, **7:**524–25
Togavirus vaccines, **7:**525
Token economy, and operant conditioning, **2:**635
Tolerance
　of barbiturates, **5:**443
　for caffeine, **2:**70
　of cocaine, **5:**441
　for deviance, **4:**85
　for drugs, **2:**514, **5:**439
　immune, **4:**325
　marijuana and cannabinoids, **4:**909
　to psychedelic drug use, **5:**448
Tomes
　granular layer of dentine, **2:**775
　process, **2:**822

Tomography, computerized
　in brain research of alcoholics, **1:**171–72
　for diagnosis of mental disorders, **4:**995, 997
　and diagnosis of strokes, **7:**304
　for localization of cognitive functions, **1:**480
Tomography, emission-computed
　brain functional area localization, **4:**202
　cerebral blood flow and, **3:**224–25
　for diagnosis of senile dementia of the Alzheimer's type, **1:**131
　and diagnosis of strokes, **7:**305
　for localization of cognitive functions, **1:**483
　Parkinson's disease and, **5:**670
　single-photon emission computed tomogram scan, **7:**305
Tomography, positron-emission. *see* Tomography, emission-computed
Tomography, radionuclide-computed. *see* Tomography, emission-computed
Tomography, ultrasonic. *see* Ultrasonography
Tomography, X-ray computed
　defined, **3:**1
　in diagnosis, **3:**5–6
　　of senile dementia of the Alzheimer's type, **1:**131
Tongue, **2:**780
　dorsal surface of, **7:**528
　muscles of, **4:**261
　and taste, **7:**527–35
Tonic cells, in saccades, **3:**558
Tonic-clonic seizures
　in epilepsy, **3:**456
　in febrile convulsions, **3:**457
　generalized, on awakening, **3:**459
Tonofilament, **2:**270, **3:**533
Tools, of hominids, **3:**496, 497–99
Tooth, **2:**599, 707, 769, 773, 789–93, 817–19, 823, 825–29. *see also* Dental enamel; Gingiva
　of anteater, **2:**707
　apical infection of, **5:**612
　bud, **2:**817, 819
　canine, **2:**770, 772
　crown, **2:**789, 792, 793, 798, 799, 819
　cusp, **2:**770, 773, 774, 819
　decay. *see* Caries, dental
　dentine, **2:**774–75, 817
　development, **2:**818, **4:**29
　and diet, **2:**798
　of Eskimos, as tool, **2:**800
　Hertwig's sheath, **2:**821
　impacted, **5:**613

Tooth (*Continued*)

incisor, **2:**770, 820, 772

labbret, **2:**801, 802

mineralization, by salivary proteins, **6:**696

mineralogenesis, **2:**828-29

molar, **2:**770, 772, 773

mutilation, cultural, **2:**802

nerves supplying, **5:**368

permanent, emergence, **4:**27, 29

and pipe-smoking, **2:**801

pulp, **2:**775-76

tissue, **2:**789

as tool, **2:**800

and toothpick, **2:**801

types, **2:**790

Tooth abrasion, **2:**779, 798, 799

Tooth calcification, **2:**823, 828-29

and hydroxyapatite, **2:**828-29

secondary, **2:**824

Tooth eruption, **2:**827-28

Tooth root, **2:**789, 792

formation, **2:**820-21, 826-27

Toothache, **2:**825, **5:**368-69

Topographic brain mapping. *see* Brain mapping

Topoisomerase, **2:**263

Torque

patterns, in bicycling, **5:**188

production, at different angles, **5:**184

Torralba, period in Spain, **2:**198

Total finger ridge count

defined, **3:**89

sexual dimorphism of, **3:**90

Total internal reflection fluorescence microscopy, **5:**581

Total lung capacity, **3:**902

Total lymphoid irradiation, **4:**364

Total tension curve, defined, **5:**181

Touch, **2:**862

active and passive conditions, **7:**53

epidermal ridges and, **3:**86

food acceptance and, **3:**644

information processing using. *see* Sensory system

object motion across the skin and sensation, **7:**59-60

by sensate robots, **7:**63

sense of, in orofacial region, **5:**362-64

sensitivity of, **7:**53-56

spatial, **7:**54-55

temporal, **7:**55-56

skin and, **7:**51-63

spatial micropatterns and texture perception and, **7:**56-58

Touch (*Continued*)

tactile character recognition and, **7:**56-58

texture perception and, **7:**58-59

Tourette's disorder. *see* Gilles de la Tourette syndrome

Toxic agents. *see* Toxins

Toxic hepatitis. *see* Hepatitis, toxic

Toxic shock syndrome, **5:**4

Toxic substances, environmental. *see* Hazardous substances

Toxic vascular headaches, **4:**54

Toxicity. *see* specific types of toxicity

Toxicokinetics, **2:**109

Toxicology, **2:**342, **7:**545-58

applied, **7:**557

concepts of, **3:**384-85

defined, **2:**342

exposure assessment, methods of, **3:**389

scope of, **5:**684

Toxin-catalyzed ADP-ribosylation

defined, **4:**13

G proteins and, **4:**14, 22, 23

gangliosides and, **3:**734

Toxins

absinthe, **2:**346

acrylamide, **2:**346

action of, route of exposure as factor in, **7:**546-47

aerial, **2:**342

algal, **7:**590-91

aluminum, in Alzheimer's disease, **1:**214

animal, **7:**595-600

atropine, **2:**346

bacterial, **4:**418, **7:**588-90

in bacterial virulence, **1:**242

benzene, **2:**346

biological, **2:**341

botulinum toxin, **2:**345

caffeine, **2:**71

carbon disulfide, **2:**346

carbon monoxide, **2:**345

chemical, **2:**341

from chemical interactions, **3:**390-91

cholera, intracellular activity of, **5:**1-3

Clostridium perfringens, **5:**3

Cyclosporin A, **4:**363

defined, **2:**341

diphtheria, intracellular activity of, **5:**2-3

drugs, **2:**342

Entamoeba histolytica, **1:**224-25

environmental, **2:**342

Escherichia coli, intracellular activity of, **5:**1-3

Toxins *(Continued)*
ethyl alcohol, **2:**346
exposure
intentional, **2:**342–43
routes of, **7:**546–47
unintentional, **2:**342
food additives, **2:**342
fungal, **7:**591–93. *see also* Mycotoxins
ganglioside receptors for, **3:**734
immunotoxins, **4:**421
inhaled
assessing risks and controlling exposures to,
7:584–85
respiratory disposition of, **7:**578–82
respiratory responses to, **7:**582–84
intracellular site of action of, **5:**1–3
invertebrates, **7:**595–98
marijuana and cannabinoids, **4:**907
mechanism of action, **7:**546–47
membrane damaging, **5:**3–4
microbial protein, **5:**1–4
natural, **7:**587–601
classification of, **7:**588
in food, **3:**672
sources of, **7:**588–600
structures of, **7:**589
uses of, **7:**600
organic chemical, **7:**551
hepatotoxicity of, **7:**550
nephrotoxicity of, **7:**550
neurotoxicity of, **7:**547–49
pulmonary toxicity of, **7:**549
reproductive effects of, **7:**551
Parkinson's disease and, **5:**667
pertussis, intracellular activity of, **5:**1–3
plant, **4:**417–18, **7:**594–95
responses of, factors influencing, **7:**546–47
snake, **7:**599–600
Staphylococcus aureus, **5:**3–4
subunit, mechanism of action of, **5:**1–3
tetanus, intracellular activity of, **5:**2–3
thiol-activated, **5:**3–4
toxic shock syndrome, **5:**4
vertebrate, **7:**598–600
Toxoplasmosis, **2:**383, **6:**277–78
drugs against, **2:**383, 385
fetal infections, **4:**456–57
as indication of fetal blood samplings, **1:**731, 732
newborn screening test for, **5:**437–38
as scalar-borne zoonosis, **7:**897–98

Trabecular meshwork, anatomy, **3:**537, 538
Trace elements, **2:**649
atomic absorption spectometry for, **2:**649
biological roles of, **7:**604
deficiency, **2:**649
definition of, **7:**603
dose-response curve for, **7:**604
essential and possibly essential, **7:**605–12
functional significance of trace elements, **5:**49
homeostatic regulation of, **7:**604–5
immunity and, **5:**513–14
importance of, **5:**51–53
metabolism in human, **2:**649
and nutrition, **7:**603–13
nutritionally essential, definition of, **7:**604
potential clinical importance of, **7:**605
toxicity of, **3:**386–87
Trachea
airway obstruction of, **6:**359–60
anatomy of, **6:**590–91
structure of, **6:**359
Tracheobronchial region, **7:**578
Traction fiber theory, **5:**71
Traction headaches, **4:**54
Tracy, **5:**716
Tragedy, human response to, **7:**323–25
Trailmaking Test, **1:**612
Traits, personality. *see under* Personality
Trance
as coping mechanism, **7:**615–23. *see also* Hypnosis
defined, **7:**615
historical perspective of, **7:**615–16
hypnotic, therapeutic use of, **7:**621–23
logic, deep hypnosis and, **7:**618
pain alleviation and, **7:**622
Tranquilizing agents, major, **2:**217, 843, 219
dopaminergic mesolimbic system and, **5:**422
as dopaminergic receptor blocker, **2:**222
for schizophrenia, **6:**774–75
side-effects, **2:**217
Tranquilizing agents, minor, **1:**135
Trans-activation (genetics)
defined, **3:**151, 701
by *fos* protein, **3:**706–7
in gene amplification, **3:**787
Transacting factors
defined, **3:**107
in gene transcription regulation, **3:**112–15
modification of, **3:**115
Transactivating domains (genetics), **3:**114–15

Transacylase, **2:**540
Transacylation, **2:**542
Transcortical sensory dysphasia. *see* Dysphasia, sensory
Transcortin, steroid transport in blood and, **7:**264
Transcription, genetic, **3:**107–16
    defined, **3:**791
    DNA binding and, **3:**117
    DNA to RNA, **2:**235, 291
        by cortical neuron, **2:**433–34
        factor, **2:**294
        factor III-A, **2:**659
           and zinc, **2:**659
        inhibition, **2:**434
        intron removal, **2:**295
        and nucleosome positioning, **2:**434
        and RNA maturation, **2:**295
        sequence deletion, **2:**435
        splicing, **2:**294
    in Epstein-Barr virus, **3:**466
    of eukaryotic genes, **3:**793–94
    of *fos* gene, **3:**702, 707
    in gene definition, **3:**792
    in muscle, **5:**165
    nucleosomes and, **3:**131–33
    upstream control sequences in regulation of, **3:**112
Transcription factors
    defined, **3:**701
    DNA methylation and, **3:**158
    eukaryotic, **3:**110–11
    nucleosome assembly and, **3:**131–32
Transcytosis, defined, **6:**63
Transducers
    defined, **3:**318, 331
    perceptual functions and, **5:**737–38
Transducin
    in amplification, **4:**15
    beta subunit of, **4:**19
    cyclic GMP phosphodiesterase and, **5:**417
    gamma subunit of, **4:**19
    in phototransduction, **4:**16
Transduction, **2:**547, 548, 552, 555
    in hair cells, **3:**235
    membranes, **4:**974–75
    signal
        at cellular surface, **7:**15–21. *see also* Signal transduction
        sensory, **7:**19
        visual, **7:**19
Transduction, genetic
    chromosome mapping and, **3:**824
    in peptides, **5:**727

Transendothelial channels, **5:**34, 35
Transexualism, **4:**997
Transfection, **2:**435
    defined, **3:**175, 801
    of mammalian cells, **3:**805
    technique, **2:**231–32, 732
    vectors in, **3:**805–6
Transfer RNA. *see* RNA, transfer
Transfer tests. *see* Retention tests
Transferases, methyl, in DNA repair, **6:**549–50
Transferrin, **2:**653–56, 659
    Sertoli cell secretion of, **7:**423–24
    targeting moiety of immunoconjugates, **4:**374–75
    transport, in brain, **1:**726
Transferrin receptors. *see* Receptors, transferrin
Transformation, cell. *see* Cell transformation
Transformation, genetic
    chromosome mapping and, **3:**824
    hyperthermia and, **4:**300
Transforming genes. *see* Oncogenes
Transforming growth factor, **2:**61, 89, 259, 260, 753, 757, 760
    effects on bone cell function, **1:**765
    molecular forms, **2:**760
    properties, **2:**757
Transforming growth factor alpha, **7:**625–29
    and epidermal growth factor, **7:**628–29
    function of, **3:**446–47
    historical aspects of, **7:**625–26
    hypercalcemia of malignancy etiology, **4:**277
    pathological roles of, **7:**628
    physiological roles of, **7:**628
    precursor, structure of, **7:**626–27
    production and estrogen, **4:**279
    receptor interactions involving, **7:**627
    structure of, **7:**626
    synthesis, cellular sources of, **7:**627
    testicle secretion of, **7:**420
Transforming growth factor alpha receptors. *see* Receptors, epidermal growth factor-urogastrone
Transforming growth factor beta, **1:**687
    active and inactive, **7:**635
    in acute phase response, **1:**28–29
    in bone and calcium metabolism, **1:**806
    bone cell function and, **1:**765
    bone resorption inhibition by, **1:**762
    bones and, **3:**485
    in cell physiology, **7:**637–38
    cell proliferation and, **7:**636
    differentiation and, **7:**637

Transforming growth factor beta *(Continued)*
  extracellular matrix and, **7:**636-37
  as factor in bone cell differentiation, **1:**788
  historical perspective of, **7:**632
  immunogenic inflammation, **4:**394
  latent form of, **7:**633-36
  metabolism in bone and calcium, **1:**806
  negative epidermal proliferation regulation and, **3:**448
  polypeptide sequences of, **7:**635
  precursor, structure of, **7:**632-33
  in regulation of beta cell response, **1:**569
  in reversing healing deficits, **7:**505-6
  role in bone cell differentiation, **1:**788
  structure of, **7:**632-33
  superfamily of genes, **7:**633
  in tissue development, **7:**637-38
  and tissue repair therapy, **7:**506-8
Transforming region. *see* Base sequence
Transfusion of blood. *see* Blood transfusion
Transgenic animals. *see* Animals, transgenic
Transgenic mice. *see* Mice, transgenic
Transient ischemic attack, **7:**303
Transition state analogue, defined, **3:**397
Transition state theory, **3:**424-25
Transketolase, **2:**528
Translation, genetic, **2:**291
  HIV-1, **1:**18
  of RNA to polypeptide, **2:**235
Translocation (genetics), **4:**951-52
  breakpoints, isochores and, **4:**255-56
  cancer and, **5:**326-27
  defined, **3:**101
  genomic mutations and, **3:**884
  insertional, chromosomal, **3:**822
  membrane proteins, **4:**972-73
  in muscular dystrophy, **5:**201
  in primate genetic relations, **3:**884
  reciprocal, chromosomal, **3:**822
Translocation (pathology), defined, **5:**139
Transmethylase, **2:**537
Transmission electron microscopy. *see* Microscopy, electron, transmission
Transpeptidases. *see* Peptidyl transferases
Transplacental exposure. *see* Maternal-fetal exchange
Transplantation
  animal models, **1:**258-59
  autologous, in Parkinson's disease, **5:**672
  brain tissue, in Parkinson's disease, **5:**671
  donor availability, **4:**365
  myoblasts, in *mdx* mice, **5:**203

Transplantation *(Continued)*
  neural, **5:**345
  solid-organ, **1:**797
Transplantation, bone marrow. *see* Bone marrow transplantation
Transplantation, heart. *see* Heart transplantation
Transplantation, kidney. *see* Kidney transplantation
Transplantation immunology, **4:**357-65
  ABO blood group matching, **4:**362
  antigens. *see* Histocompatibility antigens
  blood transfusion, **4:**363-64
  chimerism and donor-specific tolerance, **4:**364-65
  costimulatory signals, **4:**361
  cyclosporin a therapy, **4:**363
  EBV-DNA-positive lymphoproliferative disorders and, **3:**466
  effector pathway, **4:**362
  future research, **4:**365
  graft rejection, **4:**360-62
  histocompatibility antigen genetics, **4:**358-59
  hyperacute rejection, **4:**362
  immunobiology, **4:**338
  lymphohematopoietic chimerism, **4:**364-65
  making grafts nonimmunogenic, **4:**362
  MHC role, **4:**359-60
  nonspecific immunosuppression, **4:**363-64
  polyclonal and monoclonal antibodies, **4:**363
  prevention of rejection, **4:**362-65
  research goals, **4:**357-58
  sensitization pathways, **4:**360-61
  specific immunosuppression, **4:**364-65
  total lymphoid irradiation, **4:**364
Transport, axonal. *see* Axonal transport
Transport, biological. *see* Biological transport
Transthyretin. *see* Prealbumin
Transvascular exchange
  pathophysiology of, **5:**39-41
  physiology of, **5:**38-39
Transverse colon, anatomy, **3:**82
Transverse sinus, **4:**983
Transverse tubules, in muscle contraction, **5:**225
Tranylcypromine, **1:**309-10
Traub, Roger, **5:**388
Trauma. *see* Wounds and injuries
Trauma, acoustic. *see* Hearing loss
Trazodone, **1:**309
Trehalose dimycolate, **7:**753
Trematode. *see* Fluke
Tremor, **2:**221
  and dopamine, **2:**221
  ventral oral nucleus of thalamus and, **7:**442

Trenchfoot, **4:**798–99
Treponema pallidum, **1:**558–59, **4:**450
Tretinoin, **7:**68–69
Triacylglycerol lipase. *see* Lipase
Triacylglycerols. *see* Triglycerides
Triamterene, **3:**99
Triangulation, familial, **3:**245
Triatominae, and trypanosomiasis, **6:**270
Triazolam, **2:**593
Tributyrinase. *see* Lipase
Tricarboxylic acid cycle, **1:**458, **2:**240, 241
Triceps
    defined, **3:**691
    motor-unit activity in, **3:**697
    torque activation in, **3:**695
Triceps surae complex, in bicycling, **5:**187
Trichinella spiralis, **1:**265
    characteristics of, **7:**645
    encystment of, **7:**649
    scalar-borne trichinosis and, **7:**898
Trichinosis, **1:**265, **2:**387, **7:**645–57
    animals susceptible to, **7:**646–47
    clinical signs of, **7:**652–53
    control of, **7:**655–56
    diagnosis of, **7:**653–54
    epidemiology, **7:**647–49
    etiology of, **7:**647
    geographical distribution, **7:**646
    history of, **7:**646
    immunology, **7:**651–52
    lesions associated with, **7:**653
    life cycle stages in, **7:**649
    pathogenesis of, **7:**649–51
    pathology, **7:**650
    prevention of, **7:**656
    prognosis, **7:**654–55
    rodent, **7:**648
    as scalar-borne zoonosis, **7:**898
    swine, **7:**655
    sylvatic, **7:**649
    symptoms and signs of, **7:**652–53
    treatment of, **7:**655
Trichlorhinophalangeal syndrome, **2:**440
Trichomonas infections, **1:**268–69, **6:**273–74
Trichomonas vaginalis, **2:**382, 383
Trichorrhexis nodosa, **4:**48
Trichostrongylus, **2:**387
Trichothecene mycotoxicosis, **7:**659–69
    alimentary toxic aleukia disease as, **7:**665–66
    diagnosis of, **7:**667–68

Trichothecene mycotoxicosis *(Continued)*
    preventive measures for, **7:**668
    scabby wheat poisoning as, **7:**666
    therapeutic approaches to, **7:**668
Trichothecenes
    assays to detect, **7:**661
    biosynthesis of, **7:**662
    chemistry of, **7:**660–61
    conditions leading to production, **7:**661–62
    diseases attributable to, **5:**263–64
    fungi producing, **7:**661–62
    immunotoxicity of, **7:**664
    lethal dose 50 (LD$_{50}$) of, **7:**663
    metabolism of, **5:**264, **7:**666–67
    mode of action of, **7:**667
    mutagens or carcinogens, **7:**665
    natural occurrence of, **7:**662
    stability of, **7:**660–61
    structure of selected, **7:**660
    toxicological effects of, **7:**662–65
Trichotillomania, **4:**42
Trichromacy, **2:**579, 585–86
    law of, **2:**579–80
    matching experiment, **2:**580
Trichuriasis, **1:**269–70, **4:**116, 119–20
Trichuris, **2:**387
Trichuris trichiura, **4:**114
Triclabendazole, **2:**389
Tricuspid valve, **2:**180
Tricyclic antidepressive agents. *see* Antidepressive agents, tricyclic
Triethyltin, **2:**345, 346
Trigeminal nerve, **5:**363
Trigeminal neuralgia, **5:**368
Trigeminal system
    association with muscles of mastication, **7:**673
    classes of sensory receptors, **7:**675
    distribution of, **7:**672–73
    somatic sensory receptors of, **7:**671–79
Triglyceride lipase. *see* Lipase
Triglycerides, **2:**168, 170, 662
    behavior and distribution, **4:**743
    in breast milk, **4:**605
    characterization, **4:**742
    chemistry of, **3:**567–68
    defined, **3:**71, 567
    digestion of, **3:**48, 51, 80
    emulsified, **3:**587
    fatty acid distribution among, **3:**570–71
    food analysis, **4:**222

Triglycerides *(Continued)*
    in gingiva, **4:**36
    in human milk fat, **3:**48
    hydrolysis of, **3:**52
    lipase imbalance and, **3:**50
    long-chain, bile salt-stimulated lipase and, **3:**50
    metabolism, **4:**742-43
    storage of, **3:**583
Trigone, coronary, **2:**158
Triiodothyronine, **2:**20, 326, 433
    defined, **2:**321
    fetal, **3:**339
    metabolic regulation role, **4:**1006
    perinatal, **3:**348
    thyroid-stimulating hormone and, **5:**400
Triiodothyronine receptors. *see* Receptors, thyroid hormone
Trimazosine, **2:**177
Trimethoprim
    mechanism of action, **1:**326-27
    uses of, **1:**329
Trimethoprim-sulfamethoxasole, **2:**372, 383, 385
Trimethoxyphenethylamine. *see* Mescaline
1,3,7-Trimethylxanthine. *see* Caffeine
Triorthocresyl phosphate, **2:**346, 348
Triosephosphate isomerase
    deamidation of, **3:**419
    irreversible thermoinactivation rate of, **3:**421
Triosephosphate mutase. *see* Triosephosphate isomerase
Tripeptide. *see* Thyrotropin-releasing hormone
Triphosphates, deoxynucleoside. *see* Deoxynucleoside triphosphates
Triplets, Weinberg's rule and, **3:**306
Triploidy, lethal, **2:**437
Triradius
    defined, **3:**86
    extralimital, **3:**89
Trisomy, **4:**951-52
    congenital anomalies and, **3:**821-22
    defined, **3:**835
    partial, **3:**822
    rate of, **5:**247
    and spontaneous abortion, **2:**438
Trisomy 8, **2:**441
Trisomy 13, **1:**699-700, **2:**438, 441
Trisomy 14, **2:**441
Trisomy 16, **2:**438
Trisomy 18, **1:**699, **2:**438
Trisomy 19, **2:**441
Trisomy 21. *see* Down's syndrome

Trisomy X, **2:**443
Tritan axis of color, **2:**581
Tritanopia, **2:**590-91
Triticum vulgare lectins. *see* Wheat-germ agglutinins
TRNA. *see* RNA, transfer
Trochoid joints, **1:**374
Trombicula, **2:**390
Tropani alkaloids, **1:**187-79
Trophectoderm, defined, **3:**305, **6:**63
Trophoblast
    differentiation, **4:**426-27
    in embryonic development, **3:**610
    epithelial penetration, **4:**424
    plate stage, **4:**425
    previllous stages, **4:**425
Trophozoites, **1:**217-22, **2:**381
Tropical spastic paraparesis, **1:**738
Tropocollagen, **2:**204, **4:**35-36
Tropoelastin
    cross-linking of, **3:**267
    elastin and, **3:**264
Tropomyosin, **2:**92, 172, 181, **5:**232, 233, 234
    actin and, **5:**169
    Ca^{2+} and, **5:**225-26
    defined, **5:**173
    functions of, **5:**175-76, 217
    genetic analysis of, **5:**169
    as myosin substratum, **5:**165
    of nonmuscle cells, **5:**218
    in skeletal muscle, **7:**25
Troponin, **2:**3, 89, 92, 93, 172, 181, **5:**232
    actin and, **5:**169
    Ca^{2+} and, **5:**206, 225-26
    calcium and, **5:**175-76
    defined, **5:**173-74
    in skeletal muscle, **7:**25
    structure of, **5:**219
Troponin C
    as $Ca2+$ receptor protein, **3:**333
    isoforms of, **5:**220
Troponin I, **5:**220-21
Trout, rainbow, **3:**676
Truth, **2:**620-22
    and falsity, **2:**620
    and supposition, **2:**621-22
Truth disclosure, in persuasion, **5:**803
Trypanosoma, **7:**683
    African, **5:**652, **7:**682
    American, **7:**682
    antigenic variation, **4:**387-88

Trypanosoma (Continued)
classification of, **7**:681–82
fatty acid remodeling of, **5**:653
life cycle stages of, **7**:681–82
polyamine metabolism in, **6**:71–72
Trypanosoma brucei brucei, **4**:448, **5**:650
Trypanosoma brucei complex, **2**:384
Trypanosoma cruzi, **4**:448
Trypanosome lytic factor, **5**:650
Trypanosomiasis, **4**:448, **6**:269, **7**:681–85
buthionine sulfoximine for, **3**:917
and disease, **6**:269–71
Trypanosomiasis, African, **2**:383–85, **4**:448, **6**:269–70
animal, **7**:684
drugs against, **2**:383–85
human, **7**:683–84
Trypanosomiasis, American, **1**:740, **2**:383, 385, **4**:388, 448, **6**:270–71, **7**:684–85
Trypanosomiasis, Gambian, **7**:683
Trypanosomiasis, Rhodesian, **7**:683–84, 895–96
Trypanothione
Leishmania, **4**:682
as target of parasite chemotherapy, **5**:655
Trypanothione peroxidase, **4**:682
Trypomastigote, defined, **7**:681
Tryptamine receptors. *see* Receptors, serotonin
Tryptophan, **2**:13, 14, 532, 847
L-tryptophan, **3**:22
Tryptophan hydroxylase, **2**:13
Tryptophan pyrrolase, **2**:655
Tryptophan synthetase gene, of *Escherichia coli*, **3**:792
Tsetse flies, **6**:270, **7**:682
TSH. *see* Thyrotropin
Tuber cinereum, anatomy, **5**:394
Tuberculosis
antitubercular drugs, **1**:326, 329
as scalar-borne zoonosis, **7**:897
vaccines for. *see* BCG vaccine
Tuberculosis, meningeal, **5**:706
Tuberculosis, oral, **5**:614
Tuberculosis, pulmonary, **5**:706
Tuberoinfundibular neurons, **4**:307, 312
Tubular adenomas. *see* Adenomatous polyps
Tubular reabsorption
active transport and, **3**:95
defined, **3**:93
Tubular secretion, defined, **3**:93
Tubule, transverse, **2**:127, 129, 130
Tubulin, **2**:49, 164, 241
casein kinase II and, **5**:895, 896

Tubulin *(Continued)*
genetic expression of, **5**:17
in metaphase, **5**:70–71
modification of, **5**:17–18
phosphorylation of, **5**:18, 894
platelet aggregation and, **5**:901
by tyrosine-specific protein kinases, **5**:897
in vivo, **5**:898
structure of, **5**:12
Tubulin alpha
acetylated, **5**:18
detyrosination of, **5**:18
Tubulin beta, phosphorylation of, **5**:899
Tumor cells
altered adhesive properties of cells, **4**:1014
anchorage-independence, **4**:584
blood-borne, arrest, **4**:1012–13
characteristics of, **7**:687–88
entry into circulation, **4**:1011–12
epithelial, **4**:584
growth factor requirements, **4**:584
immortality, **4**:584
immunoconjugate accumulation in, **4**:377
interaction with immune system, **4**:1012
invasive properties, **4**:1013
killing of, macrophage activation and, **5**:817–18
morphological alterations, **4**:584
motility, **4**:1013
surfaces, **4**:344–45
survival in bloodstream, **4**:1012
systemic effects of peptides, **4**:275
targets for immunoconjugates, **4**:373–74
transforming growth factor-alpha synthesis in, **7**:628
Tumor clonality, **7**:687–93
animal, **7**:692
biological significance of, **7**:692–93
human, **5**:323, **7**:691–92
methods of studying, **7**:688
terminology associated with, **7**:687
X-linked markers for, **7**:690
Tumor growth factor beta, **5**:168
Tumor-infiltrating lymphocytes, **4**:1012
Tumor-like lesions, of bone, **7**:703
Tumor necrosis factor, **1**:681–82, **2**:249, 753, 757, 759
in acute phase response, **1**:28, 32
antiradiation effects of, **6**:423
and Bacillus Calmette-Guérin, **2**:759
in bone resorption, **1**:807
DNA methylation and, **3**:154
effects on bone cell function, **1**:765

Tumor necrosis factor (*Continued*)
hypercalcemia of malignancy etiology, **4:**277–78
immunotherapy with, **4:**347–48
in inflammation, **1:**313, 314, **4:**474–75
lymphotoxic, **2:**759
mediating graft injury, **4:**362
monocytopoiesis inhibition, **4:**854
and mycoplasma, **2:**759
pathogenic role, **4:**862
properties, **2:**757
and thymocyte, **2:**759
in tobacco-associated cardiovascular disease, **7:**517
types, **2:**759
Tumor necrosis factor-beta. *see* Lymphotoxin
Tumor promoters. *see* Carcinogens
Tumor rejection
B- and T-lymphocytes, **4:**340
lymphokine activated killer cells, **4:**340
macrophages, **4:**340
natural killer cells, **4:**339–40
oncogenic agents, **4:**338–39
suppressor cells and factors, **4:**341
Tumor suppressor genes. *see* Genes, suppressor, tumor
Tumorigenesis. *see* Carcinogenesis
Tumors. *see* Neoplasms
Tunicamycin, and Golgi apparatus function, **4:**12
Tuning curves
of auditory nerve fibers, **3:**238
for basilar membrane vibrations, **3:**236
central auditory system organization and, **3:**240
in cochlear pathology, **3:**241
defined, **3:**229
Turbinate, defined, **5:**523
Turkanopithecus, **3:**489
Turkey X disease, **5:**264
Turner's syndrome, **2:**443–44, **6:**859–60
genetic aspects of, **3:**821, 838
heterokaryotes and, **3:**306
muscular dystrophy and, **5:**192
Turpin, Raymond, **3:**176
Turtles, **6:**570
Twenty-four hour rhythm. *see* Circadian rhythm
Twin studies, **2:**712
of affective illness, **3:**844
in behavioral genetics, **3:**837, 842–43
of complex trait inheritance, **3:**856
and criminality, **2:**712, **3:**844
pairwise concordance, **2:**712
personality trait analysis from, **5:**771–73
Twin transfusion. *see* Fetofetal transfusion

Twins, **3:**305–15
embryology of, **3:**305–15
mothers of. *see also* Pregnancy, multiple
placentation of, **3:**305–15
"third type," **3:**306
types of, **3:**305–7
vanishing, **3:**314
Twins, conjoined, physiology of, **3:**312
Twins, dizygotic
causes of, **3:**307
vascular communications between, **3:**312
Twins, identical. *see* Twins, monozygotic
Twins, monozygotic
congenital anomalies of, **3:**313
dichorial, **3:**308–9
embryology of, **3:**307–13
fingerprints of, **3:**89
monoamniotic, **3:**312–13
monochorial, **3:**309–12
origins of, **3:**314
persuasion and, **5:**805, 806
Twisted hair, **4:**48
Twitch
in arm muscles, **3:**695
defined, **3:**691
measurement of, **5:**177–78
skeletal muscle, **7:**28
Two-dimensional chromatography. *see* Chromatography, two-dimensional
Two-dimensional electrophoresis. *see* Electrophoresis, gel, two-dimensional
Two-phase systems, liquid-liquid. *see* Aqueous two-phase systems
Tympanic membrane, **3:**230–31
Tympanic region, core temperature measurement, **4:**793
Type A personality, **4:**75
Type I (alpha) errors, **4:**62
Type II (beta) errors, **4:**62
Typhoid, **6:**709–10
parathyphoid, **6:**710
salmonella, **6:**709–10
yearly reported cases of, **6:**714
Typhoid vaccine, **6:**710
Tyramine, **1:**310
Tyrosine, **2:**12, 14
autophosphorylation, defined, **5:**725
and catecholamine, **2:**218
defined, **2:**321
MAP phosphorylation and, **5:**898
phosphatase, **2:**320

Tyrosine (*Continued*)
  phosphorylation, **2**:61
  residues, in thyroglobulin synthesis, **3**:328
Tyrosine hydroxylase, **2**:12, 224, 337
  in catecholamine synthesis, **1**:643–44
  nerve growth factor and, **5**:334
  noradrenergic nerve terminals, rat, **5**:379
  production, laminin effects, **4**:626–28
Tyrosine kinase. *see* Protein-tyrosine kinase

U

Ubiquinol cytochrome *c* oxidoreductase, **5**:58
Ubiquinone, **2**:415, 544–45
  in electron transport chain, **1**:459–60
  in oxidate phosphorylation, **5**:474
Ubiquitin, **3**:419, **4**:103
UK14,305, **2**:848
Ulceration
  defined, **3**:755
  gastric blood flow and, **3**:763
Ulna, **7**:38
Ulnar loops, dermatoglyphic, **3**:88
Ultracentrifugation, **7**:711–22
  absorption optical system in, **7**:714
  detection of macromolecules by, **7**:714
  elementary theory of, **7**:712–13
  fundamentals of, **7**:712–15
  historical aspects of, **7**:711–12
  instruments for, **7**:713–14
  Rayleigh interference optics in, **7**:715
  scope of, **7**:711–12
  sedimentation equilibrium, **7**:718–21
  sedimentation velocity, **7**:715–18
    applications of, **7**:716–18
  zone sedimentation method in, **7**:717
Ultrafiltration, **1**:708
Ultrafluorimeter, **5**:565
Ultrasonic diagnosis. *see* Ultrasonography
Ultrasonics
  cavitation by, **7**:729–30
  description of, **7**:723–24
  effect on tissue, **7**:729–30
  experimental values for velocity of, **7**:725
  finite amplitude of, **7**:728–29
  heat production by, **7**:729
  propagation in tissue, **7**:723–32
  scattering of, **7**:727–28
  specific acoustic impedance and, **7**:726
  in tissue
    absorption of, **7**:727

Ultrasonics
  in tissue (*Continued*)
    anisotropy of, **7**:725–26
    attenuation of, **7**:726–28
    average velocity of, **7**:725
    beam structure and, **7**:729
    frequency content and, **7**:728
    temperature as factor affecting, **7**:725
    tissue composition and, **7**:726
    tissue type and, **7**:726
    velocity dispersion of, **7**:725–26
  velocity of propagation, **7**:724–26
Ultrasonography
  for atherosclerosis imaging, **1**:449
  defined, **3**:1
  in detection of prostate cancer, **6**:180
  in diagnosis, **3**:6–7
  and diagnosis of strokes, **7**:305
  medical application of ultrasonic spectrum, **7**:724
  in obstetrics, **3**:7
  passage of pulse through tissue, **7**:730–31
  in pediatrics, **5**:703
Ultrasonography, prenatal
  of birth defects, **1**:704
  defined, **3**:303–4
  views of normally developing pregnancy, **1**:2–3
Ultrasound. *see* Ultrasonics
Ultrasound transducer
  in fetal blood sampling, **1**:729–30
  purity of blood samples in, **1**:730
  results of, **1**:730
  specific technical problems with, **1**:730
Ultraviolet absorption microspectrophotometer, **5**:563
Ultraviolet rays
  absorption
    in eye, **5**:594–95
    in skin, **5**:593
  acute effects on skin, **7**:66–67
  chemical reactions and, **6**:443
  chronic effects on skin, **7**:67
  detection of, in drug analysis, **3**:206
  DNA damage caused by, **6**:548
  effect on DNA, **6**:444
  effect on skin, **7**:65–69
  exposure and melanoma, **4**:888
  mutation and, **5**:252
  in protein detection, **6**:192
  skin as protection from, **7**:68–69
  systemic effects of, **7**:67–68
  tissue absorption, **5**:589

Ultraviolet rays *(Continued)*
  tumors induced by, **4**:344
  visible detection, HPLC, **4**:210
Ultraviolet therapy, **7**:68
Umami, as basic taste, **3**:644
Umbilical cord, at birth, **3**:617
Umbilicus, of monochorial twins, **3**:310
Underlying cause of death, defined, **5**:99
Underwood, Michael, **5**:699
Ungulates, gut anatomy of, **3**:60
Uniform Determination of Death, **1**:656-57
Unilamellar vesicles. *see* Vesicles, unilamellar
Unilateral spatial agnosia. *see* Agnosia, unilateral spatial
Universal eating monitor, in feeding behavior research, **3**:595
Unmarried mothers. *see* Mothers, unmarried
Unreliability problem of single neuron, **2**:544
Uphill transport. *see* Biological transport, active
Upper respiratory tract. *see* Respiratory tract
Uranyl acetate, **2**:422
Uranyl ion, **2**:726, 727
Urea
  formation, amino acid metabolism and, **6**:233
  synthesis
    bicarbonate elimination and, **5**:879
    pH$_e$ homeostasis disturbances and, **5**:880
    ph$_i$ fall and, **5**:874
  as toxin in hemodialysis, **1**:405
Ureter
  function of, **7**:734
  obstruction and cell death, **2**:251
Urethane, **2**:111
Urethra
  anaerobic bacteria found in, **1**:241
  anatomy of, **6**:561
  autonomic innervation of, **1**:543
  function of, **7**:735
  sphincter, **5**:43-44
Urethritis, chlamydial, **6**:409
Uric acid
  extrarenal excretion of, **6**:392
  production and excretion of, purine biosynthesis and, **6**:391-92
  renal handling of, **6**:392
  synthesis of, **6**:390
  thiazide therapy and, **3**:98
Uridine diphosphate glucose, **3**:921
Uridine diphosphate glucose pyrophosphorylase, **3**:921
Uridine triphosphate, **3**:921

Urinary bladder
  autonomic innervation of, **1**:543, 548
  function of, **7**:734
Urinary flow, **2**:7, 9, 10, 143-46
  carbonic anydrase inhibitors and, **3**:96
  and diuresis, **2**:143
  osmotic diuretics and, **3**:100
  in thiazide therapy, **3**:98
Urinary tract
  anatomy, **7**:733-35
  endometriosis in, **3**:369
  function of, **7**:733
Urinary tract infections
  detection of bacterial pathogens in, **1**:554
  quinolones for, **6**:407-8
Urination
  acid-base homeostasis and, **5**:828
  physiology of, **3**:93-96
Urine
  collection and storage of, **7**:734-35
  detection of bacterial pathogens in, **1**:554
  pH of, **5**:878
Urogastrone. *see* Epidermal growth factor
Urogastrone receptors. *see* Receptors, epidermal growth factor-urogastrone
Urogenital system
  detection of bacterial pathogens in, **1**:558-59
  effect of radiation on, **6**:413-14
  fetal
    development of, **3**:341-42, 343
    from fourth month, **3**:618
    at third month, **3**:611
  protozoal infections involving, **6**:273-74
Uromorun, **2**:272
Uronic acid
  carboxyl groups of, **3**:42
  in plant food residues, **3**:36-37
Urticaria, cold, **4**:797
Uterine infections, **4**:456-57
Uterine neoplasms
  DNA distribution curves in, **3**:624
  oral contraceptives and, **5**:608
  and spontaneous abortion, **1**:7
Uterus, **2**:403-5, 408
  abnormalities
    hysterosalpingogram, **1**:8
    in spontaneous abortion, **1**:7-8
  anatomy of, **6**:563
  autonomic innervation of, **1**:543
  decidual cast of, **3**:295

Uterus *(Continued)*
defined, **3:**293
endocrine regulation of function, **7:**739–40
enzyme response to estrogens, **7:**740–41
estrogen receptors in, **7:**741–42
growth factors, estrogen-induced, **7:**743
hormones controlling, **7:**737–38
implantation site, **4:**427
morphology of, **7:**738–39
nucleic acid synthesis in, **7:**740–41
oncogenes in, **7:**743–44
protein synthesis in, **7:**741
response to estrogen, **7:**737–44
retroversion, endometriosis and, **3:**363
transcriptional mechanisms in, **7:**742
translational and posttranslational effects in, **7:**742–43
UV photoelectron spectroscopy. *see under* Spectroscopy, photoelectron
Uveal tract, **3:**538–40
Uvomorulin, **2:**271, **6:**62

V

Vaccination
immune responses, **4:**325–26
for papillomavirus infection, **3:**867
for parasitic diseases, **4:**390
Vaccine therapy. *see* Immunotherapy
Vaccines. *see also* specific vaccines
anti-idiotypic, **7:**751–52
antigen, **4:**446
antimalarial, **4:**884–85, **5:**655–56
bacterial, **4:**464
bovine leukemia, **6:**646
for childhood diseases, **5:**702–3
development, principles of, **6:**642–44
gammaglobulin, **4:**466
human T cell leukemia virus, **6:**647
immunoglobulin, **4:**447
lentivirus, **6:**646–47
live bacterial, **4:**446–47
live virus, **4:**444–45, 464
nature of, in autoimmune disease, **1:**339–40
oral, in mucosal immunology, **5:**141
peptide, **7:**748–49
recombinant subunit, **7:**749–51
toxoid, **4:**445
tumor, **1:**678
viral, **6:**642, **7:**787–89
Vaccines, inactivated, **4:**445–46, 464

Vaccines, synthetic, **7:**745–55
aluminum salts as vehicles for, **7:**754
biodegradable micro-capsules as vehicles for, **7:**755
liposomes as vehicles for, **7:**755
oil emulsions as vehicles for, **7:**754–55
synthetic vs. conventional, **7:**745–46
vehicles and adjuvants for, **7:**752
Vaccines, synthetic peptide, **7:**746–47
antibody-binding B cell epitopes and, **7:**746–47
and requirement for T cell help, **7:**747
Vaccinia virus, **4:**464
VACTERL association, **1:**698
Vacuolar organelles, **4:**564
Vacuoles, **5:**875
Vagina
anatomy of, **6:**564
autonomic innervation of, **1:**543
bacteria in, **4:**457–58
protozoal infection involving, **6:**273–74
Vaginal smears, **3:**861
Vaginitis
nonspecific, effect of quinolones on, **6:**410
trichomonal, **2:**382
Vagovagal reflex, monosynaptic, **2:**31
Vagus nerve, **1:**539, **2:**31–39, 138, 186
activation, **2:**34, 35, 37
activity, **2:**138, 140
bicarbonate secretion and, **2:**37
coding of, **2:**140
digestion and, **3:**74
electrical stimulation, **2:**39
fiber, **2:**35, 36
gastric acid secretion and, **1:**547–48, **5:**715–16
gastric circulation regulation and, **3:**762
motor nucleus, dorsal, **2:**31–34
neuron, sensory, **2:**31
pathways to airway smooth muscle, **1:**546
Valine
in elastin, **3:**264
glutamic acid and, **3:**144
Valinomycin, **5:**577
Valium. *see* Diazepam
Valves, **2:**162, 180
Van Allen radiation belts, **6:**428
Vanadate, **5:**20
Vanadium
biological function of, **7:**607
toxicity of, **7:**612
Vancomycin
effect on peptidoglycan biosynthesis, **1:**280–81
mode of action, **1:**290(t)

Vanishing twins. *see under* Twins
Variable regions, antibody. *see* Antibody variable regions
Variant surface glycoproteis, trypanasoma, **5**:652
Variation (genetics), **5**:805
Varicella-zoster virus, **4**:189, 194
Varicosities, **1**:536-37, 541
Vas deferens
    autonomic innervation of, **1**:543
    sperm transport and storage of, **7**:173-74
Vascular anastomoses
    acardiac fetus and, **3**:312
    in monochorial twin placentas, **3**:310
Vascular capacitance, **7**:757-65
    active and passive changes in, **7**:761
    chemical and physical factors influencing, **7**:762
    defined, **7**:757
    measurement of, **7**:763-65
    nervous factors influencing, **7**:762
    in stress or disease, **7**:763
Vascular compliance, defined, **7**:758
Vascular diseases, peripheral, **5**:604
Vascular headaches, **4**:51-54
Vascular muscle, **5**:170
Vascular theory of emotions, **1**:104
Vascular tissue, kinin effects, **4**:594-95
Vasidilator agents, **3**:372
Vasoactive intestinal peptide, **2**:22, 26, 27, 31, 33-37, 365, 686
    alpha-chain, **2**:559
    in autonomic neurons, **1**:541
    cellular, **2**:239
    cotransmission with acetylcholine in secretory tissue, **1**:546
    functions of, **5**:718, **6**:89
    gastric, inhibitory, **2**:27
    of gray matter, **7**:205
    immune functions and, **5**:384
    mechanisms of action, **5**:715
    "pancreatic cholera" and, **5**:722
    protein 240-ka, **2**:42, 45, 46
    in rat thymic cortex, **5**:380
Vasoactive mediators, **1**:27
Vasoconstriction, **2**:365
    by angiotensin II, **5**:730
    cold-induced, **4**:796
    neuropeptide Y and, **5**:720
Vasoconstrictor agents, **2**:783, 784
Vasodilation, **2**:142, 145, 365
    in exercise, **3**:512, 519
    by kinins, **5**:730

Vasodilator agents, **2**:175, 177
Vasomotor tone, in dynamic exercise, **3**:518
Vasopressin, isoleucyl-leucyl. *see* Oxytocin
Vasopressins, **2**:20, 23, 143-46, 494
    in children, **5**:707
    effect on ingestive behavior, **4**:248
    in fetal hypothalamus, **3**:338
    functions of, **5**:397, **6**:88
    perinatal, **3**:347
    production and activity of, **6**:19-20
    regulation of, **5**:397-98
    secretion, and natriuretic factor, **1**:474
Vasospasm, cerebral, **2**:369
Vasotocin, leucyl. *see* Oxytocin
Vasovagal reaction, **2**:147
Vectors. *see* Disease vectors
Vectors, genetic. *see* Genetic vectors
Vegetable fiber
    fermentability of, **3**:44
    water-holding capacity of, **3**:41
Vegetable oils
    crop sciences and, **1**:143
    modification of, **3**:570-71
Vegetable proteins, **3**:654
Vegetables, **2**:104
    in "basic four" food groups plan, **3**:655
    nutrient composition of, **5**:496
Vegetarian diets, **3**:658-59
Vegetarians, **3**:649
Veins, **2**:185, 640. *see also* Blood vessels
    defined, **3**:755
    of spinal cord, **7**:195
    of stomach, **3**:756
    valves in, **2**:162
Velate astrocytes, **1**:436
Vena cava, **2**:179
Vending machines, food. *see* Food dispensers, automatic
Venereal diseases. *see* Sexually transmitted diseases
Venereal warts. *see* Condylomata acuminata
Venesection. *see* Bloodletting
Vengeance, media protrayal of, **1**:3, 1058
Venipuncture. *see* Bloodletting
Venous network, **2**:184-85
Venous sinuses
    blood flow in, **4**:984
    dural, **4**:982-84
Venous system, **7**:758
Venous thromboembolism. *see* Thromboembolism, venous
Ventilation, alveolar. *see* Alveolar ventilation
Ventilation, mechanical. *see* Respiration, artificial

Ventricle, heart. *see* Heart ventricle
Ventricle-assist device. *see* Heart-assist devices
Ventricular ejection fraction. *see* Stroke volume
Ventricular fibrillation, **2:**172
Ventricular function, in exercise, **3:**517–18
Ventricular nerves, **2:**140, 145, 145
Ventricular ridge, dorsal. *see* Dorsal ventricular ridge
Ventrolateral medulla, **4:**718
Venules, **2:**184–85
Venules, postcapillary, **4:**396–97
Verapamil, **2:**84–87, 96, 172, 174, 176, 392
Veratridine, **2:**144, 145
Verbal learning, **5:**124
Vergence eye movements
    characteristics of, **3:**557
    defined, **3:**553
Verner, **5:**722
Vernix caseosa, **3:**614, 615
Vertebrae. *see also* specific vertebral areas, such as Cervical vertebrae; Spine
    characteristics of, **7:**767
    structure and function of, **7:**36–37
    types of, **7:**767–69
Vertebrate body pattern formation
    molecules involved in, **3:**290–91
    morphological stages in, **3:**285–86
    spatial diversification in, **3:**289
Vertebrates
    gas convection in, mechanisms of, **6:**569–70
    nervous systems of, **5:**342
    toxins produced by, **7:**598–600
Vertebrobasilar artery, **7:**301–2
Vertical jump
    elastic energy storage in, **5:**185
    muscle interaction in, **5:**187
Very low-density lipoproteins. *see* Lipoproteins, VLDL
Vesicles, defined, **3:**47
Vesicles, intracellular, **4:**17
Vesicles, secretory
    in adrenal medulla. *see under* Adrenal medulla
    and protein sorting, **6:**252–55
Vesicles, unilamellar, **3:**54
Vesicovaginal fistula, **4:**916–18
Vesicular exocytosis, **7:**385
Vestibular organ, **2:**306
Vestibulo-ocular reflex. *see* Reflex, vestibulo-ocular
Vibrio cholera, **4:**449–50, 462
Vibrissae
    lanceolate terminals of, **7:**676
    Merkel cells of, **7:**676

Vibrissae *(Continued)*
    nerve endings on, **7:**676–77
    trigeminal innervation of, **7:**673–75
Vidarabine. *see* 9-beta-D-Arabinofuranosyladenine
vif genes. *see* Genes, *vif*
vif viral proteins. *see* Gene products, *vif*
Vigabatrin, **3:**399–400
Vigilance, cortical. *see* Arousal
Vimentin proteins, **3:**469, **5:**224
Vinblastine, **1:**181
Vincristine, **1:**181, **2:**342, 346
Vinculin, **2:**132, 270, 271, **6:**205–6
Vinyl carbamate, **2:**121
Vinyl chloride, **2:**110, 112
Vinylimine, **3:**399
Violence, **5:**445. *see also* Coercion
Violence, observed, **1:**1058
    archival analyses, **1:**7–8
    desensitization to, **1:**4
    effects on arousal, **1:**4
    field experiments, **1:**5
    habituation, **1:**4
    homocides and, **1:**8
    imitation-modeling hypothesis, **1:**2
    longitudinal studies, **1:**5–7
    morally justified vs. unjustified, **1:**2–3
    priming of aggressive associations, **1:**9
    rape, **1:**4–5
    reality of, **1:**3
    role of pornography in violence against women, **1:**4–5
    scripts, **1:**8–9
    sex differences in response to, **1:**3–4
    social comparison with observed aggressor, **1:**3
    suicides and, **1:**7–8
    Werther effect, **1:**7
Viral diseases, **7:**784
Viral envelope, **4:**481–82
Viral epidemics
    antigenic drift and, **7:**782–84
    antigenic shift and, **7:**782–84
Viral infections
    epidemiology, **7:**781–84
    mosquitoes as vectors in, **7:**895
    persistent, **7:**778–80
    scalar-borne zoonoses and, **7:**896–97
Viral infections, chronic, **7:**778
Viral latency, **2:**395, 401–2
    defined, **3:**463
Viral lytic cycle
    defined, **3:**463
    proteins of, **3:**470–72

Virilism, **3:**342, 353
Virology, **7:**771-91
Virotoxins, **7:**592
Virus diseases
    cell interactions and, **7:**777-78
    chronic, **7:**778
    control of, **7:**784-91
        chemotherapy and, **7:**789-91
        hygiene and sanitation and, **7:**784-87
        quarantine and, **7:**784
        vaccination as method of, **7:**787-89
        vector control and, **7:**787
    cytophathic effects of, **7:**785
    diagnosis of, **7:**784
    epidemiology of, **7:**781-84
    human, diagram showing spread of virions in, **7:**779
    latent, **7:**778-80
    mechanism of spread in, **7:**778
    Parkinson's disease and, **5:**667
    pathogenesis of, **7:**777-80
    persistent, **7:**778-80
    slow, **7:**778-80
    transmission of human, **7:**782
Virus replication, **6:**686-88, **7:**773-77
    double stranded DNA, **7:**775
    general features of, **7:**775
    HIV-1, **1:**16-17
    influenza viruses, **4:**486-87
    interferon effects, **4:**518-21
    strategies of, **7:**774-77
Viruses. *see also* specific viruses
    animal
        classification of, **6:**687
        shapes and sizes of, **7:**773
    budding, in studies of polarized epithelial cells, **6:**64
    and cancer, **7:**781
    classification, **7:**772-73
    definition of, **7:**771
    detection, enzyme immunoassays for, **7:**786
    DNA
        replication, **6:**686-88
        transcription of, **7:**776-77
    eradication of, **7:**789
    flow cytometry of, **3:**625
    genes
        early, **7:**773
        late, **7:**773
    genome, **7:**772
    human
        strategy of expression of families of, **7:**776
        taxonomically important families of, **7:**774

Viruses *(Continued)*
    immunity to, **7:**780-81
    inactivated, for vaccines, **4:**464
    and leukemia, **4:**691
    leukemogenic, **4:**342
    morphology, **7:**772
    multiplication cycle, **7:**773-77
    polyamine metabolism in, **6:**73
    RNA, transcription of, **7:**777
    structure and composition of, **7:**772
    symmetry, **7:**772
    transmission
        horizontal, **7:**781-82
        vertical, **7:**782
    as vectors for gene transfer, **6:**653-59
Viscera
    afferent nerves of, **3:**763, **7:**794
    afferent pathways of, **7:**798
    afferent systems of, **7:**793-99
    chemoreceptors of, **7:**794
    hypothalamus role, **4:**310-11
    inputs to hypothalamus, **4:**305-6
    pain in, **7:**797-98
    referred pain and, **7:**797-98
    sense organs of, **7:**796-97
Visceral epithelium. *see* Epithelium, visceral
Visceral mesoderm. *see* Splanchnic mesoderm
Visceral peritoneum, defined, **3:**57
Viscoelasticity, **2:**202
Viscosity
    of dietary fiber components, **3:**41-42
    fluid bilayers, **4:**971
Visible spectrum, **5:**587-88
Vision, **2:**583-89, **7:**835-44
    analysis, later cortical regions involved in, **7:**813-14
    animal, role of omega-3 fatty acids in, **5:**537
    aperture problem in processing, **7:**823
    contrast sensitivity function as measure of, **7:**817
    convergence, divergence and multiplexing information in, **7:**802
    cortical areas of, **7:**813-14
    cortical filters and measurement of, **7:**819
    cortical mapping, **4:**198
    cosine grating measurements of, **7:**817
    curvature discrimination measurements and, **7:**820
    eyeball and, **7:**835
    in infancy, development of, **5:**747
    lateral geniculate nucleus, **7:**838-40
    midbrain role in, **7:**843
    motion processing, as higher brain function, **4:**199-201

Vision *(Continued)*
  night, vitamin A and, **7**:850
  peripheral, spatial sampling and, **7**:819–21
  physiology, **7**:801–14
  principles of organization in, **7**:801–2
  retina and, **6**:615–17, **7**:836
  striate cortex processing in, **7**:806–13
  vernier acuity measurements and, **7**:820
  visual cortex and, **7**:840–43
  visual cortical efferent projections in, **7**:843
Vision, photopic, **2**:575
Vision, scotopic, **2**:575, 585
Vision, stereoscopic, **2**:581, **7**:825–26
Vision disorders
  in benign partial epilepsies, **3**:459
  perception and, **5**:742–43
Vision psychophysics, **4**:197, **7**:815–28
  concepts of, **7**:816
  depth perception and, **7**:824–26
  linear systems theory of, **7**:816
  motion perception and, **7**:822–24
  spatial filters in, **7**:817–19
  spatial pattern discrimination in, **7**:821
  vernier acuity measurements in, **7**:820–21
Visual agnosia. *see* Agnosia, visual
Visual analogue scales, in feeding behavior research, **3**:596
Visual cortex
  functions of, **3**:220
  hypercolumn in, **5**:312
  multiple representation connections in, **5**:312
  network modeling of, **5**:389
  primary, neocortical cells of, **5**:311
  saccades and, **3**:559
  stimulation of, **5**:309
Visual cortex, cat, **5**:900
Visual cortex, macaque, **5**:389–90
Visual disparity, **5**:389
Visual fields
  columnar organization of, **7**:811–12
  complex cell and, **7**:808–9
  defined, **7**:801–2
  hypercomplex cell and, **7**:809
  modular structure of, **7**:811–12
Visual illusions. *see* Illusions, visual
Visual learning, **5**:128
Visual orienting, **1**:480–81
Visual pathways
  anatomy, **3**:549–51
  smooth pursuit eye movements and, **3**:556

Visual perception, **5**:742–43
  amygdaloid nuclei and, **3**:224
  brain regions involved in, **3**:220
  cognitive aspects of, **5**:742
  disconnected left hemisphere and, **3**:223
  food acceptance and, **3**:644
  in newborn, **6**:820
  physical stimuli and, **5**:742
  positron emission tomographic studies of, **3**:225
  right occipito-parietal-(temporal) system lesions and, **3**:222
  right parieto-occipital system lesions and, **3**:223
  tertiary cortical zone and, **3**:221
  visual attributes of, **5**:742
Visual pigments
  chemistry of, **6**:618
  of retina, **6**:617–18
Visual processing of words, **4**:201–2
Visual space location. *see* Space perception
Visual-spatial dysgraphia, **3**:221
Visual-spatial dyslexia, **3**:220
Vital signs monitoring. *see* Monitoring, physiologic
Vitamin A, **7**:845–52
  absorption, **7**:847–48
  binding proteins of, **7**:857
  biochemical functions of, **7**:849–50
  chemical properties of, **7**:846–47
  compounds, structure of, **7**:854
  derivation, **4**:748
  disease states affecting, **7**:850–51
  early history of, **7**:845–46
  food analysis, **4**:219
  hepatic uptake of, **7**:854–55
  intestinal uptake of, **7**:854
  liver cell metabolism of, **7**:848
  lymphatic transport of, **7**:854
  metabolic interactions involving, **7**:850
  metabolism, **7**:847–49
  metabolites, chemical structure of, **7**:846
  in nutrition education, **3**:25
  nutritional functions of, **7**:849–50
  in prostate cancer, **6**:187
  provitamin A, **4**:219
  recommended daily allowances of, **7**:849
  regulation of, **7**:848–49
  role in vision, **6**:618
  schematic of inter- and intracellular transport of, **7**:855
  storage of, **7**:854–55
  structure of, **7**:853

Vitamin A (Continued)
  therapeutic functions of, **7**:851
  tissue distribution of, **7**:848–49
  toxicity of high doses of, **7**:851
  transport, **7**:853–58
  uptake by eye, **7**:856
  uptake by vitamin A-requiring cells, **7**:856
Vitamin A deficiency
  sensory functions and, **5**:756–57
  signs of, **7**:849
Vitamin B_1, **2**:527–28, **4**:220–21
Vitamin B_2, **4**:221
Vitamin B_6. see Pyridoxine
Vitamin B_{12}
  deficiency, **7**:608
  partial structure of, **6**:119
vitamin B complex, and immunity, **5**:513
Vitamin C. see Ascorbic acid
Vitamin D, **2**:745, **7**:859–71
  absorption of, **7**:862
  animal requirements for, **7**:869
  basic functions of, **7**:859
  bone remodeling and, **1**:822
  chemistry of, **7**:860
  derivation, **4**:748
  diseases in humans related to, **7**:869–71
  food analysis, **4**:219
  food sources of, **7**:869
  history of discovery of, **7**:860
  hydroxylation, **2**:745
  metabolism, regulation of, **7**:863
  metabolites, hypercalcemia of malignancy etiology,
    4:278
  neonatal tetany and, **3**:349
  nutritional requirements for, **7**:868–69
  in regulation of bone and calcium metabolism, **1**:801–2,
    803–4
  sterol structures of, **7**:861
  structure of, **1**:802
  in treatment of hypoparathyroidism, **5**:665
  ultraviolet radiation and, **7**:68
Vitamin D_3. see Cholecalciferols
Vitamin D deficiency, **1**:808
Vitamin deficiencies, in developing countries, **3**:26
Vitamin E, **2**:544, **7**:873–78
  absorption, **7**:875
  chemistry of, **7**:874
  derivation, **4**:748
  early history of, **7**:873
  food analysis, **4**:219–20

Vitamin E (Continued)
  forms and biological activities of, **7**:874
  and free radical scavenging, **6**:793
  longevity and, **3**:894
  in nitrosamine inhibition, **3**:679
  nutrition, **7**:875–77
  physiological functions of, **7**:875–77
  physiology of, **7**:875
  reactions and regeneration of, **7**:876
  recommended dietary allowances for, **7**:877
  sources of antioxidant activities, **7**:876
  therapeutic uses of, **7**:877–78
  tissue regulation of, **7**:875
  transport, **7**:875
Vitamin E deficiency, **7**:875–77
Vitamin K, **2**:171, 544
  dependent proteins of clotting, **4**:184
  derivation, **4**:748
  food analysis, **4**:220
Vitamin M. see Folic acid
Vitamin supplements, in Western diets, **3**:26
Vitamins
  bone cell function-related, **1**:765
  destruction, nutritional quality and, **5**:493
  estimated safe dietary intake of, **5**:486–87
  as factor in prostate cancer, **6**:187
  fat-soluble, **3**:576
  food analysis, **4**:218–21
  immunity and, **5**:512–13
  interaction with zinc, **7**:889–90
  in protein utilization, **6**:234–35
  in treatment of autism, **1**:510
Vitelline block, defined, **3**:606
Vitiligo, **5**:266
Vitreous body, anatomy, **3**:543–44
Vitreous humor, **2**:640
Vitrification, **2**:722, 725
  devitrification, **2**:724
  and ice crystal, **2**:724
Vitronectin, **2**:611
  on elastic fibers, **3**:262
Vitronectin receptors, **1**:768
VLA-2 collagen receptor, **1**:768
VNTR probes
  alleles and, **3**:138
  DNA fingerprinting and, **3**:137
Vocabulary, in older adults, **1**:120, 121
Vocal cord paralysis, **6**:359
Vocal folds, laryngeal functions, **4**:662
Vocal tract, evolution of, **4**:635–36

Vocalization, **2:**339
 and amino acid receptors, **2:**339
 role ot PAG projections to nucleus retroambiguus,
  4:718
Voice
 characteristics, of older adults, **1:**120
 normal production of, **7:**157
Voice disorders, **7:**157
Volar pads
 development of, **3:**85
 primate, **3:**92
Volatile nitrosamines. *see* Nitrosamines
Voltage clamp, **2:**306
Voluntary movement. *see under* Movement
Volvulus. *see* Intestinal obstruction
Vomiting, bulimic, **3:**247
Vomitoxin. *see* Deoxynevalenol
Von Békésy, **3:**235–36
Von Ebner glands, **2:**781
Von Ebner lines, incremental, **2:**823
Von Ebner's glands, rat, **3:**49
Von Recklinghausen neurofibromatosis. *see* Neurofibrom-
 atosis
von Willebrand factor, **4:**159–61
 deficiency, **4:**178
 in platelet activation, **6:**57–58
von Willebrand's disease, **4:**155, 159–61
Vowels, wave form of, **5:**741
vpu genes. *see* Genes, *vpu*
vpu viral proteins. *see* Gene products, *vpu*
Vulva, anatomy, **6:**564
Vulvar neoplasms, **3:**863

W

Wakefulness/sleep rhythm. *see* Sleep
Waldenström macroglobulinemia, **4:**848–49
Walk-along muscle contraction theory, **5:**226
Walking, motor pattern in, **4:**769–71
Walnut Creek Contraceptive Drug Study
 on ischemic heart disease, **5:**602
 on malignant melanoma, **5:**607
 subjects of, **5:**600
 on vascular diseases, **5:**604
Walz, E., **3:**725
Warfarin, **2:**171
Warts
 benign, papillomaviruses and, **5:**637
 description of, **5:**637
Warts, genital. *see* Condylomata acuminata

Watanabe rabbit, **2:**665
 and atherosclerosis, **2:**665
 hyperlipidemic, heritable, **2:**665
Water. *see also* Body water
 absorption
  in colon, **3:**82
  stress and intestinal absorption, **7:**288
 in cartilaginous tissue, **1:**366–67
 content values of tissues, **3:**13
 dielectric properties of, **3:**11
 enzyme-substrate binding and, **3:**426
 food spoilage and, **3:**662
 ingestion and satiety, **7:**475
 in plant cell walls, **3:**36
 retention in essential hypertension, **4:**291–92
 tertiary enzyme structure and, **3:**433
 transfer across blood-brain barrier, **1:**722–23
 transport and kinins, **4:**594
 in treatment of thermal injuries, **7:**450
Water diuretics. *see* Diuretics, water
Water-holding capacity
 defined, **3:**35
 of dietary fiber, **3:**41
 of dietary fiber polymers, **3:**37–38
Water loss, mammalian. *see* Thirst
Water-soluble vitamins, **4:**220–21
Waterhouse, Benjamin, **5:**700
Watson, James, **3:**869
Watson-Crick base pairing rule, **3:**439
Waxes
 characterization, **4:**741–42
 production of, **1:**143
Wechsler scales, **1:**611–12, **2:**709, 719, **3:**715, 838
Weed, science, **1:**145
Weight, body. *see* Body weight
Weight lifting
 aerobic exercise and, **3:**509
 cardiovascular adjustments to, **3:**520
 plasticity and, **3:**512
Weight loss
 amphetamines and, **5:**440
 and infertility, **1:**743
Weightlessness in spaceflight, **2:**147–56
 and amphetamine, **2:**149
 arm vein pressure, **2:**150
 arrhythmia, **2:**152
 autonomic outflow, **2:**152–54
 baroreceptor reflex, **2:**154
 blood volume reduction, **2:**151–52
 cardiovascular postflight changes, **2:**149

Weightlessness in spaceflight *(Continued)*
    cardiovascular response, **2:**147-56
    compliance, venous, **2:**150
    countermeasures, **2:**155
    diuresis, **2:**151
    and drug intervention, **2:**149
    early weightlessness, **2:**151
    and exercise, **2:**154-55
    experiments with humans, **2:**148
    extravehicular, activity, **2:**155
    and facial edema, **2:**149, 151
    fluid shift, headward, **2:**149-51
    and the heart, **2:**150, 152
    history, **2:**147-48
    hypotension, **2:**153
    muscular atrophy, **2:**155
    neck chamber, **2:**154
    neurotransmitter, **2:**152-54
    paucity of information on, **2:**148-49
    vasovagal reaction, **2:**147, 153
    venous pressure, **2:**150-51
Weinberg formula, **3:**306
Weizmann, Chaim, **3:**938
Well's zone of pulp, **2:**824, 825
Wernicke area. *see* Temporal lobe
Wernicke-Korsakoff's syndrome, **1:**173
Wernicke's aphasia. *see* Aphasia, Wernicke
Werther effect, **1:**7, 1058
Western Aphasia Battery, **1:**611
Western blotting, **1:**334
Westernized diets, **3:**26-28
West's syndrome. *see* Spasms, infantile
Wheal-and-flare reaction, **4:**470
Wheat, **2:**104
    in "basic four" food groups plan, **3:**656
    dietary fiber from, **3:**40
    lignified tissues of, **3:**40
    NSP content of, **3:**40
Wheat bran
    content of, **3:**36
    fermentability of, **3:**44
    phytates in, **3:**43
    water-holding capacity of, **3:**41
Wheat-germ agglutinins, **5:**454
Whipworm disease, **1:**269-70
White blood cells. *see* Leukocytes
White matter
    of brain, **2:**674-78
    dyslexia-dysgraphia and, **3:**223
    of spinal cord, **7:**196

Whole-body irradiation, **4:**364
    acute effects of, **6:**448-49
    as immunosuppression, **4:**364
Whooping cough, in infants and toddlers, **4:**460-61
Wiegandt, H., **3:**726
Wiesel, **5:**309
Wiley, Harvey W., **3:**671
Wilm's tumor, **2:**451-54
Wilms' tumor
    advances in treatment of, **5:**704
    inheritance of, **7:**707
Wilson's disease, **2:**651-52, 763
    and copper excess, **2:**651
    features, **2:**651
    inbreeding and, **4:**440
    treatment, **2:**652
Wind chill factor, **4:**794
Wine
    fermentation of, **3:**667
    types of, **1:**167
Winter depression, **2:**833
    and bright-light therapy, **2:**833
    and melatonin, **2:**833
Winterbottom's sign, **7:**683
Winterization of vegetable oils, **3:**570-71
Wisconsin Card Sorting Test, **1:**612
Wish fulfillment, in dreams, **3:**192
Wiskott-Aldrich syndrome, **1:**794
Withdrawal reaction. *see* Substance withdrawal
        syndrome
Wolf-Hirshhorn syndrome, **2:**439
Wolffian ducts, hormones in, **3:**341
Women. *see also* Females, Sexes
    body fat and menarche, **1:**741-50
    dietary cravings/aversions in pregnancy, **3:**32-33
    homozygosity, in Western countries, **3:**307
    interest patterns in, **6:**857-58
    life expectancy, in United States, **5:**102, 103
    and men
        personality differences between, **6:**857-58
        tests to determine scholastic achievement of,
            6:858-59
        tests to determine specific abilities of, **6:**858-59
    older, social status, **4:**708
    rates of psychopathology among, **6:**861-62
    reproductive system, anatomy of. *see under* Repro-
        duction
    Scandinavian, DZ twinning and, **3:**307
    sex differences. *see* Sex characteristics
    sexual response of adult, **6:**865-67

Women, black
    endometriosis and, **3**:362
    pregnancy pica of, **3**:32
Women, hominid. *see* Hominid women
Woolly monkey virus. *see* Simian sarcoma virus
Word superiority effect, **4**:201-2
Words
    antecedent for, and reading comprehension, **6**:490
    fixation, and eye movement, **6**:482-84
    identification
        and eye movement, **6**:482
        and reading, **6**:484-86
    length, and comprehension, **6**:487-88
    meaning
        and text understanding, **6**:489-91
        relation of letter identification to, **6**:484
    neural bases for, **4**:636-37
    parsing, and reading comprehension, **6**:488-89
    recognition, and reading perception, **5**:743
    repetition, and reading comprehension, **6**:490
    visual processing, **4**:201-2
Work. *see also* Multiple-task interactions, Task
        performance
    caloric costs of, **3**:507-8
    cardiac output and, **3**:516
    humanization of, **3**:473
    maximal aerobic fitness and, **3**:506-7
    oxygen consumption in, **3**:515-16
    physiology, defined, **3**:473
Work, negative muscle. *see* Negative work, muscle
Work, positive muscle. *see* Positive work, muscle
Work capacity, aerobic. *see* Fitness, aerobic
Work function (photoemission), **5**:903
World Health Organization
    diseases targeted by, **4**:447-50
    International Classification of Diseases, **5**:99-100
    on *Salmonella* food infections, **3**:665
Wormwood oil, **2**:347
Wound healing, **2**:571
    cell types and growth factors associated with, **7**:502
    and collagen, **2**:571
    fibronectin in, **3**:525
    growth factors and, **7**:499-500
    keratinocyte proliferation and, **3**:446
    psoriasis and, **3**:445-46
    TGF-alpha and, **3**:447
    thermal injuries and, **7**:454
    and vitamin C, **2**:571
Wounds and injuries
    acute, **2**:251
        and cell death, **2**:251

Wounds and injuries
    acute *(Continued)*
        radiation exposure, **2**:251
        toxic chemical exposure, **2**:251
        in ureteral obstruction, **2**:251
    of children, **5**:710
    as component of aggression, **1**:114
    dirty, thermal injuries and, **7**:454
    headaches due to, **4**:56
    prevention, **4**:74
Wrist, muscles of, **4**:264
Wrist fractures, **1**:779
Written language. *see* Language
Wuchereria bancrofti, **2**:388, **4**:448
Wurtman hypothesis
    carboydrates-craving and, **3**:689
    defined, **3**:683

X

X chromosome, **4**:253
    band Xp21, **5**:193, 194
    defined, **3**:709, 819
    DNA methylation and, **3**:160
    genetic characteristics of, **6**:838
    inactivation of. *see* Dosage compensation (genetics)
    intelligence and, **3**:840
    variation in number, **4**:956
X chromosome inactivation. *see* Dosage compensation
    (genetics)
X-linked inheritance
    disease and, **3**:820-21
    dominant, **3**:852
    muscular dystrophy, in female, **2**:481
    mutation rates of recessive traits, **5**:245
    recessive, **3**:852
    recessive disorders, genetic mapping and, **3**:828-30
X-linked lymphoproliferative syndrome. *see* Lympho-
    proliferative disorders
X-linked markers, **7**:689
X-ray, diagnostic. *see* Radiography
X-ray intensifying screens, **3**:2
X-ray photoelectron spectroscopy. *see under* Spectros
    copy, photoelectron
X-ray therapy. *see* Radiotherapy
X-rays, chemical reactions and, **6**:443
Xanthine, **3**:806
Xanthine dehydrogenase, **2**:535
Xanthine oxidase, **2**:655
Xenobiotic metabolizing enzymes
    activities in nasal cavity, **5**:286-87

Xenobiotic metabolizing enzymes *(Continued)*
    effects of inhalants and pharmaceuticals, **5:**288
    hydrolysis by, in nasal cavity, **5:**286
    inhibition and induction of, in nasal cavity, **5:**288
    oxidation or reduction by, in nasal cavity, **5:**285–86
Xenobiotics
    defined, **2:**737, **5:**285
    in upper respiratory tract, **5:**691–92
Xenogenous fertilization. *see* Gamete intrafallopian tube
        transfer
Xenophobia, aggression and, **1:**117
Xenopus laevis, **2:**659
Xenotic infections, as scalar-borne zoonoses, **7:**898
Xeroderma pigmentosum
    deficient DNA repair in, **5:**253–54
    genetic aspects of, **3:**822
Xgantigen, as genetic marker for affective disorders, **1:**98
Xylans, acidic, **3:**39
Xylitol, **2:**103, 105
Xylocaine, **2:**784
Xyloglucans, **3:**37
Xylulose, **4:**1003
XYY condition, and aggressive behavior, **4:**247

Y

Y chromosome, **4:**253
    defined, **3:**819
    genetic characteristics of, **6:**838
    male specific gene on, **6:**838–39
    in sex determination, **3:**340
    variation in number, **4:**956
Y-linked inheritance, **3:**852
Yalow, **5:**713
Yamakawa, T., **3:**726
Yeast artificial chromosome, **2:**463, 484
Yeasts
    antibody production in, **3:**805
    food spoilage and, **3:**662
    glycerol production from, **3:**938
    nucleosomes of, **3:**132
"Yellow rain," **5:**263
Yellow rice toxin. *see* Citreoviridin
Yersinia enterocolitica, **4:**450
Yersinia pestis, **7:**896
Yielding stage, of attitude change, **1:**493
Yogurt
    lactose tolerance and, **4:**621
    production of, **3:**667
Yohimbine, **2:**326, 848

Yolk sac, defined, **3:**293
Yomesan, **1:**265
Yttrium-90 isotope, **4:**372
Yu, **3:**726

Z

Z allele, in alpha-1-antitrypsin deficiency, **3:**146
Z disc
    anatomy, **5:**208
    in muscle contraction, **5:**209
Z-DNA. *see* DNA
Zearalenol. *see* Ralgro
Zearalenone toxicity, **5:**267
ZEBRA (protein), **3:**471
Zellweger's syndrome, **1:**626
Zidovudine, **2:**373, 395, 399–400
    and bone marrow toxicity, **2:**399
    and interferon synergism, **2:**373
    as nucleoside analogue, **2:**395
    in treatment of AIDS, **1:**152–53, 156
Ziehl-Nielsen stain, **1:**276
Zinc, **2:**656–60
    absorption in body, **2:**656–60, **7:**880
    antioxidant activity of, **7:**883–84
    biochemistry, **2:**657
    biological function of, **7:**607
    bodily requirements for, **7:**879
    and cell membrane structure, **7:**883
    characteristics of, **7:**884–85
    and collagen, **2:**656
    concentration in cells and tissues, **7:**881
    content of normal male, **2:**659
    deficiency, **2:**649, 657
        clinical spectrum, **2:**657
        consequences of, **7:**612, 879–80
        reported in 1963, **2:**657
        systems, **2:**657
        worldwide, **2:**649
    distribution of, **7:**880
    and DNA replication, **2:**658, 659
    endocrine system and, **7:**886–87
    excretion of, **7:**880
    and gene expression, **2:**659
    and geophagia, **2:**657
    and gonad, **2:**655, 658
    and granulocyte, **2:**659
    and growth retardation, **2:**657
    imbalance and immunity, **5:**513–14
    and immunity, **2:**657–58, **7:**885–86

Zinc *(Continued)*
importance of, **5:**51–52
and insulin, **2:**658
interaction
with metals, **7:**889–90
with vitamins, **7:**889
and lymphocyte transformation, **2:**658
metabolism, **7:**879–90
in adults and elderly, **7:**889
within cells, **7:**883–84
disorders with altered, **7:**890
human, **2:**649–60
in infants and children, **7:**888–89
model for, **7:**889
in pregnancy and fetus, **7:**888
regulation of, **7:**881
within tissues, **7:**881–83
in whole body, **7:**880–81
metalloenzymes, **7:**885
and metallothionein, **2:**659
nervous system and, **7:**887
organs rich in, **2:**659
and protein synthesis, **2:**856
regulator for 200 enzymes, **2:**649
and sex hormones, **2:**658
significance, in body, **5:**49
and T-lymphocytes, **2:**658, 659
and tandem repeats of segments of amino acids, **2:**659
and transcription factor III-A of *Xenopus laevis*, **2:**659
treatment for Wilson's disease, **2:**652
Zinc fingers, **2:**53
defined, **2:**53
in DNA-binding, **3:**112
loop protein, **2:**659
Zinc pool, **2:**660
Zinc toxicity, **7:**612, 879–80

Zollinger-Ellison syndrome
gastric mucosal hyperplasia and, **5:**721
"pancreatic cholera" and, **5:**722
secretin test for, **5:**723
Zona pellucida
acrosome reaction and, **3:**604
defined, **3:**305
in dichorial MZ twinning, **3:**308
functions of, **3:**605
loss prior to implantation, **4:**424, 429
oocyte, **5:**553–54
penetration of, **3:**605–6
penetration slit, defined, **3:**605
in twin embryogenesis, **3:**306
Zona reticularis, in puberty, **3:**355
Zonula adherens, **2:**270
Zonula occludens, **2:**271–73
Zonular fibers
anatomy, **3:**544–45
defined, **3:**533
Zoonoses, **7:**891–900
and anthroponosis, **7:**891
defined, **7:**891
flea-borne, **7:**896
fly-borne, **7:**895–96
measures for suppressing, **7:**899
scalar-borne, **7:**896–98
tick-borne, **7:**894–95
transmission, **7:**892–93
and transmission of infectious disease, **7:**643
vector-borne, **7:**893–96
Zygosity
defined, **3:**305
genetic markers for, **3:**307
Zygotes, defined, **3:**599
Zymodemes, *Entamoeba histolytica*, **1:**219
Zymogram technique, **4:**577

INDEX OF RELATED TITLES

A

Abortion, Spontaneous
Chromosome Anomalies; Embryofetopathology; Endometriosis; Implantation (Embryology); Uterus and Uterine Responses to Estrogen

Acquired Immunodeficiency Syndrome (Virology)
AIDS Epidemic (Public Health); CD8 and CD4: Structure, Function, and Molecular Biology; Cytokines in the Immune Response; Lymphocytes; Macrophages

Acute Phase Response
Cytokines in the Immune Response; Hypothalamus; Inflammation; Macrophages; Natural Killer and Other Effector Cells; Neutrophils; Phagocytes; Platelet-Activating Factor Paf-Acether; Transforming Growth Factor-β

Adaptational Physiology
Body Temperature and Its Regulation; Cardiovascular Responses to Weightlessness; Evolution, Human; pH Homeostasis

Addiction
Alcohol Toxicology; Cocaine and Stimulant Addiction; Conditioning; Marijuana and Cannabinoids; Nonnarcotic Drug Use and Abuse

Adenosine Triphosphate (ATP)
ATP Synthesis by Oxidative Phosphorylation in Mitochondria; Cell

Adipose Cell
Cholesterol; DNA Synthesis; Fatty Acid Uptake by Cells; Insulin and Glucagon; Lipids; Reticuloendothelial System

Adolescence
Development of the Self; Psychoanalytic Theory; Puberty

Adrenal Gland
Catecholamines and Behavior; Cholesterol; Hypertension; Hypothalamus; Ion Pumps; Neuroendocrinology; Plasma Lipoproteins; Steroids

Adrenergic and Related G Protein-Coupled Receptors, Structure and Function
G Proteins

Affective Disorders, Genetic Markers
Cystic Fibrosis, Molecular Genetics; Depression; DNA Markers as Diagnostic Tools; Molecular Genetics; Mood Disorders; Muscular Dystrophy; Suicide

Affective Responses
Development of the Self; Learning and Memory

Aggression
Hormonal Influences on Behavior

Aging and Language
Alzheimer's Disease; Dementia in the Elderly; Depression; Reading Processes and Comprehension; Speech and Language Pathology

Aging, Psychiatric Aspects
Alzheimer's Disease; Autoimmune Disease; Dementia in the Elderly; Depression; Depression: Neurotransmitters and Receptors; Down's Syndrome, Gerontology; Magnetic Resonance Imaging; Molecular Genetics; Mood Disorders; Suicide

AIDS Epidemic (Public Health)
Acquired Immunodeficiency Syndrome (Virology); Infectious Diseases; Leukemia; Lymphoma; Measles; Virology, Medical

Alcohol Toxicology
Brain; Central Nervous System Toxicology; Cholesterol; Cortex; Hippocampal Formation

Alzheimer's Disease
Atherosclerosis; Blood–Brain Barrier; Dementia in the Elderly; Down's Syndrome, Molecular Genetics

Amoebiasis, Infection with *Entamoeba histolytica*
Calcium Antagonists; Enzyme Inhibitors; Immunology of Parasitism; Isozymes; Lymphocytes

Amyotrophic Lateral Sclerosis
Autoimmune Disease; Gangliosides; Spinal Cord

Anaerobic Infections in Humans
Blood–Brain Barrier

Animal Models of Disease
Acquired Immunodeficiency Syndrome (Virology); Atherosclerosis; Birth Defects; Infectious Disease; Malaria; Malnutrition; Virology, Medical

Animal Parasites
Immunology of Parasitism; Malaria; Trichinosis

Antibiotic Inhibitors of Bacterial Cell Wall Biosynthesis
Antibiotics; Chemotherapy, Antiparasitic Agents

Antibiotics
Antibiotic Inhibitors of Bacterial Cell Wall Biosynthesis; DNA Synthesis; Ribosomes

Antibody–Antigen Complexes: Biological Consequences
Acquired Immunodeficiency Syndrome (Virology); Allergy; Autoimmune Diseases; B-Cell Activation, Immunology; Complement System; Inflammation; Lymphocytes; Neutrophils; Phagocytes; Reticuloendothelial System

Antidepressants
Depression: Neurotransmitters and Receptors; Mood Disorders

Anti-Inflammatory Steroid Action
Cytokines in the Immune Response; Inflammation; Interferons; Interleukin-2 and the IL-2 Receptor; Lymphocytes; Macrophages; Neural–Immune Interactions; Neuroendocrinology; Steroids

Antimicrobial Agents, Impact on Newborn Infants
Anaerobic Infections in Humans; Antimicrobial Drugs

Antimicrobial Drugs
Antimicrobial Agents, Impact on Newborn Infants; Chemotherapy, Antiviral Agents; Quinolones

Antinuclear Antibodies and Autoimmunity
Antibody–Antigen Complexes: Biological Consequences; Autoimmune Disease; DNA in the Nucleosome; DNA Markers as Diagnostic Tools

Appetite
Conditioning; Eating Disorders; Feeding Behavior; Food-Choice and Eating-Habit Strategies of Dieters; Insulin and Glucagon; Lactose Malabsorption and Intolerance; Obesity; Olfactory Information Processing; Peptide Hormones of the Gut; Salt Preference in Humans

Aqueous Two-Phase Systems
Cell; Enzyme Isolation; Histones and Histone Genes; Membranes, Biological

Articular Cartilage and the Intervertebral Disc
Articulations, Joints between Bones; Cartilage; Collagen, Structure and Function; Proteoglycans

Articulations, Joints between Bones
Articular Cartilage and the Intervertebral Disc; Cartilage; Connective Tissue

Artificial Cells
Cell

Artificial Heart
Cardiovascular System, Anatomy; Cardiovascular System, Physiology and Biochemistry; Heart–Lung Machine

Artificial Kidney
Copper, Iron, and Zinc in Human Metabolism; Hypertension

Asbestos
Pulmonary Pathophysiology; Toxicology, Pulmonary

Ascorbic Acid
Collagen, Structure and Function; Enzymes, Coenzymes, and the Control of Cellular Chemical Reactions

Astrocytes
Blood–Brain Barrier; Laminin in Neuronal Development; Lipids; Nerve Growth Factor; Neurotransmitter and Neuropeptide Receptors in the Brain; Pineal Body; Retina

Atherosclerosis
Bile Acids; Cholesterol; Obesity; Coronary Heart Disease, Molecular Genetics; Macrophages; Plasma Lipoproteins; Smooth Muscle

ATP Synthesis by Oxidative Phosphorylation in Mitochondria
Adenosine Triphosphate (ATP); Mitochondrial Respiratory Chain

Atrial Natriuretic Factor
Cardiovascular System, Anatomy; Cardiovascular System, Pharmacology; Hypertension; Hypothalamus

Attention
Cortex; Learning and Memory; Magnetic Resonance Imaging; Problem Solving; Thalamus

Attitudes as Determinants of Food Consumption
Attitude and Attitude Change; Food Acceptance: Sensory, Physiological and Social Influences

Autism
Fragile X Syndrome; Reticuloendothelial System; Schizophrenic Disorders; Seizure Generation, Subcortical Mechanisms

Autoantibodies
Antinuclear Antibodies and Autoimmunity; Autoimmune Disease; Immune Surveillance; Insulin and Glucagon; Lymphocytes; Myasthenia Gravis; Thyroid Gland and Its Hormones

Autoimmune Disease
Autoantibodies; Heat Shock; Idiotypes and Immune Networks; Immunobiology of Transplantation; Lymphocyte-Mediated Cytotoxicity; Lymphocytes; Myasthenia Gravis; T-Cell Receptors; Thymus

Autonomic Nervous System
Exercise and Cardiovascular Function; Smooth Muscle; Spinal Cord; Urinary System, Anatomy; Vascular Capacitance

B

Bacterial Infections, Detection
Anaerobic Infections in Humans; Antimicrobial Drugs; DNA Markers as Diagnostic Tools; Hemoglobin; Infectious Diseases, Pediatric; Respiratory System, Anatomy; *Salmonella;* Sexually Transmitted Diseases (Public Health)

B-Cell Activation, Immunology
Antibody–Antigen Complexes: Biological Consequences; Autoantibodies; Interferons; Interleukin-2 and the IL-2 Receptor; Mitosis; Transforming Growth Factor-β

Behavioral Effects of Observing Violence
Aggression; Crime, Delinquency and Psychopathy; Learning and Memory; Sex Differences, Psychological

Behavior: Cooperative, Competitive, and Individualistic
Altruism

Behavior Measurement in Psychobiological Research
Depression: Neurotransmitters and Receptors; Mental Disorders

Bile Acids
Cholesterol; Steroids

Bioenergetics and the Adrenal Medulla
Adenosine Triphosphate (ATP); Ascorbic Acid; Ion Pumps; Proteins

Biofeedback
Conditioning; Headache

Biological Response Modifiers
B-Cell Activation, Immunology; Cytokines in the Immune Response; Immunoconjugates; Interferons; Interleukin-2 and the IL-2 Receptor; Lymphocyte-Mediated Cytotoxicity; Lymphocytes; Macrophages; Monoclonal Antibody Technology; Natural Killer and Other Effector Cells; Thymus

Birth Control
Atherosclerosis; Cholesterol; Depression; Depression: Neurotransmitters and Receptors; Endometriosis; Estrogens and Osteoporosis; Hemostasis and Blood Coagulation; Lipids; Steroids

Birth Defects
Chromosome Anomalies; Chromosomes, Molecular Studies; Down's Syndrome, Genetic Counseling; Molecular Genetics; Stress in Families with a Handicapped Child; Teratogenic Defects

Blood Binding of Drugs
Hemoglobin; Pharmacokinetics; Plasma Lipoproteins

Blood–Brain Barrier
Blood Binding of Drugs; Brain; Central Gray Area, Brain; Plasma Lipoproteins

Blood Sampling, Fetal
Embryofetopathology; Fetus; Hemopoietic System

Blood Supply Testing for Infectious Diseases
Acquired Immunodeficiency Syndrome (Virology)

Body Fat, Menarche, and Fertility
Breast Cancer Biology; Hypothalamus; Malnutrition

Body Temperature and Its Regulation
Adaptational Physiology; Hypothalamus; Skin

Bone Cell Genes, Molecular Analysis
Collagen, Structure and Function; Cytokines in the Immune Response; Hemopoietic System; Lymphocytes; Proteoglycans; Transforming Growth Factor-β

Bone Density and Fragility, Age-Related Changes
Bone, Embryonic Development; Bone Regulatory Factors; Bone Remodeling, Human; Collagen, Structure and Function; Estrogens and Osteoporosis; Growth, Anatomical; Parathyroid Gland and Hormone; Skeleton; Thyroid Gland and Its Hormones; Vertebra; Vitamin D

Bone, Embryonic Development
Cartilage; Extracellular Matrix; Growth, Anatomical; Skeleton; Transforming Growth Factor-β

Bone Marrow Transplantation
Acquired Immunodeficiency Syndrome (Virology); Genetic Diseases; Hemopoietic System; Leukemia; Lymphocytes; Lymphoma; Macrophages; Tumors of Bone

Bone Regulatory Factors
Bone Remodeling, Human; Cartilage; Estrogens and Osteoporosis; Growth, Anatomical; Insulin and Glucagon; Insulinlike Growth Factors and Fetal Growth; Interferons; Parathyroid Gland and Hormone; Polypeptide Hormones; Thyroid Gland and Its Hormones; Tissue Repair and Growth Factors; Transforming Growth Factor-β; Vitamin D

Bone Remodeling, Human
Bone, Embryonic Development; Bone Density and Fragility, Age-Related Changes; Bone Regulatory Factors; Cartilage; Estrogens and Osteoporosis; Growth, Anatomical; Skeleton; Vitamin D

Brain
Autonomic Nervous System; Binocular Vision and Space Perception; Blood–Brain Barrier; Brain Messengers and the Pituitary Gland; Brain Spectrin; Central Gray Area, Brain; Cerebrovascular System; Cortex; Depression: Neurotransmitters and Receptors; Higher Brain Function; Hippocampal Formation; Hypothalamus; Language; Learning and Memory; Limbic Motor System; Motor Control; Neocortex; Nervous System, Anatomy; Neurotransmitter and Neuropeptide Receptors in the Brain; Schizophrenic Disorders; Speech and Language Pathology; Spinal Cord; Synaptic Physiology of the Brain; Thalamus; Visual System

Brain Messengers and the Pituitary Gland
Brain; Catecholamines and Behavior; Neuroendocrinology; Peptides; Pituitary

Brain Regulation of Gastrointestinal Function
Brain; Hypothalamus; Peptide Hormones of the Gut; Peptides

Brain Spectrin
Brain; Synaptic Physiology of the Brain

Breast Cancer Biology
Chromosome Patterns in Human Cancer and Leukemia; Cytochrome *P*-450; DNA and Gene Transcription; Metastasis; Oncogene Amplification in Human Cancer; Polypeptide Hormones; Puberty; Steroids; Transforming Factor-α; Transforming Growth Factor-β; Tumor Suppressor Genes

C

Caffeine
Abortion, Spontaneous; Cholesterol; Depression; Hypertension; Learning and Memory; Mental Disorders; Mood Disorders; Nonnarcotic Drug Use and Abuse; Sleep Disorders; Teratogenic Defects

Calcium Antagonists
Atherosclerosis; Hypertension; Ion Pumps

Calcium, Biochemistry
Adenosine Triphosphate (ATP); G Proteins; Ion Pumps; Protein Phosphorylation

Carbohydrates (Nutrition)
Atherosclerosis; Dental Caries; Insulin and Glucagon; Lactose Malabsorption and Intolerance

Carcinogenic Chemicals
Asbestos; Chromosome Patterns in Human Cancer and Leukemia; Metastasis; Neoplasms, Etiology; Oncogene Amplification in Human Cancer; Pharmacokinetics; Repair of Damaged DNA; Tobacco Smoking, Impact on Health; Tumor Clonality

Cardiac Muscle
Atrial Natriuretic Factor; Electron Microscopy; Golgi Apparatus; Mitochondrial Respiratory Chain

Cardiovascular Responses to Weightlessness
Cardiovascular System, Physiology and Biochemistry; Exercise and Cardiovascular Function

Cardiovascular System, Anatomy
Cardiac Muscle; Cardiovascular System, Physiology and Biochemistry; Respiratory System, Anatomy

Cardiovascular System, Pharmacology
Adrenal Gland; Atherosclerosis; Calcium Antagonists; Cardiac Muscle; Cardiovascular System, Physiology and Biochemistry; Catecholamines and Behavior; Cholesterol; Diuretics; Hypertension

Cardiovascular System, Physiology and Biochemistry
Adenosine Triphosphate (ATP); Autonomic Nervous System; Cardiac Muscle; Cardiovascular System, Anatomy

Carnivory

Evolving Hominid Strategies; Fats and Oils (Nutrition); Primate Behavioral Ecology; Proteins (Nutrition)

Cartilage

Articular Cartilage and the Intervertebral Disc; Articulations, Joints Between Bones; Cell Death in Human Development; Collagen, Structure and Function; Elastin; Extracellular Matrix

Catecholamines and Behavior

Antidepressants; Cortex; Depression: Neurotransmitters and Receptors; Neuropharmacology; Parkinson's Disease; Schizophrenic Disorders; Stroke

CD8 and CD4: Structure, Function, and Molecular Biology

Lymphocytes; Natural Killer and Other Effector Cells; T-Cell Receptors

Cell

Adenosine Triphosphate (ATP); Cell Death in Human Development; Cell Division, Molecular Biology; Cell Junctions; Cell Membrane Transport; Cell Nucleus; Cell Receptors; Cellular Signaling; Fatty Acid Uptake by Cells; G Proteins; Golgi Apparatus; Ion Pumps; Meiosis; Nuclear Pore, Structure and Function; Oncogene Amplification in Human Cancer

Cell Death in Human Development

Alzheimer's Disease; Cell; Cell Receptors; Embryo, Body Pattern Formation; *fos* Gene, Human; Immune Surveillance; Lymphocytes; Parkinson's Disease; Polypeptide Hormones; Radiosensitivity of the Integumentary System; Radiosensitivity of the Small and Large Intestines; Steroids

Cell Division, Molecular Biology

Cell; Cell Death in Human Development; DNA Synthesis; Hemopoietic System; Insulin and Glucagon; Insulinlike Growth Factors and Fetal Growth; Interleukin-2 and the IL-2 Receptor; Lymphocytes; Mitosis; Oncogene Amplification in Human Cancer; Receptors, Biochemistry; Retinoblastoma, Molecular Genetics; Transforming Growth Factor-α; Transforming Growth Factor-β; Tumor Suppressor Genes

Cell Junctions

Cell; Cellular Signaling

Cell Membrane Transport

Cell; Cell Receptors; Ion Pumps; Receptors, Biochemistry

Cell Nucleus

Cell; Chromatin Folding; DNA and Gene Transcription; Histones and Histone Genes; Mitosis; Nuclear Pore, Structure and Function; Repair of Damaged DNA

Cell Receptors

Autoimmune Disease; Cell; Myasthenia Gravis; Proteins

Cellular Memory

Calcium, Biochemistry; Cell; Conditioning; Eye Movements; *fos* Gene, Human; G Proteins; Hippocampal Formation; Neurotransmitter and Neuropeptide Receptors in the Brain

Cellular Signaling

Cell; Cell Junctions; Endocrine System; Myasthenia Gravis; Nerve Growth Factor; Neuroendocrinology; Neurotransmitter and Neuropeptide Receptors in the Brain; Polypeptide Hormones; Receptors, Biochemistry; Steroids; Tissue Repair and Growth Factors; Transforming Growth Factor-α

Central Gray Area, Brain

Adrenergic and Related G Protein-Coupled Receptors; Autonomic Nervous System; Brain; Brain Regulation of Gastrointestinal Function; Catecholamines and Behavior; Cortex; Hippocampal Formation; Hypothalamus; Pain

Central Nervous System Toxicology

Alcohol Toxicology; Astrocytes; Parkinson's Disease

Cerebral Specialization

Brain; Language; Evolution; Magnetic Resonance Imaging

Cerebrovascular System

Blood–Brain Barrier; Brain

Chemotherapy, Antiparasitic Agents

Amebiasis, Infection with *Entamoeba histolytica;* Leishmaniasis; Malaria; Trichinosis; Trypanosomiasis

Chemotherapy, Antiviral Agents

Acquired Immunodeficiency Syndrome (Virology); CD8 and CD4: Structure, Function, and Molecular Biology; DNA Synthesis; Herpesviruses; Influenza Virus Infection; Interferons

Cholesterol

Atherosclerosis; Lipids; Membranes, Biological; Steroids

Chromatin Folding
Chromatin Structure and Gene Expression in the Mammalian Brain; DNA in the Nucleosome; Histones and Histone Genes; Meiosis; Mitosis

Chromatin Structure and Gene Expression in the Mammalian Brain
Chromatin Folding; Chromosomes; DNA and Gene Transcription; DNA in the Nucleosome; Histones and Histone Genes

Chromosome Anomalies
Abortion, Spontaneous; Chromosome Patterns in Human Cancer and Leukemia; Chromosomes; Down's Syndrome, Molecular Genetics; Meiosis; Mitosis

Chromosome Patterns in Human Cancer and Leukemia
Chromosomes; Oncogene Amplification in Human Cancer; Retinoblastoma, Molecular Genetics; Tumor Suppressor Genes

Chromosome-Specific Human Gene Libraries
Chromosomes; DNA and Gene Transcription; DNA Markers as Diagnostic Tools; Epstein–Barr Virus; Flow Cytometry; Genes; Genetic Maps; Genome, Human; Lymphocytes; Recombinant DNA Technology in Disease Diagnosis

Chromosomes
Chromosome Anomalies; Chromosome Patterns in Human Cancer and Leukemia; Chromosome-Specific Human Gene Libraries; Genetic Maps; Telomeres, Human

Chromosomes, Molecular Studies
Chromosomes

Citrate Cycle
Adenosine Triphosphate (ATP); Cholesterol; Coenzymes, Biochemistry; DNA Synthesis; Fatty Acid Uptake by Cells; Glycolysis; Hemoglobin; Steroids

Cocaine and Stimulant Addiction
Catecholamines and Behavior; Nonnarcotic Drug Use and Abuse; Schizophrenic Disorders

Cochlear Chemical Neurotransmission
Ears and Hearing; Ion Pumps

Coenzymes, Biochemistry
Sphingolipid Metabolism and Biology; Vitamin A

Cognitive Representation in the Brain
Brain; Color Vision; Ears and Hearing; Language; Perception; Visual System

Collagen, Structure and Function
Articular Cartilage and the Intervertebral Disc; Cartilage; Golgi Apparatus; Proteoglycans

Color Vision
Vision, Psychophysics; Visual System

Comparative Anatomy
Cortex; Hair; Primates; Vertebra

Complement System
Antibody–Antigen Complexes: Biological Consequences; Autoimmune Disease; Immune System; Inflammation; Ion Pumps; Lipids; Macrophages; Neutrophils; Radioimmunoassays; Smooth Muscle

Conditional Reasoning, Development
Problem Solving; Thinking

Conditioning
Behavioral Effects of Observing Violence; Learning and Memory

Connective Tissue
Adipose Cell; Antibody–Antigen Complexes: Biological Consequences; Articular Cartilage and the Intervertebral Disc; Ascorbic Acid; Cartilage; Circadian Rhythms and Periodic Processes; Collagen, Structure and Function; Elastin; Endocrine System; Extracellular Matrix; Skin; Proteoglycans

Copper, Iron, and Zinc in Human Metabolism
Collagen, Structure and Function; Cytochrome *P*-450; DNA Synthesis; Hemoglobin; Insulin and Glucagon; Lymphocytes; Testicular Function; Zinc Metabolism

Coronary Heart Disease, Molecular Genetics
Atherosclerosis; Cholesterol; DNA Markers as Diagnostic Tools; Fatty Acid Uptake by Cells; Plasma Lipoproteins

Cortex
Brain; Comparative Anatomy; Hemispheric Interactions; Hippocampal Formation; Neocortex; Neurotransmitter and Neuropeptide Receptors in the Brain; Olfactory Information Processing; Peptides

Craniofacial Growth, Genetics
Bone, Embryonic Development; Bone Remodeling, Human; Craniofacial Growth, Postnatal; Embryo, Body Pattern Formation

Craniofacial Growth, Postnatal
Bone Density and Fragility, Age-Related Changes; Cartilage; Craniofacial Growth, Genetics; Dental and Oral Biology, Anatomy

Crime, Delinquency, and Psychopathy
Catecholamines and Behavior; Cerebral Specialization; Conditioning; Chromosome Anomalies; Epilepsy; Hormonal Influences on Behavior; Psychoanalytic Theory; Sensation-Seeking Trait

Cryotechniques in Biological Electron Microscopy
Electron Microscopy; Scanning Electron Microscopy

Cystic Fibrosis, Molecular Genetics
DNA Markers as Diagnostic Tools; Genetic Maps

Cytochrome *P*-450
Bile Acids; Breast Cancer Biology; Carcinogenic Chemicals; Cell Membrane Transport; Cholesterol; Hemoglobin; Steroid Hormone Synthesis; Steroids; Tobacco Smoking, Impact on Health

Cytokines in the Immune Response
Antibody–Antigen Complexes: Biological Consequences; B-Cell Activation, Immune System; Immunology; CD8 and CD4: Structure, Function, and Molecular Biology; Inflammation; Interferons; Interleukin-2 and the IL-2 Receptor; Lymphocytes; Macrophages; T-Cell Receptors; Transforming Growth Factor-β

D

Dementia in the Elderly
Alcohol Toxicology; Alzheimer's Disease; Brain; Depression; Mental Disorders; Huntington's Disease; Learning and Memory; Motor Control; Parkinson's Disease; Pharmacokinetics; Speech and Language Pathology

Dental and Oral Biology, Anatomy
Articulations, Joints between Bones; Bone Remodeling, Human; Cartilage; Collagen, Structure and Function; Craniofacial Growth, Genetics; Craniofacial Growth, Postnatal; Dental Caries; Dental Embryology and Histology; Gum (Gingiva); Inflammation; Lymphocytes; Salivary Glands and Saliva

Dental and Oral Biology, Pharmacology
Antibiotics; Dental and Oral Biology, Anatomy; Pain

Dental Anthropology
Dental and Oral Biology, Anatomy; Dental Caries; Evolution, Human

Dental Caries
Dental and Oral Biology, Anatomy; Salivary Glands and Saliva

Dental Embryology and Histology
Collagen, Structure and Function; Dental and Oral Biology, Anatomy; Extracellular Matrix

Depression
Adrenal Gland; Affective Disorders, Genetic Markers; Antidepressants; Depression: Neurotransmitters and Receptors; Hypothalamus; Neurotransmitter and Neuropeptide Receptors in the Brain; Pituitary; Sleep; Thyroid Gland and Its Hormones

Depression: Neurotransmitters and Receptors
Adrenal Gland; Antidepressants; Depression; Endocrine System; Hypothalamus; Neuroendocrinology; Neurotransmitter and Neuropeptide Receptors in the Brain; Pineal Body; Pituitary; Platelet Receptors; Suicide

Development of the Self
Personality Disorders (Psychiatry); Psychoanalytic Theory

Developmental Neuropsychology
Brain; Cerebral Specialization; Epilepsy; Hemispheric Interactions; Mental Disorders; Mood Disorders; Plasticity, Nervous System; Schizophrenic Disorders

Development, Psychobiology
Evolution, Human; Fertilization; Fetus; Sperm

Diagnostic Radiology
Ultrasound Propagation in Tissue

Dielectric Properties of Tissue
Hemoglobin; Membranes, Biological; Skin

Diet
Dietary Fiber, Chemistry and Properties; Eating Disorders; Food-Choice and Eating-Habit Strategies of Dieters; Food Groups; Nutrients, Energy Content; Nutritional Quality of Foods and Diets; Vitamin D

Dietary Cravings and Aversions during Pregnancy
Pregnancy, Nausea and Vomiting

Dietary Fiber, Chemistry and Properties
Diet; Proteoglycans

Digestion and Absorption of Human Milk Lipids
Lipids

Digestive System, Anatomy
Digestive System, Physiology and Biochemistry; Gastric Circulation; Intestinal Blood Flow Regulation

Digestive System, Physiology and Biochemistry
Autonomic Nervous System; Bile Acids; Brain Regulation of Gastrointestinal Function; Carbo-

hydrates (Nutrition); Digestive System, Anatomy; Fats and Oils (Nutrition); Proteins (Nutrition); Salivary Glands and Saliva; Smooth Muscle

Diuretics
Adenosine Triphosphate (ATP); Cholesterol; Ion Pumps; Pharmacokinetics; Urinary System, Anatomy

DNA Amplification
Chromosome Anomalies; DNA and Gene Transcription; Gene Amplification; Oncogene Amplification in Human Cancer

DNA and Gene Transcription
Chromatin Folding; DNA in the Nucleosome; Chromosomes; DNA Methylation and Gene Activity

DNA Binding Sites
DNA and Gene Transcription; Human Genome and Its Evolutionary Origin

DNA in the Nucleosome
Chromatin Folding; Chromatin Structure and Gene Expression in the Mammalian Brain; Chromosomes; DNA and Gene Transcription; Histones and Histone Genes

DNA Markers as Diagnostic Tools
Alzheimer's Disease; Autoimmune Disease; Coronary Heart Disease, Molecular Genetics; Cystic Fibrosis; Molecular Genetics; Genetic Diseases; Genome, Human; Hemophilia, Molecular Genetics; Huntington's Disease; Immunobiology of Transplantation; Multiple Sclerosis; Muscular Dystrophy, Molecular Genetics; Myasthenia Gravis; Phenylketonuria, Molecular Genetics; Retinoblastoma, Molecular Genetics; Sickle Cell Hemoglobin

DNA Methylation and Gene Activity
DNA and Gene Transcription; DNA Binding Sites

DNA, RNA, and Protein Sequence Analysis by Computer
Human Genome and Its Evolutionary Origin; Proteins

DNA Synthesis
DNA and Gene Transcription; DNA, RNA, and Protein Sequence Analysis by Computer

Down's Syndrome, Molecular Genetics
Alzheimer's Disease; Collagen, Structure and Function; DNA Markers as Diagnostic Tools;

Genes; Genetic Diseases; Genetic Maps; Genome, Human; Heat Shock; Influenza Virus Infection; Interferons; Recombinant DNA Technology in Disease Diagnosis

Dreaming
Psychoanalytic Theory; Sleep

Drug Analysis
High Performance Liquid Chromatography; Immunoassays, Nonradionucleotide; Radioimmunoassays

Drugs of Abuse and Alcohol Interactions
Cocaine and Stimulant Addiction; Marijuana and Cannabinoids; Pharmacokinetics

Dyslexia–Dysgraphia
Cerebral Specialization; Hippocampal Formation; Speech and Language Pathology; Thalamus

E

Ears and Hearing
Cochlear Chemical Neurotransmission

Eating Disorders
Obesity

Elastin
Extracellular Matrix; Laminin in Neuronal Development; Proteins

Electron Microscopy
Scanning Electron Microscopy

Embryo, Body Pattern Formation
DNA and Gene Transcription; DNA Binding Sites

Embryofetopathology
Abortion, Spontaneous; Birth Defects; Chromosome Anomalies; Genetic Counseling; Implantation (Embryology)

Embryology and Placentation of Twins
Birth Defects; Chromosome Anomalies; Down's Syndrome, Molecular Genetics; Follicle Growth and Luteinization; Implantation (Embryology); Pituitary

Endocrine System
Adenosine Triphosphate (ATP); Atherosclerosis; Cell Receptors; Cholesterol; Hypothalamus; Insulin and Glucagon; Neuroendocrinology; Parathyroid Gland and Hormone; Peptide Hormones of the Gut; Pituitary; Protein Phosphorylation; Radioimmunoassays; Steroids; Thyroid Gland and Its Hormones; Tissue Repair and Growth Factors

Endocrinology, Developmental
Endocrine System; Growth, Anatomical; Hypothalamus; Insulin and Glucagon; Insulinlike Growth Factors and Fetal Growth; Parathyroid Gland and Hormone; Peptides; Pituitary; Puberty; Thyroid Gland and Its Hormones; Tissue Repair and Growth Factors; Vitamin D

Endometriosis
Immune System

Endothelium-Derived Relaxing Factor
Atherosclerosis; Hemoglobin; Hypertension; Macrophages; Neutrophils; Smooth Muscle; Stroke

Energy Metabolism
Adenosine Triphosphate (ATP); ATP Synthesis by Oxidative Phosphorylation in Mitochondria; Mitochondrial Respiratory Chain

Environmental and Occupational Toxicology
Asbestos; Carcinogenic Chemicals; Tobacco Smoking, Impact on Health; Toxicology, Pulmonary

Enzyme Isolation
Chromosome-Specific Human Gene Libraries; Monoclonal Antibody Technology

Enzyme Thermostability
Proteins

Enzymes, Coenzymes, and the Control of Cellular Chemical Reactions
Adenosine Triphosphate (ATP); ATP Synthesis by Oxidative Phosphorylation in Mitochondria; Enzyme Inhibitors

Enzymology of DNA Replication
DNA and Gene Transcription; DNA Synthesis

Epidermal Proliferation
Cytokines in the Immune Response; Extracellular Matrix; Interferons; Keratinocyte Transformation, Human; Lymphocytes; Macrophages; Oncogene Amplification in Human Cancer; Transforming Growth Factor-α; Transforming Growth Factor-β

Epilepsy
Cortex; Hippocampal Formation; Neuroendocrinology; Peptides; Seizure Generation, Subcortical Mechanisms

Epstein–Barr Virus
DNA and Gene Transcription; DNA Binding Sites; Genetic Maps; Immunobiology of Transplantation; Lymphocytes; Lymphoma

Estrogen and Osteoporosis
Tissue Repair and Growth Factors; Transforming Growth Factor-β

Evolution, Human
Comparative Anatomy

Evolving Hominid Strategies
Evolution, Human; Language, Evolution

Exercise
Adenosine Triphosphate (ATP); Appetite; ATP Synthesis by Oxidative Phosphorylation in Mitochondria; Exercise and Cardiovascular Function; Fatty Acid Uptake by Cells; Insulin and Glucagon; pH Homeostasis; Skeletal Muscle

Exercise and Cardiovascular Function
Cardiovacular System, Physiology and Biochemistry; Exercise

Extracellular Matrix
Collagen, Structure and Function; Connective Tissue; Elastin; Laminin in Neuronal Development; Metastasis; Proteoglycans; Receptors, Biochemistry; Transforming Growth Factor-β

Eye, Anatomy
Eye Movements; Retina; Visual System

Eye Movements
Eye, Anatomy

F

Famine
Infectious Diseases; Nutrition and Immunity

Fats and Oils (Nutrition)
Adipose Cell; Atherosclerosis; Cholesterol; Insulin and Glucagon; Lipids; Phospholipid Metabolism; Retina

Fatty Acid Uptake by Cells
Adipose Cell; Atherosclerosis; Cholesterol; Insulin and Glucagon; Phospholipid Metabolism; Platelet-Activating Factor Paf-acether

Feeding Behavior
Appetite; Attitudes as Determinants of Food Consumption; Behavior Measurement in Psychobiological Research; Brain; Dietary Cravings and Aversions during Pregnancy; Eating Disorders; Obesity; Food Acceptance: Sensory, Physiological, and Social Influences; Food-Choice and Eating-Habit Strategies of Dieters; Malnutrition

Fertilization
Cholesterol; Phosphorylation of Microtubule Protein

Fetus

Bone, Embryonic Development; Cardiovascular System, Anatomy; Cardiovascular System, Physiology and Biochemistry; Digestive System, Anatomy; Embryo, Body Pattern Formation; Endocrine System; Implantation (Embryology); Larynx; Parathyroid Gland and Hormone; Reproductive System, Anatomy; Respiratory System, Anatomy; Respiratory System, Physiology and Biochemistry; Skin; Spinal Cord; Thyroid Gland and Its Hormones; Urinary System, Anatomy

Flow Cytometry

DNA Synthesis; Monoclonal Antibody Technology

Follicle Growth and Luteinization

Hypothalamus; Oogenesis; Pituitary

Food Acceptance: Sensory, Physiological, and Social Influences

Appetite; Olfactory Information Processing; Tongue and Taste

Food Groups

Diet; Fats and Oils (Nutrition); Nutritional Quality of Foods and Diets

Food Microbiology and Hygiene

Salmonella

Food Toxicology

Allergy; Cytochrome *P*-450; Mycotoxins; Oncogene Amplification in Human Cancer; Toxins, Natural

Food-Choice and Eating-Habit Strategies of Dieters

Eating Disorders; Obesity

Forearm, Coordination of Movements

Motor Control; Movement; Muscle Dynamics

***fos* Gene, Human**

DNA and Gene Transcription; Oncogene Amplification in Human Cancer

Fragile X Syndrome

Autism; Connective Tissue; Depression; Genetic Counseling; Seizure Generation, Subcortical Mechanisms

Fusimotor System

Motor Control; Movement; Muscle Dynamics; Muscle, Physiology and Biochemistry; Proprioceptors and Proprioception

G

Gangliosides

Cholesterol; Gangliosides and Neuronal Differentiation; Ganglioside Transport; Golgi Apparatus; Lipids; Oncogene Amplification in Human Cancer; Phospholipid Metabolism; Plasticity, Nervous System; Sphingolipid Metabolism and Biology

Gangliosides and Neuronal Differentiation

Gangliosides

Ganglioside Transport

Gangliosides; Retina; Visual System

Gastric Circulation

Digestive System, Anatomy; Digestive System, Physiology and Biochemistry

Gastrointestinal Cancer

Bile Acids; Cholesterol; Dietary Fiber, Chemistry and Properties; Digestive System, Physiology and Biochemistry; Food Microbiology and Hygiene; Lymphoma; Metastasis; Peptide Hormones of the Gut; Salt Preference in Humans

Gaucher Disease, Molecular Genetics

Bone Marrow Transplantation; DNA Markers as Diagnostic Tools; Genetic Diseases

Gene Amplification

DNA Amplification; Genes; Oncogene Amplification in Human Cancer

Genes

Chromosomes; DNA and Gene Transcription; Lymphocytes; Meiosis; Mutation Rates; Polymorphism, Genes

Genetically Engineered Antibody Molecules

Idiotypes and Immune Networks; Immune Surveillance; Monoclonal Antibody Technology; Pharmacokinetics

Genetic Counseling

Birth Defects; Chromosome Anomalies

Genetic Diseases

Atherosclerosis; Cholesterol; Chromosome Anomalies; Chromosome Patterns in Human Cancer and Leukemia; Cystic Fibrosis, Molecular Genetics; Down's Syndrome, Molecular Genetics; Gaucher Disease, Molecular Genetics; Genes; Genetics, Human; Hemophilia, Molecular Genetics; Huntington's Disease; Muscular Dystrophy, Molecular Genetics; Phenylketonuria, Molecular Genetics; Sickle Cell Hemoglobin

Genetic Maps
Chromosomes; Cystic Fibrosis, Molecular Genetics; DNA Markers as Diagnostic Tools; Genetic Diseases; Genome, Human; Hemophilia, Molecular Genetics; Huntington's Disease; Meiosis; Muscular Dystrophy, Molecular Genetics

Genetics and Abnormal Behavior
Crime, Delinquency, and Psychopathy; Depression; DNA Markers as Diagnostic Tools; Down's Syndrome, Molecular Genetics; Fragile X Syndrome; Genetics, Human; Mood Disorders; Phenylketonuria, Molecular Genetics; Schizophrenic Disorders

Genetics, Human
Chromosomes; Cystic Fibrosis, Molecular Genetics; DNA and Gene Transcription; Genes; Genetic Diseases; Genetic Maps; Hemophilia, Molecular Genetics; Huntington's Disease; Meiosis; Mitosis; Muscular Dystrophy, Molecular Genetics; Phenylketonuria, Molecular Genetics

Genital Cancer and Papillomaviruses
DNA and Gene Transcription; Keratinocyte Transformation, Human; Oncogene Amplification in Human Cancer; Sexually Transmitted Diseases (Public Health)

Genome, Human
Chromosome-Specific Human Gene Libraries; Chromosomes; DNA and Gene Transcription; Genes, Genetic Diseases; Genetic Maps; Polymorphism, Genes; Pulsed-Field Gel Electrophoresis; Telomeres, Humans

Genomes in the Study of Primate Relations
Chromosome Anomalies; Genome, Human; Human Genome and Its Evolutionary Origin

Gerontology
Autoimmune Disease; Bone Density and Fragility, Age-Related Changes; Genes; Gerontology, Physiology; Hypertension; Neuroendocrinology; Polymorphism of the Aging Immune System; Repair of Damaged DNA

Gerontology, Physiology
Aging and Language; Alzheimer's Disease; Bone Density and Fragility, Age-Related Changes; Brain; Dementia in the Elderly; Energy Metabolism; Exercise and Cardiovascular Function; Gerontology; Learning and Memory; Repair of Damaged DNA; Tobacco Smoking, Impact on Health

Glutathione
Enzymes, Coenzymes, and the Control of Cellular Chemical Reactions; Selenium in Nutrition

Glycogen
Adenosine Triphosphate (ATP); Insulin and Glucagon

Glycolysis
Adenosine Triphosphate (ATP); Glycogen; Muscle, Physiology and Biochemistry

Golgi Apparatus
Extracellular Matrix; Lipids; Proteins; Proteoglycans; Steroids

Growth, Anatomical
Embryo, Body Pattern Formation; Fertilization; Fetus; Puberty

Gum (Gingiva)
Collagen, Structure and Function; Connective Tissue; Lipids; Oral Pathology; Proteoglycans

H

Hair
Lipids

Headache
Brain; Epilepsy; Inflammation; Pain; Stroke

Health Psychology
Biofeedback; Nonnarcotic Drug Use and Abuse; Obesity; Pain; Persuasion; Psychoneuroimmunology; Stress-Induced Alteration of GI Function; Tobacco Smoking, Impact on Health

Heart–Lung Machine
Oxygen Transport to Tissue; Pulmonary Pathophysiology

Heat Shock
Genes; DNA and Gene Transcription; Immune Surveillance; Immune System; Thermotolerance in Mammalian Development

HeLa Cells
DNA Markers as Diagnostic Tools

Helminth Infections, Impact on Human Nutrition
Chemotherapy, Antiparasitic Agents; Lactose Malabsorption and Intolerance; Malnutrition

Hematology and Immunology in Space
Hemoglobin; Lymphocytes

Hemispheric Interactions
Brain; Cerebral Specialization; Motor Control

Hemoglobin
Hemoglobin, Molecular Genetics; Malaria; Sickle Cell Hemoglobin

Hemoglobin, Molecular Genetics
Hemoglobin; Sickle Cell Hemoglobin

Hemophilia, Molecular Genetics
Genetic Counseling; Genetic Diseases

Hemopoietic System
B-Cell Activation, Immunology; Cytokines in the Immune Response; Extracellular Matrix; Lymphocytes; Macrophages; Natural Killer and Other Effector Cells; Neutrophils

Hemostasis and Blood Coagulation
Hemophilia, Molecular Genetics; Tissue Factor

Herpesviruses
Chemotherapy, Antiviral Agents; DNA and Gene Transcription; Epstein–Barr Virus; Lymphocytes; Nuclear Pore, Structure and Function; Polymorphism, Genes; Virology, Medical

Higher Brain Function
Brain; Cortex; Neurotransmitter and Neuropeptide Receptors in the Brain; Reading Processes and Comprehension; Speech and Language Pathology; Vision, Psychophysics; Visual System

High Performance Liquid Chromatography
High Performance Liquid Chromatography in Food Analysis

High Performance Liquid Chromatography in Food Analysis
Ascorbic Acid; High Performance Liquid Chromatography; Lipids; Mycotoxins; Vitamin A; Vitamin D; Vitamin E

Hippocampal Formation
Alzheimer's Disease; Brain; Epilepsy; Hypothalamus; Neocortex; Thalamus

Histones and Histone Genes
Chromatin Folding; DNA and Gene Transcription; DNA in the Nucleosome; DNA Synthesis

Hormonal Influences on Behavior
Aggression; Endocrine System; Hypothalamus; Insulin and Glucagon; Peptide Hormones of the Gut; Sex Differences, Psychological; Sexual Behavior, Human; Spinal Cord; Steroid Hormone Synthesis

Human Genome and Its Evolutionary Origin
Chromosomes; DNA and Gene Transcription; Genome, Human

Human Muscle, Anatomy
Articulations, Joints between Bones; Movement; Muscle, Physiology and Biochemistry Skeletal Muscle; Tongue and Taste

Huntington's Disease
Genetic Diseases

Hypercalcemia of Malignancy
Breast Cancer Biology; Cytokines in the Immune Response; Diuretics; Lymphoma; Parathyroid Gland and Hormone; Transforming Growth Factor-α; Vitamin D

Hypertension
Cardiovascular System, Anatomy; Cardiovascular System, Physiology and Biochemistry; Smooth Muscle; Vascular Capacitance

Hyperthermia and Cancer
Body Temperature and Its Regulation; Skin, Effects of Ultraviolet Radiation; Thermotolerance in Mammalian Development

Hypothalamus
Affective Responses; Autonomic Nervous System; Blood–Brain Barrier; Body Temperature and Its Regulation; Brain; Circadian Rhythms and Periodic Processes; Feeding Behavior; Hippocampal Formation; Limbic Motor System; Neuroendocrinology; Peptides; Pituitary; Polypeptide Hormones; Thalamus; Thirst; Visceral Afferent Systems, Signaling and Integration

I

Idiotypes and Immune Networks
Autoimmune Disease; Lymphocytes; T-Cell Receptors

Immune Surveillance
Immune System; Lymphocytes; Lymphoma; Natural Killer and Other Effector Cells

Immune System
Cytokines in the Immune Response; Idiotypes and Immune Networks; Immune Surveillance; Immunogenetics; Immunology, Regional; Immunopathology; Immunotoxins; Lymphocytes; Macrophages; Natural Killer and Other Effector Cells

Immunity to Cancer
Acquired Immunodeficiency Syndrome (Virology); Epstein–Barr Virus; Herpesviruses; Idiotypes and Immune Networks; Immune Surveillance; Immune System; Immunobiology of

Transplantation; Interleukin-2 and the IL-2 Receptor; Leukemia; Lymphoma; Lymphocytes; Macrophages; Melanoma Antigens and Antibodies; Monoclonal Antibody Technology; Natural Killer and Other Effector Cells; Oncogene Amplification in Human Cancer; Papillomaviruses and Neoplastic Transformation; Retroviral Vaccines

Immunoassays, Nonradionucleotide
Radioimmunoassays

Immunobiology of Transplantation
CD8 and CD4: Structure, Function, and Molecular Biology; Cytokines in the Immune Response; Immune System; Interleukin-2 and the IL-2 Receptor; Lymphocytes

Immunoconjugates
Biological Response Modifiers; CD8 and CD4: Structure, Function, and Molecular Biology; Immune System; Interferons; Lymphocytes; Natural Killer and Other Effector Cells

Immunogenetics
Complement System; Immune System; Immunobiology of Transplantation; Lymphocytes; T-Cell Receptors

Immunology of Parasitism
Cytokines in the Immune Response; Immunopathology; Lymphocytes; Macrophages; Malaria; Monoclonal Antibody Technology; Neutrophils; Radioimmunoassays; Recombinant DNA Technology in Disease Diagnosis

Immunology, Regional
Cytokines in the Immune Response; Hemopoietic System; Immune System; Inflammation; Lymphocytes; Natural Killer and Other Effector Cells

Immunopathology
Allergy; Antibody–Antigen Complexes: Biological Consequences; Autoimmune Disease; Immune System; Inflammation; Lymphocytes; Macrophages; Natural Killer and Other Effector Cells; Smooth Muscle

Immunotoxins
Antibody–Antigen Complexes: Biological Consequences; Bone Marrow Transplantation; Cell Membrane Transport; Immune System; Immunoconjugates; Leukemia; Lymphoma; Monoclonal Antibody Technology; Ribosomes

Implantation (Embryology)
Abortion, Spontaneous; Electron Microscopy; Extracellular Matrix

Inbreeding in Human Populations
Cystic Fibrosis, Molecular Genetics; Phenylketonuria, Molecular Genetics; Population Genetics

Infectious Diseases
AIDS Epidemic (Public Health); Alzheimer's Disease; Dementia in the Elderly; Infectious Diseases, Pediatric; Influenza Virus Infection; Leishmaniasis; Malaria; Measles; Parkinson's Disease; Sexually Transmitted Diseases (Public Health); Trypanosomiasis

Infectious Diseases, Pediatric
AIDS Epidemic (Public Health); Herpesviruses; Immune System; Infectious Diseases; Lymphocytes; Macrophages; Meninges; Pathophysiology of the Upper Respiratory Tract

Inflammation
Complement System; Connective Tissue; Cytokines in the Immune Response; Kinins: Chemistry, Biology, and Functions; Lymphocytes; Macrophages; Neutrophils; Respiratory Burst

Influenza Virus Infection
DNA and Gene Transcription; RNA Replication

Insulin and Glucagon
Glycolysis; Polypeptide Hormones

Insulinlike Growth Factors and Fetal Growth
Cell Receptors; Tissue Repair and Growth Factors

Interferons
Chemotherapy; Antiviral Agents; Herpesviruses; Influenza Virus Infection; Psychoneuroimmunology

Interleukin-2 and the IL-2 Receptor
Cytokines in the Immune Response; Immune System; Lymphocytes; Macrophages; Natural Killer and Other Effector Cells

Interpersonal Attraction
Attitude and Attitude Change

Intestinal Blood Flow Regulation
Peptide Hormones of the Gut

Ion Pumps
Adenosine Triphosphate (ATP); ATP Synthesis by Oxidative Phosphorylation in Mitochondria

Isoenzymes
High Performance Liquid Chromatography; Malignant Melanoma; Metabolic Regulation; Monoclonal Antibody Technology; Population Genetics

K

Keratinocyte Transformation, Human
Carcinogenic Chemicals; Papillomaviruses and Neoplastic Transformation

Kinins: Chemistry, Biology, and Functions
Cell Receptors; High Performance Liquid Chromatography; Hypertension; Inflammation; Peptides; Radioimmunoassays; Smooth Muscle

L

Lactation
Epidermal Proliferation; Insulin and Glucagon; Lipids; Puberty; Thyroid Gland and Its Hormones

Lactose Malabsorption and Intolerance
Carbohydrates (Nutrition); Thyroid Gland and Its Hormones

Laminin in Neuronal Development
Extracellular Matrix; Peptides

Language
Aging and Language; Dyslexia–Dysgraphia; Language, Evolution; Reading Processes and Comprehension; Speech and Language Pathology

Language, Evolution
Alzheimer's Disease; Comparative Anatomy; Language; Parkinson's Disease

Laryngeal Reflexes
Larynx; Respiratory System, Anatomy; Respiratory System, Physiology and Biochemistry

Larynx
Cartilage; Connective Tissue; Laryngeal Reflexes

Learned Helplessness
Depression; Depression: Neurotransmitters and Receptors; Hippocampal Formation; Lymphocytes; Neocortex; Suicide

Learning and Memory
Alzheimer's Disease; Brain; Hippocampal Formation; Neocortex; Spinal Cord

Leishmaniasis
Chemotherapy, Antiparasitic Agents; Immunology of Parasitism; Macrophages; Parasitology, Molecular; Zoonoses

Leukemia
Bone Marrow Transplantation; Carcinogenic Chemicals; Chromosome Patterns in Human Cancer and Leukemia; Hemopoietic System; Lymphoma; Metastasis; Oncogene Amplification

in Human Cancer; Radiation Interaction Properties of Body Tissues

Life-Cycle Transitions
Estrogens and Osteoporosis

Limbic Motor System
Brain

Lipids
Bile Acids; Cholesterol; Fatty Acid Uptake by Cells; High Performance Liquid Chromatography; Sphingolipid Metabolism and Biology; Steroids

Lithium Effects on Blood Cell Production and on the Immune System
Leukemia; Lymphocytes; Macrophages; Neutrophils

Liver Cancer and the Role of Hepatitis B Virus
Epstein–Barr Virus; Oncogene Amplification in Human Cancer; Papillomaviruses and Neoplastic Transformation; Togaviruses and Flaviviruses

Locomotion, Neural Networks
Brain; Hippocampal Formation; Spinal Cord

Longevity
Atherosclerosis; Brain; Hypothalamus; Mortality; Pulmonary Pathophysiology; Thymus

Low Temperature Effects on Humans
Appetite; Stroke

Lymphocyte-Mediated Toxicity
Autoimmune Disease; Complement System; Interleukin-2 and the IL-2 Receptor; Lymphocytes; Natural Killer and Other Effector Cells; T-Cell Receptors

Lymphocyte Responses to Retinoids
B-Cell Activation, Immunology; CD8 and CD4: Structure, Function, and Molecular Biology; Cytokines in the Immune Response; Hemopoietic System; Interferons; Interleukin-2 and the IL-2 Receptor; Lymphocytes; T-Cell Receptors; Vitamin A

Lymphocytes
Autoimmune Disease; B-Cell Activation, Immunology; CD8 and CD4: Structure, Function, and Molecular Biology; Cytokines in the Immune Response; Interleukin-2 and the IL-2 Receptor; Lymphocyte-Mediated Cytotoxicity; Lymphocyte Responses to Retinoids; Phagocytes

Lymphoma
Acquired Immunodeficiency Syndrome (Virology); Carcinogenic Chemicals; Chromosomal

Anomalies; Immune Surveillance; Interferons; Leukemia; Lymphocyte-Mediated Cytotoxicity; Lymphocytes; Metastasis; Oncogene Amplification in Human Cancer

M

Macrophages
Acquired Immunodeficiency Syndrome (Virology); Autoimmune Disease; Lipids; Malaria; Phagocytes

Magnetic Resonance Imaging
Adipose Cell; Blood–Brain Barrier; Brain

Malaria
Chemotherapy, Antiparasitic Agents; Interferons; Interleukin-2 and the IL-2 Receptor; Lymphocytes; Sickle Cell Hemoglobin

Malignant Melanoma
Melanoma Antigens and Antibodies; Metastasis; Nerve Growth Factor; Transforming Growth Factor-α; Transforming Growth Factor-β; Tumor Clonality

Malnutrition
Endocrine System; Malnutrition of the Elderly; Nutrition and Immunity; Nutritional Quality of Foods and Diets

Malnutrition of the Elderly
Malnutrition

Marijuana and Cannabinoids
Natural Killer and Other Effector Cells; Radioimmunoassays

Maternal Mortality
Hypertension; Mortality

Mathematical Modeling of Drug Resistance in Cancer Chemotherapy
Leukemia

Mating Behavior
Hormonal Influences on Behavior; Primates; Reproductive System, Anatomy; Sexual Behavior, Human

Measles
Blood–Brain Barrier; Cell Receptors; Immune Surveillance; Infectious Diseases, Pediatric; Lymphocytes; Virology, Medical

Meiosis
Chromosome Anomalies; Chromosomes; Down's Syndrome, Molecular Genetics; Oncogenesis; Sperm

Melanoma Antigens and Antibodies
Gangliosides; Malignant Melanoma; Metastasis; Monoclonal Antibody Technology; Nerve Growth Factor; Proteoglycans; Tissue Repair and Growth Factors

Membranes, Biological
Adenosine Triphosphatase (ATP); ATP Synthesis by Oxidative Phosphorylation in Mitochondria; Cell Junctions; Cell Membrane Transport; Extracellular Matrix; G Proteins; Ion Pumps; Lipids; Signal Transduction at the Cell Surface

Meninges
Brain; Connective Tissue

Mental Disorders
Alzheimer's Disease; Autism; Blood–Brain Barrier; Cocaine and Stimulant Addiction; Dementia in the Elderly; Depression; Eating Disorders; Learning and Memory; Magnetic Resonance Imaging; Mood Disorders; Personality Disorders (Psychiatry); Schizophrenic Disorders; Sexual Behavior, Human; Sleep Disorders

Metabolic Regulation
Cholesterol; Circadian Rhythms and Periodic Processes; Genetic Diseases; Glycolysis; Insulin and Glucagon; Phenylketonuria, Molecular Genetics

Metastasis
Extracellular Matrix; Immune Surveillance; Lymphocytes; Lymphoma; Macrophages; Natural Killer and Other Effector Cells; Neoplasms, Etiology; Oncogene Amplification in Human Cancer; Retinoblastoma, Molecular Genetics; Transforming Growth Factor-α; Transforming Growth Factor-β

Microbial Protein Toxins
G Proteins

Microiontophoresis and Pressure Ejection
Adenosine Triphosphate (ATP); Sperm

Microtubules
Alzheimer's Disease; Mitosis; Phosphorylation of Microtubule Protein; Plasticity, Nervous System

Microvascular Fluid and Solute Exchange
Cardiovascular System, Anatomy; Cardiovascular System, Physiology and Biochemistry; Plasma Lipoproteins; Proteoglycans

Minerals in Human Life
Adenosine Triphosphate (ATP); Calcium, Biochemistry; Copper, Iron, and Zinc in Human Metabolism

Mitochondrial Respiratory Chain
ATP Synthesis by Oxidative Phosphorylation in Mitochondria

Mitosis
Chromosome Anomalies; DNA Synthesis; Down's Syndrome, Molecular Genetics; Microtubules

Molecular Recognition of Transfer RNAs
DNA and Gene Transcription; Proteins

Monoclonal Antibody Technology
Interferons; Lymphocytes; Radioimmunoassays

Mood Disorders
Adrenal Gland; Affective Disorders, Genetic Markers; Antidepressants; Appetite; Depression; Hypothalamus; Sleep Disorders; Suicide

Mortality
Longevity; Maternal Mortality

Motor Control
Brain; Locomotion, Neural Network; Motor Skills, Acquisition; Muscle Dynamics; Spinal Cord; Vision Psychophysics

Motor Skills, Acquisition
Learning and Memory; Motor Control

Movement
Motor Control; Motor Skills, Acquisition

Mucosal Immunology
CD8 and CD4: Structure, Function, and Molecular Biology; Cytokines in the Immune Response; Immune System; Lymphocytes; T-Cell Receptors

Multiple Sclerosis
Autoimmune Disease; CD8 and CD4: Structure, Function, and Molecular Biology; Immune System; Lymphocytes; Macrophages; Myelin and Demyelinating Disorders; Neural–Immune Interactions; T-Cell Receptors

Muscle Development, Molecular and Cellular Control
Atherosclerosis; Cardiac Muscle; Cell Junctions; DNA and Gene Transcription; DNA Binding Sites; DNA Synthesis; Muscle, Physiology and Biochemistry; Skeletal Muscle; Smooth Muscle; Thyroid Gland and Its Hormones

Muscle Dynamics
Muscle, Physiology and Biochemistry; Skeletal Muscle

Muscle, Molecular Genetics
Muscle Development, Molecular and Cellular

Muscle, Physiology and Biochemistry
Adenosine Triphosphate (ATP); Cardiac Muscle; Motor Control; Human Muscle, Anatomy; Protein Phosphorylation; Skeletal Muscle; Smooth Muscle

Muscular Dystrophy, Molecular Genetics
DNA Markers as Diagnostic Tools; Genes; Genetic Diseases; Skeletal Muscle

Mutation Rates
Carcinogenic Chemicals; Chromosome Anomalies; DNA Synthesis; Lymphocytes; Muscular Dystrophy, Molecular Genetics; Repair of Damaged DNA; Retinoblastoma, Molecular Genetics; Sickle Cell Hemoglobin

Myasthenia Gravis
Autoantibodies; Autoimmune Disease; Thymus

Mycotoxins
Alkaloids in Medicine; DNA Synthesis; Food Microbiology and Hygiene

Myelin and Demyelinating Disorders
Autoimmune Disease; Cholesterol; Electron Microscopy; Lipids; Multiple Sclerosis; Proteins

N

Nasal Tract, Biochemistry
Cytochrome P-450; Olfactory Information Processing

Natural Killer and Other Effector Cells
Herpesviruses; Immune Surveillance; Immune System; Interferons; Interleukin-2 and the IL-2 Receptor; Lymphocyte-Mediated Cytotoxicity; Lymphocytes; Macrophages; T-Cell Receptors

Neocortex
Brain; Cortex; Cerebral Specialization; Hemispheric Interactions; Seizure Generation, Subcortical Mechanisms; Somatosensory System; Thalamus; Visual System

Neoplasms, Biochemistry
Isoenzymes; Metastasis; Neoplasms, Etiology; Tumor Clonality

Neoplasms, Etiology
Gastrointestinal Cancer; Lymphoma; Neoplasms, Biochemistry; Oncogene Amplification in Human Cancer; Retinoblastoma, Molecular Genetics; Tumor Suppressor Genes

Nerve Growth Factor
Cognitive Representation in the Brain; Hippocampal Formation; Learning and Memory; Neurotransmitter and Neuropeptide Receptors in the Brain; Oncogene Amplification in Human Cancer

Nerve Regeneration
Astrocytes; Plasticity, Nervous System; Laminin in Neuronal Development; Nerve Growth Factor; Nervous System

Nervous System, Anatomy
Adrenal Gland; Autonomic Nervous System; Brain; Central Gray Area, Brain; Cortex; Hypothalamus; Meninges; Pineal Body; Pituitary; Spinal Cord; Synaptic Physiology of the Brain; Thalamus

Neural Basis of Oral and Facial Function
Autonomic Nervous System; Cortex; Muscle Dynamics; Muscle, Anatomy; Pain; Somatosensory System; Thalamus; Tongue and Taste

Neural–Immune Interactions
Autoimmune Disease; Blood–Brain Barrier; Cytokines in the Immune Response; Hemopoietic System; Hypothalamus; Interleukin-2 and the IL-2 Receptor; Lymphocytes; Macrophages; Pituitary

Neural Networks
Cortex; Hippocampal Formation; Olfactory Information Processing; Retina; Somatosensory Systems; Vision (Physiology); Visual System

Neuroendocrinology
Endocrine System; Hypothalamus; Neuropharmacology; Peptide Hormones of the Gut; Pituitary; Puberty; Steroids; Thyroid Gland and Its Hormones

Neurology of Memory
Hippocampal Formation; Learning and Memory

Neuropharmacology
Adrenergic and Related G Protein-Coupled Receptors; Antidepressants; Autonomic Nervous System; Epilepsy; Nerve Growth Factor; Neurotransmitter and Neuropeptide Receptors in the

Brain; Pain; Parkinson's Disease; Schizophrenic Disorders; Stroke

Neurotransmitter and Neuropeptide Receptors in the Brain
Alzheimer's Disease; Brain; G Proteins; Mood Disorders; Parkinson's Disease; Schizophrenic Disorders

Neutrophils
Hemopoietic System; Inflammation; Macrophages; Respiratory Burst

Newborn Screening
Acquired Immunodeficiency Syndrome (Virology); Cystic Fibrosis, Molecular Genetics; Muscular Dystrophy, Molecular Genetics; Phenylketonuria, Molecular Genetics; Sickle Cell Hemoglobin

Nonnarcotic Drug Use and Abuse
Cocaine and Stimulant Addiction; Marijuana and Cannabinoids

Nuclear Pore, Structure and Function
Adenosine Triphosphate (ATP); Antibody–Antigen Complexes: Biological Consequences; Cell Membrane Transport; DNA and Gene Transcription; Electron Microscopy; Lymphocytes; Mitosis; Monoclonal Antibody Technology

Nucleotide and Nucleic Acid Synthesis, Cellular Organization
DNA and Gene Transcription; DNA Synthesis; HeLa Cells; Metabolic Regulation; Mutation Rates; Purine and Pyrimidine Metabolism; Ribonucleases, Metabolic Involvement

Nutrients, Energy Content
Adenosine Triphosphate (ATP); ATP Synthesis by Oxidative Phosphorylation in Mitochondria

Nutritional Modulation of Aging Processes
Atherosclerosis; Cholesterol; Lipids; Longevity; Malnutrition

Nutritional Quality of Foods and Diets
Carbohydrates (Nutrition); Cholesterol; Diet; Dietary Fiber, Chemistry and Properties; Fats and Oils (Nutrition); Food Groups; Lipids; Minerals in Human Life; Oral Pathology; Phospholipid Metabolism; Proteins (Nutrition)

Nutrition and Immunity
Ascorbic Acid; Autoimmune Disease; Cytokines in the Immune Response; Immune System; Interferons; Interleukin-2 and the IL-2 Receptor; Lip-

ids; Proteins (Nutrition); Trace Elements (Nutrition); Vitamin A; Vitamin E

O

Obesity
Eating Disorders; Hypothalamus; Insulin and Glucagon; Satiating Power of Foods

Olfactory Information Processing
Alzheimer's Disease; Hippocampal Formation; Hypothalamus; Nasal Tract, Biochemistry; Neocortex; Thalamus

Omega-3 Fatty Acids in Growth and Development
Fats and Oils (Nutrition); Nutrition and Immunity

Oncogene Amplification in Human Cancer
Breast Cancer Biology; Chromosome Anomalies; DNA Synthesis; Gene Amplification; Malignant Melanoma

Oogenesis
Cell Junctions; DNA Synthesis; Meiosis; Steroids; Tissue Repair and Growth Factors

Optical Microscopic Methods in Cell Biology
Scanning Optical Microscopy

Optical Properties of Tissues
Collagen, Structure and Function; Eye, Anatomy; Hemoglobin; Skin; Skin, Effects of Ultraviolet Radiation

Oral Contraceptive and Disease Risk
Breast Cancer Biology; Cholesterol; Hypertension; Malignant Melanoma; Stroke

Oral Pathology
Acquired Immunodeficiency Syndrome (Virology); Dental Caries; Herpesviruses; Papillomaviruses and Neoplastic Transformation; Sexually Transmitted Diseases (Public Health)

Oxygen Transport to Tissue
Hemoglobin; Red Cell Membrane; Respiratory System, Physiology and Biochemistry

P

Pain
Spinal Cord; Thalamus

Papillomaviruses and Neoplastic Transformation
Keratinocyte Transformation, Human; Recombinant DNA Technology in Disease Diagnosis; Retinoblastoma Molecular Genetics; Skin, Effects of Ultraviolet Radiation; Tumor Suppressor Genes

Parasitology, Molecular
Leishmaniasis; Malaria; Trichinosis; Trypanosomiasis

Parathyroid Gland and Hormone
Bone Remodeling, Human; Circadian Rhythms and Periodic Processes; Vitamin D

Parkinson's Disease
Genetic Diseases

Pathology, Comparative
Acquired Immunodeficiency Syndrome (Virology); Atherosclerosis; Bone Marrow Transplantation; Retroviruses as Vectors for Gene Transfer

Pathology, Diagnostic
Blood Supply Testing for Infectious Diseases; Electron Microscopy; Gas Chromatography, Analytical; Genetic Counseling; High Performance Liquid Chromatography; Lymphocytes; Radioimmunoassays

Pathophysiology of the Upper Respiratory Tract
Allergy; Cytochrome P-450; Immune System; Olfactory Information Processing; Pulmonary Pathophysiology; Repair of Damaged DNA; Respiratory System, Anatomy; Respiratory System, Physiology and Biochemistry

Peptide Hormones of the Gut
Insulin and Glucagon; Peptides; Radioimmunoassays

Peptides
Adrenal Gland; Catecholamines and Behavior; Cytokines in the Immune Response; Hypothalamus; Insulin and Glucagon; Interferons; Interleukin-2 and the IL-2 Receptor; Kinins: Chemistry, Biology, and Functions; Peptide Hormones of the Gut; Pituitary

Perception
Cochlear Chemical Neurotransmission; Dyslexia–Dysgraphia; Ears and Hearing; Eye, Anatomy; Eye Movements; Language; Learning and Memory; Olfactory Information Processing; Proprioceptors and Proprioception; Reading Processes and Comprehension; Retina; Skin and Touch; Sweetness, Chemistry; Tongue and Taste; Vision (Physiology); Visual System

Perception, Food and Nutrition
Dietary Cravings and Aversions during Pregnancy; Olfactory Information Processing; Perception; Perception of Complex Smells; Salivary Glands and Saliva; Salt Preference in Humans; Tongue and Taste

Perception of Complex Smells
Gas Chromatography, Analytical; Olfactory Information Processing; Perception

Personality
Behavioral Measurement in Psychobiological Research; Hippocampal Formation; Personality Disorders (Psychiatry)

Personality Disorders (Psychiatry)
Depression; Depression: Neurotransmitters and Receptors; Psychoanalytic Psychotherapy; Psychoanalytic Theory

Phagocytes
CD8 and CD4: Structure, Function, and Molecular Biology; Cytokines in the Immune Response; Inflammation; Lymphocytes; Macrophages; Neutrophils

pH and Carbohydrate Metabolism
Adenosine Triphosphate (ATP); Cell Membrane Transport; Glycolysis; Insulin and Glucagon; pH Homeostasis

Pharmacogenetics
Cytochrome *P*-450; Pharmacokinetics

Pharmacokinetics
High Performance Liquid Chromatography; Membranes, Biological; Pharmacogenetics; Radioimmunoassays

Phenylketonuria, Molecular Genetics
DNA Markers as Diagnostic Tools; Genetic Diseases; Population Genetics

pH Homeostasis
Glycolysis; Golgi Apparatus; Ion Pumps; Magnetic Resonance Imaging; Neutrophils; Respiratory System, Physiology and Biochemistry

Phospholipid Metabolism
Cell Membrane Transport; Platelet-Activating Factor Paf-acether

Phosphorylation of Microtubule Protein
Alzheimer's Disease; Isoenzymes; Microtubules; Mitosis; Nerve Growth Factor; Plasticity, Nervous System

Photoelectric Effect and Photoelectron Microscopy
Electron Microscopy

Pineal Body
Adenosine Triphosphate (ATP); Circadian Rhythms and Periodic Processes; Salivary Glands and Saliva

Pituitary
Cholesterol; Golgi Apparatus; Growth, Anatomical; Hypothalamus; Peptides; Steroids

Plasma Lipoproteins
Bile Acids; Cholesterol; Fatty Acid Uptake by Cells; Lipids; Proteins

Plasticity, Nervous System
Hippocampal Formation

Platelet-Activating Factor Paf-acether
Immune System; Inflammation; Lymphocytes; Macrophages; Neutrophils; T-Cell Receptors; Thymus

Platelet Receptors
Adenosine Triphosphate (ATP); Collagen, Structure and Function; Platelet-Activating Factor Paf-acether

Polarized Epithelial Cells
Embryo, Body Pattern Formation; Golgi Apparatus; Neoplasms, Etiology

Polyamine Metabolism
Cell Membrane Transport; Malaria; Protozoal Infections, Human; Virology, Medical

Polymorphism of the Aging Immune System
Immune System; Lymphocytes; Thymus

Polymorphism, Genes
Evolution, Human; Genes; Population Genetics

Polypeptide Hormones
Atrial Natriuretic Factor; DNA and Gene Transcription; Hypothalamus; Insulin and Glucagon; Peptide Hormones of the Gut; Radioimmunoassays

Population Differentiation and Racial Classification
Stereotypes and Prejudice

Population Genetics
Genes; Genetics, Human; Genome, Human; Mutation Rates

Porphyrins and Bile Pigments, Metabolism
Reticuloendothelial System

Pregnancy, Nausea and Vomiting
Abortion, Spontaneous; Steroids

Primates
Comparative Anatomy

Problem Solving
Thinking

Procrastination
Psychoanalytic Theory

Proprioceptive Reflexes
Fusimotor System; Proprioceptors and Proprioception; Skin and Touch; Spinal Cord

Proprioceptors and Proprioception
Fusimotor System; Human Muscle, Anatomy; Proprioceptive Reflexes; Skeletal Muscle; Skin and Touch

Prostate Cancer
Breast Cancer Biology; DNA and Gene Transcription; Endocrine System; Fats and Oils (Nutrition); Metastasis; Obesity; Vitamin A

Protein Detection
Proteins

Protein Phosphorylation
Insulin and Glucagon; Muscle, Physiology and Biochemistry; Oncogene Amplification in Human Cancer; Phosphorylation of Microtubule Protein; Proteins

Proteins
Adenosine Triphosphate (ATP); Collagen, Structure and Function; DNA and Gene Transcription; DNA Binding Sites; DNA Synthesis; Elastin; Enzymes, Coenzymes, and the Control of Cellular Chemical Reactions; Genetic Diseases; Ion Pumps; Membranes, Biological; Phosphorylation of Microtubule Protein; Protein Detection; Protein Phosphorylation; Proteoglycans

Proteins (Nutrition)
Energy Metabolism; Malnutrition

Protein Targeting, Basic Concepts
Golgi Apparatus; Histones and Histone Genes; Nuclear Pore, Structure and Function; Polarized Epithelial Cell; Proteins; Ribosomes

Protein Targeting, Molecular Mechanisms
Adenosine Triphosphate (ATP); Cell Membrane Transport; Golgi Apparatus; Heat Shock; Nuclear Pore, Structure and Function; Proteins; Protein Targeting, Basic Concepts; Ribosomes

Proteoglycans
Alzheimer's Disease; Atherosclerosis; Cartilage; Collagen, Structure and Function

Protozoal Infections, Human
Amoebiasis, Infection with *Entamoeba histolytica;* Malaria; Trypanosomiasis

Pseudoautosomal Region of the Sex Chromosomes
Chromosomes; Meiosis; Schizophrenic Disorders; Telomeres, Human

Psychoanalytic Psychotherapy
Psychoanalytic Theory

Psychoanalytic Theory
Aggression; Psychoanalytic Psychotherapy

Psychoneuroimmunology
Autoimmune Disease; Depression; Endocrine System; Lymphocytes; Neuroendocrinology

Psychophysiological Disorders
Atherosclerosis; Conditioning; Hypertension; Psychoanalytic Theory; Pulmonary Pathophysiology

Puberty
Adolescence

Pulmonary Pathophysiology
Allergy; Larynx; Macrophages; Pathophysiology of the Upper Respiratory Tract; Respiratory System, Physiology and Biochemistry

Pulsed-Field Gel Electrophoresis
Cystic Fibrosis, Molecular Genetics; Genetic Maps; Muscular Dystrophy, Molecular Genetics

Purine and Pyrimidine Metabolism
Adenosine Triphosphate (ATP)

Q

Quinolones
Antimicrobial Drugs; Pharmacokinetics; Sexually Transmitted Disease (Public Health)

R

Radiation, Biological Effects
Hemopoietic System; Lymphocytes; Macrophages; Radiation Interaction Properties of Body Tissues; Radiobiology; Radiosensitivity of the Integumentary System; Radiosensitivity of the Small and Large Intestines; Skin, Effects of Ultraviolet Radiation

Radiation Interaction Properties of Body Tissues
Adipose Cell; Connective Tissue; Skeletal Muscle; Lipids; Malnutrition; Obesity

Radiobiology
Diagnostic Radiology; Hemopoietic System; Mutation Rates; Radiosensitivity of the Integumentary System; Radiosensitivity of the Small and Large Intestines; Repair of Damaged DNA; Tobacco Smoking, Impact on Health

Radioimmunoassay
Immunoassays, Nonradionucleotide; Monoclonal Antibody Technology; Parathyroid Gland and Hormones

Radiosensitivity of the Integumentary System
Collagen, Structure and Function; Hair; Hyperthermia and Cancer; Keratinocyte Transformation, Human; Radiation, Biological Effects; Skin; Skin, Effects of Ultraviolet Radiation

Radiosensitivity of the Small and Large Intestines
Radiation, Biological Effects

Reading Processes and Comprehension
Eye Movements

Receptor Molecules, Fertilization
Fertilization; Sperm

Recombinant DNA Technology in Disease Diagnosis
Breast Cancer Biology; Chromosome Patterns in Human Cancer and Leukemia; Cystic Fibrosis, Molecular Genetics; DNA Markers as Diagnostic Tools; DNA, RNA, and Protein Sequence Analysis by Computer; Leukemia; Lymphoma; Muscular Dystrophy, Molecular Genetics; Oncogene Amplification in Human Cancer; Papillomaviruses and Neoplastic Transformation; Phenylketonuria, Molecular Genetics; Polymorphism, Genes; Tumor Suppressor Genes

Red Cell Calcium
Adenosine Triphosphate (ATP); Calcium, Biochemistry; Sickle Cell Hemoglobin

Red Cell Membrane
Lipids

Repair of Damaged DNA
DNA and Gene Transcription

Reproductive Processes, Neurochemical Control
Hypothalamus; Neuroendocrinology; Pituitary; Reproductive System, Anatomy

Reproductive System, Anatomy
Cardiovascular System, Anatomy; Puberty; Sperm; Uterus and Uterine Responses to Estrogen

Respiration, Comparative Physiology
Respiratory System, Anatomy; Respiratory System, Physiology and Biochemistry

Respiratory Burst
Inflammation; Lymphocytes; Neutrophils; Phagocytes

Respiratory System, Anatomy
Macrophages; Respiratory System, Physiology and Biochemistry; Skeleton

Respiratory System, Physiology and Biochemistry
ATP Synthesis by Oxidative Phosphorylation in Mitochondria; Respiratory System, Anatomy

Reticuloendothelial System
Hemopoietic System; Inflammation; Lymphocytes; Macrophages; Neutrophils; Phagocytes; Respiratory Burst

Retina
Cell Junctions; Color Vision; Eye, Anatomy; Vitamin A

Retinoblastoma, Molecular Genetics
Retina

Retroviral Vaccines
Acquired Immunodeficiency Syndrome (Virology); Lymphocytes

Retroviruses as Vectors for Gene Transfer
DNA and Gene Transcription; Oncogene Amplification in Human Cancer

Ribonucleases, Metabolic Involvement
DNA and Gene Transcription; HeLa Cells; Interferons; Nucleotide and Nucleic Acid Synthesis, Cellular Organization

Ribosomes
DNA Synthesis

RNA Replication
Chromosomes; DNA and Gene Transcription; DNA Synthesis; Retroviruses as Vectors for Gene Transfer

S

Salivary Glands and Saliva
Autonomic Nervous System; Cystic Fibrosis, Molecular Genetics; Dental and Oral Biology, Anatomy; DNA Synthesis

Salmonella
Antibiotics; Food Microbiology and Hygiene; Platelet-Activating Factor Paf-acether

Salt Preference in Humans
Hypertension; Tongue and Taste

Satiating Power of Food
Appetite; Attitudes as Determinants of Food Consumption; Dietary Fiber, Chemistry and Properties; Fats and Oils (Nutrition); Obesity; Proteins (Nutrition); Sweetness, Chemistry

Scanning Optical Microscopy
Optical Microscopic Methods

Schizophrenia, Psychosocial Treatment
Schizophrenic Disorders

Schizophrenic Disorders
Depression; Magnetic Resonance Imaging; Mental Disorders; Psychoanalytic Theory; Schizophrenia, Psychosocial Treatment; Suicide

Seizure Generation, Subcortical Mechanisms
DNA and Gene Transcription; Epilepsy; Hippocampal Formation; Limbic Motor System

Self-Monitoring
Development of the Self

Sensory-Motor Behavioral Organization and Changes in Infancy
Cortex; Motor Skills Acquisition; Perception

Serotonin in the Nervous System
Cellular Signaling; Depression; Eating Disorders; Headache; Hypothalamus; Neuroendocrinology; Neurotransmitters and Receptors; Obesity; Pain; Sleep

Sex Differences, Biological
Sex Differences, Psychological

Sex Differences, Biocultural
Eating Disorders; Primates; Sex Differences, Psychological

Sex Differences, Psychological
Depression; Fragile X Syndrome; Mental Disorders; Nonnarcotic Drug Use and Abuse; Sex Differences; Sexual Behavior, Human

Sexual Behavior, Human
Conditioning; Cortex; Hormonal Influences on Behavior

Sexuality, Anthropological and Evolutionary Perspectives
Bony Pelvis of Archaic Homo Sapiens; Primates; Puberty; Sexual Behavior, Human

Sexually Transmitted Diseases (Public Health)
AIDS Epidemic (Public Health); Herpesviruses; Papillomaviruses and Neoplastic Transformation

Sickle-Cell Hemoglobin
Electron Microscopy; Hemoglobin; Hemoglobin, Molecular Genetics

Signal Transduction at the Cell Surface
Cell; G Proteins; Insulin and Glucagon; Ion Pumps; Olfactory Information Processing; Receptors, Biochemistry; Visual System

Skeletal Muscle
Adenosine Triphosphate (ATP); Connective Tissue; Muscle, Physiology and Biochemistry; Muscular Dystrophy, Molecular Genetics; Myasthenia Gravis

Skeleton
Articular Cartilage and the Intervertebral Disc; Articulations, Joints Between Bones; Hemopoietic System; Vertebra

Skin
Collagen, Structure and Function; Connective Tissue; Elastin; Hair; Inflammation; Malignant Melanoma; Skin, Effects of Ultraviolet Radiation; Vitamin A; Vitamin D

Skin and Touch
Skin; Somatosensory System

Skin, Effects of Ultraviolet Radiation
Epidermal Proliferation; Inflammation; Malignant Melanoma; Optical Properties of Tissue; Skin; Vitamin D

Sleep
Circadian Rhythms and Periodic Processes; Cortex; Learning and Memory; Sleep Disorders; Thermotolerance in Mammalian Development

Smooth Muscle
Cardiac Muscle; Cell Junctions; Muscle, Physiology and Biochemistry; Skeletal Muscle

Social Categorization
Stereotypes and Prejudice

Soft Drugs
Antimicrobial Drugs; Pharmacokinetics; Steroids

Somatosensory System
Hemispheric Interactions; Pain; Spinal Cord; Thalamus

Spaceflight Health Care Systems
Space Travel, Biochemistry and Physiology

Space Travel, Biochemistry and Physiology
Bone Remodeling, Human; Cardiovascular System, Anatomy; Cardiovascular System, Physiology and Biochemistry; Muscle, Physiology and Biochemistry; Respiratory System, Anatomy; Respiratory System, Physiology and Biochemistry; Skeletal Muscle

Speech and Language Pathology
Language

Sperm
Adenosine Triphosphate (ATP); Meiosis; Puberty; Reproductive System, Anatomy

Sphingolipid Metabolism and Biology
Ganglioside Transport; Gangliosides and Neuronal Differentiation; Leukemia; Lipids; Malignant Melanoma; Protein Phosphorylation

Spinal Cord
Meninges; Motor Control; Pain; Peptides; Proprioceptive Reflexes; Proprioceptors and Proprioception; Spinal Cord Dysfunction

Spinal Cord Dysfunction
Neuropharmacology; Plasticity, Nervous System; Spinal Cord

Stereotypes and Prejudice
Aggression; Psychoanalytic Theory

Steroid Hormone Synthesis
Cholesterol; Cytochrome *P*-450; DNA and Gene Transcription; Polypeptide Hormones; Steroids

Steroids
Adrenal Gland; Anti-Inflammatory Steroid Action; Cholesterol; Cytochrome *P*-450; Endocrine System; Estrogens and Osteoporosis; Hypertension; Lymphocytes; Reproductive Processes, Neurochemical Control; Steroid Hormone Synthesis

Stress-Induced Alteration of Gastrointestinal Function
Brain Regulation of Gastrointestinal Function; Peptides; Salivary Glands and Saliva

Stroke
Autoimmune Disease; Cerebrovascular System; Magnetic Resonance Imaging

Studying and Learning Strategies
Learning and Memory; Problem Solving

Sudden Infant Death Syndrome
Brain; Cortex

Suicide
Aggression; Depression; Depression: Neurotransmitters and Receptors; Nonnarcotic Drug Use and Abuse

Sulfites in Foods
Food Microbiology and Hygiene

Superovulation and Ovulation Induction
Abortion, Spontaneous; Follicle Growth and Luteinization

Superoxide Dismutase
Radioimmunoassays

Sweetness, Chemistry
Perception; Tongue and Taste

Synaptic Physiology of the Brain
Brain; G Proteins; Neurotransmitter and Neuropeptide Receptors in the Brain; Plasticity, Nervous System

T

T-Cell Receptors
CD8 and CD4: Structure, Function, and Molecular Biology; Immune System; Lymphocytes; Thymus

Telomeres, Human
Chromosomes; DNA and Gene Transcription

Teratogenic Defects
Chromosome Anomalies; Embryo, Body Pattern Formation; Teratology, Human

Teratology, Human
Birth Defects; Embryo, Body Pattern Formation; Embryofetopathology; Teratogenic Defects

Testicular Function
Cellular Signaling; Chromosomes; Macrophages; Meiosis; Puberty; Sperm; Steroids

Thalamus
Cortex; Parkinson's Disease

Thermal Injuries
Immune System; Inflammation; Ion Pump; Skin

Thermotolerance in Mammalian Development
Heat Shock

Thinking
Learning and Memory; Neural Networks; Problem Solving

Thirst
Blood–Brain Barrier; Salt Preference in Humans

Thymus
Acquired Immunodeficiency Syndrome (Virology); Autoimmune Disease; CD8 and CD4: Structure, Function, and Molecular Biology; Immune System; Lymphocytes; T-Cell Receptors

Thyroid Gland and Its Hormones
Hypothalamus; Pituitary; Plasma Lipoproteins

Tissue Factor
Atherosclerosis; Cell Receptors; Cytokines in the Immune Response; Hemostasis and Blood Coagulation; Lymphocytes

Tissue Repair and Growth Factors
Cartilage; Collagen, Structure and Function; Extracellular Matrix; Inflammation; Macrophages;

Muscle, Physiology and Biochemistry; Neutrophils

Tobacco Smoking, Impact on Health
Asbestos; Atherosclerosis; Macrophages

Togaviruses and Flaviviruses
Virology, Medical

Tongue and Taste
Perception; Salivary Glands and Saliva; Signal Transduction at the Cell Surface

Total Body Calcium Measurement
Bone Density and Fragility, Age-Related Changes; Skeleton

Toxicology
Asbestos; Central Nervous System Toxicology; Immunotoxins; Pulmonary Pathophysiology

Toxicology of Pesticide Residues in Foods
Carcinogenic Chemicals; Food Toxicology; Teratology, Human

Toxicology, Pancreatic
Alcohol Toxicology; Endocrine System; Insulin and Glucagon; Peptide Hormones of the Gut; Toxicology

Toxicology, Pulmonary
Asbestos; Macrophages; Neutrophils; Pathophysiology of the Upper Respiratory Tract; Tobacco Smoking, Impact on Health; Toxicology

Toxins, Natural
Alkaloids in Medicine; Antibiotics; Bacterial Infections, Detection; Mycotoxins

Trace Elements (Nutrition)
Copper, Iron, and Zinc in Human Metabolism; Dental Caries; Selenium in Nutrition; Zinc Metabolism

Transforming Growth Factor-α
Keratinocyte Transformation, Human; Transforming Growth Factor-β

Transforming Growth Factor-β
Cell Receptors; Extracellular Matrix; Transforming Growth Factor-α

Transmission of Infectious Diseases
Herpesviruses; Infectious Diseases; Influenza Virus Infection; Measles; Zoonoses

Trichinosis
Inflammation

Trichothecene Mycotoxicoses
DNA Synthesis; Food Microbiology and Hygiene; Hemopoietic System; Immune System; Lymphocytes; Mycotoxins; Radioimmunoassays

Trigeminal System, Somatic Sensory Receptors
Hair; Muscle, Physiology and Biochemistry; Skin; Skin and Touch

Trypanosomiasis
Parasitology, Molecular

Tumor Clonality
Chromosome Anomalies; DNA Markers as Diagnostic Tools; Leukemia; Lymphoma; Metastasis; Retinoblastoma, Molecular Genetics

Tumors of Bone
Lymphoma; Neoplasms, Etiology

Tumor Suppressor Genes
Mutation Rates; Oncogene Amplification in Human Cancer; Retinoblastoma, Molecular Genetics

U

Urinary System, Anatomy
Reproductive System, Anatomy

Uterus and Uterine Response to Estrogen
DNA and Gene Transcription; *fos* Gene, Human; Tissue Repair and Growth Factors

V

Vaccines, Synthetic
AIDS Epidemic (Public Health); Idiotypes and Immune Networks; Immune Surveillance; Infectious Diseases

Vertebra
Skeleton

Virology, Medical
Acquired Immunodeficiency Syndrome (Virology); AIDS Epidemic (Public Health); Chemotherapy, Antiviral Agents; Herpesviruses; Immune System; Influenza Virus Infection; Interferons; Measles; Oncogene Amplification in Human Cancer; *Salmonella*

Visceral Afferent Systems, Signaling and Integration
Atrial Natriuretic Factor; Hypothalamus; Pain; Peptides

Vision (Physiology)
Binocular Vision and Space Perception; Color Vision; Retina; Visual System

Vision, Psychophysics
Binocular Vision and Space Perception; Color Vision; Retina; Visual System

Visual Illusions
Vision Psychophysics; Visual System

Visual System
Binocular Vision and Space Perception; Color Vision; Eye, Anatomy; Retina; Vision (Physiology)

Vitamin A
Keratinocyte Transformation, Human; Visual System; Vitamin A Transport, Inter- and Intracellular

Vitamin A Transport, Inter- and Intracellular
Retina; Vitamin A

Vitamin D
Cytochrome *P*-450; Immune System; Insulin and Glucagon; Keratinocyte Transformation, Human; Leukemia; Lymphocytes; Macrophages; Nutritional Quality of Foods and Diets; Skeletal Muscle; Skin, Effects of Ultraviolet Radiation; Steroids

Vitamin E
Cholesterol; Fatty Acid Uptake by Cells; Nutrition and Immunity

Z

Zinc Metabolism
Cytokines in the Immune Response; Endocrine System; Fetus; Hippocampal Formation; Immune System; Insulin and Glucagon; Interleukin-2 and the IL-2 Receptor; Membranes, Biological; Testicular Function; Thymus; Vitamin A

Zoonosis
Protozoal Infections, Human; Trichinosis